REHABILITATION ENGINEERING APPLIED TO MOBILITY AND MANIPULATION

Other books in the series

The Physics and Radiobiology of Fast Neutron Beams
D K Bewley

Biomedical Magnetic Resonance Technology
C-N Chen and D I Hoult

Introductory Medical Statistics
R F Mould

Radiation Protection in Hospitals
R F Mould

RPL Dosimetry–Radiophotoluminescence in Health Physics
J A Perry

Physics of Heart and Circulation
J Strackee and N Westerhof

The Physics of Medical Imaging
S Webb

The Physics of Three Dimensional Radiation Therapy
S Webb

Mathematical Models in Cancer Research
T E Wheldon

Medical Science Series

REHABILITATION ENGINEERING APPLIED TO MOBILITY AND MANIPULATION

Rory A Cooper

University of Pittsburgh, Pittsburgh, USA
and
Highland Drive Veterans Affairs Medical Center,
Pittsburg, USA

Taylor & Francis
Taylor & Francis Group
New York London

Published in 1995 by
Taylor & Francis Group
270 Madison Avenue
New York, NY 10016

Published in Great Britain by
Taylor & Francis Group
2 Park Square
Milton Park, Abingdon
Oxon OX14 4RN

Printed in the United States of America on acid-free paper
10 9 8 7 6 5 4 3

International Standard Book Number-10: 0-7503-0343-3 (Hardcover)
International Standard Book Number-13: 978-0-7503-0343-9 (Hardcover)
Library of Congress catalog number: 95-38066

Library of Congress Cataloging-in-Publication Data

Catalog record is available from the Library of Congress

The Medical Science Series is the official book series of the International Federation for Medical and Biological Engineering (IFMBE) and the International Organization for Medical Physics (IOMP)

IFMBE

The IFMBE was established in 1959 to provide medical and biological engineering with an international presence. The Federation has a long history of encouraging and promoting international cooperation and collaboration in the use of technology for improving the health and life quality of man.

The IFMBE is an organization that is mostly an affiliation of national societies. Transnational organizations can also obtain membership. At present there are 33 national members with a total membership in excess of 12 000. An observer category is provided to give personal status to groups or organizations considering formal affiliation.

Objectives

- To reflect the interests and initiatives of the affiliated organizations.
- To generate and disseminate information of interest to the medical and biological engineering community and international organizations.
- To provide an international forum for the exchange of ideas and concepts.
- To encourage and foster research and application of medical and biological engineering knowledge and techniques in support of life quality and cost-effective health care.
- To stimulate international cooperation and collaboration on medical and biological engineering matters.
- To encourage educational programmes which develop scientific and technical expertise in medical and biological engineering.

Activities

Publication of the journal *Medical and Biological Engineering and Computing*, *MBEC News*, regional conferences and a world congress every three years with IOMP in the context of IUPESM. Working Groups and Divisions are active in specific fields of medical and biological engineering.

IOMP

The IOMP was founded in 1963. The membership includes 52 national societies, two international organizations and 10 000 individuals. Membership of IOMP consists of individual members of the Adhering National Organizations. Two other forms of membership are available, namely Affiliated Regional Organizations and Corporate Members. The IOMP is administered by a Council, which consists of delegates from each of the Adhering National Organizations; regular meetings of Council are held every three years at the International Conference on Medical Physics (ICMP). The Officers of the Council are the President, the Vice-President and the Secretary-General. IOMP committees include: developing countries; education and training; nominating; and publications.

Objectives

- To organize international cooperation in medical physics in all its aspects, especially in developing countries.
- To encourage and advise on the formation of national organizations of medical physics in those countries which lack such organizations.

Activities

Official publications of the IOMP are *Physiological Measurement*, *Physics in Medicine and Biology*, and the *Medical Science Series*, all published by Institute of Physics Publishing. The IOMP publishes a bulletin *Medical Physics World* twice a year.

Two Council meetings and one General Assembly are held every three years at the ICMP. The most recent ICMPs were held in Kyoto, Japan (1991) and Rio de Janeiro, Brazil (1994), and a future conference is scheduled for Nice, France (1997). These conferences are normally held in collaboration with the IFMBE to form the World Congress on Medical Physics and Biomedical Engineering. The IOMP also sponsors occasional international conferences, the most recent being in Seoul, Xian and Guangzhou (1991) and Pakistan, India and Argentina (1992).

For further information contact: Hans Svensson, PhD, DSc, Professor, Radiation Physics Department, University Hospital, 90185 Umeå, Sweden, Tel: (46) 90 103891, Fax: (46) 90 101588.

This book is dedicated to my students and to all my friends with physical impairments who have provided me with valuable insight. Furthermore, this book would not have been possible had it not been for the support of my wife, Rosemarie Cooper, and my grandfather and best friend, Roy Munn.

Rory A Cooper
March 1995

CONTENTS

PREFACE

The field of rehabilitation engineering is continually evolving. The pace of the progress in rehabilitation engineering is so rapid that technologies and techniques seem to become obsolete before texts are ever written. This book attempts to provide a foundation for learning rehabilitation engineering as applied to mobility and manipulation. The primary audience for this book is beginning graduate biomedical or rehabilitation engineering students. It is inherently difficult to write a book that serves different readers coming to the subject with various backgrounds and degrees of technical knowledge. Rehabilitation engineering is a young discipline. The combination of disabilities and technology necessitates rehabilitation engineering being a markedly interdisciplinary field. Rehabilitation engineers must be knowledgeable about the possible outcomes of the rehabilitation process, and with the technical potentials offered by engineering. The rehabilitation medicine physician and therapist must be aware of the abilities technology has to offer, whether available or impending, in order to select and apply assistive technology and therapies that will optimize the patient's rehabilitation. The rehabilitation engineer must have thorough knowledge of available technologies and current research, and have the ability to design technical solutions to ameliorate disability.

This book reflects the current status of rehabilitation in that there is a lot of information presented at various levels. Some material is presented in great detail, whereas other material is treated at lesser depth. This reflects the current status of the discipline, as well as my personal biases. One of the difficulties in writing this book was the broad variety of backgrounds which students and professionals in rehabilitation have. This means that some areas of the book may appear simplistic for one reader and overwhelming for another. I have tried to place enough background and motivational material in the book to help readers learn what they need to understand the material or to have sufficient references to relevant work to be able to develop sufficient knowledge for a specific topic. Many books related to assistive technology or rehabilitation engineering do not provide sufficient depth to be considered as engineering textbooks. I have attempted to provide some clarification of difficult topics and to define rehabilitation engineering in terms familiar to those who have studied traditional engineering and who have some medical background. The book includes over one thousand reference listings to offer a path to greater depth in each topic.

Students and professionals come to rehabilitation engineering courses with a variety of backgrounds. Most biomedical engineering programs require that their students take a course in rehabilitation engineering. This leads to some students being very well prepared in a classical engineering discipline (e.g., electrical engineering, mechanical engineering, civil engineering, chemical engineering) or in a life science discipline (e.g., biology, physiology, biochemistry). This book assumes that the students have taken courses in human anatomy and human physiology or have working knowledge of these fields. Students should also have senior or graduate level engineering knowledge for some sections of this book.

Classical engineering disciplines incorporate substantial analysis and design in their curricula, but focus little attention on consumers and the legislative process. Rehabilitation engineering is a discipline which provides service to people with impairments. This book would be incomplete without addressing some of the legislation which influences the practice of rehabilitation engineering. Chapter 1 provides a definition of rehabilitation engineering and describes some of the activities performed by rehabilitation engineers. It should give readers an overview of the environment in which rehabilitation engineering is practiced. Legislative issues, consumer concerns, and practice approaches are presented. This chapter can be supplemented with local legislation and policies. It should set the tone that the consumers, i.e., people with disabilities, are the focus of rehabilitation, and that consumer involvement is critical to successful rehabilitation engineering.

Chapter 2 describes how consumer input can be incorporated into rehabilitation engineering. The goal is also to make readers aware of the multidisciplinary nature of rehabilitation engineering. The remainder of the chapter is intended as reference or refresher material. For graduate classes, instructors can simply state that students are responsible for knowing the material in Chapter 2. Students will then have an idea of the minimum basic engineering material required to understand this text.

Rehabilitation engineering applied to mobility and manipulation relies on biomechanics to obtain a basis for the person's abilities. Biomechanical analyses are used to design mobility and manipulation devices as well as to analyze such devices. Chapter 3 provides an overview of kinesiological biomechanics applied to the design and analysis of mobility and manipulation devices. Kinetic and kinematic analysis techniques are described. These tools are applied to wheelchair propulsion, prosthesis ambulating, and prosthesis manipulation. This chapter provides supporting material for Chapters 5, 9, 10, and 12, and is suitable for rehabilitation scientists and engineers.

Universal design and rehabilitation engineering are related fields. The primary goal of universal design is to develop environments and systems that incorporate knowledge about people. Universal design incorporates the needs of a broad spectrum of society. Rehabilitation engineering tends to focus on people with disabilities and on how best to use technology to re-integrate a person into the mainstream of society. Chapter 4 starts by providing motivation for and describing universal design. The concept of the enabler is introduced as a tool for universal design. The latter part of the chapter is dedicated to some specific design situations, including accessibility legislation. Legislation is presented from an international perspective. This chapter is appropriate for people from a variety of professions.

Chapter 5 provides some guidance on the selection, design, and testing of restraints, lifts, and vehicle controls. Personal transportation provides greater independence and autonomy for many people with mobility impairments. The biomechanics of restraints are described. Chapter 5 provides basic information about international standards with regard to the safe transportation of people with disabilities. Engineers and clinicians should benefit from the material presented here.

Chapter 6 focuses on safety, testing, and international standards. Stability of the wheelchair and occupant is presented. Static, impact, and fatigue strength requirements, results, and test methods are presented. Test dummies are described. Consideration is given to both manual and power wheelchairs. Communication standards for integrated controls are also described. This chapter includes some advanced mathematics. Some of the concepts require substantial engineering knowledge. However, sections of this chapter can be skipped by those who desire only a cursory overview of the standards. For those interested in wheelchair design and testing, the chapter provides considerable depth of information.

Chapter 7 describes the process of manual wheelchair design. The information presented can be understood independently of Chapter 6. However, the information in both Chapters 6 and 7 taken together gives a much greater understanding of wheelchair design. Chapter 7 focuses on definitions of manual wheelchairs and their components, and provides some explanation for their use. Ergonomics of wheelchair propulsion and selection are also described. This chapter is suitable for engineers and clinicians. It is not intended to provide all of the information necessary to prescribe wheelchairs, but to provide information about their design.

Chapter 8 describes power wheelchairs from an engineering perspective. Power wheelchairs are prescribed for people with severe mobility impairments, of whom many also have manipulation impairments. An entire chapter could be dedicated to the selection of power wheelchairs, and quite possibly to how impairment and disability etiology affect power wheelchair design. Chapter 8 does not attempt to provide in-depth information in the areas described above, but presents an engineering approach to power wheelchairs. Criteria for selection of motors, microprocessor controls, batteries, gear trains, and control input devices are described. Chapter 8 is intended for graduate level engineers. Chapters 6 and 7 should be read prior to Chapter 8. The information relating to structural and performance considerations in Chapters 6 and 7 is for the most part applicable to power wheelchairs, and provides a basis for Chapter 8.

Chapter 9 on postural support and seating is closely related to power wheelchair design. Many people who require specialized seating and postural support use power wheelchairs. Chapter 9 provides basic information about seating and postural support for wheelchair users. Concepts of seating and postural support are presented along with current theory on cushion and postural support design. Some advanced topics are presented, but most of the material is appropriate for clinicians and engineers. Chapters 6–9 provide in-depth information about wheelchairs. Wheelchairs are possibly the most prolific assistive devices in use to compensate for mobility impairments. Seating and postural support involve substantial knowledge of biomechanics and functional anatomy. Chapter 3 provides the requisite background in biomechanics. Readers must be familiar with basic functional anatomy or this material must be provided by the instructor.

Chapter 10 combines the basic concepts of prosthetic and orthotic design. This chapter provides only a brief overview of this field. Entire books and curricula are dedicated to prosthetics and orthotics, and many countries require certification to practice the application and design of these devices. For these reasons, biomedical and rehabilitation engineering programs often provide only an overview of prosthetics and orthotics. This chapter focuses on the engineering involved in prosthesis and orthosis design. People interested in further information on prosthetics and orthotics should refer to one of the many excellent textbooks in this area. Particular attention is given to electromyographic (EMG) control of prostheses. This is because there are very few reference books accessible to students on this topic (most books simply state it is beyond their scope), and many engineering students seem to discover this area each time they take a course in rehabilitation. This emphasis will help to address their curiosity and provide enough background to focus the students' interest. The basic principles of prosthetics and orthotics can still be obtained even if the section on EMG control is ignored. This chapter includes information about walking aids (e.g., walkers, crutches, canes). I have included functional neuromuscular stimulation (FNS) as part of this chapter. FNS could easily have been included in a separate chapter, and some people will surely argue that this should have been the case. FNS remains primarily a research area. Widespread clinical and consumer use of FNS in mobility and manipulation devices has yet to occur. There are some regional examples of the possibilities of FNS, but they have yet to be implemented in most rehabilitation clinics. FNS has also received substantial attention in several other texts. These texts are better suited to provide the in-depth neuroscience, physiology, anatomy, and engineering required to develop a complete understanding of FNS. Students and instructors interested in greater coverage of FNS should use a supplementary text specializing in this area.

Chapter 11 describes several of the numerous recreational activities pursued by people with disabilities. Recreation is an important modality in peoples' lives. This chapter focuses on popular recreational activities which use specialized assistive devices. People with disabilities participate in nearly every recreational and sporting activity that has been devised. Many activities do not require specialized equipment. This chapter should be accessible to all readers. Chapters 1, 2, 3, 7, 9, and 10 provide valuable background information which will make reading Chapter 11 more meaningful.

Chapter 12 gives an overview of the theory and design of rehabilitation robotics. Rehabilitation robotics is an area where rehabilitation engineers have attempted to transfer technology designed for industrial automation to rehabilitation. Robotics technology has not enjoyed the success that many originally anticipated. Some people attribute the lack of success to the cost of robotic systems, while others blame the inflexibility of current robotic technology. These factors surely contribute to the lack of success robotic systems have experienced with consumers, but consumers have reported that the user interface must be natural for them to use, and must allow them greater productivity. Interface and user-friendly software design are likely to be areas of research that will yield greater acceptance of robotic technology in the future. Chapter 12 is intended for students with an engineering background, including dynamic systems and controls. Robots are complex systems, and some mathematical and engineering sophistication is required to understand their design and function.

This book attempts to couch rehabilitation engineering as a participatory process involving consumers, physicians, therapists, and engineers. It is intended to convey a consumer-centered approach to rehabilitation engineering. This approach assumes that each consumer sets her or his goals for the rehabilitation process. The consumer's goals set the direction for the rehabilitation process and to a large extent define the rehabilitation engineering process. The success of the rehabilitation engineering is based upon its ability to meet the goals defined by the consumer. This book focuses on rehabilitation engineering and is not intended to replace books related to assistive technology selection, rehabilitation medicine or engineering. Some readers will undoubtedly need to refer to supplementary texts on medicine, rehabilitation or engineering to fully benefit from the information contained here.

This book is the result of six years of teaching and fifteen years of research and design in rehabilitation and biomedical engineering. Portions of this book were developed while teaching at California State University in

Sacramento and at the University of Pittsburgh. The goal for this book was to develop a graduate level text suitable for teaching rehabilitation engineering as it applies to mobility and manipulation. The focus is clearly on motor impairment. Sensory impairment also influences mobility and manipulation, but specific considerations for sensory impairment are better left to another text. This book grew from lecture notes, laboratory exercises, and research projects related to rehabilitation engineering and science. I have attempted to include material meaningful to engineers, clinicians, therapists, and health-related scientists. Material from this text has been used to teach graduate and undergraduate level biomedical and rehabilitation engineering students. I have also used portions of the text to teach undergraduate and graduate therapy students. More recently, material from the text has been used to teach graduate rehabilitation science and technology students. I am grateful to these universities and to my colleagues for providing an environment for personal and professional growth, which enabled me to write this textbook.

I am indebted to a number of people in academic, government, and industrial circles who have contributed in different, but important, ways to the preparation of this book. In particular, I wish to extend my appreciation to R N Robertson, J F Ster, C J Robinson, C E Brubaker, M L Boninger, P Axelson, D Counts, G Stout, R Weigle, G Bardsley, F D Baldini, J T Manns, P Stankovic, T Detre, T Davey, J Harris, W E Langbein, L Miller, and D Van Sickle. I would also like to acknowledge the students at California State University, Sacramento and the University of Pittsburgh for their contributions to shaping this textbook. Special thanks go to J A Revill, S Pidgeon, and P Edge of Institute of Physics Publishing for their commitment to excellence in all aspects of the publication of this book. I also wish to acknowledge the individuals and organizations cited in the captions of numerous figures throughout the book for their permission to use that material. I would especially like to thank the Paralyzed Veterans of America for providing me with overwhelming support and opportunities.

This book is a reflection of the significant progress that has been made in the field of rehabilitation engineering. As is usual in a project such as this, progress continues after work on the manuscript stops. I have attempted to place emphasis on fundamental concepts, an approach that, among other things, will hopefully provide a measure of constancy in a rapidly evolving body of knowledge.

Rory A Cooper
March 1995
Pittsburgh

CHAPTER 1

INTRODUCTION

1.1. REHABILITATION ENGINEERING

Rehabilitation requires the expertise of numerous professionals working in concert (see figure 1.1). The person with a disability is the focal point and the driving force behind the rehabilitation process. It is the consumer's goals that are implemented, and the consumer's abilities that are enhanced. Rehabilitation engineers play an important role in the rehabilitation process, and contribute unique knowledge which is important to the consumer, and the other rehabilitation professionals. Rehabilitation engineering can be one of the most rewarding careers an engineer can choose. The lives of many people can be improved through the careful and considerate application of engineering to meet human needs (McFarland and Scadden 1986). Although engineers have contributed to the rehabilitation of persons with disabilities for over 50 years, a generally accepted definition of rehabilitation engineering remains to be developed. Rehabilitation engineering can be defined as the application of science and engineering to the design and development of assistive (adaptive) technology and rehabilitation techniques. This definition implies some degree of technical competence. In classical engineering fields (e.g. electrical, mechanical, chemical, industrial, civil) the baccalaureate or equivalent is generally considered the minimum qualification to practice in the field. Some have argued that areas such as biomedical engineering require a graduate degree to ensure competency.

As with biomedical engineers, rehabilitation engineers must interact with people trained in other disciplines. The contributions of rehabilitation engineers must be recognized for their uniqueness and their value (Gaster and Gaster 1992, Enderle 1992). Rehabilitation engineering requires knowledge of engineering fundamentals, as well as some degree of specialization within civil, electrical, mechanical or chemical engineering. Rehabilitation engineering also requires specialized knowledge related to (re)habilitation: physiology, anatomy, biomechanics, neuroscience and kinesiology (Levine 1990). Rehabilitation engineers must be capable of communicating with other rehabilitation professionals: physicians and therapists (Foort 1985).

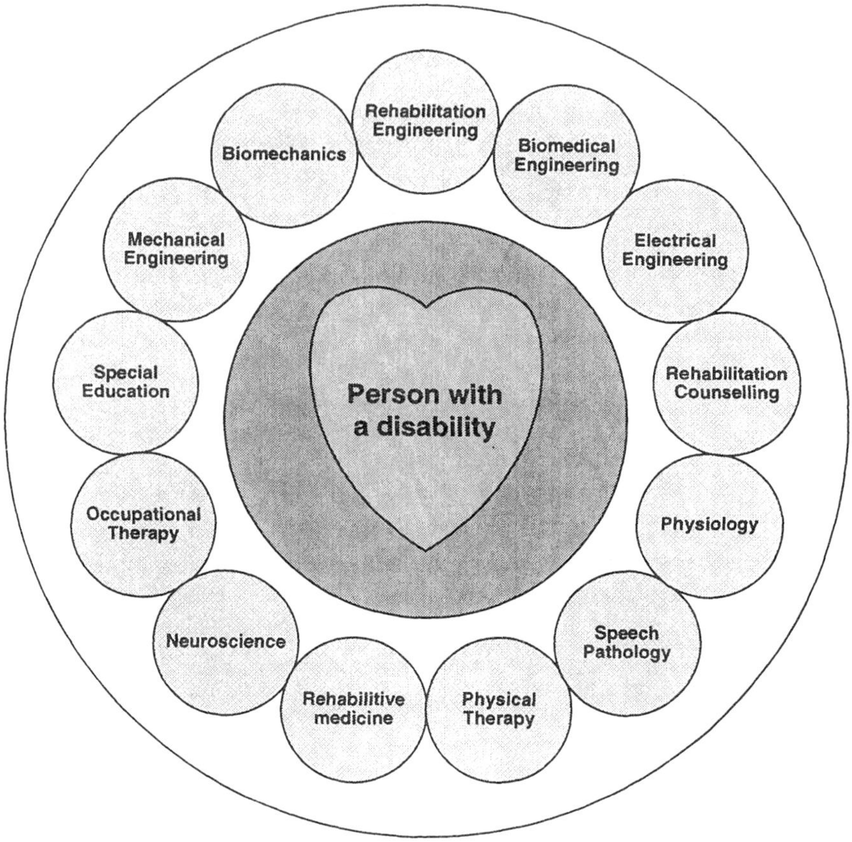

Figure 1.1. *Many professionals are involved in the successful rehabilitation of an individual with a disability. The consumer is at the center of the rehabilitation process. Several other professionals may also be involved.*

Successful rehabilitation engineering requires matching the appropriate technology with the proper techniques to the person with a disability to achieve the goals set forth by that person (Cooper *et al* 1991, Hale *et al* 1991). This requires specialized knowledge as well as the ability to work within a (re)habilitation team (Hyman and Miller 1988, Mann 1986).

1.1.1. Engineering expertise

Assistive technology is varied and complex. Few people can be experts in all areas of engineering. However, all rehabilitation engineers should have basic knowledge of electrical and electronic circuit design, machine design, signals and systems, mechanics of materials, and computers. Engineering design is a primary responsibility of the practicing rehabilitation engineer. Greater depth may be required depending upon the degree of specialization and the work setting (i.e. research versus a clinical setting) (Burgess 1984, Childress 1984). Typically, degree granting programs in rehabilitation engineering are graduate programs (Cooper *et al* 1991). Students entering these programs generally possesses an undergraduate degree in electrical or mechanical engineering, or they must enroll in a substantial number of undergraduate engineering courses. Rehabilitation engineers specialize in a number of assistive technology fields: mobility, sensory aids, robotics, augmentative and alternative communication, computer access, architectural access, rural rehabilitation, cognitive aids, quantitative assessment, prosthetics, and orthotics. Each requires specialized training and knowledge (Potvin *et al* 1986).

1.1.2. Scientific expertise

Rehabilitation engineering is a people-oriented field, more so than most fields of engineering. This makes the field attractive to some engineers, but it also requires the acquisition of specialized skills. Because of the optimal interaction required between the technology or technique and the person whose (re)habilitation goals must be met, specialized scientific techniques are required to evaluate the efficacy of the technology or technique (Cole and Edgerton 1990, Pope and Tarlov 1991). Rehabilitation engineers must rely on scientific methods and user feedback to evaluate the effectiveness of the technology or technique. Therefore rehabilitation engineers must be knowledgeable about experiment design and statistical analysis. This can help to avoid the temptation to apply the same technology or technique to all people equally. Rehabilitation engineers must also be familiar with assessment techniques used by therapists and physicians in order to make appropriate technology recommendations.

1.1.3. Assistive (adaptive) technology expertise

Knowledge of existing assistive technology and techniques is paramount to success for rehabilitation engineers. They must be knowledgeable of available devices and resources. Rehabilitation engineers must be capable of evaluating, modifying and integrating assistive technology. They must also be capable of taking a systems perspective which integrates people and technology. Where possible cost–benefit considerations must be made without compromising the goals of the individual (Mallik and Tech 1990).

1.1.4. Rehabilitation engineering in practice

There are five primary employment settings for rehabilitation engineers: (1) research/training in a university or government setting; (2) product development by a manufacturer; (3) delivery of rehabilitation engineering services in a clinical setting; (4) rehabilitation engineering services for a state department of rehabilitation; and (5) private consulting services. The majority of rehabilitation engineers work in hospital or university settings (Trachtman 1990, 1991), with the primary responsibility of the majority of rehabilitation engineers being service delivery or research and development. Many rehabilitation engineers work alone or in groups containing three or fewer engineers. The responsibilities of the rehabilitation engineer vary depending upon job setting although they are not mutually exclusive. Rehabilitation engineers may have to fulfill several of these roles. A rehabilitation engineer working in research may develop a device and then work with clients, physiatrists, and therapists in the the delivery of the device, and may have to provide assistance to obtain funding for the device. In a clinical setting a rehabilitation engineer may work with a client, physiatrist, therapist and a durable medical equipment supplier to determine the best device to meet the client's needs. The device may require modification by the rehabilitation engineer or it may require integration with other devices the client uses. Rehabilitation engineers working for manufacturers in product development often work closely with therapists, suppliers and clients to develop the best possible product at a reasonable price. This task is made more difficult by the diversity of the needs of various clients, which makes standardization difficult (increasing manufacturing costs) and flexibility a must. In addition, most assistive technology is funded by third-party providers and not by the end user. This requires alternative approaches to marketing and design. Moreover, third-party providers generally approach funding of assistive technology from the perspective of being necessary for the rehabilitation of the client and not for the comfort or convenience of the client. This has been interpreted in some cases to mean that the least expensive functional technology should be purchased rather than the best available technology. With this model all assistive technology must be medically justified. This mind set has made it difficult in some instances to find funding for rehabilitation engineering services in a clinical setting. The goal of the rehabilitation engineer regardless of the setting is to promote the development and application of the most appropriate and cost-effective technology with which to meet the (re)habilitation goals of the person with a disability.

1.2. US LEGISLATION INFLUENCING REHABILITATION ENGINEERING

On 26 July 1990, President George Bush signed into law *The Americans with Disabilities Act* (ADA) (*PL 101-336*). This law gives civil rights protection to persons with disabilities, similar to that provided to individuals on the basis of race, gender, national origin and religion (Buckeye Chapter 1992). The ADA provides equal opportunity for people with disabilities in the areas of employment, state and local government services, public transportation, privately operated transportation available to the public, places of public accommodation and communication services offered to the public. The ADA may arguably be the sweeping piece of civil rights legislation since the *Civil Rights Act* of 1964 (Eastern Paralyzed Veterans 1992). The ADA may be the broadest civil rights law enacted to date, because of the widespread physical and social barriers it could remove (Parrino 1991, Wilson 1992).

Civil rights legislation for people with disabilities has been long in coming. The first major US legislation providing services to persons with disabilities was the *Smith–Fess Vocational Rehabilitation Act* signed into law in 1920. Since that time people with disabilities have gradually won more recognition under the law. Some of the most critical pieces of legislation were introduced following major US military conflicts. The Smith-Fess act followed World War I, the *Vocational Rehabilitation Act Amendment (PL 78-113)* followed World War II, the *Vocational Rehabilitation Act Amendment (PL 83-565)* followed the Korean Conflict, and the *Vocational Rehabilitation Act Amendment (PL 89-333)* was prompted by the Vietnam Conflict. These acts laid the ground work for the disability rights struggle of the 1970s and 1980s. Foremost in this movement has been the desire and ability of diverse individuals with disabilities from a broad spectrum of arenas to work together, both formally and informally, towards the goal of full participation in American (and some would say World) society for all people with disabilities. The strong desire for full participation and the conviction that this ideal is morally correct is undeniable, and is indeed the very foundation upon which the freedom and opportunity of all citizens of the United States is based (US General Services Administration 1992).

The first major successes of the disability rights movement in America was the passage of the *Rehabilitation Act* of 1973 *(PL93-112)* and the *Education For All Handicapped Children Act* of 1975 *(PL94-142)*. These acts did not come without struggle. The *Rehabilitation Act* of 1973 was passed in September 1973 after two vetoes by President Nixon in October 1972 and March 1973. Section 504 of this act contains the regulations for implementation. On 28 April 1977 the US Department of Housing, Education and Welfare (HEW) Secretary Califano signed section 504 after extensive sit-ins at HEW offices in Washington and San Francisco which garnered wide public and local political support. These laws prohibited discrimination on the basis of disability in local programs and activities benefiting from federal financial assistance. Their enforcement has led to improved program accessibility for people with disabilities to health care, social programs, recreation, housing, and transportation. Perhaps most importantly, these laws began to open educational opportunities at all levels for people with disabilities. This contributed to the further empowerment of people with disabilities. Another important aspect of this legislation was the establishment of National Rehabilitation Engineering Centers with the appropriate funding, and the recognition of rehabilitation engineering as an important discipline within the rehabilitation field. In 1976, the Veterans Administration (now Department of Veterans Affairs) established the Rehabilitation Research and Development (RR&D) service (RR&D Annual Report 1991). The RR&D service was to be dedicated to the application of engineering science and technology to rehabilitation. An important component of the RR&D service was the establishment of centers of excellence where members of major university communities could collaborate with the VA research community. The *Education For All Handicapped Children Act* of 1975 *(PL94-142)* pushed further, requiring mainstreaming of children with disabilities into regular classrooms if appropriate, and the establishment of individualized educational programs.

Despite the advances of the 1970s, people with disabilities were far from fully integrated into society (Ficke 1992, Zola 1982). In 1985 a poll conducted by the Louis Harris Company showed that the common thread of people with disabilities in America was unemployment (Eastern Paralyzed Veterans 1992). This poll showed that two-thirds of all Americans with disabilities between the ages of 16 and 64 were unemployed. These results confirmed those of the 1980 census which found that non-disabled men participated in the labor force at 88% whereas men with disabilities participated at only 42%. Non-disabled women participated in the labor force at 64% whereas women with disabilities participated at only 24%. Although access to public assisted programs had improved, architectural and technology related access was not keeping pace.

In 1981 the Architectural and Transportation Barriers Compliance Board adopted the final format for its 'Minimum Guidelines and Requirements for Accessible Design' which was published in the Federal Register on 27 January 1982. The *Handicapped Infants and Toddlers Act (PL 99-457)* of 1986 extended the *Education of All Handicapped Children Act* of 1975 to children from birth to age 3. This law established early intervention programs, which granted federally funded assistive technology and educational media to qualified individuals. The 1986 amendments to the *Rehabilitation Act* of 1973 *(PL 99-506)* required all states to include provisions for acquiring assistive technology devices and services. The *Assistive Technology Related Assistance for Individuals with Disabilities Act* of 1988 *(PL 100-407)* was signed into law. This act recognized the important role assistive technology and rehabilitation engineering play in the full integration of people with disabilities into American society. The act provides funds to states to develop programs to increase the availability and quality of assistive technology. Grants are awarded to states on a competitive basis. This act has had favorable impact on the employment of rehabilitation engineers and has increased the awareness of rehabilitation engineering services among state agencies.

A primary force behind the *Americans With Disabilities Act* was that during the two decades prior, education and vocational training program opportunities had dramatically improved, while employment opportunities in the private sector, and access to places of public accommodation to include public and privately operated transportation services, had not improved. The ADA was not without opposition (Weaver 1990). Lobbyists for business interests argued that the act would inflict undue hardship on business. They argued that in times of limited fiscal resources, intense international competition and demands for many new or expanded social programs—including other programs to help people with disabilities live independently or to move from education to employment—the impact of this legislation would be too great. It was argued that the ADA goes too far by requiring firms with 15 employees or more to make reasonable accommodations to hire people with disabilities who can perform the essential task for a particular job. Business lobbyists stated that this was imposing costs on an employer to hire an employee of lesser ability, and that the legislation went far beyond equal pay for equal work. Advocates for the rights of persons with disabilities were successful in convincing the legislature and president that the benefits would outweigh initial investments, and that if the ADA were not passed business would continue to apply discriminatory practices. The ADA pulls together the experience of the previous two decades to provide a foundation for full participation in American society.

1.3. INTERNATIONAL POLICY INFLUENCING REHABILITATION ENGINEERING

In 1950, the United Nations estimated that 3% of any country's population had a disability. In 1983, the United Nations estimated that more than 500 million people had a mental, physical or sensory impairment. Based upon 1983 estimates, the United Nations suggested that 25% of a country's population may be affected by a person with a disability. The World Health Organization projected that these percentages were unlikely to decline for the next 15 years.

The United Nations started a concerted effort to assess the worldwide status of people with disabilities as part of the International Year of Disabled Persons in 1980. A survey was distributed to member nations in March 1980 which had elicited 64 responses by May 1981: 22 from African countries, 15 from European countries, 11 from American countries, 11 from Asian and Pacific countries and five from west Asian countries. The results of industrialized and developing countries were compared. Approximately 80% of all people with disabilities are located within rural areas of developing countries. Some countries estimate that 20% of their population may have some form of disability. This means that disability may influence 50% of the population in developing countries (e.g. parents, children, close relatives). Industrialized countries were more likely to report diabetes, cancer, disease or trauma to the nervous system, genito-urinary diseases, and allergies as causes of disability. Developing countries were more likely to report inadequate nutrition, prenatal care, lack of immunizations, and violence as the causes of disability. The Economic Commission for Africa and the Organization of African Unity (OAU) in 1980 summarized that disability in developing countries could be traced to socio-economic conditions resulting from an unjust economic order. Furthermore, the causes of disability are often imposed on Africa by external phenomena such as decolonization, struggles for freedom, and aggression. Conflict and violence are often underplayed when investigating disability etiology. In Zimbabwe, approximately 13% of all disability is caused by violence. In Japan, almost 5% of disability among people over 18 years of age was caused by war. In the United States violence is the fastest growing cause of disability.

A country's social policy and resources can alleviate the effects of disability. Sweden has one of the most comprehensive programs of financial support and rehabilitation for people with disabilities. However, a relationship prevails between disadvantageous economic conditions and disability. Of Swedish residents with a mobility or severe visual impairment, 52% were hourly working class employees, whereas only 17% were salaried professional employees. In developing and industrialized countries, people with disabilities have worse working conditions and lower incomes than the general population of the same age. The relationship between socio-economic circumstances and the likelihood of a person developing a disability are strongest in countries with inadequate free public services (e.g. medical, sanitary).

The United Nations, through its World Health Organization, has proposed the World Program of Action (WPA) for the prevention of disability. The WPA involves activities for the improvement of hygiene, education, nutrition, better access to food and health care. The program places special emphasis on mother and child care, counseling parents, immunization, disease/infection control, accident prevention, and environmental intervention. The components directed specifically at disability aim to prevent disability, develop rehabilitation for the realization of full participation of people with disabilities in social, vocational and educational life, and achieve equality. The WPA outlines a plan for reducing disability and improving rehabilitation:

1. early detection, diagnosis and intervention;
2. medical care and treatment;
3. social, psychological and other types of counseling and assistance;

4. training in self-care activities, including mobility, communication, and daily living skills;
5. provision of technical and mobility aids and other assistive devices;
6. specialized education services;
7. vocational rehabilitation services, guidance, training and placement;
8. follow-up.

While most countries agree with the importance of disability legislation, some countries have not implemented laws regarding disability. For example Spanish legislators have expressed concern that special legislation for people with disabilities may be seen as discriminatory. There has been considerable discussion in the United Kingdom concerning disability legislation because such legislation could face formidable problems of definition; it may have to make provision for affirmative action programs in favor of people with disabilities, and it could lead to increased complexity and cost for the bureaucracy to ensure compliance. Portugal has taken the position that disability legislation should be minimized, and only used when regional issues can not be resolved. In Luxembourg, administrative policy and precedents are used in place of legislation. Administrators have sufficient authority to institute and enforce policies to assist people with disabilities.

Many developing countries do not have comprehensive disability policies. A common problem is that rehabilitation resources are often unavailable where needed most. Most rehabilitation centers are located in cities near large hospitals rather than in rural areas. This often restricts access to these facilities by the most needy. In developing countries cultural and family beliefs must be carefully considered. Rehabilitation must include understanding the meaning of disability, and involve family members. In Nigeria, like many developing countries, a cultural stigma is attached to disability which inhibits reintegration into society. Disability in newborns may be blamed on the parents', ancestors' or child's ill behavior in a previous life. If disability occurs later in life it is viewed as punishment for inappropriate behavior. For these reasons, employers are reluctant to hire people with disabilities for fear they may scare customers away. In a study by Osuala, less than 10% of 96 employers in Nigeria hired workers with disabilities. A disproportionate number of people with disabilities were employed by government agencies due to government laws regulating employment quotas. However, supervisors and co-workers within the government held the same fears for their colleagues with disabilities as did their private industry counterparts.

Jamaica began to formalize its rehabilitation services in 1954. In 1972, formal education programs for rehabilitation professionals were developed, and in 1977 programs were established to provide care and services for children with disabilities. In 1981, the Combined Disabilities Association Ltd was formally organized to gain greater accessibility to public buildings and schools throughout the Caribbean.

Many European countries operate within a framework of collective security. This focuses the costs of chronic illness, unemployment, and accidents from the individual to the community. Several European countries offer national health insurance (e.g. France, Germany, the Netherlands, Sweden, the United Kingdom) which includes disability coverage. National insurance in these countries provides assistive technology deemed medically necessary for rehabilitation.

According to Copeland, there are four primary components which define a quota system: eligibility requirements to qualify as a worker with a disability; stipulations regarding which employers are obliged to meet the quota; declaration of the proportion of the staff that is to consist of people with disabilities, and stipulation of penalties for non-compliance. Quota systems presuppose a precise definition of disability. Disability is often defined in terms of etiology, expected duration, complicating vocational factors which may reduce earnings, and extent of potential or actual earning loss. Rehabilitation services are often combined with evaluations for quota systems.

France, Germany, Italy, Luxembourg, and the United Kingdom have quota systems for employment of people with disabilities. Germany requires an estimated 50% reduction in earning capacity to be eligible for the quota system. German companies are required to participate in the quota system if they have at least 16 employees. The German quota system employs a range from 5 to 10% employment of people with disabilities, depending upon demand, and the size of the company. The government collects monthly fines for each position that goes unfilled. The funds collected through the quota process are used for vocational rehabilitation services, and to provide technical aids to workers who need them. German employers are also encouraged, through tax deductions and credits, to provide assistive devices to their disabled workers. Some workers with a disability may be eligible for additional paid leave, and no overtime.

The British parliament has made significant efforts to contribute to the social enhancement of people with disabilities. The *Disabled Persons Act* of 1944 empowered the Minister of Labour to provide industrial rehabilitation for people with disabilities who experienced difficulty in entering the workforce. The *Chronically Sick and Disabled Persons Act* of 1970 has drawn greater attention to the needs of people with disabilities. The British have a proud history of volunteer service to people with disabilities, with many individuals over the years receiving royal recognition. In 1981, many of the separate disability organizations were united to form the British Council of Organizations of Disabled People. The United Kingdom quota system is set at 3% for all employers with a minimum of 21 workers. To be eligible, workers must demonstrate difficulty in obtaining work due to their disability. The worker's age, experience, and training are also considered. An employer must apply for a waiver to hire a unimpaired employee if their quota has not been met. However, the United Kingdom quota system is rarely enforced and has become ineffective. Economic factors have influenced the availability of rehabilitation services, technical aids and assistant services. Economics has also influenced the ability of agencies to enforce quota systems.

The Dutch government has developed a *bonus malus* system which includes a paid bonus to employers who employ a worker with a disability for more than one year. The government may subsidize up to 20% of the employee's income. Employers are required to pay fines for employees who are injured or who develop a disability through their work. Employees with a disability receive vocational rehabilitation services; however, they have no effect on the employee's official disability status. A medical disability rating is used to determine the extent of services for which the employee is eligible.

Australia and New Zealand have also had a long history of providing services to people with disabilities. The Technical Aid to the Disabled Organization (TAD) was founded in Sydney in 1970 as a multi-disciplinary organization aimed at providing volunteer technical assistance to people with a disability. TAD has received some government recognition, but not government protection. TAD and its individual volunteers are privately insured because they may be held liable for legal action due to inappropriately applying assistive devices or for ill advising a client. Some service can only be provided by a licensed professional. The Crippled Childrens Society of New Zealand was formed in 1935 to provide rehabilitation services. During the 1960s several government funded welfare agencies were established to provide services to children with disabilities. In 1972, New Zealand enacted the *Accident Compensation Act* which provides compensation to those with job-related disability of up to 80% of their previous earnings, medical expenses, and rehabilitation assistance. Services vary by region, and are often restricted by difficulty in obtaining financial support for technology and services.

Despite the efforts of countries around the world attempting to legislate or entice integration of people with disabilities into various facets of society, only moderate success has been obtained. A key to reintegration of people with disabilities is empowerment and self-direction. Until people with disabilities are given the technical and legal means of influencing government and business decisions, they will be unable to obtain full integration into society.

1.4. ASSISTIVE TECHNOLOGY

For a person with a newly acquired disability, adaptation to the realities of a new life is usually the first step made toward re-entry into mainstream society

(Kyes 1990). During this exploration period, people express two factors that can make transition easier: the will to remain as active as possible and the availability of expert advice on both coping strategies and the specific equipment and accessories available to meets the person's goals (Oullette-Tries 1988). Much of this phase is experienced during the person's in-patient rehabilitation. During this time it is critical that therapists, physiatrists, counselors and rehabilitation engineers provide meaningful and useful advice (English *et al* 1985). The primary product of rehabilitation engineering is appropriate assistive technology. In some ways the rehabilitation engineer is continuously involved in the rehabilitation process, even when there is no specific client. The rehabilitation engineer must use experience and training to anticipate the technology demands to meet the user's rehabilitation goals. The will of the client comes from within, but it can be intensified by interaction with positive role models. The advice and expertise comes from many sources including therapists, doctors, rehabilitation engineers, equipment dealers, and other users (Zejdlik 1992). When informative assistance and adaptive equipment are available, people with disabilities find the physical aspects of returning to work, school, home, and social activities less formidable than many initially expect.

1.4.1. Evaluation of assistive technology

The evaluation of assistive technology is critical to its proper application. Assistive technology must be evaluated to determine the potential contribution of the device of technology to the rehabilitation process (Galvin 1991). The value of the technology is dependent upon the person for whom it is being evaluated and the user's intended use of the device. The goal of an evaluation is the satisfaction of the user (Galvin and Phillips 1991).

Technology can be evaluated from several perspectives: (1) technical, (2) functional, (3) comparative, and (4) user perception (McFarland 1990). A technical evaluation takes an engineering approach. Rehabilitation engineers evaluate devices to determine whether they were designed using sound engineering principles, and whether there were any design or manufacturing flaws, to determine the quality of the components used in the design, and to determine the adequacy of devices' safety features. A technical evaluation must be made within the framework of the intended user performing a predefined set of activities.

A functional evaluation determines the effectiveness of the device or technique in ameliorating an aspect of a persons disability. This type of evaluation usually involves testing the device or technique with people. The goal of functional evaluation is to determine whether the device or technique significantly enhances the ability of a person to perform a certain task. Another goal of functional evaluation is to determine the population of potential users that the device or technique will serve best. The process of functional evaluation may be lengthy, requiring up to several years for some complex devices and techniques.

An example of functional evaluation is in the area of rehabilitation robotics where there has been much speculation about the potential benefits of robots to people with upper limb impairments. Several robots have been developed or modified for the purpose of assisting persons with limited or no upper limb function in performing a specific task. Several tasks must be performed to operate a computer workstation. Some of these tasks, such as use of the keyboard, are best performed by specialized computer access devices, whereas other tasks, such as changing the paper in a printer or loading and unloading floppy disks, may be best performed by a robot. The case of using a robot to change and remove paper from a printer and loading and unloading floppy disks from a computer requires some complex motions. Robots have been shown to be programmable to perform both of these tasks. This eliminates the need for two specialized devices. However, does the robot perform these tasks better? i.e. does the robot give the user greater independence? are the tasks relatively simple for the user to perform independently? does the robot perform the tasks reliably and repeatably? These questions must be asked and answered to perform a functional evaluation.

A comparative evaluation examines the properties of a specific device or technique and compares and contrasts it with the properties of other devices or techniques designed to perform similar tasks or ameliorate similar functional impairments. Product comparisons require that the consumer should have some prior knowledge of the quality of existing technology. A comparative evaluation is a relative measure of the properties of a specific device or technique as compared and contrasted with existing technology. Product comparisons are generally less comprehensive than are technical or functional evaluations. However, comparative evaluations are often more readily comprehensible to the consumer and provide information as to where the products of various manufacturers stand. Comparisons may include side-by-side listings of information provided in manufacturers' published marketing literature. However, one must be careful to ensure that all of the information is reported in a consistent manner. For example, one manufacturer may report the weight of their wheelchair without armrests, footrests, and wheel-locks whereas another may report the the weight of the wheelchair including all of these components. The reader of such a comparison may be misled if the reporting is not consistent. A product comparison must provide the information necessary to enable a person to survey the technology which may meet their needs.

Satisfaction of the consumer with a disability is the most important basis for evaluating assistive technology. However, factors consumers consider important are often overlooked or not considered during the development of new assistive technology. Much of the scientific literature in rehabilitation engineering contains technical descriptions of assistive technology, but most lack comprehensive analyses of the needs of the consumer. In more established engineering disciplines there are well developed design procedures which reflect the needs of the consumer. In general all consumers desire a quality product (quality is always defined by the consumer) which is simple (in this case simple is a relative term), elegant, cost effective, and functional. Because the needs of consumers with disabilities vary greatly, it is more difficult to develop assistive technologies which work well for a broad spectrum of people with disabilities. A valuable model to use when developing assistive technology is that there is a person doing something, somewhere. In other words there is a person, a task, and an environment (human, context, activity). All too often technology has been developed in a vacuum void of substantive consumer input or the consumer input has been severely limited. This results in a common aggravating cycle for consumers.

1. The consumer is provided an assistive device for use and evaluation based upon some clinical or personal selection process. This is often done with all parties only having limited information about the device.
2. The consumer uses the assistive device and determines its inadequacies for his or her needs. The device often undergoes moderately successful modification or alteration.
3. The consumer continues to use the device, although not completely satisfied, until the device is obsolete or the consumer abandons the device.
4. The consumer selects another device which meets some of the needs unmet by the previous device, but which has shortcomings of its own.

Some of these problems can be avoided by working with consumers early in the design process. Ideally, consumers would be involved in developing the design concept and would work as full partners with the design team. Some problems may never be solved by manufacturers. Because of the specific needs that some people with disabilities have, there will likely always remain the need for custom, one-off, products produced by rehabilitation engineering services. The design of these devices must rely heavily on consumer involvement. An area of assistive technology development which has largely been overlooked is the integration of several assistive devices to meet the needs of some consumers. For example, to best meet the needs of a consumer with upper and lower limb impairments, an augmentative communication device, a power wheelchair, and environmental controls may be recommended. Each of these devices affects the performance of the others, yet they are often developed and applied separately. This frequently results in frustration and anger on the part of the consumer. Although some clinics have specialized in the integration of several assistive devices for a single consumer, they are handicapped by the lack of coordination and standardization among manufacturers of various devices. It is also very important for the rehabilitation engineer to understand that the needs of an individual change as they adapt to their disability, as their disability changes or through experience and aging (Blackman 1991, Coogler 1992, Rioux 1992).

1.5. REHABILITATION TECHNOLOGY SUPPLIERS

Consumers and rehabilitation professionals have relied heavily on the knowledge of rehabilitation technology suppliers for product information. The National Association of Medical Equipment Suppliers (NAMES) has recognized the need to develop standards of practice for rehabilitation technology suppliers (NAMES 1992). Standards of practice help to ensure quality service from NAMES members. NAMES is made up of manufacturers, dealers, clinicians and consumers. Rehabilitation technology suppliers provide assistive technology and services to people with disabilities. Rehabilitation equipment is different from other durable medical equipment in that assessment and evaluation of the technology to meet the client needs is an integral part of the specification process; the equipment and/or components may be customized to meet the client's goals. Rehabilitation technology suppliers are commonly an integral part of the rehabilitation team. Often the rehabilitation technology supplier is responsible for making the methods and terms of payment clear to the consumer, and for providing service and maintenance for the technology.

1.6. ENGINEERING ACCEPTABLE PERFORMANCE

Assistive technology is designed to assist people to carry out some purpose (Blackman 1991). Assistive technology can be approached from a systems perspective to mean that there are people, devices, and activities in various environments. The system may include one person working with one device in a single environment or it can be several people operating several devices in many environments. The objective of the rehabilitation engineer is to work with the team (consumer, therapist, physician, etc) to determine what must be done to perform the given tasks (as defined by the consumer) and the best way to do it. The rehabilitation engineer is responsible for ensuring the technology achieves an acceptable level of performance for potential users. The rehabilitation engineer engineers human performance into the assistive technology. Engineering refers to translating scientific findings into applied technology. Although engineers should not be confined purely to application of technology, it is an important component of engineering. Engineering implies deliberate planning by the designer to make things happen the way the team wants them to happen; this requires active involvement to create a product that meets the needs of the consumer to the best of his/her ability, and not to passively wait to see what develops.

Acceptable performance in the context of rehabilitation engineering is defined as the result of a pattern of actions carried out to satisfy an objective according to some standard (Mueller 1990). The actions may affect observable physical behavior or non-observable cognitive processing. The performance of the system depends upon the technology, the person, and the interaction between the person and the technology. Any performance objective must be defined according to some predefined standard. The two most common standards are speed and accuracy. The quality of the performance is often defined by the standard set. It is often too easy for rehabilitation engineers to simply let standards evolve, much to the dismay of consumers. Without performance standards virtually all performance becomes acceptable. An example of a human performance standard is the common test used to evaluate and tune power wheelchair controllers by maneuvering into a toilet stall. The controller is tuned until the person using the chair can maneuver into the stall close enough to use the toilet in a specified amount of time. Several other such standards can be applied.

To be of any practical use standards must be measurable. Simple standards such as weight (mass) can be measured with a scale and be recorded in pounds (kilograms). Human performance of the system is often measured using standards of accuracy, speed, training time, and consumer satisfaction. Measuring human performance is critical to the successful design of assistive technology. The testing of human performance often requires a statistical approach when dealing with devices not designed for a specific individual. Rehabilitation engineers must also have experience in developing surveys to determine consumer satisfaction and acceptance. Rehabilitation engineers must be aware that they must face trade-offs to optimize the results of all performance measures. Generally, changes to improve the results of one standard will affect another standard. For example, if one wishes to improve accuracy some speed may be sacrificed or complexity and training time will increase. The skill of the rehabilitation engineer is affected by his/her ability to make trade-offs that lead to the best overall human performance.

1.7. REHABILITATION ENGINEERING DESIGN

Design is deliberate purposeful planning. Rehabilitation engineering design may be defined as deliberate purposeful planning applied to assistive technology. In many senses rehabilitation engineering itself may best be defined as rehabilitation engineering design. The design skills of the rehabilitation engineer are what distinguishes him or her from the other members of the rehabilitation team. Heuristics, one of the oldest design methods, is often applied to assistive technology. When using heuristics, the engineer imagines that the device is completed and that the consumer is using it. By imagining that the problem has been solved certain conditions are met. By working one's way backwards from the solution one can develop the criteria necessary for the design. This approach has been applied to solve many complex design problems.

For example, let us say we would like to design a power wheelchair for use by people with quadriplegia. We start by saying the power wheelchair will be used indoors and outdoors. Hence, we choose to use electricity to power the wheelchair so that the noise and fumes will not present a problem indoors. We choose to use batteries (Gel Cell Marine batteries for cost and durability) because mains electricity may not always be available outdoors, but because we use batteries we have limited power. We do not want anybody to get stuck somewhere so we want to minimize the energy drain on the batteries. For this reason we choose to use a switching servo amplifier in a full H-bridge (so the chair can go forwards and backwards). We can get simplicity, high torque, and eliminate the need for a negative supply voltage by using a permanent magnet d.c. motor. This allows us to work on a 24 V system (two 12 V batteries in series). We use an oil-filled, sealed gear box to keep noise and maintenance down (we do not want dirt to damage our wheelchair). Since we selected a high-torque motor, the frame needs to be designed to accommodate it. We select a chromium molybdenum steel alloy for the frame because of its high strength and good machinability. Wide-profile pneumatic tires were selected to provide the user with some shock absorption. A simple linear seating system was selected because it can accommodate a variety of user needs. The wheelchair was designed to be controlled with a joystick because of the user's limited upper-limb function. By working backwards each previous step can be worked out; from there the design criteria can be developed for individual components.

One of the first steps in the design process is to define when the problem is solved. Generally, the rehabilitation engineer will sit with the consumer and they will discuss the problems the consumer has identified. The rehabilitation engineer will assist in defining the problem and then translate the problem into design criteria. At this point it is best if the rehabilitation engineer, the consumer and others on the design team meet to refine the problem and the design criteria. Once the problem has been defined, then ideas need to be generated to solve the problem. Typically there are numerous solutions to a single problem. All solutions have their merits; a key issue is finding a solution that best meets the consumer's needs.

To be successful at creating innovative solutions to problems one must be able not to rely too heavily on past experience. Although past experience is valuable, it often limits our perspectives and it may hide important solutions to the problem. A popular method for forming new ideas is called brainstorming. Brainstorming simply requires assembling a group of people with some knowledge in the problem area and letting them present numerous solutions to the problem. There are only two rules to brainstorming: (1) the ideas should meet the need and (2) criticism of any sort is forbidden.

Once brainstorming is complete, there should be many more ideas for solutions than are needed. The best ideas must be selected for further development. This is critical to the success of the design. It is important to remember that all are solutions to the problem; the process is to select the best solutions. Once the most promising ideas have been selected, details are added until a final design is selected. After the final design is selected, either the device is built, another idea is selected or the brainstorming begins again. If the final design is built, it must be evaluated and then refined or discarded based upon the results.

1.8. ASSISTIVE TECHNOLOGY DESIGN CRITERIA

A substantial portion of time and effort for many rehabilitation engineers is spent in the evaluation or design of assistive technology. The design of assistive technology is critical to its performance and acceptance by consumers. There are 17 design factors which have been identified as being of significant importance to consumers (Batavia and Hammer 1990). These factors form a guideline for the design of assistive technology.

1.8.1. Affordability

Consumers are concerned that the purchase, maintenance and repair of the device will cost too much. Because of the dependence upon assistive technology to perform major life functions and the discouraging fact that many people with disabilities must live on very limited incomes, real expense over the usable life of the product is of great concern to consumers. This is true despite the fact that most assistive technology is being purchased by third-party providers.

1.8.2. Compatibility

The ability to integrate devices is of concern to consumers. It is also discouraging that some devices require much specialized hardware to function properly. Consumers are concerned that some devices are sold independently even though they require integration with another device for them to function. Rehabilitation engineers need to be aware of various standards for interfaces and need to use them in their designs. There are international standards for serial communication (RS232) and for general purpose interface buses (GPIB, IEEE-488). These and other standards are important for applications involving power wheelchairs, robotics, and communications. Engineers need to standardize components and hardware whenever possible. Two major problems, among others, persist in the design of power wheelchairs: (1) there are no standards for type and location of mounting hardware for seating systems, and (2) there are no standards for the location and type of locking mechanisms for restraints used for public transportation. There needs to be some compatibility among various manufacturers and across product types. In other words, wheelchairs of various manufacturers need to share some common components or design features, and other products such as communication and environmental control devices must readily interface to wheelchairs.

1.8.3. Consumer repairability

Assistive technology must be designed such that whenever possible the consumer or an assistant can perform simple repair and maintenance to their personal equipment. For more advanced technologies the consumer should be able to operate a self-diagnostics system to assist with the repair and maintenance of their personal equipment. For example, if a power wheelchair were to incorporate a watchdog microprocessor that could be queried by the user to determine the probable problem with his/her chair then he/she could inform the repair center so that proper arrangements could be made for repair or maintenance.

1.8.4. Dependability

Many consumers depend upon their assistive technology for their independence and optimal performance. Because alternative technologies often are not available when a device is inoperable, devices must be extremely dependable. Not only should the device be functional; it must also function repeatably and predictably. Consumers need to be aware of the restrictions in the use of the device, and restrictions should be as minimal as possible. Devices need to be designed to be fail-safe and be robust to continuous severe use.

1.8.5. Durability

Assistive devices must remain operable over extended periods of time. Consumers should clearly understand the life expectancy of all of their devices. Engineers should design products to meet the life expectancy needs of consumers. Consumers need to be aware of the care and maintenance required to get the maximum life expectancy from their devices.

1.8.6. Ease of assembly

Devices should be designed to be shipped such that they can be reassembled by consumers and/or their assistants whenever possible. Manufacturers should pay attention to packaging as well as to the design of the device.

Ease of assembly also plays an important role in the independent travel of people. Often airline travel requires the disassembly and assembly of assistive devices. Quite often this must be done rapidly with few if any tools. It is in the best interest of all involved if critical components can be removed and replaced by the consumer and/or assistant. If tools are required, then they should be provided by the manufacturer and a storage place for them should be designed into the product.

1.8.7. Ease of maintenance

It is critical that all assistive technology be maintained in an operable and safe condition. Devices should be designed so that the user can conduct maintenance and clean the device. Obviously, the maintenance of some devices requires specialized training. However, in such cases the consumer should be aware of when and where this maintenance is to be performed. Devices should be designed for minimum maintenance. Manufacturers should provide equipment with clear maintenance instructions, and a maintenance and repair log.

1.8.8. Effectiveness

If an assistive device is not effective in improving the consumer's living situation, as perceived by the consumer, then it will not be used. Effectiveness must evaluated in terms of enhancing functional capability or increasing independence. The device must live up to the claims of the manufacturer as perceived by the consumer. The device must meet the specific needs of the consumer or its shortcomings must be clearly defined.

1.8.9. Flexibility

All consumers' needs are not alike and the needs of a particular consumer may change with time or environment. Therefore, it is useful to provide some adjustment (tuning) of the device for each person. It is also important to provide some options to the consumer upon purchase of the device. Many people can be accommodated by a single basic design if accessories are provided to allow some degree of customization. The number and type of options should be determined with input from consumers.

1.8.10. Learnability

When designing or specifying an assistive device, one must decide how the tasks will be divided between the device and the user. With some devices a great deal of technology is used so that the user applies simple commands to make the device perform complicated tasks. By doing this some of the flexibility of control by the user is sacrificed. Another approach is to give the user numerous options. However, the latter approach makes learning to use the device more difficult and time consuming. Upon initially receiving the device, the user should be able to learn to use the device and start independently using the device within a reasonable period of time regardless of whether specialized training is required or not. The definition of reasonable depends upon the complexity of the tasks the device is to perform, and the abilities and motivation of the user. The engineer should design the device to allow maximum flexibility with minimum training time. However, this is not always possible. Therefore, the consumer must be made fully aware of the typical training time and the commitment required. All instructions should be clearly written in terms consumers can understand and any training should be convenient and readily available to consumers. The cost of specialized training should not be prohibitive. The consumer should have the option of increasing the complexity of tasks while gaining experience and familiarity with the device.

1.8.11. Operability

An assistive device is of no value to a consumer if s/he can not operate it. All assistive devices must be readily operated by the experienced user, and must respond to the consumer's operative commands according to the consumer's standards. All controls provided to the user must be easily accessible and safe. The start-up time for each command must be reasonable from the consumer's perspective. The rehabilitation engineer is to ensure that all controls provide clear feedback to the user as to their mode of operation and are appropriate for the person's disability. Any set-up routines should be easy to use and not require an excessive amount of time. Potential glitches or pitfalls must be avoided and minimized. The rehabilitation engineer must ensure that all voice systems are clearly understandable and reflect the pitch of the user's gender. The rehabilitation engineer must ensure that all audible tones/alarms are acceptable in terms of harshness, loudness, length, and frequency. Visual displays should be selected based upon acceptable viewing angles, colors, shapes, and types.

1.8.12. Personal acceptability

Assistive technology must be designed to augment the abilities of the individual using it. It is important to consider the psychosocial impact of the assistive technology on the user. Assistive devices should not only be functional, but they must also be aesthetically pleasing. Devices should be designed or specified to be congruent with the user's personality and self-image. Devices should not be embarrassing in appearance or tone. The self-image and position within the societal framework can be influenced by assistive technology.

1.8.13. Physical comfort

Assistive technology should not cause the user physical discomfort or pain. Rehabilitation engineers should be aware of consumers' concerns about comfort, and should avoid using materials that may cause skin irritation, or noises that may cause irritation. Comfort should be a primary consideration when designing assistive technology.

1.8.14. Physical security

The consumer must be able to operate his/her assistive technology without significant risk of injury or irritation. All assistive devices should incorporate safety features in their design. Devices should be designed not to disturb the normal physiological function of the user. Users should be protected from electric shock, abrasion, puncture, cuts, burns, allergic reactions, and excessive force or pressure.

1.8.15. Portability

The ability to transport or move an assistive device may play an important role in the restoration of function and independence for a person with a disability. Devices should be made light and transportable without compromising function. Portable power supplies are recommended for electronic devices. Batteries should be easily removed and replaced. Devices should be made so that they can fit in compact spaces.

1.8.16. Securability

The loss of an assistive device can have devastating effects. Assistive technology must be designed so that it can be kept within the physical control of the consumer to reduce the likelihood of theft or vandalism. Devices should be made such that they are easily secured or that they require a password or code to operate them. Rehabilitation engineers should consider security measures in their design.

1.8.17. Professional repairability

All assistive devices eventually need repair or replacement. Some repair and maintenance should be performed by the consumer. More complicated repairs should be possible by local repair facilities. Local facilities should be able to make common repairs within a reasonable period of time. Manufacturers should produce and make available replacement parts to local repair facilities. If repairs must be made by the factory, the turn-around time must be minimized. If required the manufacturer or local repair facility should provide temporary replacements.

1.9. PRODUCT TESTING

Product testing is an important part of the development of a quality product. Rehabilitation engineering demands total quality management. Product testing differs from product evaluation in so far as the goal of product testing is to develop information useful for the engineer in improving product design, whereas product evaluation focuses more on information for the consumer. Product testing occurs at three basic levels: (1) the testing of individual components of the system, (2) testing of preproduction prototypes, and (3) testing from the production line. Testing helps to ensure a quality product. Testing is beneficial to the consumer in that better products become available. Third-party providers benefit from testing by being able to quantitatively evaluate the quality of a product. Manufacturers can use testing to improve their products, to develop new designs, and to demonstrate the quality of their products. Testing consists of several important parts: measurement of the characteristics of the product and its key components, measurement of performance of the product and key components, measurement of the robustness of the product and key components, and durability of the product and key components.

1.9.1. Measurement of system characteristics

All assistive devices must be made to some tolerances and specifications. These specifications may be for electrical/electronic and/or mechanical components of the system. It is important that the final product and its components meet specifications. Testing for tolerances simply requires measuring the properties of interest. This may simply entail measuring the weight and a few key dimensions. It could entail testing the power consumption of electrical components or testing software for fault tolerance. Testing system characteristics helps to ensure at least minimum performance on a component scale. It also important to hold the device and its components within reasonable tolerances to ensure similar function for different units, and to make replacement parts.

1.9.2. Measurement of performance

Testing should also ensure that the product meets prespecified performance standards. Performance standards vary greatly with product type. However, there are some common characteristics. All products should meet some form of safety, and controllability standard. Many products should also meet some speed standard. Safety standards should minimize the discomfort and risk of injury to the consumer, assistants, and repair personnel. All assistive devices need to be controllable by the consumer and/or their assistant. The degree of freedom and autonomy of the consumer is related to the degree of controllability of the device. For some automated devices, robots, and power wheelchairs, a speed standard is also required to ensure minimum safe performance. Performance tests are often standardized tests that most or all products undergo in the factory. For example, the maximum speed, maximum acceleration/deceleration, turning acceleration, and directional control may be tested in the factory.

1.9.3. Measurement of robustness

Assistive devices have the potential to be subject to impacts, spills, heat, and cold. Because of the important benefits assistive technology can bring to the user, it must be designed to withstand reasonable mishaps. Commonly products and their components are tested for their impact strength. This can be simply accomplished by dropping from a specific height on to a hard surface. Mechanical and electronic devices require testing for robustness. Electronic devices should undergo operation at both extremes of their operating range for periods of at least 24 hours. When testing for robustness the surface finish (appearance) of the device should also be considered. Rehabilitation engineers can spend time observing the use or using assistive technology themselves to develop an understanding of the types of event the device may be exposed to. The tests for robustness must be at least as severe as the effects of actual use.

1.9.4. Measurement of durability

Durability is often the most time consuming and difficult testing to perform. Fatigue testing is commonly used to measure the durability of a device. However, fatigue testing for both electronic and mechanical systems is a statistical process. In other words, it is difficult to gather much information from a single test. If, for example, three wheelchairs with a 200 pound dummy were tested by pulling them back and forth over a half inch door threshold 100 thousand times, two wheelchairs might break where the cross frame member meets a side frame member, whereas the third may show no damage. If only one chair had been tested the potential problem might not have been identified. Durability is tested by performing tests during continuous operation with a number of units. From these tests information is gathered to help improve on weaknesses in the design or to verify the quality of the design.

1.10. SUMMARY

Rehabilitation engineering is a many-faceted discipline, blending engineering, science, and medicine. The tasks of the rehabilitation engineer are in

Table 1.1. Competencies taught in rehabilitation engineering curricula.

MEDICAL FOUNDATION FOR REHABILITATION ENGINEERING

Functional human anatomy: A basic understanding of human anatomy, and how the body functions from a mechanical and kinesiological perspective.

Human physiology: A basic understanding of major body processes, bioenergetics, and neurophysiology.

Biomechanics: Basic kinesiological biomechanics, analysis of physical impairment, and biomechanical analysis of adaptations for physical impairment.

Neuroscience: Basic neuroscience of unimpaired and impaired neurological function. Neuropathology and disability. Injury and disease processes which can lead to impairment and disability.

Psychological and sociological effects of disability: Understanding and assessment of the impact disability has on the individual and society. Study of the affects rehabilitation has on the psychological well-being of the individual, and on the individual's potential contribution to society.

BASIS FOR REHABILITATION ENGINEERING IN THE REHABILITATION PROCESS

Assistive technology: Introduction to the function and application of existing assistive technology.

Rehabilitation engineering service delivery: Instruction in the rehabilitation process, in legal and legislative aspects of service delivery, and in management of a rehabilitation engineering service.

Clinical practice in rehabilitation engineering: Internship in a rehabilitation service with assistive technology component. Students learn teamwork, assessment, fabrication, consumer training, billing, and ordering.

REHABILITATION ENGINEERING DESIGN AND ANALYSIS

Rehabilitation engineering applied to mobility and manipulation: Analysis and design of devices and systems for ameliorating and assessing physical impairment. Addresses the design and testing of wheelchairs, prosthetics, orthotics, robotics, transportation, and architecture.

Rehabilitation engineering applied to information access and communication: Analysis and design of devices and systems for ameliorating and assessing sensory or communication impairment. Addresses the design and testing of computer access devices, specialized software, communication devices, hearing aids, and visual aids.

Rehabilitation engineering design: Provides a basis for working in multi-disciplinary teams, implementing total quality management, human subject's and experimental device approval, and quantitative/qualitative evaluation.

Rehabilitation engineering research: Participation in a rehabilitation engineering research project which contributes new knowledge to the rehabilitation of people with disabilities, ameliorates disability, or helps to prevent disability.

many ways similar to those of other engineering fields. However, the settings and the applications vary significantly from those of other engineers. The contributions rehabilitation engineering makes to the lives of people with disabilities are substantial. Rehabilitation engineering is a formal engineering discipline with unique areas of knowledge. Specialized training and experience is required to perform rehabilitation engineering. Curricula in rehabilitation engineering are based upon competencies. The competencies addressed by most rehabilitation engineering training programs are presented in table 1.1. Additional courses are often offered to provide students with greater opportunity to specialize or to learn more about other areas of interest.

FURTHER READING AND REFERENCES

Angerman B and Dreschler H 1991 Results of statutory rehabilitation measures in Germany *Int. J. Rehab. Res.* **14** 195–202

Batavia A I and Hammer G S 1990 Toward the development of consumer-based criteria for the evaluation of assistive devices *J. Rehab. Res. Dev.* **27** 425–36

Berkowitz E D 1979 *Disability Policies and Government Programs* (New York: Praeger)

Berkowitz E D 1987 *Disability Policy* (New York: Cambridge University Press)

Blackman A 1991 Machines that work miracles *Time* 18 February pp 70–1

Buckeye Chapter Paralyzed Veterans of America 1992 *Opening New Doors for Paralyzed Veterans and the Physically Challenged* (OH: Euclid)

Buehler C and Schmidt M 1993 User involvement in evaluation of assessment of assistive technology. *Proc. Second Eur. Conf. on the Advancement of Rehabilitation* pp 30.1

Burgess E M 1984 The road ahead *J. Rehab. Res. Dev.* **21** 10–3

Childress D S 1984 Rehabilitation engineering *Mech. Eng.* **106** 56–65

Clywd A 1981 The European parliament and the disabled *Br. Dental J.* **151** 265–6

Cole T M and Edgerton V R (ed) 1990 *Medical Rehabilitation Research* (Washington, DC: National Institute on Child and Human Development, National Institutes of Health)

Coogler C E 1992 Falls and imbalance *REHAB Management* **5** 53–7, 79, 116–77

Cooper R A, Cook A M and Grey T 1991 Development of a model rehabilitation engineering program *Proc. RESNA 14th Annual Conf. (Kansas City, MO, 1995)* pp 82–4

Copeland L S 1979 Social security abroad: mandatory employment of the handicapped *Soc. Secur. Bull.* **42** 23–4

Cornes P 1984 The future of work for people with disabilities: a view from Great Britain *Int. Exchange of Experts and Information in Rehabilitation* (New York: World Rehabilitation Fund)

Cornes P and Hunter J 1985 *Eur. Conf. on Research in Rehabilitation, Work, Disability and Rehabilitation (East Lansing, University Center for Int. Rehabilitation, Michigan State University, 1985)*

Couch R H 1993 Rehabilitation innovations in Central America *Int. J. Rehab. Res.* **16** 13–22

De Jong G 1984 Independent living and disability policy in the Netherlands: three models of residential care and independent living *Int. Exchange of Experts and Information in Rehabilitation* (New York: World Rehabilitation Fund)

Department of International Economic and Social Affairs 1986 *Disability: Situation, Strategies and Policies, United Nations Decade of the Disabled Person (1983–1992)* (New York: United Nations)

Driedger D 1989 *The Last Civil Rights Movement* (New York: St. Martin's)
Eastern Paralyzed Veterans of America 1992 *Understanding the Americans with Disabilities Act* (New York: Jackson Heights)
Editorial 1981 Disability in the developing world. *Br. Med. J.* **282** 928
Editorial 1988 Disability in the third world *Lancet* **1** 253–4
Enderle J D (ed) 1992 *Engineering Senior Design Projects to Aid the Disabled* (Washington, DC: National Science Foundation)
English J B, Knorr K H and Schuch J Z 1985 Rehabilitation engineering technology applied in a total care facility: a feasibility study *Proc. RESNA 8th Ann. Conf. (Memphis, TN, 1955)* pp 148–50
Ficke R C 1992 *Digest of Data on Persons with Disabilities* (Washington, DC: National Institute on Disability and Rehabilitation Research, US Department of Education)
Fisher P, Toczek M and Seeger B R 1993 Technology for people with disabilities: a survey of needs. *J. Assistive Technol.* **5** 106–18
Foort J 1985 Comments for a new generation of rehabilitation engineers *J. Rehab. Res. Dev.* **22** 2–8
Galvin J C 1991 *Evaluation of Assistive Technology* (REquest, Rehabilitation Engineering Center, National Rehabilitation Hospital)
Galvin J C and Phillips B 1991 *What is Appropriate Technology?* (REquest, Rehabilitation Engineering Center, National Rehabilitation Hospital)
Gaster L S and Gaster J C 1992 Acceptance of rehabilitation engineering: a survey of healthcare administrators *Proc. RESNA Int. '92 (Toronto, 1992)* pp 196–8
Gostin L O and Beyer H A 1993 *Implementing the Americans with Disabilities Act: Rights and Responsibilities of All Americans* (Baltimore, MD: Brookes)
Hale P N, Schweitzer J R and Puckett F D 1991 Training in rehabilitation engineering/technology *Proc. RESNA 14th Ann. Conf. (Kansas City, MO: 1991)* pp 295–7
Hudson B J 1991 Preproduction quality assurance: quality intrinsic to design *Med. Des. Mater.* April pp 34–40
Hunt S 1991 Book reviews: is the law fair to the disabled? a European survey *Soc. Sci. Med.* **33** 745–8
Hyman W A and Miller G E 1988 Rehabilitation engineering service delivery through engineering student design projects *Proc. ICAART (Montreal, 1988)* vol 3 pp 16–7
Kyes K 1990 *User Perspective: Hope, Direction and Confidence* (Los Angeles, CA: Q. Allied Health Care Publications) pp 22–5
Leavitt R L 1992 *Disability and Rehabilitation in Rural Jamaica* (London: Associated University Presses)
Levine S P 1990 Position paper: certification (credentialing) of rehabilitation engineers *RESNA 13th Ann. Conf. (Washington, DC, 1990)*
Mallik K and Tech M 1990 Rehabilitation engineering: who pays for it? *Proc. RESNA 13th Ann. Conf. (Washington, DC, 1990)* pp 317–8
Mann R W 1986 Selected perspectives on a quarter century of rehabilitation engineering *J. Rehab. Res. Dev.* **23** 1–6
McFarland S 1990 *A Guide to Making Informed Choices about Assistive Technology* (REquest, Rehabilitation Engineering Center, National Rehabilitation Hospital)
McFarland S R and Scadden L A 1986 Marketing rehabilitation engineering *SOMA Eng. Human Body* **1** 19–23
Mitchell R G 1988 Disability in the third world *Dev. Med. Child Neurol.* **30** 1–2
Mittler P 1993 Political and legislative conditions for the successful education of children with special needs *Int. J. Rehab. Res.* **16** 289–97
Mueller J 1990 Toward universal design: an ongoing project on the ergonomics of disability *Am. Rehab.* Summer 15–19, 33

NAMES Rehab Section 1992 *Standards of Practice and Protocol for Rehabilitation Technology Suppliers* (Alexandria, VA: National Association of Medical Equipment Suppliers)

Newsome L R 1988 Disability in New Zealand: a study of rehabilitation and disability organizations *Int. Exchange of Experts and Information in Rehabilitation* (New York: World Rehabilitation Fund)

Noble J H 1979 Rehabilitating the severely disabled: the foreign experience *J. Health Politics, Policy Law* **4** 221–49

Osuala J D C 1993 Employment problems of the adults disabled in Nigeria *Int. J. Rehab. Res.* **16** 343–5

Ouellette-Tries H 1988 A new approach to rehabilitation: a consumer's perspective *Proc. ICAART (Montreal, 1988)* vol 3 178–9

Parrino S S 1991 In the spirit of the ADA *Focus, Natl Council Disability* Summer 1

Pope A M and Tarlov A R (ed) 1991 *Disability in America: Toward a National Agenda for Prevention* (Washington, DC: National Academy)

Potvin A R, Crosier W G, Fromm E, Lin J C, Neuman M R, Pilkington T C, Robinson C J, Schneider L W, Strohbehn J W, Szolovitis P and Tompkins W J 1986 Report of an IEEE task force—an IEEE opinion on research needs for biomedical engineering systems *IEEE Trans. Biomed. Eng.* **BME-33** 48–59

Rehabilitation Research and Development Center, Palo Alto, Annual Report 1991

Rioux M H 1992 A culture of diversity *Abilities* Spring 60–1

Robinson C J 1993 What is 'Rehabilitation Engineering'? *IEEE Trans. Rehab. Eng.* **RE-1** 1

Shown D 1991 Evaluation of rehabilitation services in Nigeria *Int. J. Rehab. Res.* **14** 182–4

Symington D C 1994 Megatrends in rehabilitation: a Canadian perspective *Int. J. Rehab. Res.* **17** 1–14

Thompson D L, Thomas K R and Fernandez M S 1994 The Americans with Disabilities Act: social policy and worldwide implications for practice *Int. J. Rehab. Res.* **17** 109–21

Trachtman L H 1990 *Who is a Rehabilitation Engineer?: Report on a Survey of RESNA's Rehabilitation Engineering Professional Specialty Group* (Washington, DC: RESNA)

Trachtman L H 1991 Who is a rehabilitation engineer? *Proc. RESNA 14th Ann. Conf. (Kansas City, MO, 1991)* pp 190–2

United Nations 1986 *Disability: Situation, Strategies and Policies (United Nations Decade of Disabled Persons 1983–1992)* (New York: United Nations)

US General Services Administration 1992 *Council On Accessible Technology*

Warren C G 1993 Cost effectiveness and efficiency in assistive technology service delivery *J. Assistive Technol.* **5** 61–73

Weaver C L 1990 The ADA: another mandated benefits program? *Am. Enterprise* May/June 81–4

Wilson D L 1992 New federal regulations on rights of the handicapped may force colleges to provide better access to technology *Chron. Higher Ed.* **38** A1,A21–2

Zeitzer I R 1983 Social security trends and developments in industrialized countries *Soc. Sec. Bull.* **46** 52–62

Zeitzer I R 1994 Recent European trends in disability and related programs *Soc. Sec. Bull.* **57** 21–6

Zejdlik C P 1992 Managing SCI *REHAB Management* **5** 59–62, 121

Zola I K 1982 Denial of emotional needs to people with handicaps *Arch. Phys. Med. Rehab.* **63** 63–7

CHAPTER 2

FUNDAMENTALS OF REHABILITATION ENGINEERING DESIGN

2.1. DESIGN CONSIDERATIONS

2.1.1. The device

Proper materials for an assistive devices cannot be selected without understanding some basic characteristics of the final product and the user's intentions for the product. The rehabilitation engineer must ask: how will the device interact with the user? Material selection and the design itself are affected by the answer to this question. If the device is to come into contact with the user it should not cause abrasion, allergic reaction, or discomfort. The device may come into contact with body fluids, which must not create a hazard for the consumer or for the device. Devices must also consider the cognitive abilities of the prospective user. Cognitive ability will affect material and component selection, and may require greater safety consideration.

Rehabilitation engineers must be able to determine or reasonably estimate the intended or expected loads and stress the device will experience. The environment in which the device will be used and the frequency of use will affect the evaluation of the required mechanical and electrical properties of the device. Depending on the intended use and the environment a device may require a metal or plastic case and be battery operated. A wheelchair frame may be made of aluminum for active users. The intended use will also affect the design as well as the materials.

Often people use multiple assistive devices to perform functions of daily living and for their vocation. The design of a device must consider how it will be integrated into the users existing technology. Such considerations as power consumption, size, weight, and flexibility are affected by the technology with which the device or devices are to be integrated.

Material selection varies depending upon the type of device to be designed. Mechanical, electronic, electrical, and electromechanical devices all have properties to be considered when designing an assistive device. Many devices integrate components from each of these areas. Mechanical design may incorporate the structural components of a device as well as the packaging. Mechanical design may also involve heat flow considerations. Electrical considerations may include supplying and controlling power to the device or communication with other devices. Electronic design is involved with coordination and control of electronic devices. Electronic design can be used for computer control of devices or for digital communication. Electromechanical design involves the conversion of electrical energy into mechanical energy and vice versa. The design and selection of motors and their drive circuitry is the primary application of electromechanical design to assistive technology.

Assistive technology must assist a person to perform a task prescribed by that person. However, merely meeting user-designed performance goals will not necessarily be met with satisfaction: as assistive technology often becomes an integral part of one's personal appearance, and when successfully designed can become an expression of self-image. Therefore aesthetic considerations have to be evaluated throughout the design process. A good 'rule of thumb' to follow is to make the device as unobtrusive as possible. When viewing a person with a disability the assistive technology should not detract from the person. There are two approaches to this: (1) the device is designed to be small and essentially transparent or (2) the device is blended into the user's persona. In the second case the concept is to integrate the user and the device such that the device becomes a reflection of the user. This is an effective means of integrating larger devices such as wheelchairs.

2.1.2. The user

Assistive technology cannot be properly designed without input from the user. Input from the user, in many cases, is not purely setting functional and aesthetic goals. Often more information about the user's characteristics is required to optimize the performance of the device. The anthropometry, cognitive abilities, and motivation of the consumer must be considered for successful design.

People are not all alike. Anthropometry is the study of the physical characteristics of people. When designing assistive technology age and gender characteristics must be determined. Age and gender can affect body size, body type, and body shape. Age may also affect cognitive ability. Age and gender may also affect personal aesthetic considerations. Other factors that may need to be considered are weight, height, strength, and body type. These factors may affect size and strength requirements.

Disability etiology may also affect design of assistive technology. The extent and type of disability affect the abilities of the user. The rehabilitation engineer must consider the extent of the person's proprioceptive perception. The engineer must be aware of the existence and degree of spasticity. The degree of impairment also affects the relationship between the assistive technology and the user.

The independence level and the attitude of the person towards assistive technology will affect the user. Some people are not receptive to technology and must therefore be approached differently than others. Some people never achieve the same level of independence as other people with similar abilities, whereas others far exceed what is expected of them. This is only a reflection of people themselves. However, the danger lies in forgetting that everyone is an individual and has individual needs. It is important for the rehabilitation engineer to be an astute observer and a good listener.

People are motivated to differing degrees. Assistive technology often requires a significant time commitment for set-up and training. The greater the time demands before the user masters the technology, the greater the returns should be. It is critical that the user defines the goals, because when the goals of the user do not match the abilities of the technology the imposition placed on the user can have long-term negative effects. The rehabilitation engineer has the responsibility to clearly define the limitations of the assistive technology. Technology which requires major time commitments often meets with opposition from consumers and it is difficult to get them to comply with training.

2.1.3. Intended use

Assistive technology has no intrinsic value. What makes assistive technology worthwhile are the achievements facilitated with assistive technology. For assistive technology to be successful it has to incorporate information about the user and the intended use. The desired task(s) will affect the design, the person for whom the device is designed will influence the design and so will the location. With assistive technology there is always someone doing something somewhere.

Rehabilitation engineering considers the nature of the activities for which the device will be used and anticipates possible abuses or use for unintended purposes. Equipment for recreation may require greater durability than that for home use. Communication devices require different design criteria than do prosthetics. Often the nature of the activity determines a significant amount of the design.

Assistive technology is used somewhere. The environment is which the design is to be used will affect its design. A device designed for indoor use will differ from one design for outdoor use. Combined indoor and outdoor use may require different specifications than either those used for indoor or outdoor devices. Proper engineering calls for the consideration of the type of circumstances the device may be exposed to. These may include dust, fluids, moisture, shock, humidity, vibration, heat, cold, and a number of other hazards. It is impractical to design devices such that every foreseeable hazard can be withstood by the device. It is prudent to carefully consider the environment in which the device will be used and then design the device to operate safely and reliably within that environment.

Some devices are used more often than others. Frequency of use is an important consideration when designing assistive technology. Cost and performance can be optimized by considering frequency of use. The frequency of use of the device itself and of specific features of the device are important to design. For example, many power wheelchairspower] have programmable maximal speed, maximal forward, reverse and yaw acceleration, and filter parameters. Some of these features are adjusted infrequently. The engineer must decide whether these features should be adjustable from an input on the wheelchair or from an external device. If the device itself is used infrequently then cost can be reduced by making it less durable than would be necessary for the same type of device if it were used constantly. Also, infrequently used devices may require special power considerations.

2.1.4. Engineering design criteria

Assistive technology requires engineering design as well as human performance design. Classical measures of engineering design quality must also be applied to assistive technology. Rehabilitation engineers must evaluate the effectiveness of a device based upon a cost to benefit ratio over the life of the product. Determining the life of an assistive device requires looking beyond profit and considering the rate of development of improved technology to meet the needs of the client. With consideration for the needs of the consumer, classical engineering design criteria can assist with the development of practical assistive technology.

Function and cost are often the dominant design criteria for assistive technology. As reported earlier, people with disabilities have some of the lowest income levels recorded. Related to their lower income, most assistive technology is purchased by a third-party provider. Third-party providers purchase the cheapest assistive technology which meets the users needs. Over the long term, this may not be the most cost-effective approach. However, it is the system in place. The rehabilitation engineer is challenged to provide quality equipment based upon cost and meeting the user's needs. Because cost is amortized over the life of the product, more expensive technology must be justified based upon longer life and/or greater function. Often funding decisions are based upon meeting a person's minimal needs, not upon comfort or convenience. This presents three basic paths for the engineer: (1) sophisticated solutions that perform important tasks other systems cannot perform; (2) products that last longer than competitors; and

(3) products that are least expensive. Developing advanced technology is expensive and may be difficult to justify. Developing products which last must be balanced with advances in technology. Developing a quality product at a low price promotes acceptance for payment.

Although quality is a somewhat nebulous term defined by the consumer, it has some characteristics that can be described by engineers. Engineers define quality of products based upon function, durability, reliability, robustness, repairability, and manufacturability. Function has been discussed previously and will not be expanded upon here.

Durability defines a product's resistance to wear, fatigue, and other forms of damage or decay. It can be tested by simulating use of the device under harsh environments. The International Standards Organization (ISO) develops standards related to durability for assistive technology. Assistive devices should provide maximal durability without limiting function or exceeding cost.

Reliability is a measure of the amount of time the product functions properly over its lifetime or it may be defined as the percentage of time the product functions properly over a given interval. Because people use their assistive technology to achieve some functional goal, reliability is related to the percentage of time the person is not able to reach that goal due to the technology. Nearly one hundred per cent reliability is desired within the given budget. In many cases reliability must take precedence over cost.

Robustness determines the degree to which a system is able to function when components vary in tolerance or fail. Robustness can also be used to mean the degree of variability in the environment that a system can tolerate. Robustness is important from a manufacturing perspective because robust designs permit normal operation with some variation in component tolerances. Robustness is important to the consumer because robust designs will function properly longer as components wear, deteriorate or in some cases fail. Robust designs also permit systems to operate acceptably in environments for which they may not have been intended.

Some assistive devices are one-off, while some are made in quantity. If a rehabilitation engineer is going to make a single device for a specific client then available fabrication resources must be considered. If an assistive device is to be manufactured in quantity than the rehabilitation engineer must consider the appropriate manufacturing methods for the device. Because nearly all assistive technology requires some degree of customization, designs for manufacture must account for this in thorough product design and manufacturing process design.

Once a product has been developed it will invariably require repair or replacement at some time. The ease and cost of repair are heavily influenced by the design of the system. However, system reliability and repairability often conflict, which requires making trade-offs. Typically it is best to use modular design with assistive technology. This allows repairs to be made at multiple levels, relatively quickly and easily. When a part no longer functions properly, the module in which it is installed can be replaced. The damaged module can then be sent to the factory or service center for repair. Hence field sites do not have to invest in expensive test equipment.

People with disabilities rely on assistive technology to assist them in performing tasks of their choosing. Because of this, people may use their devices or systems in places engineers may never have imagined. Many devices require a substantial period of time for accommodation. Therefore it is important to use parts and materials that are readily available. Components that must be custom ordered or which come from a single supplier (sole source) may cause undue hardship on the user. It is critical that rehabilitation engineers use widely available, high-quality parts from reliable and competent sources.

Some assistive devices are required to be powered by an external source, i.e. battery, or powersupply] distribution network. The freedom of mobility of the user will depend upon the source of power and its availability. Generally it is best to minimize power consumption. Minimal power consumption will lead to less heat being generated, and longer operation at less cost. Battery power is preferred for many devices so that they can be used in a variety of locations.

Assistive devices work with people and when machines and people work together there is shared function and control. Some machines are purely passive and convert the input of the user into another form of motion or form of energy. Other machines have some degree of 'intelligence'. In the latter case the rehabilitation engineer must decide which tasks are to be the responsibility of the machine and which tasks are to be the domain of the user. The engineer must also decide whether the roles are fixed or if control can be relinquished or acquired by the user. It may be desirable to have a robot function as a teleoperator until it has 'learned' certain characteristics of the user's behavior, at which time it can suggest to the user that it is capable of handling that task. Other devices may not be capable of learning, but may be switched to require a greater or lesser degrees of input from the user. Even purely passive systems utilize shared control and the abilities of the user must be considered when determining the extent of control to be delegated to the machine.

2.2. TOTAL QUALITY MANAGEMENT IN REHABILITATION ENGINEERING

Total quality management is a process and atmosphere for promoting quality as defined by the consumer. Total quality management is accomplished by treating people as intelligent colleagues who strive to do a good job. Rehabilitation engineers often work in teams with other rehabilitation professionals, and consumers. The knowledge and experiences of these people need to be applied to rehabilitation engineering tasks. Rehabilitation engineers cannot place the blame on consumers for not using or accepting their devices or techniques. Dr W Edwards Deming (1982) founded the principles of total quality management. Deming outlined 14 points to promote quality and productivity. These have been summarized as they apply to rehabilitation engineering.

1. Consistently strive to improve the product and service, with the aim of becoming or remaining competitive and to create job opportunities.
2. Clearly understand your responsibilities, and take a leadership role for a change.
3. Build quality into devices and systems from the onset. Cease dependence upon final inspection.
4. Develop long-term relationships with suppliers and other rehabilitation professionals, based upon loyalty and trust.
5. Constantly and forever improve every process for planning, production, and service, to achieve quality and productivity.
6. Institute and participate in training.
7. Adopt and institute leadership. Help people, machines, and devices do a better job.
8. Drive out fear from the working environment so that consumers and rehabilitation professionals can work effectively together.
9. Break down barriers between professions. People of various professions and consumers must work together as a team to foresee problems and to develop solutions.
10. Teams cannot improve quality or productivity through slogans or productivity targets. Quality and productivity require cooperative interaction between people, agencies, offices, and departments. Avoid creating adversarial relationships.
11. Substitute leadership for management by objectives or quotas. Participate as a team member. Lead the team to achieve goals.
12. Remove barriers which hinder pride in individual workmanship. Recognize people for their contributions to quality.
13. Institute a vigorous program of self-education and self-improvement. Become active in professional activities and professional societies. Participate in consumer and advocacy group activities.
14. Everybody needs to participate in developing quality and responsibility, and must assume responsibility.

Rehabilitation engineering design and total quality management are interrelated. Total quality management provides a framework in which quality rehabilitation engineering services and design can be attained. Total quality management tools can assist rehabilitation teams in developing optimal solutions for challenges faced by people with disabilities. Rehabilitation engineering service delivery requires interaction between consumers and

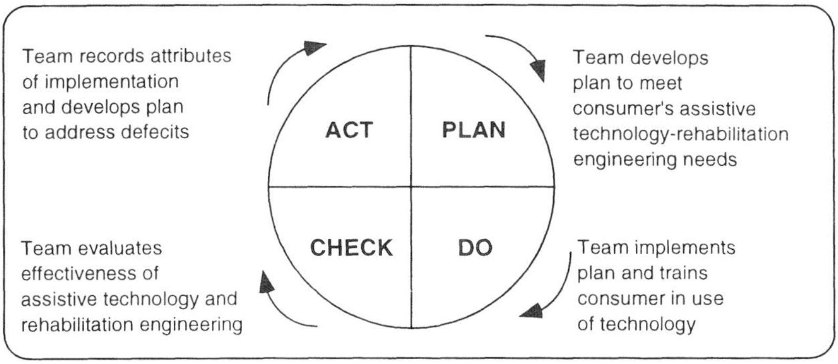

Figure 2.1. *Example of Deming plan–do–check–act cycle applied to rehabilitation engineering service delivery. Definitions for each task can be used to describe research, design, or development tasks.*

rehabilitation professionals. Dr Deming developed the plan–do–check–act cycle to describe how productivity and quality can be maximized while maintaining a healthy working environment. The plan–do–check–act cycle as applied to rehabilitation engineering service is illustrated in figure 2.1.

The plan–do–check–act cycle emphasizes a team approach. Teams of people with various experiences and perspectives are more apt to address all of the critical issues required to define optimal solutions to meet the consumer's goals. The planning cycle entails rehabilitation professionals and consumer(s) working together to articulate problems, to define goals, and to describe solutions. During the do cycle, rehabilitation professionals and consumers purchase and/or design the appropriate assistive technology. The consumer also receives training in the use and care of the technology. Training is a very important component of assistive technology service delivery, and of rehabilitation engineering. Rehabilitation engineers have had numerous good products rejected by therapists or consumers because they were not sufficiently trained to realize the benefits of the devices. A consumer is presented with two power wheelchairs, one which uses four switches for foot control (simple switch control), and one which uses a proportional foot control (foot joystick) and several switches. The consumer can get into the simple switch control power wheelchair and can learn to perform simple driving maneuvers within a very short time (usually within a few hours). However, simple switch control makes it difficult to maneuver in close quarters (e.g. offices, small rooms, rooms with much furniture). A foot joystick and several switches often takes longer to learn to drive—often several sessions of a couple of hours each. Joystick control is proportional, which gives the user greater control over the wheelchair, but requires superior skill for common driving maneuvers than simple switch control. Therefore, the learning curve is steeper. Once the consumer masters proportional control then s/he will be able to maneuver in more complex environments more effectively than with switch control. Of course some people do not have the abilities required to operate proportional control devices, and can benefit from switch-operated controllers. The check cycle is a quantitative and qualitative evaluation of the effectiveness of the consumer's current means of functioning, including assistive technology, and of the new approach developed during the plan and do cycles. Prior to and after training the abilities of the consumers to perform the tasks to be enhanced must be measured. The satisfaction of the user with the current and proposed method for enhancing task performance must also be measured. Training can have a tremendous influence on the check cycle, and thus affect decisions about the assistive technology or rehabilitation intervention. It is important to obtain quantitative data, because decisions should be based upon statistical analysis as well as opinion. The act cycle is used to record the attributes of the process (i.e. assistive technology, evaluation, design, training) and to develop a plan to address the deficits. The cycle assumes that the process of providing rehabilitation engineering and assistive technology services is iterative. This is quite different than the approach taken by many institutions and funding agencies. Without incorporating an act cycle, consumers are apt to be dissatisfied or to abandon the solution, even if they were part of developing the solution. During the act cycle a plan must be developed to address the deficits and to make continuous progress towards obtaining the consumer's goals. Third-party providers are often skeptical when asked to pay for several assistive devices for evaluation purposes. However, in some cases assistive technology can be rented, until a satisfactory solution is obtained. Some institutions have large selections of assistive devices, and broad rehabilitation engineering capabilities. These institutions can implement a number of iterations of the plan–do–check–act cycle before all purchases must be made.

The plan–do–check–act cycle outlines a process for implementing quality. Several other aids can be used to help focus on problems so that solutions can be developed. The fishbone diagram is a useful tool for identifying components which contribute to a problem. Once components have been identified solutions can be developed. A common problem facing assistive technology service delivery and rehabilitation engineering service programs is the abandonment of assistive technology by consumers. A fishbone diagram for an assistive technology service delivery program with a problem of consumers abandoning assistive technology solutions is illustrated in figure 2.2. In the fishbone diagram the problem is placed at the end of the arrow along the primary axis. Major components leading to the problem flow into the primary axis arrow. Sub-components of the problem flow into the major components. A single problem can have multiple layers.

Although fishbone diagrams focus on problems they are not meant as a tool to punish people, but as a tool to develop solutions. Problems are a result of the system and not of the people. Organizations generally

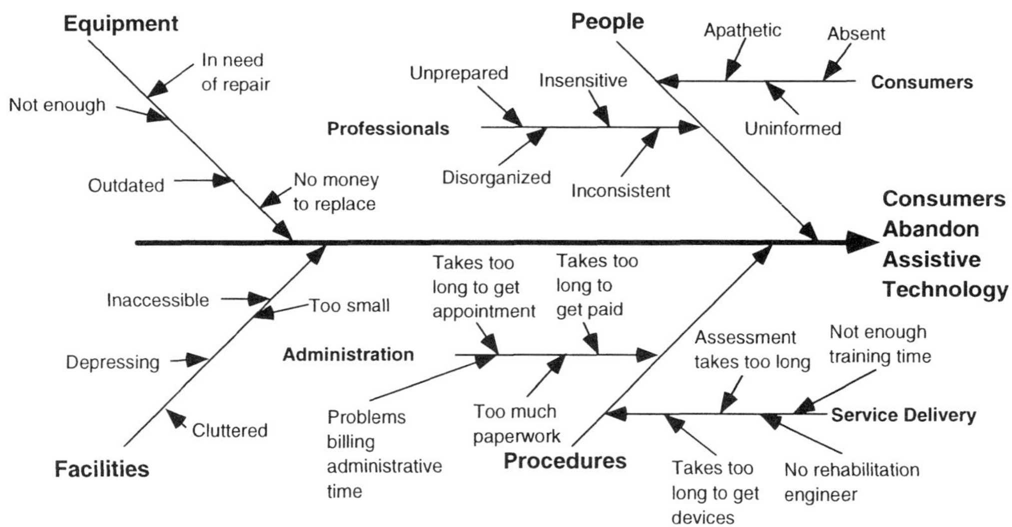

Figure 5.11. *Left-hand control for personal motor vehicle with automatic transmission. Pushing forward on handle operates brake, whereas rotating the handle about the steering column operates accelerator. (Photograph courtesy Mobility Products and Design, Inc.)*

employ qualified people who are supportive of quality when working in the proper environment. The fishbone diagram does not illustrate the interaction between various components which may contribute to problems within the process. Flow charts are useful for illustrating interactions between components, and for identifying feedback loops. Figure 2.3 illustrates a flow chart for a rehabilitation engineering service. The flow chart is divided into three categories: rehabilitation engineering; rehabilitation medicine; physical therapy; occupational therapy; and consumer. Tasks or processes are distributed among the groups. The process is described in sequence from top to bottom. Several steps may be interconnected while others incorporate feedback loops. Problems with the process can be identified or illustrated using flow charts by showing weaknesses in the process flow, feedback loops, or with interactions. Arrows are used to illustrate the direction of information flow, and line types are used to discriminate between forward and feedback paths. Tasks are placed in rectangular boxes and decisions are placed in diamond shapes.

The house of quality (HOQ) is a tool used to design systems and devices of high quality. The house of quality can include several layers, and provides a means of incorporating ratings from several groups (e.g. consumers, therapists, engineers, physicians). Figure 2.4 demonstrates how three layers can be incorporated into a house of quality. Customer requirements are listed down one side of the HOQ, and assistive technology attributes are listed across the top. The interaction between the customer requirements and the assistive technology attributes forms a matrix within the house. Values are assigned to each box within the internal matrix which reflect the quality of the interaction or the benefit which is expected to be obtained. The next layer reflects the interaction between rehabilitation interventions (e.g. therapy, medical rehabilitation, strength, fitness, training) and assistive technology. The physical, emotional and cognitive abilities of the consumer must be appropriate for assistive technology to be acceptable. Many people who have never used a computer may feel very uncomfortable with high-technology assistive devices. Rehabilitation interventions also interact with several processes (e.g. administration, funding, referrals, assessments). The quality of the interaction between the rehabilitation interventions and processes is then placed in a matrix within the third house. The HOQ allows processes to be decomposed into basic components. Each component of the process relates directly or indirectly to quality as defined by the consumer.

Figure 2.5 illustrates how a house of quality can be used to evaluate a robotic workstation for assisting a person to perform several tasks. Substantial research has been performed to develop robotic workstations for people with severe mobility impairments. Few robotic workstations have been accepted by consumers. Here a consumer has listed several tasks the s/he would like to be able to perform independently. Several factors related to the function of the robotic workstation are listed vertically to the right of the

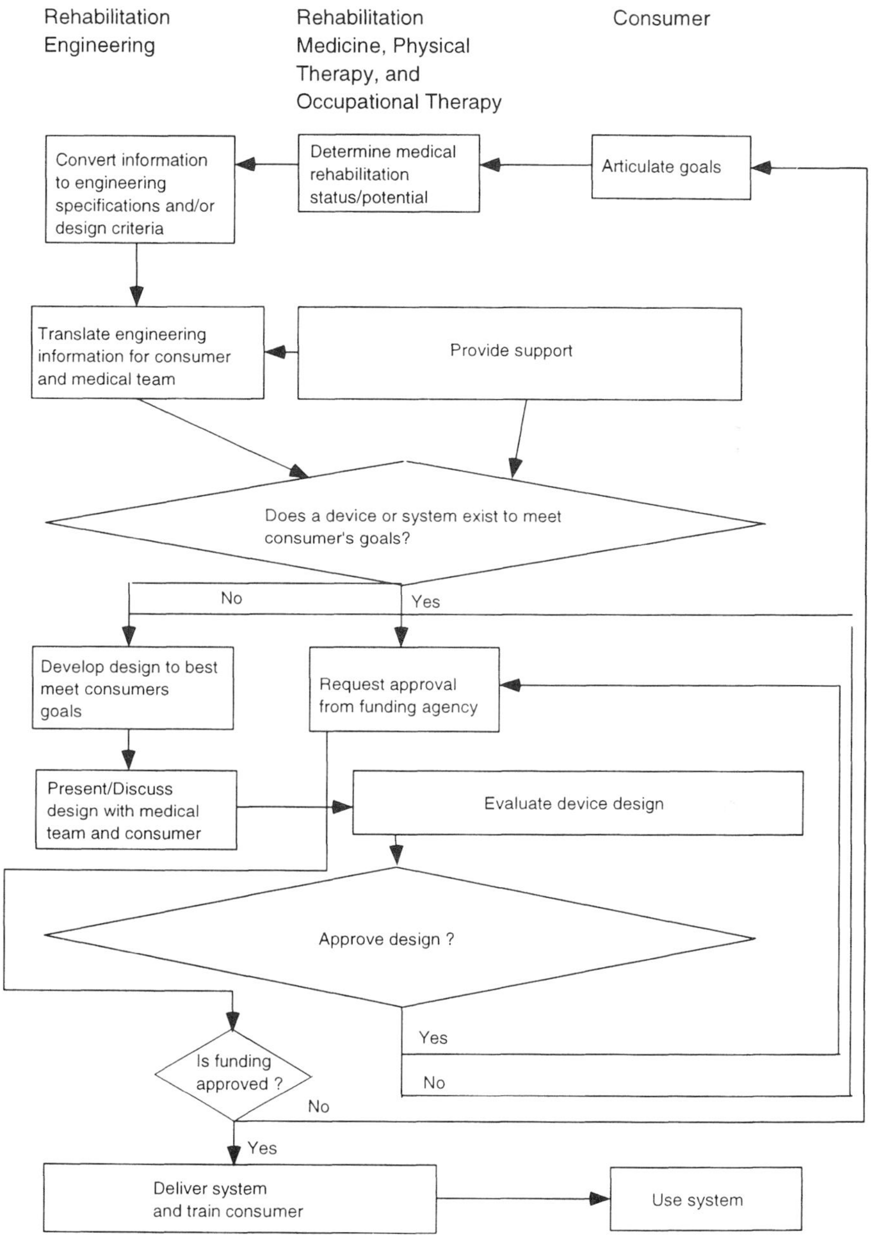

Figure 2.3. *Simple block diagram for total quality management in a rehabilitation engineering service. Several feedback loops illustrate the importance of cross-communication between members of the rehabilitation team. Quality design is dependent upon meaningful feedback.*

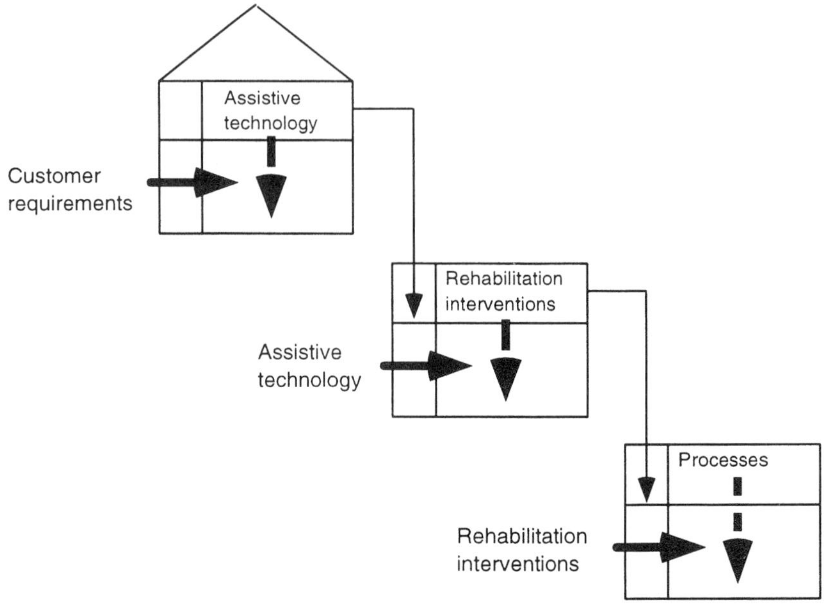

Figure 2.4. *Example of three layers of the house of quality (HOQ) for an assistive technology application process.*

consumer's list of desired tasks. The consumer rates the robotic workstation's ability to assist performing each of the tasks independently (i.e. without a human assistant). A clinician rates the robotic workstation's ability to meet the customer's requirements for the various categories. A score of 1 means that the interaction was of high quality, whereas a score of 5 represents poor quality. The scores of the consumer and the clinician are added and placed in the appropriate box within the matrix. Boxes with high scores correlate with problem areas, and boxes with low numbers represent high performance. With this information the system can be refined to provide the consumer greater quality.

The house of quality provides some numerical feedback as to problem areas, and areas of high quality. The house of quality also provides feedback as to the quality of interaction between various factors. Multiple layers can be used for complex systems and complex interactions. Once problem areas are identified and their interactions with other factors are understood, solutions can be developed.

2.3. STEEL AS A STRUCTURAL MATERIAL

Steel is primarily a combination of iron and carbon. Steel also contains varying amounts of other materials. The percentage of carbon and other

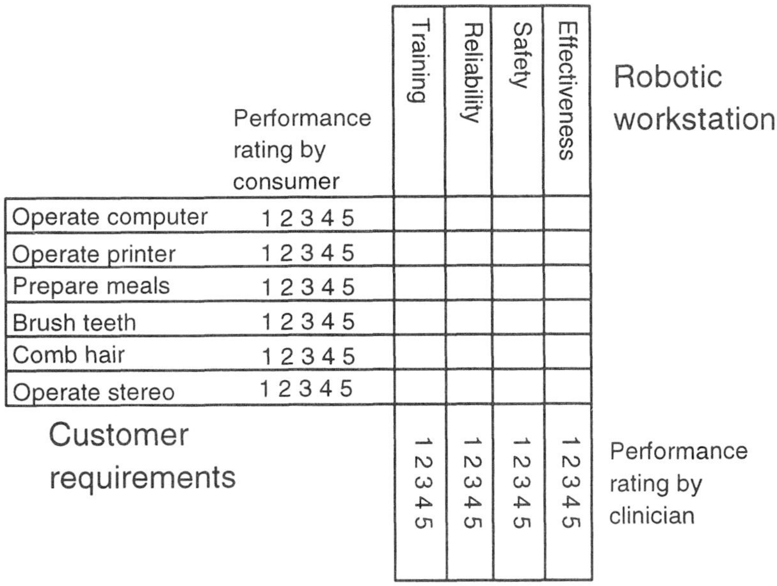

Figure 2.5. *House of quality applied to using a robotic workstation to assist with some activities of daily living. The HOQ could be expanded to incorporate the assistant's evaluation, and other rehabilitation interventions required to make using a robot possible. HOQ data for various interventions can be compared to determine the best solution to the consumer's needs.*

materials mixed with the steel affects the properties and characteristics of the resultant steel. Steel is an important structural material for assistive technology design.

The principle raw materials for steel manufacture are iron ore, iron, steel scrap, coal and limestone. Coal is converted into coke, gas and chemicals in the coke ovens. Coke is used in the blast furnace as a fuel and reducing agent, the gas is burned in heating units and the chemicals are processed into various organic materials. Limestone is employed as a flux in both the blast furnace and steel making furnace where it serves to remove impurities form the melt.

The base material for making steel is pig iron. Pig iron is the product of the blast furnace. Pig iron contains large amounts of carbon, manganese, phosphorous, sulfur, and silicon. In its raw form, pig iron is hard and brittle. Steel making is the process of removing undesirable elements from pig iron and replacing them with desirable elements in the proper proportions. The final elements may be the same as those originally removed, but in different proportions. Most steel is manufactured using open-hearth, basic oxygen or electric arc processes.

Open-hearth furnaces are used for a wide range of steels which need only nominally restrictive control. They can be used to create high-quality basic steels. Basic oxygen furnaces (BOF) involve the same chemical reactions as open-hearth furnaces. However, a BOF introduces nearly pure oxygen into the process at key times. This increases the rate at which steel can be produced. Specialty steels, such as stainless, tool steel, and high-alloy, are made in electric arc furnaces. Electric arc furnaces provide much greater temperature control which is required for making complex steels.

2.3.1. *Properties of steel*

The percentages of various alloying elements has an effect on the properties of the steel. The steel industry has studied the properties of steel with different alloying elements in varying amounts. The purpose of these studies was to develop some standard steels which engineers could use in their designs. These standards were designed to meet significant metallurgical and engineering needs of steel users. The primary organizations responsible for steel standards are the Society of Automotive Engineers (SAE) and the American Iron and Steel Institute (AISI).

A number series is used to identify the chemical composition of carbon steel. The last two numbers of an identifier indicate the approximate mid-range of percentage carbon content. For example, steel designated 1035 has a carbon content ranging from 0.32–0.38%. It is sometimes necessary to deviate from this rule for some carbon ranges or for variations in manganese, phosphorous, or sulfur within a given carbon range. When a five-digit series is used, the last three digits designate the carbon range.

The prefix 'E' in the AISI numbering system is used to designate steels made in an electric furnace. Steels without a prefix are normally manufactured using an open-hearth or basic oxygen process.

2.3.2. *Standard carbon steels*

Steel is defined to be carbon steel when no minimum aluminum, boron, chromium, cobalt, columbium, molybdenum, nickel, titanium, tungsten, vanadium, or zirconium content is specified. No other elements may be added for the purpose of alloying either. Carbon steel has less than 0.60% copper, 1.65% manganese, 0.60% silicon. Standard carbon steels are commonly used for structural components which require moderate strength or substantial machining or forming. They generally have excellent properties for welding, and standard carbon steels can typically be welded by all processes. Table 2.1 shows the first two digits of the AISI numerical code for some standard carbon steels and their properties.

The most common standard carbon steels used in the design and construction of assistive devices are AISI numbers 1010, 1018 and 1020. These are all readily machinable, formable and weldable. They have good strength properties and are readily available.

Table 2.1. *Standard carbon steel designations.*

10xx	Non-resulferized carbon steel grades
11xx	Resulferized carbon steel grades
12xx	Rephosphorized and resulfurized carbon steel grades
15xx	Non-sulfurized high-manganese carbon steel grades

Table 2.2. *Some standard alloy steel designations.*

13xx	Manganese 1.75%
41xx	Molybdenum 0.20 or 0.25%
43xx	Nickel 1.83%, chromium 0.50 or 0.80%, molybdenum 0.25%
47xx	Nickel 1.05%, chromium 0.45%, molybdenum 0.20 or 0.35%
51xx	Chromium 0.80, 0.88, 0.93, 0.95 or 1.00%
61xx	Chromium 0.60 or 0.95%, vanadium 0.13 or 0.15%
86xx	Nickel 0.55%, chromium 0.50%, molybdenum 0.20%
92xx	Silicon 2.00%

2.3.3. Standard alloy steels

Steel is designated to be alloyed if the maximum range for the content for alloying elements meets or exceeds one or more of the following limits: manganese, 1.65%, silicon, 0.60%, or copper, 0.60%. Steel is also considered an alloy if a minimum is specified for any of the following elements: aluminum, boron, chromium, cobalt, columbium, molybdenum, nickel, titanium, tungsten, vanadium, zirconium, or any other alloying element. Table 2.2 shows the first two digits of the AISI numerical code for some standard alloy steels and their properties.

Varying alloying elements and percentages affect different engineering properties. Steel by definition is a combination of iron and carbon. Steel is alloyed with various elements to improve physical properties and to produce special properties. Table 2.3 lists some common alloying elements and the properties they elicit.

Assistive devices are commonly constructed using AISI number 4130, 4340, and 8620 common alloy steels. AISI 4130 is a chromium–molybdenum alloy which is widely used because of its strength, weldability, and ease in fabrication. It can be treated for higher strength and to resist abrasion. AISI 4340 is a chromium–nickel–molybdenum alloy which possesses remarkable ductility and toughness. It has high fatigue strength and is good for highly stressed parts. AISI 8620 is a chromium–nickel–molybdenum alloy which has excellent machinability, and responds well to polishing. It should be stress-relieved after welding.

Table 2.3. *Some common alloying elements and the physical properties they induce.*

Aluminum (Al)	Deoxidizes and degasifies.
Carbon (C)	Raises tensile strength, hardness and resistance to wear and abrasion. Lowers ductility, toughness and machinability.
Chromium (Cr)	Increases tensile strength, hardness, hardenability, toughness, resistance to wear and abrasion, resistance to corrosion.
Cobalt (Co)	Increases strength and hardness. Intensifies effects of other major elements.
Lead (Pb)	Improves machinability
Manganese (Mn)	Deoxidizes, degasifies and reacts with sulfur to improve formability. Increases tensile strength, hardness, hardenability, and resistance to wear. Decreases scaling and distortion.
Molybdenum (Mo)	Increases strength, hardness, hardenability, and toughness. Improves machinability, resistance to corrosion and intensifies the effects of other alloying elements.
Nickel (Ni)	Increases strength and hardness without sacrificing ductility and toughness. Increases resistance to corrosion when introduced in suitable quantities in high-chromium (stainless) steels.
Phosphorus (P)	Increases strength, hardness, and improves machinability. Adds substantial brittleness.
Sulfur (S)	Improves machinability at low heat, can produce brittleness without sufficient manganese at red heat. Decreases weldability, impact toughness and ductility.
Silicon (Si)	Deoxidizes and degasifies. Increases tensile and yield strength. Increases hardness, forgeability and magnetic permeability.
Tungsten (W)	Increases strength, hardness and toughness.
Vanadium (V)	Increases strength, hardness, and resistance to impact. Intensifies effects of other major elements.

2.4. ALUMINUM FOR ASSISTIVE TECHNOLOGY DESIGN

Aluminum is the world's most abundant metal. At present, the only practical source of aluminum is bauxite ore which is 45–65% aluminum oxide. Bauxite ore is carefully treated through a series of chemical processes until alumina (aluminum oxide) is produced. The alumina, cryolite and aluminum fluoride are mixed in an electric smelting furnace where the aluminum is freed and siphoned off to form ingots of usable aluminum.

The aluminum industry uses a four-digit index system for the designation of aluminum and its alloys. The first digit indicates the alloy group according to the major alloying elements.

2.4.1. Common grades of aluminum

The 1xxx series is the basic designation for commercially pure aluminums. In this group the aluminum content is 99%, and there are no major alloying elements.

The second digit indicates modifications in impurity limits. If the second digit is zero, there are no special controls or impurities. Numbers 1 through

Table 2.4. *Major alloying elements.*

1xxx	No major alloying elements
2xxx	Copper
3xxx	Manganese
4xxx	Silicon
5xxx	Magnesium
6xxx	Magnesium and silicon
7xxx	Zinc
8xxx	Other element
9xxx	Unused

9 are used as the second digit to indicate special control of one or more impurities.

The last two digits indicate specific minimum aluminum content. The absolute minimum aluminum content in this group is 99%. The minimum for certain grades may exceed 99%, and these last two digits represent the hundredths of a percent over 99. For example, 1030 would designate 99.30% minimum aluminum content, without special controls. Also 1130, 1230 or 1330 indicate the same minimum level of aluminum content, 99.30%; however, each has special control on one or more impurity. Similarly 1100 designates 99.00% aluminum with control of one impurity.

SAE 1100 is commercially pure aluminum. It is soft, ductile and has excellent workability. It is ideal for intricate form where low strength is acceptable. It is the most weldable aluminum. It has excellent resistance to corrosion and can be given a decorative finish. It is excellent for food handling equipment or for heat sinks.

2.4.2. Aluminum alloy designations

If the first number in the designation is any number between 2 and 9 the material is made of an aluminum alloy. In other words anything but a 1 in the first position of the SAE number designates an alloy. For alloys the second digit of the designation numbers indicates alloy modification. If the second digit is zero, it indicates the original alloy. Other numbers in the second digit indicate alloy modifications. The last two digits serve to identify the different alloys in the group and follow no prescribed pattern. Table 2.4 presents major aluminum alloying elements and their designations.

Experimental alloys are designated according to the four-digit system, but they are prefixed by the letter X. Three aluminum alloys are commonly used within rehabilitation engineering: SAE 2024, SAE 6061, and SAE 7075. These are commonly used as structural materials. SAE 2024 is a common high-strength aluminum alloy. It is used for structural parts which require high fatigue resistance, good strength to weight ratio, high machining, but do not require welding. SAE 2024 does not have good corrosion resistance so

Table 2.5. *Temper designations for non-heat-treatable alloys.*

xxxx	-F	As-fabricated
xxxx	-O	Annealed
xxxx	-H	Strain hardened (followed by 2 or 3 digits)
	-H1	Strain hardened only
	-H2	Strain hardened, partially annealed
	-H3	Strain hardened then stabilized
		Tempering (indicated by second digit)
xxxx	-Hy2	1/4 hard
xxxx	-Hy4	1/2 hard
xxxx	-Hy6	3/4 hard
xxxx	-Hy8	Full hard
xxxx	-Hy9	Extra hard

Table 2.6. *Temper designations for heat-treatable alloys.*

xxxx F		As-fabricated
xxxx-O		Annealed
xxxx-T		Heat treated (always followed by one or more numbers)
	T351:	Solution heat treated, stress relieved, cold worked
	T4:	Solution heat treated, naturally aged
	T5:	Artificially aged only
	T6:	Heat treated, artificially aged
	T10:	Artificially aged, cold worked

it is commonly anodized or used with Alclad. SAE 6061 is an inexpensive and versatile structural aluminum alloy. Most aluminum wheelchair frames are made of this alloy. It offers good mechanical properties and corrosion resistance. It can be welded using most common methods. SAE 7075 is one of the highest strength aluminum alloys. It is ideally suited for high-stress parts. It is not recommended for welded parts.

2.4.3. Aluminum property designations

Aluminum alloy temper designations consist of suffixes to the numeric alloy designations. For example, with 6061 T6, 6061 denotes the alloy and T6 denotes the temper or degree of hardness. The temper designation also indicates the method by which the hardness was obtained. Temper designations differ for heat-treatable and non-heat-treatable alloys, as shown in tables 2.5 and 2.6.

Alloys commonly used for assistive technology typically use T4 or T6 temper designations. The tempering process which yields the best results is dependent upon the type of aluminum alloy. SAE 2024 is commonly specified as 2024 T4, and SAE 6061 T6 is used as 6061 T6. Specifying the proper alloy and temper designation can help to make designs successful.

2.5. USE OF COMPOSITES FOR ASSISTIVE TECHNOLOGY DESIGN

2.5.1. *Plastics*

Plastics is the common name for high polymers. High polymers are complex chains of polymers. Polymers are combinations of two or more monomers. Monomers are basic building blocks made of elements which have an affinity for linking. The bonding of two monomers is called copolymerization. Under the proper circumstances monomers copolymerize and begin to form chains. These chains can begin to build and form more polymers, leading to the development of plastics. In this sense plastics are composite materials.

Despite having shiny, smooth and homogeneous appearances, plastics do not exhibit homogeneity down to the molecular level. The properties of plastics are determined largely by how the monomers are arranged. Thermoplastics have linearly linked bonds, whereas thermosetting plastics have cross-linked bonds. The links of a thermoplastic are free to rotate. The side branches and rotatable backbones of thermoplastics entangle with each other giving the structure flexibility. The entanglements and a relatively strong van der Waals force give many thermoplastics their stiffness. If heat is added the increased motion of the molecules weakens the van der Waal bonds and the plastic can be molded or shaped.

High polymers can be made which have very strong bonds. Three-dimensional linkages of carbon atoms can form covalent bonds. Thermosetting plastics are made by creating extensive three-dimensional covalent linkages between molecules. This process is called cross-linking, and often occurs at elevated temperatures. Ultraviolet radiation may break the covalent bonds in the main polymer chains, reducing strength. Manufacturers use additives to reduce this effect. Table 2.7 lists some common plastics used in rehabilitation engineering.

2.5.2. *Fiber-reinforced composites*

Fiber-reinforced composites are deliberate compositions of strong, stiff fibers in a matrix of a softer material. Glass spun to long, consistent, and strong filaments called fiberglass is the most widely used material in composites. Fiberglass has high tensile strength and low cost. Polyester plastic is widely used as a matrix. However, epoxies have become the standard for engineering applications because of their superior properties. A significant advantage of epoxies is that they are addition polymers, i.e. nothing is given off during the curing process. Condensation polymers give off moisture during curing which must be vented.

The principal drawback of fiberglass as a reinforcement material is its low stiffness. Graphite, Kevlar and boron fibers have high stiffness and

Table 2.7. *Some common plastics for rehabilitation engineering.*

	Specific gravity	Tensile strength (MPa)	Young's modulus (MPa)	Distortion temperature (°C)
Thermoplastics				
ABS	1.05	41	2206	90
Nylon	1.14	81	2758	75
Polycarbonate	1.2	62	2275	130
Polyethylene	0.95	18	1379	49
Thermosets				
Polyester	1.31	59	2344	54
Epoxy	1.1–1.4	28–90	2413	46–280
Epoxy glass	1.6–2	69–138	20960	120–260

Table 2.8. *Properties of some high-performance fibers.*

	Density (kg m^{-3})	Elastic modulus (MPa)	Tensile strength (MPa)
S glass	2491	89 640	3999
Kevlar	1439	124 110	2758
Steel	7750	206 850	3999
Boron	2519	448 180	6412
Graphite	1661	979 090	20 685

nearly the same tensile strength as glass fibers. Other more exotic materials (tungsten and sapphire) have several desirable properties, but their costs are prohibitive.

Kevlar is an organic fiber which is yellow in color and soft to the touch. It is extremely strong, tough and forms a very lightweight composite. Kevlar is resistant to impact, but has rather low compression strength.

Graphite fibers are created by extreme stretching and heating of rayon fibers which changes their molecular structure. Graphite fiber has very low density, a high modulus of elasticity, and a high tensile strength. Table 2.8 presents engineering properties of some common high-performance fibers.

2.6. DESIGN WITH ENGINEERING MATERIALS

Materials have different properties and these properties can be taken advantage of to yield various devices. The properties of the materials must be selected based upon knowledge of the final product and its uses. From

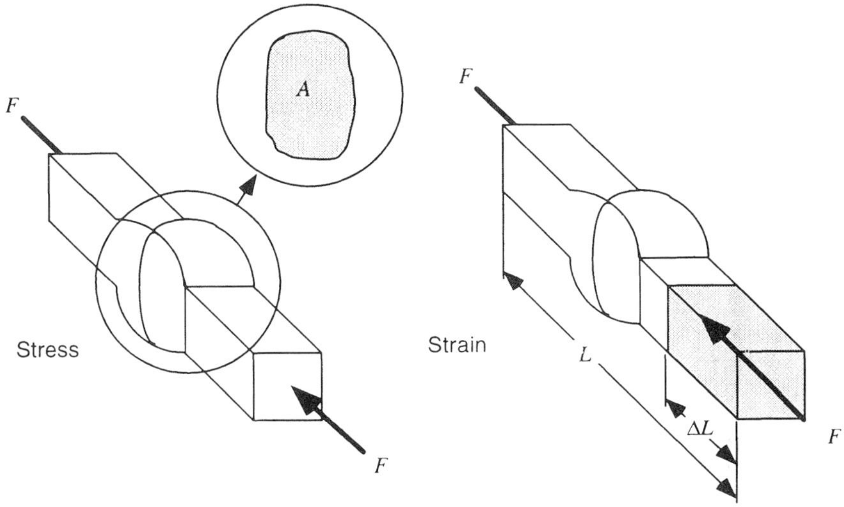

Figure 2.6. *When force is applied to a material it experiences stress and strain. In this illustration compression forces are applied to the end of a beam. The center section of the beam bulges, increasing the maximum cross-sectional area of the beam. The compression forces also cause the beam to shorten from its resting length. The change in the resting length is related to the strain.*

this information the expected loads on the systems can be estimated and used to determine the materials most appropriate for the device final design. Stress and strain within engineering materials and within structures are important for designing safe and effective devices. Stress and strain have simple definitions which are often confused. Stress is the applied force divided by the cross-sectional area of the deformed section. Strain is the ratio of the change in length due to an applied force to the original length, as shown in figure 2.6

2.6.1. Principal stresses

A force (F) acting upon an object will cause the object to deform. The deformation, called the strain, is usually small and disappears when the force is removed. Elastic deformation occurs when the strain returns to zero once a force has been removed. When an object remains in its deformed state it is said to have undergone plastic deformation. Many forming processes take advantage of this behavior. The point at which the deformation changes from elastic to plastic is called the yield point. Stress (σ) is the reaction of the atomic structure of the material in response to an external force

$$\text{stress} = \sigma = \frac{F}{A} \text{ kilopascals} \tag{2.1}$$

where A is the cross-sectional area of the material. Stresses may be perpendicular to the force; these stresses are called tensile or compression forces. Shear stresses act in a plane parallel to the external force.

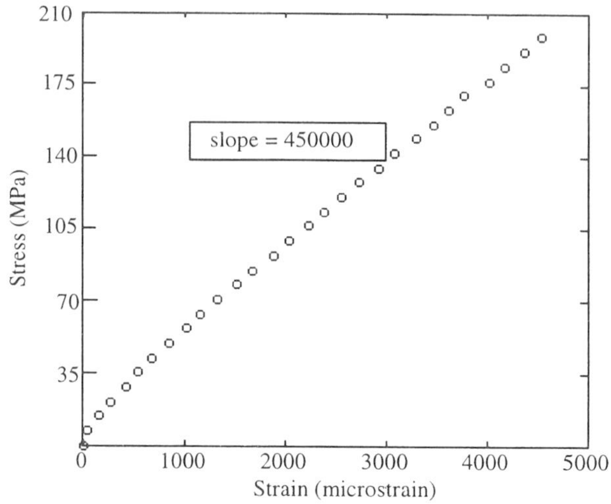

Figure 2.7. *Stress and strain data for a carbon fiber test specimen. The modulus of elasticity is equal to the slope. Note that actual data only approximate a line.*

2.6.2. *Modulus of elasticity*

The modulus of elasticity is the ratio of stress to strain. Strain (ε) is defined as the ratio in the change in length to the total length

$$\text{strain} = \varepsilon = \frac{\Delta L}{L}. \tag{2.2}$$

The modulus of elasticity (E) is a measure of stiffness for a material, and defines the slope of the stress–strain curve in the elastic region during tension (see figure 2.7) or compression loading

$$\text{modulus of elasticity} = E = \frac{\sigma}{\varepsilon} \text{ kilopascals.} \tag{2.3}$$

The modulus of elasticity in shear (G) is the shear stress (σ_s) divided by the elastic deformation (γ)

$$\text{shear modulus of elasticity} = G = \frac{\sigma_s}{\gamma} \text{ kilopascals.} \tag{2.4}$$

2.6.3. *Safety factor*

It is often prudent to use lower permissible stress values than are calculated or found in tables. A safety factor is a number (N) multiplied by the load force or divided by the stress

$$\text{safety factor} = SF = \frac{\sigma}{N} A \quad \text{newtons.} \tag{2.5}$$

The magnitude of the safety factor is based upon our knowledge of variability of stress values, the magnitude of residual stresses, environmental influences, accidental loads, quality of inspection and servicing, and the danger to life and property.

2.6.4. *Bending stresses*

The general equation for bending stresses (tensile and compression) is dependent upon the bending moment (M Nm), the distance from the center of pressure (point at which net force is applied) to the plane in which the stress is under consideration (c m), and the area moment of inertia (I m^4)

$$\text{bending stresses} = \sigma_t \quad \text{or} \quad \sigma_c = \frac{Mc}{I} \text{ kilopascals.} \tag{2.6}$$

2.6.5. *Torsional stresses*

Torsional stresses occur when an element is twisted by shear stresses. Torsional stresses are dependent upon the torsional moment (T Nm), the distance from the point of stress to the neutral axis (c m), and the torsional moment of inertia (J m^4)

$$\text{torsional stress} = \sigma_{tor} = \frac{Tc}{J} \text{ kilopascals.} \tag{2.7}$$

2.7. FABRICATION

Parts must be fabricated and materials worked to develop a device or product. The quality of the final product is a combination of the design, the selection of appropriate fabrication processes, and the skill of the workmanship with which the components and total device are made.

2.7.1. *Welding*

A weld is defined as a local union of materials accomplished by applying heat and pressure with or without fill material being added. The strength of the weld depends upon how the heat and pressure are applied as well as the materials welded and any fill material. Welding permits complex devices to be constructed from readily formable components, i.e. tubes and plates. Welding encompasses three common processes—welding, brazing and soldering. Brazing refers to the use of brass as a filler material when bonding two metals. Soldering refers to the process of bonding two metals at low temperatures ($< 425°$ C), and is commonly used for the assembly of electric circuits. Welding implies that there is no filler material or that the melting point of the filler material is near that of the materials being bonded. Welding includes a number of high-temperature bonding processes.

Torch welding is a process whereby the adjoining edges of the materials to be bonded are heated until they reach a molten state. The materials flow together and form a local bond. The flame, usually oxygen–acetylene, is adjusted to a narrow point, at which the temperature is nearly 6000 °F. A high level of skill is required to get full penetration and proper flow of the fill material. The greatest advantage of torch welding is that the equipment is inexpensive, portable and requires no external power. Torch welding works well for most common standard carbon steels. Torch welding does not work well with many steel alloys or with other metals, i.e. aluminum or titanium.

Electric arc welding using either d.c. or a.c. electric current to create localized heating at the points to be joined. Electric arc welding does not require oxygen to create heat and can thus be used for a greater variety of metals. Shielded metal arc welding (stick welding) is the most common form of electric arc welding. Stick welding uses a consumable electrode. As the weld progresses the electrode melts into the weld to provide filler material. The electrode can be covered with flux to help remove impurities and oxidants from the weld, in which case the electrodes are short 'sticks', or the flux can be in the center of the electrode which permits use of a spool of cored wire for the electrode. Most common steels can be stick welded with the proper electrode.

Gas metal arc welding, also known as metal inert gas (MIG), or CO_2 welding, uses a continuous bare metal wire electrode. The welding zone is shielded by an inert gas, such as CO_2 or argon. This method requires less skill by the user and is suitable for most common aluminums and steels. It is only recommended for pieces less than or equal to 0.25 inches thick.

Gas tungsten arc welding, also known as tungsten inert gas (TIG) welding, uses a non-consumable electrode shielded by an inert gas, helium or argon, and a separate filler rod. TIG can be used with most weldable metals, i.e. steel, aluminum, or titanium. It requires a substantial degree of skill to achieve good welds.

Electric arc welding processes create high localized concentrations of heat. This property permits the welding of materials with high coefficients of thermal conductivity. The shielding permits welding materials which oxidize rapidly.

Thermal plastics can also be welded by heating with electrical filaments. Typically, thermal plastics are welded by applying a heated metal surface to the materials to be bonded and then applying moderate pressure.

2.7.2. Fasteners

Because joints assembled with screws, bolts, and studs can be readily disassembled and reassembled, threaded fasteners have been standardized in such a way that the selection of the proper fastener is resolved by determining critical dimensions.

There is a wide variety of screw types: flat, round, oval, pan, binding, truss, fillister, and hexagon. The selection of the proper screw type depends upon the amount of space available, the probability that someone will come in contact with it and the amount of force the screw must support. The heads may be slotted or recessed (Phillips, frearson or torque). Machine screw points are not finished, i.e. they are flat. They are used in threaded holes or with nuts. Machine screws are commonly used on thin metal sections like those found on electrical equipment.

Slotted cap screws come with round, flat and fillister head styles. Capscrews are available in fluted socket, as used in a typical wrench, and recessed socket, allen capscrews. Capscrews have fine tolerances and finishes. They are typically used on tapped holes or with nuts. They are used in applications where high reliability and strength are required.

Setscrews are used to set the position of collars or pulleys on soft shafts. They are available with a variety of recessed heads. Setscrews come with flat, cone, oval, cup, dog, and half-dog points.

Machine bolts are available with square, hex, rounded, or flat countersunk heads. They make snug, full fit connections and are usually held in place with a nut. They are used for general fastening.

Tap bolts are regular hexhead bolts with fully threaded shanks. They are used in threaded holes wherever a machine bolt or capscrew would be impractical.

2.7.3. Working with adhesives

Epoxy systems differ greatly in their working properties. Useful epoxies require the proper combination of workability and strength. Good epoxies must have very low toxicity. Other desirable characteristics are low water absorption, good material wetting properties without drainage, and low density for a lighter laminate. Composite materials are bonded by wetting the cloth to be used with epoxy. Just enough epoxy to wet all of the fibers should be used. Excess epoxy will make the end product heavier. Once the cloth has been wetted it is layed-up on the mold. For small quantities a styrofoam mold covered in plastic wrap works well. The wetted cloth should be layed in criss-crossing (quasi-isotropic) layers until the desired number of layers is achieved. All materials have to be free from oil, grease and dirt or a strong bond will not be achieved.

2.8. BASIC ELECTRIC CIRCUITS

One of the most attractive properties of electricity compared with other power sources is its mobility. Electrical energy can be transported to any point along a couple of wires and with the proper equipment be converted into light, heat, or motion. An electric circuit is a closed path of electrical elements which permits current to flow. Current through an element is defined as the time rate of change of charge, and its units are amperes

$$i = \frac{\mathrm{d}q}{\mathrm{d}t}. \tag{2.8}$$

Current is represented by i and charge by q. Charge is the fundamental unit of matter responsible for electric phenomena. The voltage across an element is defined as the work required to move a positive one coulomb charge from one side of the element to the other

$$v = \frac{\mathrm{d}w}{\mathrm{d}q}. \tag{2.9}$$

Voltage is represented by v and energy by w. The units of voltage are volts. Power is the time rate of expending or absorbing energy

$$p = \frac{\mathrm{d}w}{\mathrm{d}t}. \tag{2.10}$$

The units of power are watts, and energy is expressed in joules. The power associated with the current through an element can be derived from these basic definitions

$$p = \frac{\mathrm{d}w}{\mathrm{d}t} = \frac{\mathrm{d}w}{\mathrm{d}q}\frac{\mathrm{d}q}{\mathrm{d}t} = vi. \tag{2.11}$$

Using this definition power absorbed by the element is positive. The use of electric power is very important to the design of assistive technology. Many lives have been significantly improved through its proper use.

2.8.1. Linear models

Engineers use models to represent the elements of an electric circuit. Models are generated for most manufactured devices in order to manipulate parameters of the models and to establish bounds on the circuit's operating characteristics. Often simple models are used to make problems more manageable. Models serve to help engineers understand the response of real circuits, but are not substitutes for them. A good model serves to further understand and enhance predictions. Many circuit elements have well defined circuit models. Linearity forms the basis for linear models. Linearity is defined by the excitation and response of the element and on a larger scale by the circuit. Two basic tests are applied to determine linearity: they are the principle of superposition, and the property of homogeneity. These are illustrated in the following equations

Superposition

$$\begin{aligned} i_1 &\Rightarrow v_1 \\ i_2 &\Rightarrow v_2 \\ i_1 + i_2 &\Rightarrow v_1 + v_2. \end{aligned} \tag{2.12}$$

Homogeneity

$$\begin{aligned} i_1 &\Rightarrow v_1 \\ ki_1 &\Rightarrow kv_1. \end{aligned} \tag{2.13}$$

An element or circuit model must pass both of the tests to be linear, otherwise it is said to non-linear. The property of a material to resist the flow of current is called resistivity, denoted by ρ. Resistance (R) is the physical property of an element that impedes the flow of current. The units of resistance are ohms

$$i = \frac{Av}{\rho L} = \frac{v}{R} \Rightarrow v = iR. \tag{2.14}$$

A is the cross-sectional area and L the length of the resistor. Practical resistors are characterized by their nominal resistance, accuracy, power dissipation, and stability as a function of temperature and humidity. The accuracy determines how well the actual resistor can be expected to approximate the nominal value. Accuracy is defined by a tolerance. The most common type of resistor found in electric circuits is made of carbon film. Carbon film resistors are available which dissipate between $\frac{1}{8}$ and 2 W of power. Other materials are used for high-power resistors. Multicolored bands are painted on the resistor body to indicate the nominal value of the resistance, and the tolerance, as shown in table 2.9. Bands a, b, and c give the nominal resistance, and the tolerance band gives the percent the resistor may vary from the nominal value.

A capacitor is a two-terminal element that can be modeled by two conducting plates separated by a dielectric material. Electric charge can be stored on the plates. The capacitance is proportional to the dielectric constant (ε) which determines the energy stored per unit volume per unit voltage difference across a capacitor

$$C = \frac{\varepsilon A}{d} = \frac{q}{v} \tag{2.15}$$

where A represents the area of plates, and d the space between the plates. The units of capacitance are farads. The current through a capacitor can be found by differentiating equation (2.16) below. The capacitor type is determined by its dielectric material. As a practical matter, capacitors are given a voltage rating which is the maximum voltage that can safely be applied to the capacitor. Practical capacitors exhibit leakage currents due to their internal

Table 2.9. *Color code for carbon resistors.*

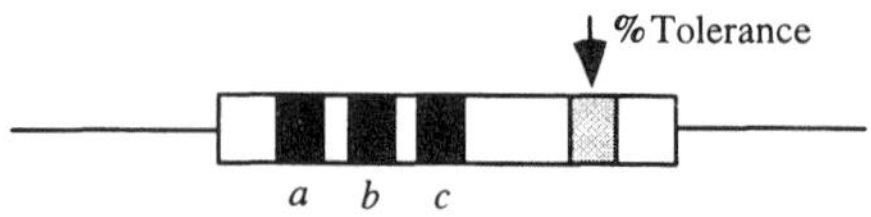

$R = (10a + b)10^c \pm$ %tolerance

Color	Value
Coding for bands a, b and c	
Gold*	−2
Silver*	−1
Black	0
Brown	1
Red	2
Orange	3
Yellow	4
Green	5
Blue	6
Violet	7
Gray	8
White	9
Tolerance band	
Gold	±5%
Silver	±10%

* These are seldom used for coding bands.

leakage resistance. The product of the leakage resistance and capacitance is often given by manufacturers as a specification capacitor loss. Common types of capacitors include ceramic (barium titanate), Mylar, Teflon, polystyrene, and tantalum. These types of capacitors range in value from 100 pF to 1 μF having tolerances 3, 10, and 20%. Resistance–capacitance products are from 10^3 ΩF (ceramic) to 2×10^6 ΩF (Teflon). Larger values of capacitance require the use of electrolytic capacitors. Electrolytic capacitors are constructed of polarized layers of aluminum oxide or tantalum oxide and can yield values of 1 to 100 000 μF. Electrolytic capacitors have high leakage currents with resistance–capacitance products from 10 to 10^3 ΩF. Because electrolytic capacitors are polarized, they must be connected to the circuit with the proper voltage polarity

$$q = Cv \Rightarrow \frac{dq}{dt} = i = C\frac{dv}{dt}. \tag{2.16}$$

A conductor may be shaped into a multiturn coil. The voltage across the coil is proportional to the rate of change of the current. The inductance (L)

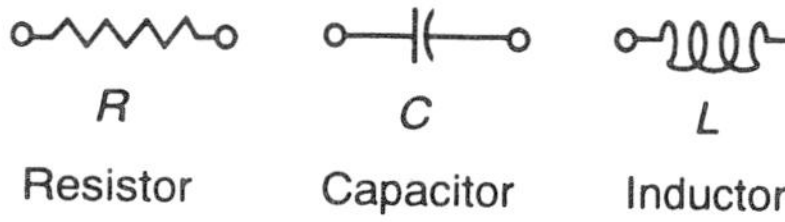

Figure 2.8. *Engineering symbols for a resistor, capacitor, and inductor.*

is the constant of proportionality, which is measured in henrys

$$v = L\frac{\mathrm{d}i}{\mathrm{d}t}. \tag{2.17}$$

Inductors are commonly available with values ranging from 1 μH to 100 H. Practical inductors dissipate small amounts of power due to ohmic losses associated with the wire making the inductor, and core losses due to the induced currents within the core about which the inductor is wound. Large inductors are obtained by using many turns and a ferrous (iron) core; the combination of these two characteristics often increases the power dissipation.

2.8.2. Linear circuits

A closed path in a circuit is a traversal through a series of nodes which ends at the starting node without encountering a node more than once. A closed path is a loop. Ohm's law gives the voltage and current relationships for a single resistor, figure 2.8.

Current and voltage in a circuit are related by Kirchhoff's laws. Kirchhoff's current law states that the algebraic sum of the currents entering a node is identically zero for all time instants. Kirchhoff's voltage law states that the algebraic sum of the voltage drops around any closed path in a circuit is identically zero for all times. The voltage appearing across one of a series of resistors connected in series with a voltage source (see figure 2.9) will be the ratio of its resistance to the total resistance,

$$v_n = \frac{R_n}{R_1 + R_2 + R_3 + \cdots + R_n}. \tag{2.18}$$

This principle is called voltage division. The denominator of equation (2.18) is said to be an equivalent resistance for a series circuit. The current is identical in circuit elements connected in series. Circuit elements are connected in parallel when the voltage across each element is identical (figure 2.10).

The current through one of a set of resistors connected in parallel will be the ratio of its conductance (G_k) to the total conductance (G_{eq})

$$\begin{aligned} i_k &= \frac{G_k}{\sum_{i=1}^{N} G_i} = \frac{(1/R_k)i_s}{1/R_1 + 1/R_2 + 1/R_3 + \cdots 1/R_N} = \frac{G_k i_s}{G_{eq}} \\ &= \frac{(1/R_k)(v_s/R_{eq})}{1/R_{eq}} = \frac{v_s}{R_k}. \end{aligned} \tag{2.19}$$

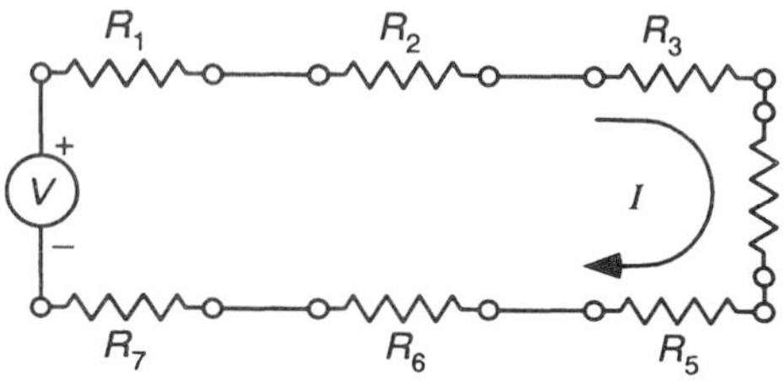

Figure 2.9. *Example of a series resistance circuit. The current (I) is identical in each resistor (R_1, R_2, R_3, R_4, R_5, R_6, R_7). The equivalent resistance is $V/I = R_1 + R_2 + R_3 + R_4 + R_5 + R_6 + R_7$, and from the equations given the voltage across each resistance can be found.*

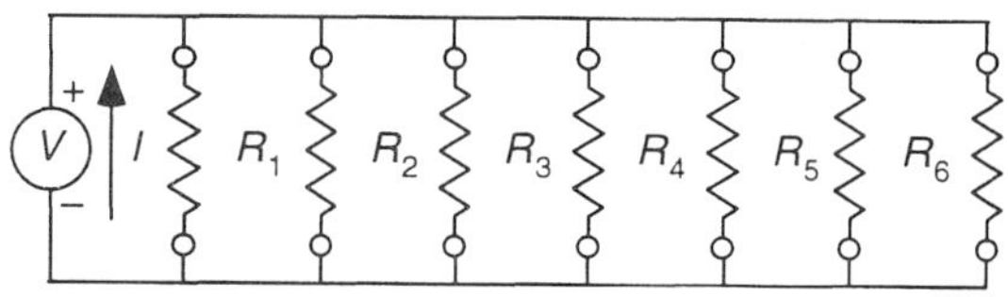

Figure 2.10. *Parallel resistance circuit. The voltage across each resistor is identical. The equivalent resistance for this parallel circuit is* $R_{eq} = \frac{1}{1/R_1+1/R_2+1/R+\cdots 1/R_7}$.

The equivalent capacitance of a set of N parallel capacitors (C_k) is simply the sum of the individual capacitors (C_{eq})

$$C_{eq} = C_1 + C_2 + C_3 + \cdots + C_N = \sum_{k=1}^{N} C_k. \tag{2.20}$$

The equivalent capacitance (C_{eq}) of a set of N series capacitors is equal to the sum of the inverse of each individual capacitance (C_k)

$$C_{eq} = \frac{1}{C_1} + \frac{1}{C_2} + \frac{1}{C_3} + \cdots + \frac{1}{C_N} = \sum_{k=1}^{N} \frac{1}{C_k}. \tag{2.21}$$

The equivalent inductor (L_{eq}) for a set of N inductors connected in parallel is the sum of the individual inductors (L_k)

$$L_{eq} = L_1 + L_2 + L_3 + \cdots + L_N = \sum_{k=1}^{N} L_k. \tag{2.22}$$

The equivalent inductance (L_{eq}) of a set of N series inductors is equal to the sum of the inverse of each individual inductors (L_k)

$$L_{eq} = \frac{1}{L_1} + \frac{1}{L_2} + \frac{1}{L_3} + \cdots + \frac{1}{L_N} = \sum_{k=1}^{N} \frac{1}{L_k}. \tag{2.23}$$

The analysis of many analog circuits can be accomplished by nodal voltage or mesh current analysis. These methods provide systematic procedures for obtaining simultaneous equations describing the circuit behavior. In a planar circuit a branch is a path that connects two nodes. A planar circuit with N nodes will require $N-1$ equations to obtain the voltages at the $N-1$ nodes, since one node serves as the reference node. The voltage across each branch is related to a reference node. The voltage at a node is determined using Kirchhoff's current law at each of the circuit's nodes, with the exception of the reference node. The voltage across each node ($\boldsymbol{v}$) and the source currents ($\boldsymbol{i}_s$) in the circuit are linearly related by the circuit conductance matrix ($\mathbf{G}$) when only independent sources are in the circuit

$$\boldsymbol{i}_s = G\boldsymbol{v}$$

$$\boldsymbol{v} = \begin{bmatrix} v_a \\ v_b \\ \vdots \\ v_N \end{bmatrix} \qquad \boldsymbol{i}_s = \begin{bmatrix} i_{s1} \\ i_{s2} \\ \vdots \\ i_{sN} \end{bmatrix} \qquad \mathbf{G} = \begin{bmatrix} \sum_a G & -G_{ab} & -G_{ac} \\ -G_{ab} & \sum_b G & -G_{bc} \\ -G_{ac} & -G_{bc} & \sum_c G \end{bmatrix}. \tag{2.24}$$

In the conductance matrix, $\sum_k G$ is the sum of the conductances at node k, and G_{ij} is the sum of conductances connecting nodes i and j. If there are $\mathbf{N}$ node voltages to be solved for, the pattern of symmetry remains for $\mathbf{G}$ and the matrix is $N \times N$. If the kth source is not present, then $i_{sk} = 0$. Circuits with dependent sources may be solved similarly with a few modifications.

Kirchhoff's voltage law (KVL) around a closed path can be used to determine the behavior of many planar circuits. A loop is a closed path starting at a node and tracing a path which returns to the original node without passing an intermediate node more than once. A mesh is a loop that does not contain any other loops within it. The mesh current is the current that flows through the elements constituting the mesh. For circuits with only independent voltage sources (v_s) the mesh current i_k in the kth mesh can be found with a general linear equation

$$\sum v_s = \sum_{p=1}^{P} R_p i_k - \sum_{q=1}^{Q} R_l i_q. \tag{2.25}$$

For mesh k, the mesh current i_n is multiplied by all resistances R_j around the mesh. The terms due to the connecting resistances R_i multiplied by the mesh current in the adjacent mesh i_q for all Q adjacent meshes are subtracted from the sum for the mesh of interest. The current and voltage relations for mesh analysis with independent voltage sources can be described by a set of linear equations

$$v_s = \mathbf{R}\boldsymbol{i}. \tag{2.26}$$

In the above equation, $\mathbf{R}$ is a matrix with a diagonal consisting of the sum of resistances in each mesh, and the off-diagonal elements are the negative of the resistances connecting two meshes. The vector $\boldsymbol{i}$ consists of the N mesh currents. The source vector $\boldsymbol{v}_s$ consists of the sum of sources in the jth mesh with the appropriate sign assigned to each source

$$\boldsymbol{v}_s = \begin{bmatrix} v_{s1} \\ v_{s2} \\ \vdots \\ v_{sN} \end{bmatrix}. \tag{2.27}$$

Circuits with independent current sources or dependent voltage sources may be solved using similar methods with accommodations being made for these differences.

2.8.3. *Steady-state sinusoidal analysis*

Sinusoidal forcing functions are particularly important in the analysis of power sources, and communication signals. A forcing function causes a forced response, whereas the natural response is caused by the internal dynamics of the circuit. The natural response in many cases decays with time, while the forced response continues as long as the forcing function is present. Sinusoids are an important class of forcing functions

$$v_s = V_m \sin(\omega t + \phi). \tag{2.28}$$

The amplitude of the sinusoid is V_m, the radian frequency is ω (rad s^{-1}), and the phase shift is ϕ (radians). A sinusoid is a periodic function

$$v_s(t + T) = v_s(t). \tag{2.29}$$

The period of oscillation is T, and the reciprocal of T defines the frequency (f) or number of cycles per second (hertz)

$$f = \frac{1}{T}. \tag{2.30}$$

The angular (radian) frequency (ω) is often more useful for analysis of sinusoidal functions

$$\omega = 2\pi f = \frac{2\pi}{T}. \tag{2.31}$$

A sinusoidal current or voltage at a given frequency can be characterized by its amplitude and phase angle. A real sinusoidal current with magnitude I_m, frequency ω, and phase angle θ can be represented by a simple phasor

$$\begin{aligned} i(t) &= I_m \cos(\omega t + \theta) = \mathrm{Re}\left\{I_m \mathrm{e}^{\mathrm{j}(\omega t+\theta)}\right\} \\ v(t) &= V_m \cos(\omega t + \phi) = \mathrm{Re}\left\{V_m \mathrm{e}^{\mathrm{j}(\omega t+\phi)}\right\}. \end{aligned} \tag{2.32}$$

The notation Re, which represents the real part of the equation, and the redundant $e^{j\omega t}$ can be dropped without loss of generality

$$\begin{aligned} I &= I_m e^{j\theta} = I_m \angle\theta \\ V &= V_m e^{j\phi} = V_m \angle\phi. \end{aligned} \tag{2.33}$$

The abbreviated representation is the phasor notation. Using this notation circuit responses can be calculated in the frequency domain. Analysis in the frequency domain permits many circuits to be solved using algebra and then transformed back into the time domain instead of using calculus in the time domain. When using steady-state frequency domain analysis, many electric components can be transformed into their frequency domain equivalent impedance

$$\begin{aligned} &R \Leftrightarrow R \\ &L \Leftrightarrow j\omega L \qquad j = \sqrt{-1} \\ &C \Leftrightarrow \frac{1}{j\omega C}. \end{aligned} \tag{2.34}$$

The impedance of inductors and capacitors is frequency dependent. Impedances exhibit similar properties to resistors in direct current circuits. However, complex arithmetic must be used when working with impedances. The appearance of inductors or capacitors in a circuit can affect the magnitude and introduce a phase shift in the voltage and current relationships. The impedance of an element (Z) is the ratio of the phasor voltage (V) and the phasor current (I)

$$Z = \frac{V}{I} = \frac{V_m \angle\phi}{I_m \angle\theta} = \frac{V_m}{I_m} \angle\phi - \theta = |Z| \angle\varphi = Z_m e^{j\varphi} = R_m + jX_m. \tag{2.35}$$

Nodal and mesh analysis techniques can be used with impedances just as they are with resistances. Complex arithmetic must be used when working with sinusoidal excitations. The instantaneous power ($p(t)$) delivered to a circuit element is the product of the voltage ($v(t)$) and the current ($i(t)$)

$$\begin{aligned} p(t) &= v(t)i(t) \\ P &= VI = (V_m \angle\phi)(I_m \angle\theta) = V_m I_m \angle\phi + \theta = P_m \angle\beta. \end{aligned} \tag{2.36}$$

The average power (P_{avg}) delivered to a circuit element can be found by integrating the instantaneous power ($p(t)$) over a single period, and dividing by the period (T)

$$\begin{aligned} &Z = Z_m \angle\varphi \qquad v(t) = V_m \cos\omega t \qquad i(t) = I_m \cos(\omega t - \varphi) \\ &P_{\text{avg}} = \frac{V_m I_m}{T} \int_0^T \cos\omega t \cos(\omega t - \varphi)\, dt = \frac{V_m I_m}{T} \left(\frac{T}{2} \cos\varphi \right) \\ &\quad = \frac{V_m I_m}{2} \cos\varphi. \end{aligned} \tag{2.37}$$

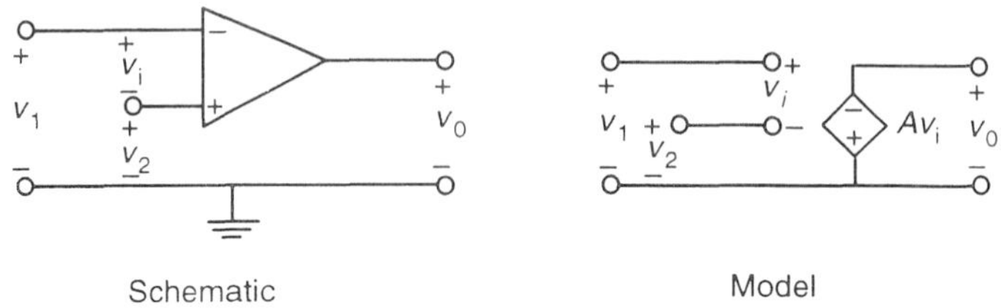

Figure 2.11. *Operational amplifier schematic and model with common reference.*

The average power delivered by a sinusoidal current is related to the effective value of the current (I_{eff}). The average power delivered to an element can be represented in terms of effective voltage (V_{eff}) and current

$$P_{\text{avg}} = \frac{V_m I_m}{2} \cos\varphi = \frac{V_m}{\sqrt{2}} \frac{I_m}{\sqrt{2}} \cos\varphi = V_{\text{eff}} I_{\text{eff}} \cos\varphi. \tag{2.38}$$

The product of the effective voltage (V_{eff}) and effective current (I_{eff}) is called the apparent power and has units of volt-amperes (V A). When dealing with motors the efficiency of power transmission is important. The ratio of the average power to the apparent power is defined as the power factor (pf)

$$pf = \frac{P_{\text{avg}}}{V_{\text{eff}} I_{\text{eff}}} = \cos\varphi. \tag{2.39}$$

The angle φ is often referred to as the power factor angle. Maximum power is transferred from the source to the load if the power factor angle is zero.

2.8.4. *Operational amplifiers*

An operational amplifier is an active element with a high gain ratio designed to be used with other circuit elements to perform a specific function. An operational amplifier is illustrated in figure 2.11.

The input terminal identified by the + sign is the non-inverting terminal, and the terminal identified by the − is the inverting terminal. An operational amplifier is defined by its input–output relationship

$$v_0 = -Av_i = -A(v_1 - v_2). \tag{2.40}$$

The open loop gain (A) is by design large, on the order of 10^6. In an ideal model of an operational amplifier (figure 2.11), the impedance between terminals 1 and 2 is infinite (input impedance), and the amplifier acts like a voltage controlled voltage source. The current flowing between inputs 1 and 2 is quite small, therefore $v_i = v_1 - v_2$ is essentially zero. To be a useful device circuit elements are used in conjunction with an operational

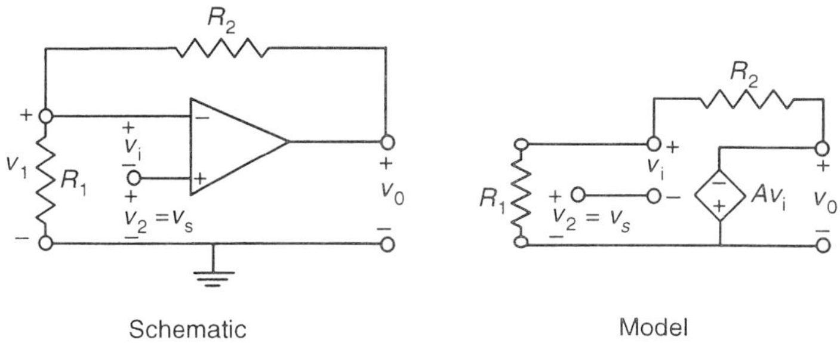

Figure 2.12. *Non-inverting amplifier.*

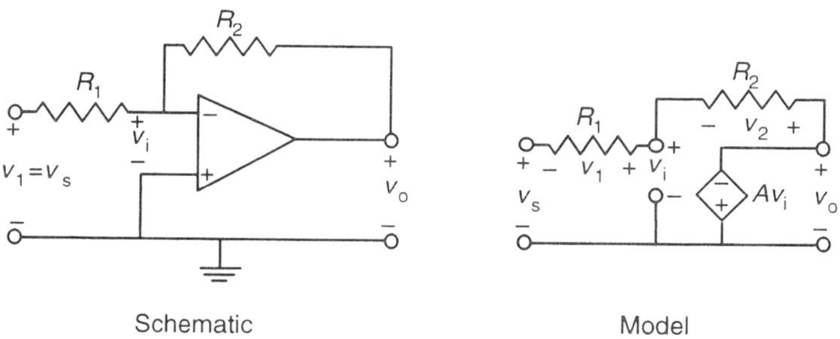

Figure 2.13. *Diagrams for a simple inverting amplifier.*

amplifier. Operational amplifiers are commonly used as feedback devices. A simple non-inverting amplifier is shown in figure 2.12.

This circuit will be analyzed for its input–output properties.

$$\begin{aligned} v_0 &= -Av_i = -A(v_1 - v_2) = -A(v_1 - v_s) = -A(kv_o - v_s) \\ v_o(1 + Ak) &= Av_s. \end{aligned} \tag{2.41}$$

Using voltage division, the relationship between v_1 and v_o can be determined

$$v_1 = \frac{R_1}{R_1 + R_2} v_o = kv_o. \tag{2.42}$$

If A is much larger than one, the non-inverting amplifier gain is a function of the external resistors

$$\frac{v_o}{v_s} = \frac{A}{1 + Ak} \approx \frac{1}{k} = \frac{R_1 + R_2}{R_1}. \tag{2.43}$$

Feedback from the output terminal to the negative input terminal is achieved by connecting the resistor R_2. Feedback elements from the output should normally only be connected to the negative input terminal. An inverting amplifier will yield a response proportional to the negative of the input, as shown in figure 2.13.

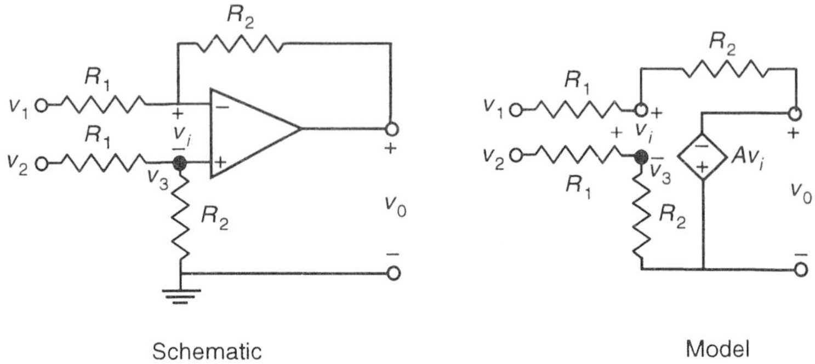

Figure 2.14. *Circuit for a simple differential amplifier.*

The input–output relation for this circuit follows from the ideal model, and voltage division

$$
\begin{aligned}
v_i &= v_s + v_1 \\
v_1 &= \frac{R_1}{R_1 + R_2}(v_o - v_s) = k\,(v_o - v_s) \\
v_o &= -Av_i = -A\,(v_s + v_1) = -A\,[v_s + k\,(v_o - v_s)] \\
v_o(1 + Ak) &= -Av_s(1 - k).
\end{aligned}
\tag{2.44}
$$

The transfer (gain) function is easily derived from these equations

$$
\frac{v_o}{v_s} = \frac{-A(1-k)}{1 + Ak} \approx \frac{-(1-k)}{k} = -\left(\frac{R_2}{R_1}\right). \tag{2.45}
$$

The transfer function illustrates the sign-changing (inverting) property of this operational amplifier configuration. Operational amplifiers can also be analyzed in the frequency domain. Consider the inverting operational amplifier just described with impedance elements substituted for the external resistors. The input–output characteristics can be derived in a similar manner using complex arithmetic

$$
\begin{aligned}
&- V_o + I(Z_1 + Z_2) + V_s = 0 \\
&I = \frac{V_o - V_s}{Z_1 + Z_2} \qquad V_i = IZ_1 + V_s \qquad V_o = -AV_i \\
&SR = \left.\frac{\mathrm{d}v_o}{\mathrm{d}t}\right|_{\max}.
\end{aligned}
\tag{2.46}
$$

Sometimes the differential voltage across an element or between two nodes is required. Under these circumstances, an amplifier referenced to ground will not yield the appropriate measurement. A differential amplifier is used to amplify the difference between the voltages applied to the input (figure 2.14).

The gain of this circuit can be found by using voltage division and the properties of operational amplifiers

$$v_3 = \frac{v_2 R_2}{R_1 + R_2}$$
$$i = \frac{v_1 - v_3}{R_1} = \frac{v_3 - v_o}{R_2}. \tag{2.47}$$

These equations can be manipulated to determine the differential gain for this circuit

$$v_o = \frac{(v_2 - v_1) R_2}{R_1}. \tag{2.48}$$

This differential amplifier has a low input impedance which may not be desirable in many cases. Additional operational amplifiers can be used to increase input impedance, and enhance performance. In most cases only resistors and capacitors are used in operational amplifier circuits. The practical operational amplifier is less than ideal, and requires some special consideration for each application. In some cases, the output voltage is not zero when the input terminals are shorted; this is called the input offset voltage. In some operational amplifiers an external voltage can be applied to zero the input offset voltage. A physical operational amplifier has slight asymmetries that make the output voltage have a small component due to the average of the input voltages (v_1 and v_2 from figure 2.11)

$$v_o = -Av_i + B\left(\frac{v_1 + v_2}{2}\right) = -Av_i + Bv_c. \tag{2.49}$$

The average component (v_c) is called the common mode component. The differential component is $-v_i = v_2 - v_1$. A small common mode component is desirable, the figure of merit called the common mode rejection ratio ($CMRR$) is used to characterize this component of amplifiers

$$CMRR = \left|\frac{A}{B}\right|. \tag{2.50}$$

Typical values range from 10^3 to 10^6. Amplifiers are only linear devices over a limited range. If the input is too great the output will saturate. The slew rate (SR) is the maximum rate at which the output can change when in saturation

$$SR = \left.\frac{\mathrm{d}v_o}{\mathrm{d}t}\right|_{\max}. \tag{2.51}$$

The slew rate determines how fast an amplifier can respond to changes in the input. Amplifiers also exhibit a frequency dependence. The high open loop gain of operational amplifiers rolls off with increasing frequency. The closed loop frequency response is also related to gain. The data sheet for the amplifier should be consulted for high-gain or high-frequency applications. Operational amplifiers are important for many rehabilitation applications.

FURTHER READING AND REFERENCES

Blanchard K and Johnson S 1986 *The One Minute Manager* (Glasgow: Collins)
Cornesky R 1991 *The Quality Professor* (Madison, WI: Magna)
Crosby P B 1988 *The Eternally Successful Organization: The Art of Corporate Wellness* (New York, NY: McGraw-Hill)
Crosby P B 1989 *Let's Talk Quality: 96 Questions You Always Wanted to Ask Phil Crosby* (New York, NY: McGraw-Hill)
Dally J W, Riley W F and McConnell K G 1993 *Instrumentation for Engineering Measurements* (New York, NY: Wiley)
Deal T E and Kennedy A A 1982 *Corporate Culture* (Reading, MA: Addison-Wesley)
Deming W E 1982 *Out of the Crisis* (Cambridge, MA: Productivity)
Dummer G W A and Winton R C 1990 *An Elementary Guide to Reliability* (Oxford: Pergamon)
Dorf R C 1993 *Introduction to Electric Circuits* (New York, NY: Wiley)
Edwards K S and McKee R B 1991 *Fundamentals of Mechanical Component Design* (New York, NY: McGraw-Hill)
Johnson D E, Hilburn J L and Johnson J R 1978 *Basic Electric Circuit Analysis* (Englewood Cliffs, NJ: Prentice-Hall)
Kolstee H M 1984 *Machine Design for Mechanical Technology* (New York, NY: Holt, Rinehart and Winston)
Kotter J P 1990 What leaders really do *Harvard Business Rev.* May–June
Mabie H H and Reinholtz C F 1987 *Mechanisms and Dynamics of Machinery* (New York, NY: Wiley)
Norman R A 1988 *Principles of Bioinstrumentation* (New York, NY: Wiley)
Pietrocola F 1991 Plastics: Will their success story continue? *Med. Des. Mater.* April 42–9
Van Vlack L H 1980 *Elements of Materials Science and Engineering* (Menlo Park, CA: Addison-Wesley)
Waintraub J L and Brumgnach E 1989 *Electric Circuits for Technologists* (St Paul, MN: West)
Webster J G 1992 *Medical Instrumentation: Application and Design* 2nd edn (Boston, MA: Houghton Mifflin)

CHAPTER 3

BIOMECHANICS OF MOBILITY AND MANIPULATION

3.1. HUMAN MOTION ANALYSIS

Human motion analysis is the discipline which describes, analyzes, and assesses human movement and posture. Specific terminology is used to describe human movement and the process of analyzing and assessing it. A quantitative assessment must be preceded by a measurement, description, and analysis process. The primary assessment level uses direct observation. Astute observers document what they observe, monitoring changes, analyzing information, and diagnosing the causes. Quantitative measurements can present data in a convenient manner so that comparisons can be made readily.

Several steps must be completed to achieve an understanding of the biomechanics of a task or posture. The basis for a biomechanical analysis lies in the measurement, description, and monitoring process. Measurement is the quantification of variables related to human movement or posture. Description is the presentation of measured data in a manner which assists in understanding movement or posture. Monitoring is the measurement and description of data over time. Accurate and reliable measurements are required to assess the efficacy of intervention strategies. Calibrated data which have been conditioned to minimize the effects of noise and artifacts can be used for analysis. Analysis involves the mathematical or statistical manipulation of the data to extract information. Analysis may also be used to combine data to create a new variable which provides information not directly measurable. A common important analysis technique involves the calculation of net joint moments and forces from kinetic, kinematic, and anthropometric data.

The most complex task in a biomechanical evaluation is assessment and interpretation. Assessment and interpretation are used to develop and implement positive intervention strategies. Assessment may be used to suggest strategies for correcting or preventing a debilitating posture or movement pattern. Assessment may also be used to determine an individual's progress after intervention.

Biomechanics is closely related to anatomy, physiology, and neurology. The neuromuscular system acts as a control of the release of metabolic energy for the purpose of generating controlled patterns of muscle tension in the presence of passive structures. Muscles act at joints to maintain posture or create movements, which results in the generation or absorption of mechanical energy. At the motor unit level one observes the response to central nervous system signals with the convergence of many excitatory and inhibitory responses. Each muscle demonstrates a second level of convergence in the summation of all motor unit forces in a process described as recruitment. At the joint level the moment force is an algebraic summation of all local agonist/antagonist muscle tensions acting through the moment arm lengths. Integrated central nervous system (CNS) control manifests itself as coordinated control of limbs or the entire body. If performance degrades because of metabolic response (e.g., fatigue) it will manifest itself in changes in muscle forces, moments, and mechanical powers. If the neurological system is deficient biomechanical and electromyographic changes will occur.

3.1.1. Kinematics

Kinematics is the study of motion independent of the forces which cause movement. Kinematic variables include linear and angular displacements, velocities, and accelerations. Data are typically recorded from limb centers of mass, joint centers, or anatomical landmarks, and pertinent points on any devices the person is interacting with. Kinematic data can be related to an absolute (e.g., fixed point on the horizon) or relative (e.g., a point on the body) reference frame. Kinematics can illustrate differences in motion patterns for pathologies.

Kinematic conventions are used so that common references are used throughout the world. Anatomical conventions have existed throughout the medical literature for years. The body and its motions can be accurately described using terms like distal, extension, and abduction. These are relative terms; a convention for an absolute spatial reference frame has been established as well.

A commonly used absolute spatial reference frame denotes the vertical direction as y, the direction of progression (anterior-posterior) as x, and the medial-lateral movement as z, as shown in figure 3.1. Absolute angles in the xy-plane are measured from the horizontal, with positive angles increasing counter clockwise. Velocities are defined as positive for increasing position (e.g., if x is growing larger in the direction of forward progression than the derivative of x or the x-velocity is positive). Acceleration is positive when velocity is increasing. A counter-clockwise angular increase results

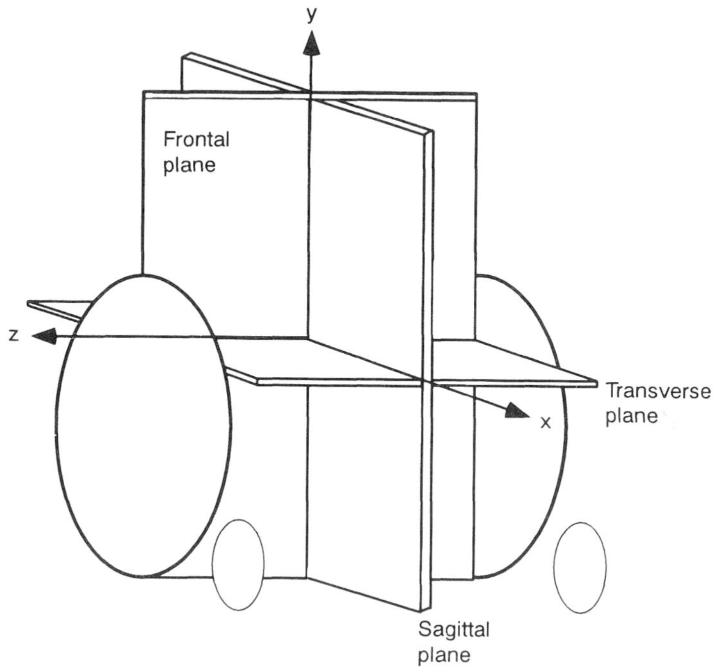

Figure 3.1. *Spatial coordinate system.*

in a positive angular velocity. When angular velocity is increasing then a positive angular velocity results.

Kinematic data can be measured in several ways. Goniometers are devices which attach directly to the body to measure angular or linear position or displacement. Goniometers may be simple mechanical measuring devices such as protractors or calipers used to measure static segment angles, positions, or lengths. More commonly goniometers use an electrical or electronic transducer to convert a linear or angular position to a proportional electrical signal. A variety of transducers can be used to create a goniometer (e.g., strain gage, potentiometer, optical encoder, fiber optic cable). The appropriate goniometer depends upon the task being studied and the information required. Goniometers are generally attached directly to the segment being measured, and hence care should be taken not to impair or alter normal movement. Goniometers are appropriate for measuring small movements (e.g., motions of the hands and fingers) or when desiring a simple immediate relative measure (e.g., knee flexion–extension angle). An advantage of goniometers is that they can measure motion in the plane of the joint regardless of absolute position. There are some disadvantages to goniometers: absolute angle and displacements are not readily available, proper fitting can be difficult and time consuming, movement may be encumbered, and measuring complex joints may be difficult.

Accelerometers are sometimes used to directly measure acceleration as their name implies. Accelerometers are typically small force transducers which incorporate a fixed mass. When applying Newtonian mechanics force equals mass times acceleration, thus when there is a fixed mass, changes in force are proportional to acceleration. Several types of force transducers are suitable for measuring acceleration: metal foil strain gages, piezoelectric material, and piezoresistive strain gages. Triaxial (single-axis accelerometers mounted along orthogonal axes) are required to determine the resultant acceleration vector. If the limb rotates then multiple triaxial accelerometers are required to determine resultant acceleration. Accelerometers are typically electrically configured in a Wheatstone bridge to convert changes in the transducer to an electrical signal. Typically, the output from the bridge is quite small (e.g., on the order of a few millivolts) thus requiring amplification. Many suitable instrumentation amplifiers exist for this purpose. Accelerometers are best for measuring vibration or impact where large accelerations may occur with small motions. Image-based techniques are best when measuring gross motion.

Because of the complexity of some motions an imaging system is the most appropriate means of capturing the data. To gather dynamic data many images sequenced in time must be gathered. Images may be collected using film or electronic media (e.g., video). From the point of the optics, film and electronic cameras are essentially the same. Using an electronic media has several advantages (e.g., lower cost, electronic image stabilization, automatic digitizing, rapid sampling rates) over film, and thus dominates the motion analysis market. A single camera can capture motion in its plane of view. Therefore motion in a plane requires only a single camera. Motion in three-dimensional space requires at least two cameras. Multiple cameras are often used because each camera is only capable of tracking markers in its view plane. Multiple cameras minimize the possibility that a marker may be obstructed from one of the camera's view planes. Multiple cameras produce redundant data which can be combined to create a simulated 3D view using a homogeneous transformation. Data is collected at a fixed frequency. Therefore the camera provides a 2D image of its view for each sampling time. This provides time–motion data which can be used to evaluate the movement of posture.

The converted coordinate frame data are called raw data. Filtering and/or smoothing are often applied to the raw data to minimize the error (i.e., maximize the signal to noise ratio). Kinematic data are often corrupted by noise. Noise may come from digitization error, shuttering, quantization error, obscuring of markers, and a number of other sources. One of two assumptions are made about the noise spectrum: either the noise is of higher frequency than the signal or it is white noise (i.e., constant power across the frequency spectrum). If the noise frequency is correctly assumed higher than the signal, then an appropriate low-pass filter will remove much of the noise. If the noise power is assumed white, then more sophisticated techniques are required.

Several techniques may be applied to create suitable *conditioned* data. Much of the noise which pollutes kinematic data is of higher frequency than the signal. This noise can be easily removed using a suitable low-pass filter (e.g., Butterworth, Chebychev). There are several ways to choose the appropriate cut-off frequency. The appropriate cut-off frequency for the filter can be determined by using an estimate of the signal with noise power spectral density. Many software packages (e.g., Matlab, Mathematica) provide routines for calculating power spectral density estimates. A decision can be made based upon the power spectral density as to which frequency should be the cut-off, then an appropriate order low-pass filter can be used to condition the data. Another method is to perform a residual analysis of the difference between the filtered and unfiltered data over a wide range of cut-off frequencies. The inflection point of the residual curve can be used to select the appropriate cut-off frequency. The residual at any cut-off frequency is given by

$$R_x = \sqrt{\frac{1}{N}\sum_{i=1}^{N}(X_i - \hat{X}_i)^2} \qquad R_y = \sqrt{\frac{1}{N}\sum_{i=1}^{N}(Y_i - \hat{Y}_i)^2}. \tag{3.1}$$

Researchers and clinicians often attempt to optimize the sampling frequency of the motion (kinematic) and force/torque (kinetic) data. This results in two different sampling rates. To calculate multimodal data (e.g., net joint moments, net joint forces), the two sampling rates must be synchronized. This has been accomplished using various spline techniques. However, polynomial spline methods only provide local synchronization, are heavily dependent upon boundary conditions, and do not guarantee stable results. The periodicity of the kinematic data may be exploited to develop the optimal cosine series representation for smoothing and synchronization.

Kinematic data are often cyclic. Each marker of the kinematic model (e.g., shoulder, elbow, wrist, knuckle) of each subject generates x and y sets of data which may be periodic. The frequencies of the x and y data are dependent upon the anthropometry of the individual, and the speed of movement.

The periodic nature of the kinematic model can be exploited to develop a cosine series representation of x and y for each marker from a set of kinematic data including several trials

$$\begin{aligned} \hat{x}(t) &= \sum_{i=1}^{N} A_i \cos(\omega_i t + \theta_i) \\ \hat{y}(t) &= \sum_{j=1}^{N} B_j \cos\left(\bar{\omega}_j t + \varphi_j\right). \end{aligned} \tag{3.2}$$

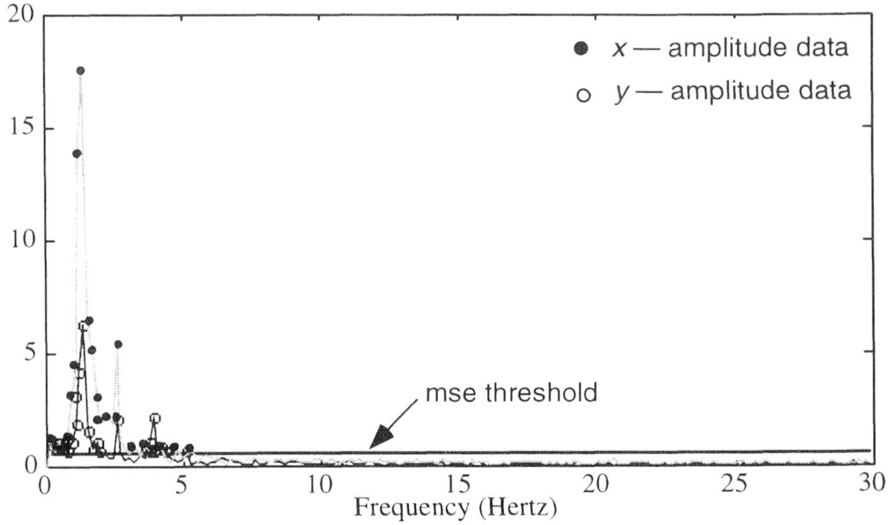

Figure 3.2. *Amplitude spectral density with threshold for kinematic data from a manual wheelchair propulsion wrist marker.*

In order to determine the optimal cosine series representation of the data, the proper amplitude threshold must be selected. The d.c. component was removed from the x and y data for each kinematic marker to avoid biasing the results based upon marker spatial location (i.e., the shoulder motion may be small relative to its location with respect to the hub)

$$\tilde{x}(t) = x(t) - \bar{x} \qquad \tilde{y}(t) = y(t) - \bar{y}. \tag{3.3}$$

The frequency, amplitude, and phase for the cosine series representation (3.1) were selected using an estimate of the amplitude spectrum of the x and y data. The length of the fast Fourier transform (FFT) is denoted by n.

$$\begin{aligned} A(\omega) &= \frac{2}{n}\text{FFT}(\tilde{x})\text{FFT}(\tilde{x})* \qquad & \theta(\omega) &= a\tan\left(\frac{\text{Im FFT}(\tilde{x})}{\text{Re FFT}(\tilde{x})}\right) \\ B(\bar{\omega}) &= \frac{2}{n}\text{FFT}(\tilde{y})\text{FFT}(\tilde{y})* \qquad & \varphi(\bar{\omega}) &= a\tan\left(\frac{\text{Im FFT}(\tilde{y})}{\text{Re FFT}(\tilde{y})}\right). \end{aligned} \tag{3.4}$$

The slower discrete Fourier transform (DFT) is used, instead of the (FFT), for data which were not radix 2. The amplitude threshold can be taken as the mean divided by the maximum of the amplitude spectrum for x and y respectively.

The cosine series estimates of x and y were calculated in a loop with the base amplitude threshold being multiplied by the iteration variable each pass, as shown in figure 3.2

$$K_x = \frac{1/n\sum A(\omega)}{\max(A(\omega))} \qquad K_y = \frac{1/n\sum B(\bar{\omega})}{\max(B(\bar{\omega}))}. \tag{3.5}$$

The maximum inflection point of the residuals, equation (3.1), can be used to select the appropriate amplitude threshold for smoothing. Smoothing or filtering of data is very important before derivatives are calculated as noise may be amplified by the differentiation process. Given the smoothed coordinate data from each of the anatomical markers, the limb-segment angles can be calculated.

If markers are in line with the long bone axis it is simple to calculate the absolute and relative segment angles. The absolute angles are calculated using the absolute positions for the segment proximal (P) and distal (D) markers.

$$\theta_{D,P} = \tan^{-1} \frac{y_P - y_D}{x_P - x_D}. \tag{3.6}$$

The segment angles are absolute in the defined external spatial reference frame. The joint angles can be calculated using the absolute angles from two adjacent segments. Flexion and extension as well as adduction and abduction are defined relative to each joint. One must be aware of the convention for all link-segments being studied. The joint angles can be calculated using the difference between the absolute segment angles

$$\phi_J = \theta_{D-1,P-1} - \theta_{D,P}. \tag{3.7}$$

Some precaution must be taken to account for the joint-angle boundary. This will prevent joint angles from being defined outside their anatomical range. Once data have been suitably filtered and/or smoothed, velocity and acceleration can be calculated from position data. When determining the velocity and acceleration it is important to maintain synchronization with position data. A simple method for differentiating position with the resultant velocity and acceleration occurring at the same time is to use a basis of $2\Delta t$ sampling periods. Other more sophisticated reports can be used as well

$$\begin{aligned}
v_{x_i} &= \frac{x_{i+1} - x_{i-1}}{2\Delta t} & v_{y_i} &= \frac{y_{i+1} - y_{i-1}}{2\Delta t} \\
a_{x_i} &= \frac{v_{x_{i+1}} - v_{x_{i-1}}}{2\Delta t} & a_{y_i} &= \frac{v_{y_{i+1}} - v_{y_{i-1}}}{2\Delta t} \\
a_{x_i} &= \frac{x_{i+2} - 2x_i + x_{i-2}}{2\Delta t} & a_{y_i} &= \frac{y_{i+2} - 2y_i + y_{i-2}}{2\Delta t}.
\end{aligned} \tag{3.8}$$

Velocity and acceleration of markers can lead to further insight into the underlying features of the motion. Angular velocities and accelerations can be calculated in a similar manner.

3.1.2. Kinetics

The study of forces and the resultant energetics of the human body are referred to as kinetics. An understanding of forces can lead to an understanding of the origins of human movement. Moments and forces can be measured directly by surgically implanting transducers. Typically, implanting a transducer is only possible under special circumstances. One case where implantable transducers are particularly valuable is with implantable prostheses. Instrumented prostheses can give clinicians valuable information about their use.

For most applications, indirect estimates of internal forces are calculated from measurements of external forces, kinematic data, and anthropometric data. Calculating joint moments of the critical joints (e.g., wrist, elbow, knee, hip) during wheelchair propulsion or ambulation may help to further the understanding of possible injury mechanisms. A general link-segment model can be developed using an inverse solution with the Newton–Euler method based on a variable degree-of-freedom body coordinate system (i.e., the number of body segments is not fixed). A recursive matrix back-propagation algorithm can be derived for determining three-dimensional net joint moments from kinetic, kinematic, and anthropometric data. The algorithm is computationally efficient and numerically stable.

Evaluating joint moments during wheelchair propulsion of the shoulder via kinetic and kinematic analysis may identify causes of musculoskeletal overloading and rotator cuff injuries. Carpal tunnel syndrome and other wrist and hand injuries are also experienced by wheelchair users. Research in this area suggests that carpal tunnel syndrome may be the result of nerve compression which occurs during forceful exertions with the hand and wrist in hyperflexion or hyperextension. Repetitive strikes of the heel of the hand against the pushrim may cause pain and numbness of the thumb and fingers. The ability to accurately calculate joint moments for the wrist, elbow, and shoulder may help in establishing preventative measures for wheelchair users.

Net joint moments may also be helpful in determining effective intervention strategies of people with gait impairments. Net force and torque patterns can assist a physician or therapist in evaluating the rehabilitation of a person who has had a total hip replacement. These techniques can also be applied to developing surgical intervention strategies for children with crouched gait. Joint moment and force information is important for the surgeon, therapist, and kinesiologist in their diagnostic assessments.

Knowledge of the time profiles of joint forces and moments are necessary for an understanding of the cause and implications of any movement. Muscle and tendon forces of humans most often must be determined indirectly from available kinematic, anthropometric, and reaction force data. Rigid link-segment models are often used to calculate net joint reaction forces and net muscle moments. Useful calculation of net joint reaction forces and net muscle moments requires a full kinematic description, accurate anthropometric measures, and a full description of the external forces.

Several assumptions are made when using a link-segment model to estimate net joint moments. Accurate measures of segment masses, centers of mass, joint centers, and moments of inertia must be available. Each segment is assumed of fixed mass which is modeled as a point mass located at the center of mass (COM). The segment COM remains stationary with respect to the segment, i.e., there is no relative motion. The mass moment of inertia of each segment about its proximal end remains constant during the motion. The length of each segment remains constant during the motion, i.e., no migration. Accurate kinematic and external force-torque (kinetic) data must be available which are temporally and spatially synchronized. All body joints are considered either hinge or ball and socket joints.

Four basic types of forces act upon link-segment models: (1) gravitational forces; (2) external forces; (3) muscle and ligament forces; and (4) inertial forces. Gravitational forces act downwards through the COM of each segment and are equal to the mass times acceleration due to gravity. External forces must be measured using an external sensor. Such forces typically act over some area and are modeled by a force vector acting upon the center of pressure. Muscle and ligament forces are computed in terms of net muscle moments and net joint forces. Co-contraction, joint friction, and passive properties of muscle and tendons prevent determining actual muscle forces. Inertial forces always oppose motion and are a function of each limb-segment's resistance to change in motion.

Each body segment acts independently under the influence of reaction forces, muscle moments, inertial forces, and force due to gravity. A feature of link-segment models is that the free-body diagrams for each segment are similar. This feature can be exploited to develop a single recursive algorithm with back-propagation to calculate net muscle moments, and net joint forces.

To develop a three-dimensional net muscle moment model and net joint force model, three free-body diagrams will be used. Muscle moments and joint forces are determined from the free-body diagrams using force and moment balance equations.

$$\begin{aligned} &\sum F_x = Ma_x \qquad \sum F_y = Ma_y \qquad \sum F_z = Ma_z \\ &\sum M_{Px} = I_{Px}\alpha_{Px} \qquad \sum M_{Py} = I_{Py}\alpha_{Py} \qquad \sum M_{Pz} = I_{Pz}\alpha_{Pz}. \end{aligned} \tag{3.9}$$

Net joint moment equations can be derived from the free-body diagrams of link-segment models. However, the equations of motion can be programmed in any number of forms, some more efficient than others. A numerically stable algorithm for efficiently computing three-dimensional net muscle moments and net joint forces for n-degree-of-freedom link-segment models will be developed. A matrix approach will be used to develop an algorithm for computing three-dimensional net muscle moments and net joint forces. The similarity of the structure of the free-body diagrams for link-segment models can be exploited.

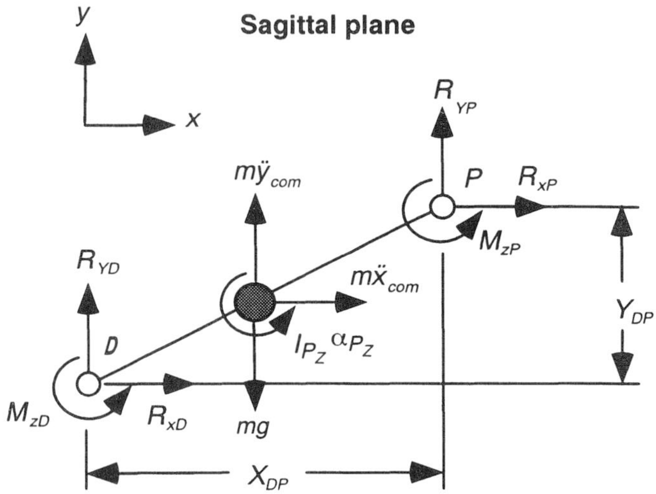

Figure 3.3. *Complete sagittal plane free-body diagram of a single segment.*

The moments and forces acting in the sagittal plane were determined with vector addition using force and moment balance equations (figure 3.3)

$$\begin{aligned} &R_{xP} + R_{xD} - m\ddot{x}_{com} = 0 \\ &R_{yP} + R_{yD} - m\ddot{y}_{com} - mg = 0 \\ &M_{zP} + M_{zD} + R_{xD}y_{DP} - R_{yD}x_{DP} + mg(x_{DP} - x_{com}) - I_{Pz}\alpha_z = 0. \end{aligned} \tag{3.10}$$

The moments and forces acting in the frontal plane can be determined with vector addition using force and moment balance equations (figure 3.4)

$$\begin{aligned} &R_{zP} + R_{zD} - m\ddot{z}_{com} = 0 \\ &R_{yP} + R_{yD} - m\ddot{y}_{com} - mg = 0 \\ &M_{xP} + M_{xD} + R_{zD}y_{DP} - R_{yD}z_{DP} + mg(z_{DP} - z_{com}) - I_{Px}\alpha_x = 0. \end{aligned} \tag{3.11}$$

The moments and forces acting in the transverse plane can be determined with vector addition using force and moment balance equations (figure 3.5).

$$\begin{aligned} &R_{zP} + R_{zD} - m\ddot{z}_{com} = 0 \\ &R_{xP} + R_{xD} - m\ddot{x}_{com} = 0 \\ &M_{yP} + M_{yD} + R_{zD}x_{DP} - R_{xD}z_{DP} - I_{Py}\alpha_y = 0. \end{aligned} \tag{3.12}$$

The following definitions simplify transforming equations (3.10)–(3.12) into matrix form: for $x_{com} = K_{x_{DP}}$, $y_{com} = K_{y_{DP}}$, $z_{com} = K_{z_{DP}}$, then

$$a_{xP} = K\,(\ddot{x}_P - \ddot{x}_D) \qquad a_{yP} = K\,(\ddot{y}_P - \ddot{y}_D) \qquad a_{zP} = K\,(\ddot{z}_P - \ddot{z}_D)\,. \tag{3.13}$$

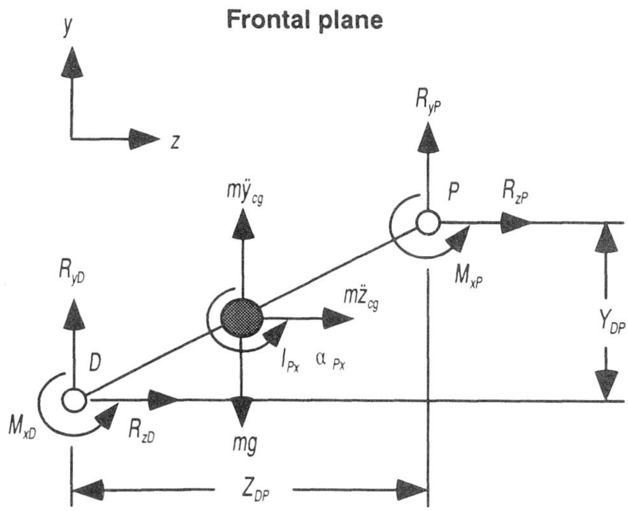

Figure 3.4. *Complete frontal plane free-body diagram of a single segment.*

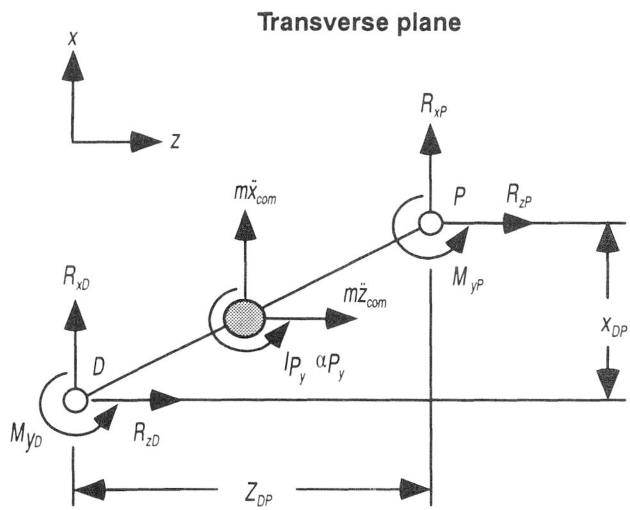

Figure 3.5. *Complete transverse plane free-body diagram of a single segment.*

The inertias (I_p) about the x, y, and z axes of the fixed reference frame are related to the inertias (I_s) of the segment about the axes fixed to the limb by a homogeneous rotation matrix. By selection of the Eulerian angle description the finite rotation sequence can be made commutative. The vector triad ($\boldsymbol{I}$, $\boldsymbol{J}$, $\boldsymbol{K}$) is fixed to the reference along X, Y, Z axes and another triad ($\boldsymbol{i}$, $\boldsymbol{j}$, $\boldsymbol{k}$) is fixed to the moving segment along the x, y, z axes; any arbitrary finite rotation can be expressed in terms of the Euler angles (ψ, θ, ϕ).

$$I_p = T I_s T^{-1} = T I_s T^T = \begin{bmatrix} I_{xx} & I_{xy} & I_{xz} \\ I_{xy} & I_{yy} & I_{yz} \\ I_{xz} & I_{yz} & I_{zz} \end{bmatrix}$$

$$T = \begin{bmatrix} c\phi c\theta & s\phi c\theta & -s\theta \\ -s\phi c\psi + c\phi s\theta s\psi & c\phi c\psi + s\phi s\theta s\psi & c\theta s\psi \\ s\phi s\psi & -c\phi s\psi + c\psi s\theta c\psi & c\theta c\psi \end{bmatrix}. \tag{3.14}$$

The angular accelerations about the proximal end, $\alpha_x, \alpha_y, \alpha_z$, are the second derivatives of each segment absolute angle about the appropriate axis. The centripetal acceleration component of the moments is found from the cross product of the inertia matrix with the angular velocity vector.

$$\begin{bmatrix} \tilde{M}_x \\ \tilde{M}_y \\ \tilde{M}_z \end{bmatrix} = I_p \begin{bmatrix} \alpha_x \\ \alpha_y \\ \alpha_z \end{bmatrix} + (\omega \times I_p \omega)$$

$$= I_p \begin{bmatrix} \alpha_x \\ \alpha_y \\ \alpha_z \end{bmatrix} + \begin{bmatrix} 0 & -\omega_z & \omega_y \\ \omega_z & 0 & -\omega_x \\ -\omega_y & \omega_x & 0 \end{bmatrix} \begin{bmatrix} I_{xx} & I_{xy} & I_{xz} \\ I_{xy} & I_{yy} & I_{yz} \\ I_{xz} & I_{yz} & I_{zz} \end{bmatrix} \begin{bmatrix} \omega_x \\ \omega_y \\ \omega_z \end{bmatrix}.$$

Equations (3.10)–(3.14) can be written as a matrix equation using the following definitions

$$\mathbf{r}_P = \begin{bmatrix} r_{xP} \\ r_{yP} \\ r_{zP} \\ M_{xP} \\ M_{yP} \\ M_{zP} \end{bmatrix} \quad \mathbf{r}_D = \begin{bmatrix} r_{xD} \\ r_{yD} \\ r_{zD} \\ M_{xD} \\ M_{yD} \\ M_{zD} \end{bmatrix} \quad \mathbf{a} = \begin{bmatrix} a_{xD} \\ a_{yD} \\ a_{zD} \\ \alpha_x \\ \alpha_y \\ \alpha \end{bmatrix} \quad \mathbf{M} = \begin{bmatrix} 0 \\ m \\ 0 \\ m(1-K)z_{DP} \\ 0 \\ m(1-K)x_{DP} \end{bmatrix}$$

$$\mathbf{I} = \begin{bmatrix} m & 0 & 0 & 0 & 0 & 0 \\ 0 & m & 0 & 0 & 0 & 0 \\ 0 & 0 & m & 0 & 0 & 0 \\ 0 & 0 & 0 & I_{xx} & I_{xy} & I_{xz} \\ 0 & 0 & 0 & I_{xy} & I_{yy} & I_{yz} \\ 0 & 0 & 0 & I_{xz} & I_{yz} & I_{zz} \end{bmatrix}$$

$$\mathbf{\Phi} = \begin{bmatrix} -1 & 0 & 0 & 0 & 0 & 0 \\ 0 & -1 & 0 & 0 & 0 & 0 \\ 0 & 0 & -1 & 0 & 0 & 0 \\ 0 & z_{DP} & -y_{DP} & -1 & 0 & 0 \\ z_{DP} & 0 & -x_{DP} & 0 & -1 & 0 \\ -y_{DP} & x_{DP} & 0 & 0 & 0 & -1 \end{bmatrix}$$

$$\Omega = \begin{bmatrix} 0 & 0 & 0 & 0 & 0 & 0 \\ 0 & 0 & 0 & 0 & 0 & 0 \\ 0 & 0 & 0 & 0 & 0 & 0 \\ 0 & 0 & 0 & 0 & -\omega_z & \omega_y \\ 0 & 0 & 0 & \omega_z & 0 & -\omega_x \\ 0 & 0 & 0 & -\omega_y & \omega_x & 0 \end{bmatrix} \qquad \omega = \begin{bmatrix} 0 \\ 0 \\ 0 \\ \omega_x \\ \omega_y \\ \omega_z \end{bmatrix}.$$

The variables defined above can be combined to form a single matrix equation

$$\mathbf{r}_p = \Phi \mathbf{r}_D + \mathbf{Ia} + \Omega \mathbf{I} \omega + \mathbf{M} g. \tag{3.15}$$

Equation (3.15) can be used to calculate 3D net muscle moments and joint reaction forces for a particular 'snap-shot' of a single segment.

The particular application determines the implementation of equation (3.15). For real-time implementation, the joint moment vector is calculated for the most distal end, the point where the reaction vector is measured, and then its negative is used as the reaction vector for the next most proximal link-segment model. This process is repeated until all desired joint moment vectors for a particular snap-shot have been calculated. With the anthropometric, kinematic, and most distal joint-segment kinetic data defined, a real-time algorithm for each snap- shot can be outlined as follows

$$\begin{aligned} &\textit{for } j = 1, n \\ &\mathbf{r}_p(j,t) = \Phi(j,t)\mathbf{r}_D(j,t) + \mathbf{I}(j,t)\mathbf{a}(j,t) \\ &\qquad + \Omega(j,t)\mathbf{I}(j,t)\omega(j,t) + \mathbf{M}(j,t)g \\ &r_D(j+1,t) = -r_p(j,t) \\ &\textit{end} \end{aligned} \tag{3.16}$$

where j is the joint-segment index, and n is the number of joint-segments of interest in the model. The most distal joint-segment is defined as joint 1 (figure 3.6). The algorithm in equation (3.16) would be called for each snap-shot (time sample) to compute the joint reaction vector for joint-segments of interest.

Currently, post-processing is the only means of determining muscle moments and joint reaction forces. This is due to limitations in image-processing systems, and the integration of kinetic and kinematic data. When post-processing all of the anthropometric, kinematic and most distal link-segment, kinetic data are available. Most systems store these data in matrix format.

The kinetic data set for the most distal link-segment is defined as the reaction matrix

$$\mathbf{R}_D(j) = \left[r_j(t_0) r_j(t_1) \cdots r_j(t_N) \right] \qquad \mathbf{R}_D = R^{6xN+1}.$$

The accelerations (linear and angular) for each joint segment can be written as an acceleration matrix

$$\mathbf{A}(j) = \left[a_j(t_0) a_j(t_1) \cdots a_j(t_N) \right] \qquad \mathbf{A} = R^{6xN+1}.$$

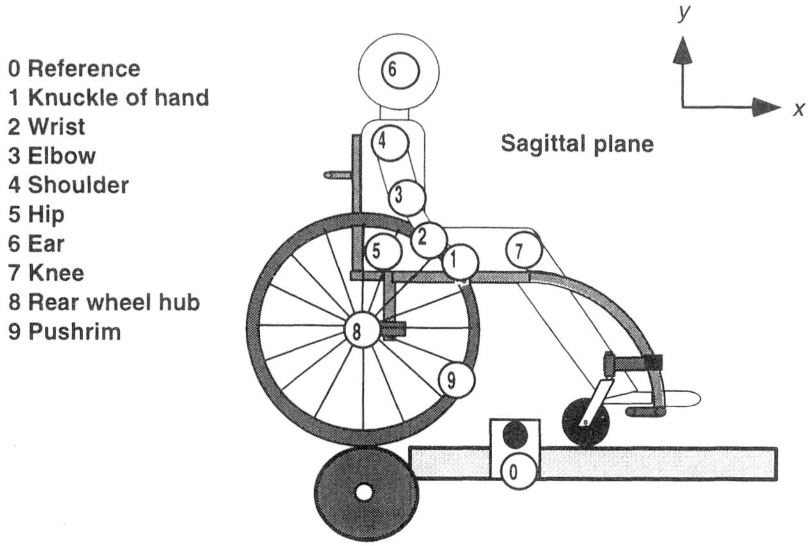

Figure 3.6. *Marker locations.*

The mass and moment arms of the Earth's gravitation can be combined into a single matrix for each joint-segment

$$\tilde{\mathbf{M}}(j) = \left[M_j(t_0)M_j(t_1)\cdots M_j(t_N)\right] \qquad \tilde{\mathbf{M}} = R^{6xN+1}.$$

Selecting the form of the kinematic data matrix can improve the efficiency of the algorithm

$$\mathbf{X} = \begin{bmatrix} x_{1DP}(t_0) & y_{1DP}(t_0) & z_{1DP}(t_0) & \cdots & x_{nDP}(t_0) & y_{nDP}(t_0) & z_{nDP}(t_0) \\ x_{1DP}(t_1) & y_{1DP}(t_1) & z_{1DP}(t_1) & \cdots & x_{nDP}(t_1) & y_{nDP}(t_1) & z_{nDP}(t_1) \\ \vdots & \vdots & \vdots & \ddots & \vdots & \vdots & \vdots \\ x_{1DP}(t_N) & y_{1DP}(t_N) & z_{1DP}(t_N) & \cdots & x_{nDP}(t_N) & y_{nDP}(t_N) & z_{nDP}(t_N) \end{bmatrix}.$$

All of the matrices are of the proper dimension to calculate the muscle moments and joint reaction forces. However, the reaction matrix requires some preconditioning which makes the post-processing algorithm slightly more complex than the real-time algorithm.

$$
\begin{aligned}
&\textit{for } j = 1, n \\
&\textit{for } t = 0, N \\
&\tilde{R}_D(j,t) = \Phi(j,t)\mathbf{R}_D(j,t) \\
&I_p(j,t) = T(j,t)I_s(j)T^T(j,t) \\
&\tilde{\Omega}(j,t) = \Omega(j,t)I(j,t)\bar{\omega}(j,t) \\
&\tilde{I}(j,t) = I(j,t)A(j,t) \\
&\textit{end} \\
&\mathbf{R}_p(j) = \tilde{R}_D(j) + \tilde{I}(j) + \tilde{\Omega}(j) + \mathbf{M}(j)g \\
&\mathbf{R}_D(j+1) = -R_p(j) \\
&\textit{end}
\end{aligned}
\tag{3.17}
$$

This approach is computationally efficient because the matrix structure takes advantage of computer data structures. The algorithms only require simple matrix multiplication and addition. The matrices are triangular or block diagonal which simplifies computation. All transformations are numerically stable. The algorithms can be implemented with any number of languages (C, Basic, or FORTRAN) or software packages (Matlab, Mathcad, or Mathematica). Computations of muscle moments and joint reaction forces show a substantial speed-up with the matrix approach compared with directly programming model equations. In addition, the algorithm is independent of model order. This makes the algorithm useful for any number of biomechanical models.

Link-segment models assume that each joint is a hinge or ball joint and that the joint moment is due to a pure joint torque. If joint moments were due to a pure joint torque, then reaction forces at each joint would be the same as the bone-on-bone forces. However, muscles do not exert pure torque, they act more like a linear motor which produces additional compressive and shear forces across the joint surfaces. For an accurate representation of the joint bone-on-bone forces, the muscle-induced forces must be overlaid onto the free-body diagram. At the extreme ranges of motion of joints tendon and ligament forces must also be included.

3.1.3. Anthropometry

Anthropometry can be simply defined as the physical measurement of the human body. Anthropometric data can be used to assist rehabilitation engineers in the design of work spaces, controls, and other human interface devices. Human measurements are used to develop and prescribe products which will be functional for a large portion of the population. A useful rule is to design for 95% of the target population. There are two basic types of anthropometric measures which are commonly made. Static measurement is the characterization of fixed properties of the body (e.g., mass, density, limb lengths). Functional measurements are those taken with people assuming common movement positions or performing movement tasks which help to determine the properties of the body (e.g., range of motion, reach).

Often when making a biomechanical analysis the body is considered as a system of mechanical links. The physical size and form of the body segments are some of the most common anthropometric measures. Segment lengths are determined using anatomical landmarks, commonly readily identifiable joints. Scientists have carefully dissected cadavers and estimated joint centers of rotation and have determined model sets of data lengths as functions of total body length. Drillis and Contini showed that measurements based upon palpaple bony landmarks are highly correlated with measurements from cadaver studies.

The properties of segments are important to biomechanical analyses. Each segment has a unique composition of bone, muscle, fat, and other tissue. Gravity influences the motion of segments as well as external and internal forces. Control and support mechanisms for people with severe mobility impairments must consider the effect of gravity. One need only hold his/her arms elevated in front of them for any length of time to experience the effect of gravity. Cadaver studies have been used to determine estimates of segment densities. Segment densities are related to their masses and volumes

$$\text{density} = \frac{\text{mass}}{\text{volume}}. \tag{3.18}$$

Body-segment volumes can be determined on living subjects, hence segment mass on living people can be estimated using their actual segment volume and a segment density estimate. Segment densities are typically closely related to total body density. Segment center of mass is also an important anthropometric measure. The mass of a segment can be represented by the sum of the mass of small portions of the limb

$$m_{\text{segment}} = \sum_{i=1}^{n} m_i = \sum_{i=1}^{n} \delta_i V_i \approx \delta \sum_{i=1}^{n} V_i \tag{3.19}$$

where m_i is the mass, δ_i is the density, and V_i is the volume of the ith portion of the segment. The center of mass must create the same moment due to gravity with respect to any point along the segment as the original mass distribution. The center of mass can be defined as the distance q from the proximal end of the segment

$$mq = \sum_{i=n}^{n} m_i q_i \Rightarrow q = \frac{1}{m} \sum_{i=1}^{n} m_i q_i. \tag{3.20}$$

When performing many biomechanical analyses the center of mass of multiple segments is required. When a person is moving one or more segments, the body center of mass location is changing, therefore the center of mass must be calculated as a time series

$$x_0(t) = \frac{m_1x_1(t) + m_2x_2(t) + m_3x_3(t) + \cdots m_nx_n(t)}{m}$$
$$y_0(t) = \frac{m_1y_1(t) + m_2y_2(t) + m_3y_3(t) + \cdots m_ny_n(t)}{m}. \tag{3.21}$$

The center of mass location is most useful for helping to provide support for posture and stability. If rotational accelerations are involved in movements then inertial resistance must be factored into analyses. Inertia is a measure of the ability of the segment to resist changes in angular velocity. The magnitude of inertia depends upon the axis about which rotation takes place, and is minimum about the center of mass. The moment of inertia is a constant, and can be calculated with respect to the segment's distal end given the mass of each small portion

$$I = \sum_{i=1}^{n} m_i q_i^2 \quad \text{or} \quad I = \int q^2 \, \mathrm{d}m. \tag{3.22}$$

For most analyses the moments of inertia are considered to act around either the joint or mass center. The moment of inertia about a joint center, when multiplied by the segment angular acceleration, provides an estimate of the instantaneous joint moments and forces. This information can be combined with static measurements for functional postures to produce useful kinetic analysis. In cadaver studies or with paralyzed limbs, segments can be swung as a pendulum around the appropriate joint axes. The period of oscillation of the pendulum can be used to estimate moment of inertia

$$I = \frac{WL}{4\pi^2 f^2} \tag{3.23}$$

where W is the weight of the segment, L is distance from pivot to segment center of mass, and f is the frequency of oscillation. Several other effective techniques have been developed to determine moments of inertia on living people. Many tables of anthropometric values use the radius of gyration (ρ) as a means of presenting moment of inertia data.

$$I_0 = m\rho_0^2. \tag{3.24}$$

Often it is convenient to calculate the moment of inertia about some axis parallel to the center of mass, but located at some other point in the plane. This is extremely useful in stability and postural control analyses. The parallel axis theorem makes translating the moment of inertia about the center of mass to another axis in the plane simple

$$I = \frac{m}{2}(x - \rho_0)^2 + \frac{m}{2}(x + \rho_0)^2 = m\rho_0^2 + mx^2 = I_0 + mx^2. \qquad (3.25)$$

The distance x can be any direction from the center of mass as long as the axis of rotation is parallel to the axis of the moment of inertia about the center of mass.

Anthropometric measures are useful in evaluating simple measurements of an individual's ability to perform some tasks. Strength, reach, and dexterity tests are required to assess a person's ability to operate devices or perform job functions. Engineers and scientists must be aware that well motivated and/or trained individuals can voluntarily improve their performance on functional tests. It is important to make multiple measurements to assure reliable data. Segmental mass determinations, moments of inertia and center of mass locations are dependent on knowing the density and volume of various tissues within the segment and how they are distributed throughout the segment. For instance, we know that cortical bone has a specific gravity of greater than 1.8, fat is less than 1.0 and muscle is greater than 1.0. Additionally, the distribution of these tissues within the segment is not symmetrical or homogeneous. Both of these factors contribute to the difficulty of determining body-segment parameters for the living person. Segment lengths must also be measured in order to determine relative locations of centers of mass. Determinations of joint centers and end-of-segment locations must be determined for external anatomical features of the living subject and decisions made regarding where one segment ends and the next begins.

There are a number of techniques which have been used to determine body-segment parameters. Information from a few cadaver studies utilizing small numbers of specimens from a restricted population (generally male, Caucasian specimens) have been widely used by researchers. The limited diversity and numbers of cadavers used for these determinations presents problems when a different living population is being studied. This is particularly true for studies requiring estimations of body-segment parameters for the spinal cord injured, as this population through disuse and overuse of different areas of the body has proportionally different mass and density distributions of body tissues than the able-bodied population.

Cadaveric studies are difficult to do because of the length of time required to dissect and measure segmental information, therefore making it hard to obtain information on large numbers for representative populations. In fact, Miller and Nelson noted that fewer than 50 cadavers had been studied over the past century. In their chapter on body-segment parameters, Miller and Nelson provide a summary of cadaver studies. Most of the studies involved very few specimens and most were adult male Caucasians. Additionally, the majority of the work has focused on determining segment masses proportional to whole body mass and the location of segmental centers of mass. Clauser *et al* had the largest sample ($n = 14$), but they only reported group statistics. Other populations were studied by Meeh, male and female infants, while Mori and Yamamoto, and Fujikawa characterized specific body-segment parameters for male and female Japanese.

Different body-segmental parameters were determined in each of the studies. Dempster and Clauser *et al* performed the most comprehensive measurements including anthropometric measurements, segmental masses, centers of gravity, and density. In addition to providing the data on his specimens, Dempster also determined moments of inertia and presented one of the most rigorous methodologies. He employed a balance plate to measure segmental centers of gravity, a free-swinging pendulum technique for moment of inertia calculations, and volumes were determined by immersion.

A number of tables have been developed relating segmental masses to mass of the entire body. Barter combined the data of Braune and Fischer, Fischer and Dempster to increase the sample size in order to develop regression equations relating segment masses to total body mass. Clauser *et al* also developed regression equations from their data predicting segment masses from total body mass. Immersion techniques have been used to determine segmental masses. One of the difficulties is that it is difficult to use the technique for the head and trunk. The determination is based on the fact that a segment immersed in water will displace an amount of fluid equal to its volume. If a certain density is assumed—which in most studies was the same as water, 1, then mass can be determined. Zook measured 164 boys by slowly immersing them into a tank of water. The volume of the overflow was measured when the water levels reached the ankle, knee, crotch, navel, bottom of sternum, top of sternum, auditory canal and top of the head. Dempster used a similar technique to measure segment masses except that he used different tanks to accommodate the different segments. Clauser *et al* refined these techniques and used a two tank closed system for their measurements which did not rely on spillage or the subjects immersing their limbs. This technique reduced errors due to spillage and subject movement.

Locations of segment center of masses have been estimated assuming fixed relationships between the length of the segments and location of the center of mass. Harless, Braune and Fischer, Dempster, and Clauser *et al* determined the location of segment centers of mass as a proportion of total segment length. As a general note, Braune and Fischer stated that extremity segment center of mass locations were approximately 44% from the proximal end while Dempster indicated that they were 43%.

Few studies have been completed on computing moments of inertia for body segments. Hay discussed the two techniques primarily used—the compound pendulum and mathematical modeling. The compound pendulum is based on measuring the period of oscillation of a body segment hung from a fixed point. The following expression allows the moment of inertia to be calculated

$$I = \frac{W_j T^2}{4p^2} \tag{3.26}$$

where W is the weight of the object, p is the distance from the center of gravity to the fixed suspension point, and T is the period of oscillation. This technique was used by Fischer and Dempster to determine moments of inertia for various cadaver body segments. Santschi *et al* determined whole body moments of inertia using this technique for eight different body positions using 66 subjects. Drillis and Contini made plaster casts of limb segments and measured the moments of inertia using the compound pendulum technique. In order to determine the moments of inertia, a correction factor was incorporated relating the relative weights of the actual segment and the cast.

Some researchers have determined radii of gyration from cadaveric moment of inertia information. Fischer determined that generally for the extremity segments, radius of gyration for moments of inertia for all axes passing through the mass center and aligned perpendicular to the long axis were 0.3 of the segment length. Plagenhoef used the data of Dempster to determine radii of gyration for moments of inertia in relationship to axes perpendicular to the long segment axes and passing through either the distal or proximal end.

Other methods for determining segmental parameters have been used with varying degrees of success. Williams and Lissner, and Drillis and Contini estimated segmental weights by utilizing a reaction-change technique or as it is sometimes referred to a board and scale technique. Subjects lie on a board which is supported on a scale at one end and a narrow ridge on the other end is supported by the floor. By applying static principle of moments, a determination of center of mass of the whole body can be made. Subsequent movements of segments from the anatomical position result in a determination of segmental weight proportion. The drawbacks of this system are that the segment center of mass locations must be known and segmental moments of inertia cannot be determined.

Photogrammetry has been used to estimate segmental volumes. Pierson describes this technique. Essentially, equal width colored bands of opaque material were projected onto the subject. The individual was then photographed from different positions. Calculations of volumes were made by determining the areas occupied by the projected bands on the person's body. Digitization of the areas occupied by each band is a cumbersome process and limits the ability of the researcher to collect large samples.

High-energy gamma rays were used by Caspar to determine the mass, center of mass and moments of inertia for non-biological materials. The principle of this technique was that the transmitted photons picked up by the detector were directly related to the mass of the substance. Brooks used the same technique to determine body-segment parameters for a leg of lamb. The gamma ray technique was found to be accurate when compared with other standard techniques. In living subjects brief scanning times and low radiation levels would be required before the technique could be used safely.

Drillis and Contini describe a quick-release technique for estimating the moment of inertia of a segment with respect to a horizontal axis passing through the proximal joint. The analysis is based on the idea of producing an acceleration of the limb independent of muscle contraction. If the force at the time of release is known and the angular acceleration calculated, the moment of inertia can be determined based on the relationship

$$\tau = I\alpha \tag{3.27}$$

where τ = torque—force applied through a transducer times its distance to the axis of rotation, I is the unknown moment of inertia and α is the angular acceleration determined from the accelerometers. If two accelerometers are used, the effect of gravity is eliminated by using the differential output from the accelerometers. The limitations with this technique include the inability to obtain the moments of inertia about the three principle axes of all the segments, isolating individual segments, and the difficulty of obtaining data on the head and trunk.

Several mathematical models have been developed which determine body-segment parameters. Whitsett used 14 rigid, homogeneous segments in his model and estimated all segment masses, except the head and trunk, for utilizing Barter's regression equations. Dempster's data were used to find center of mass locations for the upper and lower extremities, except for the feet and hands. Mid-points along the symmetrical axes were used to approximate mass center locations for these segments and for the head and trunk. Segmental dimensions were measured on a group of living subjects. Hanavan developed a 15-segment model based on modeling segments as an ellipsoid for the head, elliptical cylinders for the upper and lower trunk, solid spheres for the hands, and frustra of right circular cones for the the remaining portions of the upper extremity and all segments of the lower extremity. Hanavan also used Barter's regression equations to develop a set of equations for determining individual segment center of mass locations and moments of inertia about the three principal axes. A set of anthropometric data including lengths, diameters and circumferences were required inputs to determine segment parameters for individual subjects. Hanavan compared his results with those determined experimentally on the data from 66 subjects collected by Santschi *et al*. Hanavan's model parameters compared favorably with the experimental results, the moment of inertia calculations falling within 10% of those values. Miller and Morrison compared two versions of Hanavan's model to predict segment weights, specific gravities and principal moments of inertia. It was suggested that Hanavan's model utilizing Clauser *et al*'s multi-step weight distribution regression equations should be the model used in biomechanical investigations, although the authors recommended that efforts should be continued to refine Hanavan's model.

3.2. GAIT ANALYSIS

Unimpaired human walking can be defined as a method of locomotion involving the use of two legs, alternately, to provide both support and propulsion with at least one foot being in contact with the ground at all times. Gait describes the manner or style of walking. There are several important temporal and spatial factors which are involved in gait analysis and which must be defined in order to discuss types of gait. The gait cycle is divided into the stance phase and swing phase. During the stance phase, a foot is in contact with the ground as the body passes over it. During the swing phase, a leg moves forward through the air. Single support describes the period when only one foot is in contact with the walking surface. During the dual stance period (i.e., double-support period), both feet are concurrently in contact with the walking surface. The temporal sequencing of unimpaired gait is presented in figure 3.7.

The basic spatial gait parameters most frequently used are stride length, step length, walking base, and toe-out angle (see figure 3.8). Stride length is the distance between two successive placements of the same foot. Step length is the distance by which the chosen, marked, foot moves forward in front of the other foot. Walking base is the width of the stride (i.e., the side-to-side distance between the centerline of each foot). Toe-out angle is the angle between the direction of progression and the centerline, midline, of the foot.

The basic temporal gait parameters most frequently used are velocity, cadence (step frequency), stride time, and step time. Velocity is the rate of linear motion along the line of progression of the body's center of mass. Cadence (step frequency) is the number of steps taken per minute, thus it is a measure of half cycles. Stride time is the amount of time required to complete an entire gait cycle (e.g., from heel strike to heel strike). Step time is the amount of time between heel strikes or in some cases foot contacts (e.g., as in toe walking) with the designated, marked, foot.

A single normal gait pattern is impossible to define as there is too much variation among individuals. Measurement of gait is affected by anthropometric (e.g., gender, size, mass, age) and experimental (e.g., indoors versus outdoors, length of walkway, floor surface) conditions. There are several features that are characterisitc of unimpaired gait. Unimpaired gait is symmetrical (i.e., equal left and right step lengths and step times). There is consistency between successive stance phases as well as between successive

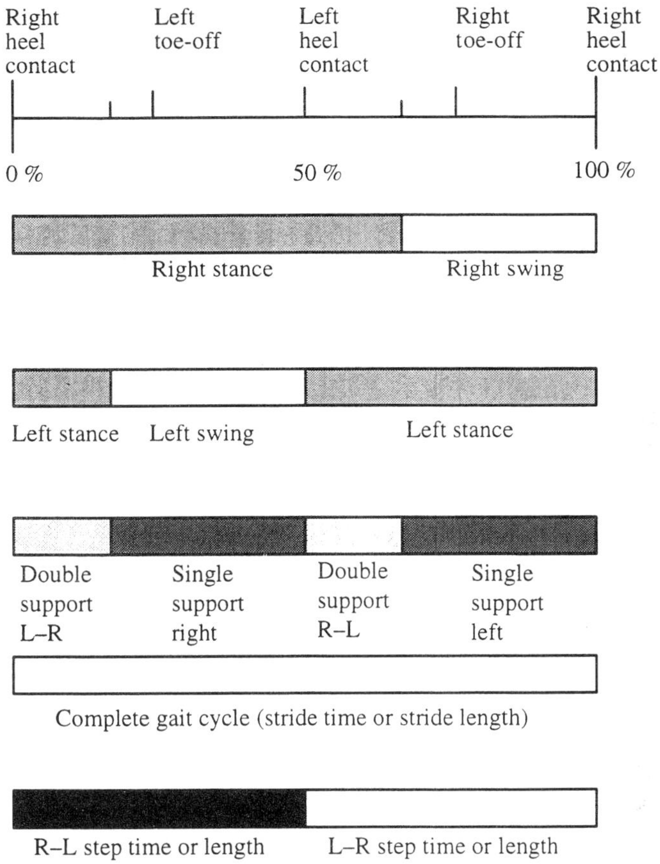

Figure 3.7. *Basic temporal and spatial parameters of an unimpaired gait cycle.*

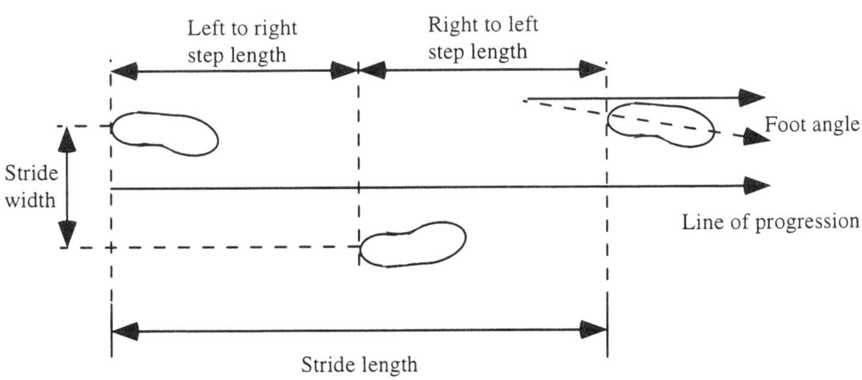

Figure 3.8. *Basic spatial parameters used in gait analysis.*

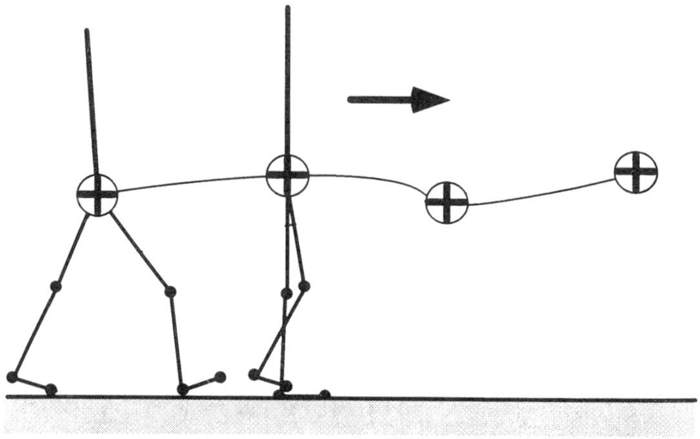

Figure 3.9. *Motion of the body center of mass during unimpaired walking.*

swing phases. In unimpaired gait, the stance phase consists of 60% of the gait cycle, whereas the swing phase accounts for the other 40%. Double-limb support may consume 10 to 30% of the gait cycle.

The body center of mass (COM) is an imaginary point about which the mass of the body is concentrated at a given time instant. The precise point of the COM is variable. The COM during erect posture is typically located an inch anterior to the second sacral vertebra within the pelvic basin. The COM usually has maximum vertical oscillation of about 50 mm during unimpaired walking (see figure 3.9). Peak COM height occurs during mid-stance whereas minimum COM height coincides with the dual-support phase.

Raising and lowering the body COM consumes energy (i.e., work is approximately change in COM height times the body weight). Control of the COM requires precise coordination of hip, knee, and ankle motion. Impairments to or the absence of the coordination of these joints often results in increased energy cost.

3.2.1. Walking efficiency

There are many mechanical aspects to consider when analyzing the efficiency of gait. Efficient gait minimizes the forces outside of the line of progression (i.e., most of the force is directed opposite to the direction in which the person is walking). This helps to conserve energy by minimizing forces and moments which do not contribute to motion. Inefficiencies include excessive hip sway, exececessive trunk rotation (e.g., arm swing and/or pelvic rotation), and lateral deviations (hip, knee, and foot movements perpendicular to the line of progression); see figure 3.10. During unimpaired gait, the COM displays a horizontal amplitude oscillation of about 50 mm. People must shift their COM from side to side during normal gait in order to advance.

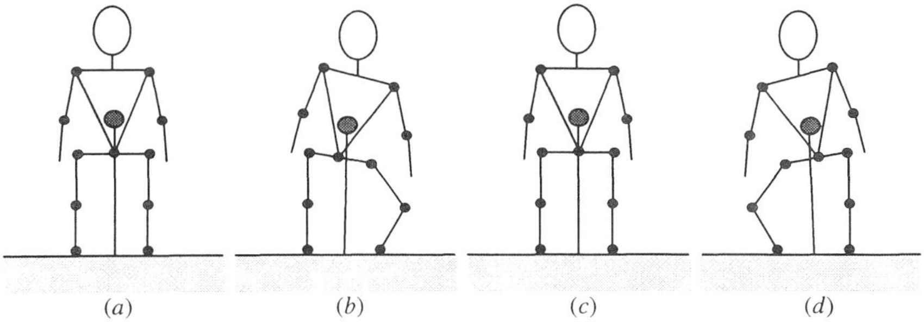

Figure 3.10. *Front view of person walking with a sway inefficiency. In part (a) the person is upright, in (b) the person takes a step forward and sways to the right and then comes to the center again. When leading with the next foot the person sways to the left. Some sway is natural, but excessive sway uses additional energy and may tax body structures over long periods. Sway is not necessarily symmetrical.*

During walking the COM is positioned almost directly over the supporting leg with each step in order to minimize the moments around the hip which leads to a tendency to drop the unsupported side of the pelvis. A narrow base width (e.g., within 50 to 100 mm) helps to minimize the horizontal excursion of the body COM.

Stride length can also influence efficiency. Stride length and cadence vary with walking speed. As speed increases stride length and cadence increase. Stride length affects the rise and fall of the torso (i.e., the center of mass rise and fall) which requires work against gravity. Cadence requires the primary muscles for locomotion to fire in sequence at a rate sufficient to maintain cadence for a given stride length. If stride length is too long then the work against gravity can be great enough to reduce efficiency. To complicate matters muscle contractile forces are rate and length dependent. If stride length is too long the muscles may not produce force efficiently. Similarly if the cadence is too fast the rate of muscle contraction may be too high to produce force efficiently. Changes in cadence and stride length may also influence efficiency through braking during foot contact. If a large braking force is applied, then a large propulsive force is required to accelerate the body to the desired speed.

Injury, impairment, or inflexibility of muscles and connective tissues may also influence gait efficiency. If muscles (antagonists) and ligaments are inflexible, agonist muscles must exert more force to bring the joint through normal ranges of motion. For example, tight hamstrings may cause knee extensors to create a larger than normal force in order to bring the knee joint into full extension during heel-strike, toe-off and swing phases. Excessive contraction of agonist muscles or insufficient contraction of anatagonist muscles during some periods of the gait cycle can lead to increased energy expenditure. For example, when excessive vertical toe-off force is exerted, a bounce is added to the gait pattern. Bounce increases the vertical component of the motion of the body's center of mass which increases energy cost through the increased work against gravity.

Environmental factors also affect the efficiency of gait. Ground reaction forces are required to generate propulsive forces during gait. The coefficient of friction and the damping properties of the walking surface influence efficiency. Surfaces with low coefficients of friction exert small reaction forces and thus reduce propulsion efficiency. For example, walking with leather-soled shoes on ice requires a high amount of energy. This is because greater force is applied to maintaining balance, and only a small percentage of the propulsion force generated by the body is applied towards forward motion. However, if a person wears ice-skates then the situation can be reversed. Ice-skates restrict motion in all direction except forward (i.e., the blades dig into the ice to reduce lateral motion, but allow the person to glide forward with much less effort than when walking). Hence, the low coefficient of friction is used to increase efficiency. Surfaces with high damping properties absorb large amounts of energy. For example, when walking on tiled floor the damping coefficient is low and the energy efficiency should be high. However, the damping coefficient of loose sand is much higher, absorbing more energy, and the energy efficiency is lower. The principles of coefficient of friction and coefficient of damping have been used quite successfully to improve shoe design.

3.2.2. Force plate analysis of gait

The forces exerted by the foot against the ground during walking are of major interest to biomechanists performing gait evaluations. By understanding how shoe soles, floor materials, prosthetics, and orthotics affect foot forces, improved designs to increase efficiency, reduce risk of accidental injury, and decrease chronic biomechanical stresses that lead to lower extremity and back disorders can be developed. To acquire data regarding ground reaction forces, stiff platforms supported by sensitive force transducers have been developed (figure 3.11). Typical force platforms are stationary and measure about 40×60 cm. Force plates or platforms are mounted level with a floor and permit the measurement of three orthogonal forces and moments exerted on the platform during walking, running, or jumping.

Force transducers can be linear variable differential transformers (LVDT), resistive strain gages, or piezoelectric crystals. A force platform must be a mechanically stiff system so that the natural frequency (a resonant frequency of 1 kHz is common) of the platform is well above the frequencies of the ground reaction forces being measured. The platform and force transducers must be sufficiently strong to withstand a 5 kN force. The sensitivity of a force platform is about 1 N (sensitivity of 0.2 N is attainable). Stable high-gain instrumentation amplifiers make force plates readily available and quite reliable.

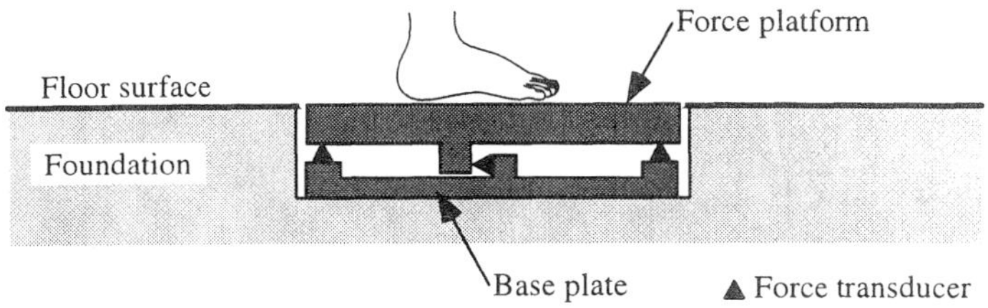

Figure 3.11. *Schematic of a force platform for measuring ground reaction forces during ambulation. Each force transducer measures one force. A force transducer is placed in each corner of the force platform. Two additional force transducers are used to measure anterior–posterior (A–P) forces. Two more force transducers are located along the centerline of the force platform and orthogonal to the A–P force transducers in order to measure medio–lateral (M–L) force.*

Force platforms can be used to estimate the instaneous coefficient of friction necessary to stop the foot from sliding. The damping properties of shoes and prosthetic feet can also be measured. Another application is the determination of the dynamic effects of ambulation with various assistive devices or surgical interventions. Combined kinematic and kinetic (e.g., force platform) analysis can assist in determing net joint forces and moments at various body joints. This method has been applied to estimate the yield and fatigue strength of prosthetic hip implants. The frequency, amplitude, and cumulative load can be estimated from net joint moment and force data.

The superior–inferior (S–I) component of ground reaction forces for unimpaired subjects shows a double-hump pattern during walking (figure 3.12). The S–I forces are related to the changes in the body center of mass position. During the initial contact stages the stance leg is extended with the body center of mass (COM) posterior to the foot contact point. The negative anterior–posterior (A–P) force causes the body COM to rotate forward about the foot contact point. At this time the body COM is accelerated upward causing an increase in the S–I force. The acceleration tapers off as the body COM is directly over the foot contact point (i.e., the body COM has reached a dynamic equilibrium). As propulsion begins, the body COM is accelerated upward and forward, causing another rise in the S–I forces. The upward acceleration of the body COM helps to create clearance for the swing leg, and allows more leg muscles to contribute to forward propulsion. The body COM falls and slows as the stance leg becomes fully extended and the swing leg makes contact with the floor.

The A–P forces show an alternating pattern related to propulsion and braking (figure 3.13). The A–P forces are usually about one-tenth the amplitude of the S–I forces. The A–P forces start with a braking phase which coincides with foot contact (i.e., heel-strike). During the roll-over phase the body center of mass is elevated over the stance foot and there is a corresponding shift from a braking (i.e., negative force) to a propulsive (i.e.,

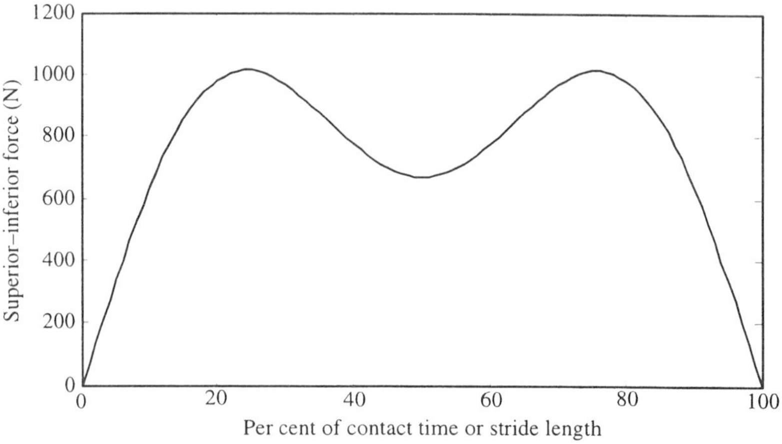

Figure 3.12. *Superior–inferior force during unimpaired walking. This figure shows the characteristic double hump of walking S–I force.*

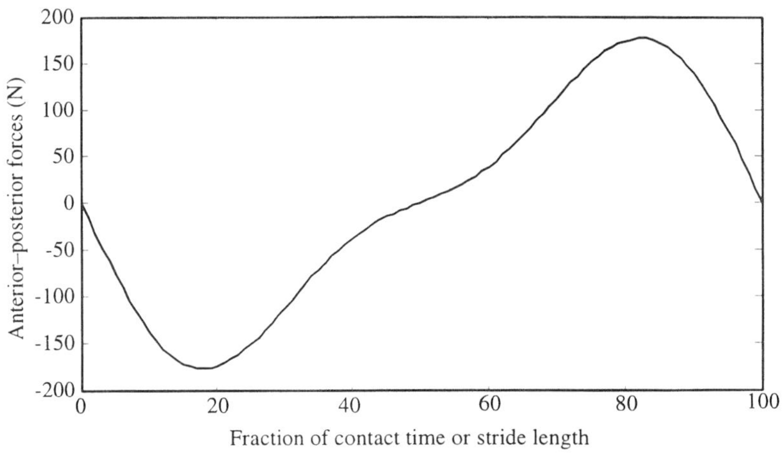

Figure 3.13. *A–P ground reaction forces applied to a force platform during walking. The negative forces at the beginning of the stride represent braking, whereas the positive forces at the end of the stride represent push-off.*

positive force). When the body center of mass is directly over the stance foot the A–P forces are nearly zero. As the body center of mass shifts forward of the stance foot, propulsion begins. The propulsion force reaches a maximum with maximum acceleration of the body center of mass. The A–P force declines as the stance leg extends and the swing leg prepares for contact. Ideally, the A–P forces of the stance leg reach zero force coincident with foot contact of the opposite leg.

Recently, pressure-sensitive insoles, which use piezoelectric or conductive polymer sensors, have been developed to measure pressure distribution over the bottom of the foot caused by ground reaction forces. This technique can be used to design shoes and orthotic devices.

A problem that faces clinicians when using force plates is the variability of data over consecutive trials. Clinical force-plate forces are divided in three orthogonal axes: A–P; medio–lateral (M–L); and S–I. Several attempts have been made to measure ground reaction forces on a motor driven treadmill. S–I forces can be measured by placing a standard force plate into the frame of the treadmill, under the moving belt. A longer than standard force platform (1.2×0.5 m^2) is typically used for this application. A longer plate is used to accommodate the sliding of the stance foot across the treadmill as the belt moves. Rosenrot *et al* developed the following relationship between contact distance (D_c) and belt speed (V_b) for adult males while walking

$$D_c = 0.655 + 0.25V_b. \tag{3.28}$$

Jansen reported on a design which implemented a dual-belt treadmill with separate force platforms for each belt. The force platforms were mounted as the supporting structure of the belt. Again, only the S–I forces could be recorded. Dual belts permitted measurement of the S–I force profiles without the superimposed force profile signal for periods when both feet were on the force platform. Weidenhielm *et al* used two force platforms which extended across their entire laboratory. They were able to collect six contiguous gaits cycles. However, there were some concerns about variation in speed, and for the natural frequency of the force platforms. Stewart *et al* have developed a composite treadmill with instrumented legs. Their treadmill is capable of measuring S–I and A–P forces. Preliminary data for this instrumented treadmill show promise for clinical applications. With a force-sensing treadmill, multiple strides can be studied and energy cost measurements can be combined with kinetic and kinematic analysis. This should lead to a more complete understanding of gait.

3.2.3. Gait analysis with lower limb prostheses

Walking with a prosthesis alters gait. The more proximal the amputation the greater the alteration of gait patterns. Thus, the gait parameters for a below-knee (BK) amputee are closer to unimpaired gait than those for an above-knee (AK) amputee. Murray *et al* have shown that mean common gait parameters for people with AK amputations are about two standard deviations from unimpaired parameters, while people with BK amputations are about one standard deviation different. People with single leg amputations at any level demonstrate asymmetrical gait patterns. Walking with a prosthesis lowers normal walking speed.

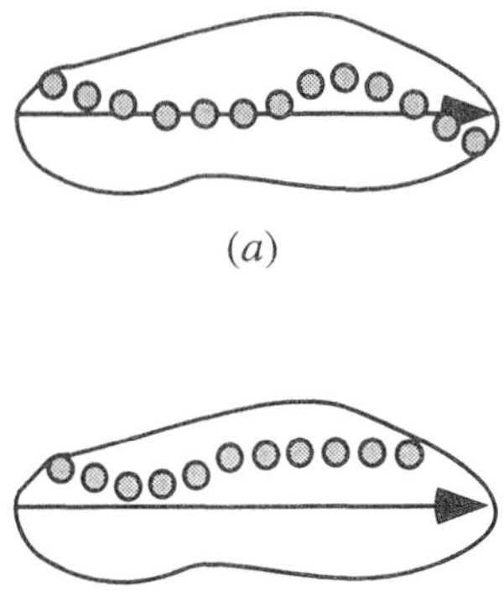

(*a*)

(*b*)

Figure 3.14. *Center of pressure measurements for the left foot of an unimpaired person (a), and a person with a left AK amputation (b).*

Absence of knee and ankle flexion in early stance increases the excursion of the body center of mass. The observable effect is a gait deviation known as vaulting (i.e., vaulting refers to excessive vertical movement of the torso). Vaulting is regularly observed by people with AK amputations. Vaulting can be caused by prosthetic ankle joints aligned in excessive plantar flexion, prosthetic feet that are too long, or misalignment of the knee and ankle. Vaulting can lead to uneven step length, irregular timing, abnormal side sway, and other inefficiencies. Ambulation with an AK prosthesis requires substantially more energy than unimpaired ambulation. Energy costs can range from 30 to 70% higher than for unimpaired gait. The combination of reduced speed, decreased stability, and higher energy cost often make people with AK amputations reluctant to rely on their prosthesis for ambulation. This is especially true for people with double leg amputations. In the case of double leg amputations many people opt to use a wheelchair for mobility outside of the home or office. During ambulation with an AK prosthesis the support time during the stance phase is shortened. This has the effect of shortening the swing phase as well, because of the difficulty in maintaining single-limb support on the prosthetic leg. The step length of the prosthetic leg is longer than the step length for the intact leg. This results in a longer prosthetic leg swing phase. Gait with an AK prosthesis usually results in decreased cadence, longer cycle duration, and shorter stride lengths. In some instances wider stride widths, and less prosthetic foot toe-out have been recorded. The center of pressure (COP) (i.e., the point at which force is applied to the force platform) reveals kinetic differences between unimpaired and AK prosthetic gait. In unimpaired gait, the COP shows a transition from heel to the great toe as well as internal rotation (figure 3.14). This is characterized by a smooth roll-off of the force from the toes. During AK gait the center of pressure often remains either internal or external to the line of progression along the foot.

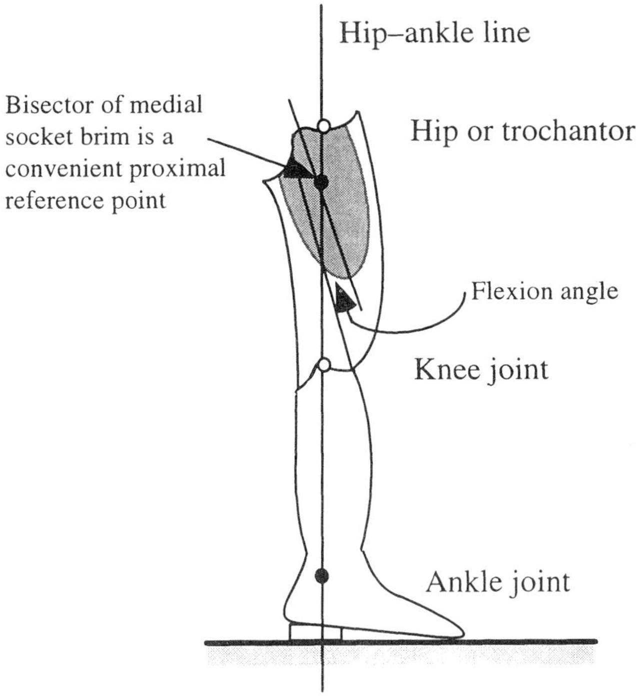

Figure 3.15. *Alignment of above-knee prosthesis to provide appropriate knee stability. The position of the knee joint relative to the hip–ankle line is adjusted to accommodate each individual's residual limb length.*

Stability of the knee joint at initial foot–ground contact and during stance is of paramount concern for people with AK amputations (see figure 3.15). A person with an AK amputation does not have the knee musculature for fine control of position or speed of the lower leg. When a simple hinge joint or four-bar linkage system is used as a replacement for a human knee, the person's upper leg and hip muscles must control the prosthetic knee and shank motions. These muscles also play an important role in controlling the motion and location of the pelvis and thigh. Changes in the angular accelerations at the hip control the damped pendulum-like swing of the prosthetic shank and foot. In order to maintain a stable and comfortable gait, the person must minimize the risk of uncontrolled foot contact during the swing phase, reach near full extension of the knee previous to heel strike, and maintain knee extension until heel-strike of the contralateral leg.

Fitting the anterior brim of the prosthetic socket well above the ischial seat helps to provide efficient voluntary knee control. Adjusting the knee joint center of rotation with respect to the hip–ankle line can provide additional knee control. The knee joint center can be adjusted to place the socket in flexion. For people with short residual limbs 5 to 20 degrees of flexion are used, and the knee joint is typically placed about 10 mm behind the hip–ankle line. For individuals with medium residual limb length about 5 degrees of flexion is used, with the knee joint center placed along the hip–ankle line. When the individual has a long residual limb, less than 5 degrees of flexion is required and the knee joint center is aligned with or slightly in front of the hip–ankle line. This method of alignment places the extension musculature of the hip in a favorable position for controlling the knee. This alignment also removes some of the support from the ischial tuberosities and transfers it to the hamstring tendons at the ischium. The gluteus maximus is placed to encourage sitting on the posterior brim of the socket which increases the tendency to hold the ischium on the ischial seat at push-off phase of the gait cycle. This method of alignment also minimizes lumbar lordosis in the presence of flexion contracture. This alignment makes developing an aesthetic prosthesis difficult.

The directions, magnitudes, and points of application of ground reaction forces substantially limit prosthetic limb stability during walking. The considerable concern for stability, and the energy required for AK ambulation, contribute to the alterations in the ground reaction force profiles (see figure 3.16). For an adult with an AK amputation, the propelling force for the prosthetic limb is about 33% less than in an unimpaired limb. Children with AK or knee disarticulate amputations are more asymmetrical with their prosthetic limb, developing about 53% less propulsive force than with their natural limb. Knee stability can be obtained by aligning the prosthesis so that the resultant ground reaction force vector is anterior to the knee axis. Locating the knee anteriorly with respect to the knee axis is a method of increasing knee stability during the period immediately following foot–ground contact. However, too much alignment stability, from a hyperextended prosthetic knee, can result in involuntary knee control. Although involuntary knee control provides extreme stability, it also makes it very difficult to obtain knee flexion at push-off which can have a dramatic effect on energy cost.

Prosthetic knee mechanisms are designed to complement hip musculature in controlling prosthetic limb dynamics. Ideally, the prosthetic knee mechanism would simulate the dynamics of normal knee musculature. Swing-phase control is obtained by timed dissipation and storage of energy within the knee mechanism. Energy dissipation is accomplished via a Coulomb friction damper with a constant friction mechanism. Knee mechanisms with constant friction may also incorporate an energy storage component (e.g., spring) to inhibit knee flexion and assist knee extension. Viscous damping has also been used with pneumatic or hydraulic knee mechanisms.

The patellar-tendon-bearing (PTB) BK prosthesis was the culmination of years of research at the University of California. The aim in providing a prosthesis for the person with a BK amputation is to make the fullest use of

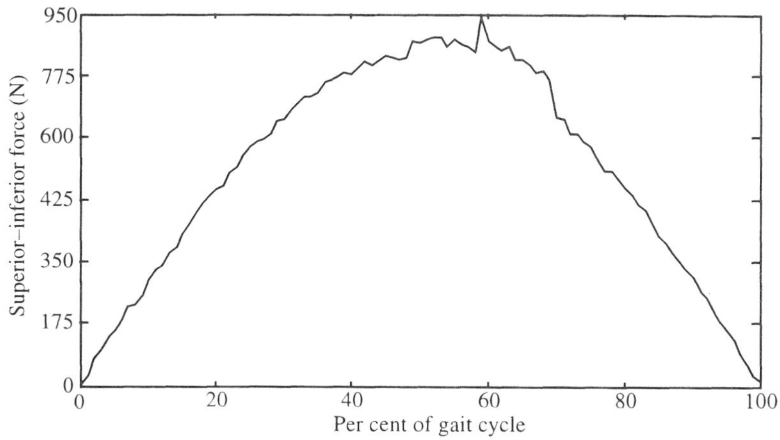

Figure 3.16. *Example of S–I ground reaction forces for a person with an AK amputation using a SACH (solid ankle cushioned heel) foot. In this case the double-hump profile typical of unimpaired gait is not present.*

the knee joint. During surgery the residual limb is designed to minimize the functional degradation of the anatomical knee. When successful, the residual limb BK can be used for weight bearing and dynamic load transfer to the ground. Scientists, engineers, and clinicians assume that the person with a BK amputation who has a healthy residual limb of average, sufficient, length will be able to walk in a manner similar to people without a lower-limb impairment. This implies making full use of the knee and employing a narrow-based gait. Rigid-body models can be used to estimate the forces applied to the residual limb while walking (see figure 3.17).

Forces S, L and M represent the net forces acting between the prosthetic socket and and the residual limb in the vertical, lateral and medial directions, respectively. The force of gravity acting upon the person's center of mass is represented by W. The ground reaction force, R, acts along a line connecting the center of pressure and the body's center of mass. The body's lateral inertial forces are represented by I acting at the body's center of mass. The forces S, L and M are of considerable imporatance in socket design and alignment. These forces and the distribution of pressure must be adjusted based upon residual limb length, and the residual limb's ability to sustain distal and lateral pressure.

The force patterns acting upon the residual limb influence socket design. The socket must pre-compress the residual limb in areas where force is transmitted. Pre-compression is used to reduce pistoning, that is motion of the socket relative to the skeletal structure, which can cause discomfort and tissue breakdown. Vertical support can largely be borne by the patellar tendon. The tip of the residual limb is unsuited for load-bearing. Areas which can sustain some load-bearing are illustrated in figure 3.18. The socket must

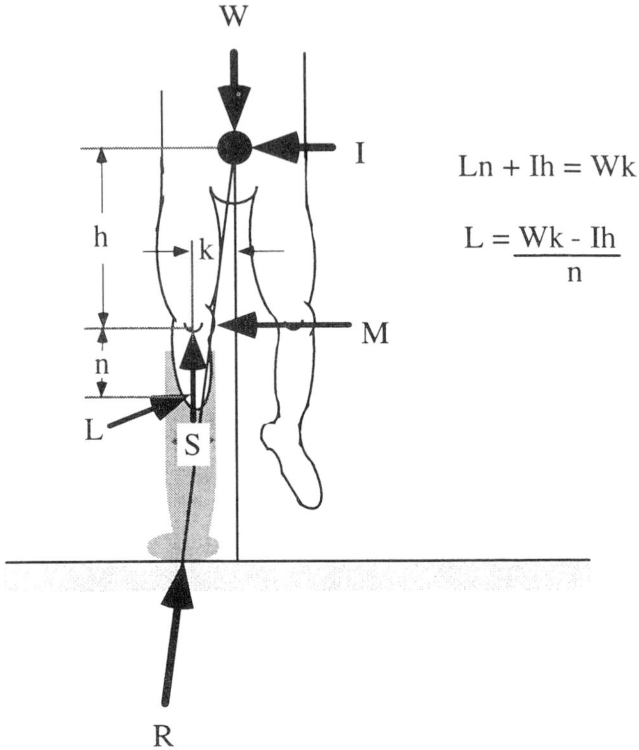

Figure 3.17. *Rigid-body model for estimating the forces applied to the residual limb of a person with a BK amputation. This diagram illustrates the anterior forces during mid-stance.*

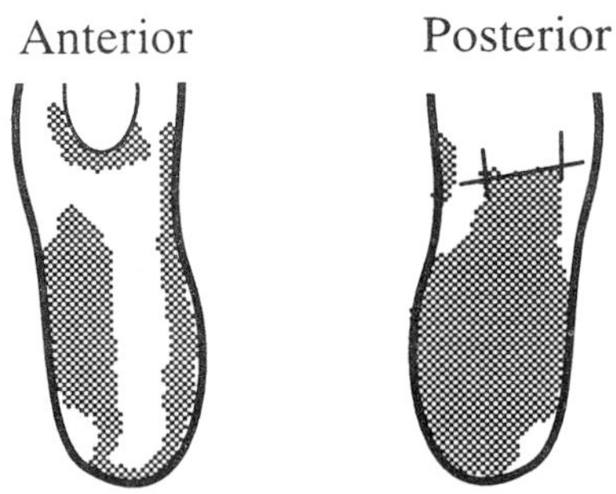

Figure 3.18. *Pressure tolerant areas for BK residual limb.*

be designed to protect pressure-sensitive areas by loading pressure-tolerant areas.

Ambulation with a BK prosthesis has many of the same characteristics as ambulation with an AK prosthesis. However, the differences from unimpaired gait are smaller. Velocity, cadence, and stride length are lower than for unimpaired gait, and gait cycle duration is longer. For people with single-leg amputation, the prosthetic limb stance phase is shorter and the swing phase is longer than for the intact limb. The prosthetic leg step length is longer than for the intact limb, and the prosthetic foot shows less toe-out. Widened stride width is also common. The dual-stance phase is longer for a person with a BK amputation (e.g., 17% compared with 10% for unimpaired gait). A BK prosthesis only requires an ankle mechanism. With a BK prosthesis, the ankle and part of the tibia/fibula are replaced. The biological control of knee and shank angles and velocities make the gait pattern more symmetrical than for prosthetic knee control.

Waters *et al* assessed the energy cost of ambulation by measuring the amount of oxygen consumed per meter traveled normalized by body weight. They found that control subjects (unimpaired gait) consumed $0.16\,\text{ml}\,\text{kg}^{-1}\,\text{m}^{-1}$, whereas subjects with (BK) amputations consumed $0.20\,\text{ml}\,\text{kg}^{-1}\,\text{m}^{-1}$, and subjects with (AK) amputations consumed $0.25\,\text{ml}\,\text{kg}^{-1}\,\text{m}^{-1}$. Waters and colleagues also found that cadence fell from 116 steps min^{-1} (unimpaired) to 99 steps min^{-1} for people with BK amputations, and to 87 steps min^{-1} for people with AK amputations. Stride length followed a similar pattern: 1.50 m (unimpaired), 1.44 m (BK), and 1.20 m (AK). The speeds of forward progression for the three groups were also different: $82\,\text{m}\,\text{min}^{-1}$ (unimpaired), $71\,\text{m}\,\text{min}^{-1}$ (BK), and $52\,\text{m}\,\text{min}^{-1}$ (AK).

3.2.4. Gait analysis with lower-limb orthoses

Lower-extremity functional orthoses may be characterized as exo-skeletons which provide support during standing or ambulation for people with lower-limb impairments or muscle imbalance. There are two basic types of functional lower-extremity orthoses: ankle—foot orthoses (AFO); and knee–ankle–foot orthoses (KAFO). The AFO provides support for the foot and ankle. The ankle may permit limited or no rotation. The KAFO provides support and limited or no movement at the ankle and knee joints.

During KAFO gait the knee is often locked, and ankle motion is limited to within about 15 degrees in the sagittal plane. This often leads to increased energy expenditure because the entire leg must be accelerated and decelerated. KAFO ambulation can be performed using a reciprocating or swing-through gait. With a recriprocating gait, the hip must be raised and rotated forward to initiate and control the leg during the swing phase. The hip must be sufficiently contracted to yield adequate ground clearance for the swing leg. Once the swing leg contacts the floor, the person raises the opposite hip and brings the opposing leg in-line with the lead leg. This leads to a pattern of single-leg stance, swing forward, single-leg stance, dual-leg stance. The dual-stance phase is substantially longer than during normal

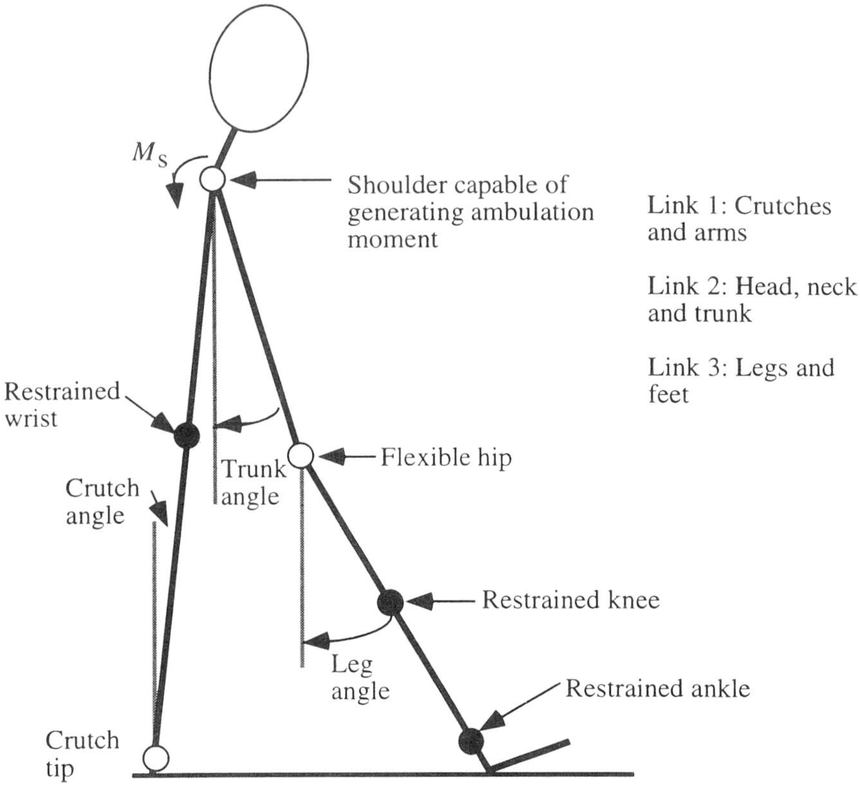

Figure 3.19. *Sagittal plane swing-through gait three-link model for kinematic analysis. This model can be used to determine the characteristics of the gait, and to measure the time spent in each phase. The KAFO restrains the ankles and knees, whereas the crutch restrains motion of the wrist.*

walking. KAFO ambulation is too slow and energy inefficient to be practical for most people with complete paralysis or with progressive neuromuscular diseases.

The swing-through gait allows individuals with paraplegia to ambulate using crutches and KAFOs (see figure 3.19). The KAFOs provide a rigid, stable support during the stance phase. Forward motion is obtained by using the upper extremities and torso to place the hips in extension. Once the person is balanced over the lower extremities, the walker or crutches can be moved forward. This causes the hips to go into flexion. The torso weight is supported by the walker or crutches through the arms during this time. The arms and torso are extended, which lifts the legs and causes them to swing forward. The pendulum action of the lower limbs can cause them to rotate forward beyond the crutch tips with minimal shoulder moment. The legs are non-weight bearing until heel contact. Upon heel contact a shoulder

For equilibrium: (a) sum of all forces must equal zero.

(b) sum of all moments about the ankles must equal zero.

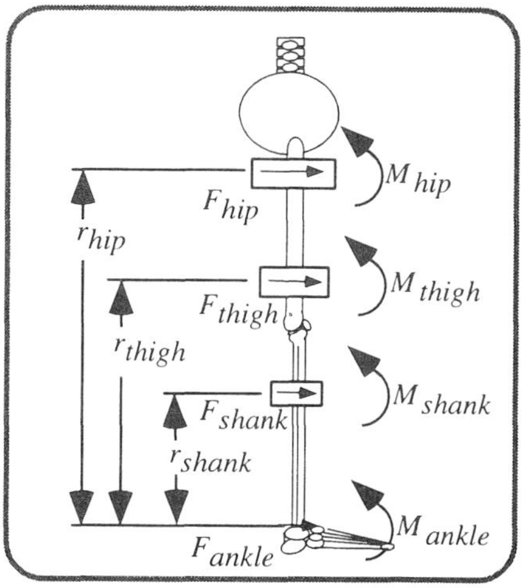

$$F_{hip} + F_{thigh} + F_{shank} + F_{ankle} = 0$$

$$M_{hip} + M_{thigh} + M_{shank} + M_{ankle} - F_{hip}\, r_{hip} - F_{thigh}\, r_{thigh} - F_{shank}\, r_{shank} = 0$$

Figure 3.20. *Sagittal plane force and moment equilibrium analysis for a knee–ankle–foot orthosis. This diagram can be used to interpret the forces applied, by the KAFO restraints, to the leg during ambultation. Any two forces and moments can be determined provided the others have been measured.*

and hip moment is used to bring the hips into extension and to bring the crutch tips forward. Energy expenditure during the swing phase is high because the body must be continuously lifted during this phase. Rovick and Childress studied timing parameters for swing through gait with a KAFO. They determined that the first double-support phase consumed 12% of the gait cycle, while the second double-support phase, coincident with heel-strike, consumed about 10%. The body stance phase consumed 30% of the gait cycle, while the swing phase consumed about 48% of the gait cycle. Gait velocity was measured to be $0.9\,\mathrm{m\,s^{-1}}$ with an average stride length of 1.5 m.

The forces that stabilize the knee are conventionally applied by straps or moulded bands. The positioning of these straps is important to providing a safe and effective orthosis. Occasionally, KAFO straps can cause the knee to shift backwards (a phenomena called *genu recurvatum*). KAFOs often rely on three-point force application to stabilize the knee (see figure 3.20). Straps must be designed to minimize knee force, and to distribute the contact force to mimize risk of shear and skin breakdown. The design of a KAFO can be posed as a set of four design criteria: (1) minimize the total force required for knee stabilization; (2) minimize force per unit square area at restraints; (3) minimize anatomical knee shear; and (4) apply forces only to tolerant body tissue. Lehman and Warren measured the forces applied to various strap locations and sizes during ambulation with KAFOs. They found that several common configurations produced acceptable results. The maximum force applied by a restraining strap was found to vary between 100 and 170 N with a strap width of 30 mm. A simple pressure calculation would indicate that strap pressures exceed capillary pressure. Skin breakdown probably does not occur more frequently because of the variations in pressure due to the cyclic nature of ambulation and upright posture. Wider restraints are recommended for people who ambulate frequently or who are prone to skin breakdown. Particular attention must be paid to the superior edge of the posterior thigh band. If the orthosis is too long or the band placed too high, the person may sit on the band and shut-off blood flow or produce abrasions.

For people with good upper limb coordination and strength the most preferred option for orthosis ambulation is reciprocal gait. Reciprocal gait can enable people to ambulate in a wide variety of situations, and more closely resembles normal gait. The more efficient and flexible the ambulation, the more likely it is that peole will adopt it. Reciprocal walking refers to placing one foot in front of the other with at least one leg always bearing weight. Successful reciprocal gait requires: clearance of the swing leg from the ground; ability to move the hip joint from extension to flexion; and the ability to move the trunk forward over the stance leg. Most people with thoracic level lesions require orthotic compensation for the hip and legs or a hip–knee–ankle–foot orthosis (HKAFO). Two basic methods have been used to design HKAFOs. One method uses relative hip movements and an orthotic mechanism to ensure reciprocal gait. Cables are used to actuate the reciprocal motion. This type of orthosis has been named the reciprocating gait orthosis (RGO). Douglas *et al* give a detailed description of the development of the RGO. Another method is to minimize the frictional losses at the hip to achieve the swing phase through a pendulum effect. The difficulty with the second approach is in attaining sufficient ground clearance. Conceptually, ground clearance can be obtained by introducing structural rigidity in the coronal plane in order to maximize crutch force efficiency. The ORLAU ParaWalker uses the second approach and has been widely used clinically. Proponents of imposed reciprocation and free hip orthoses both claim higher efficiency and cosmesis. A study by Whittle and Cochrane showed that there was no significant difference in energy cost between the RGO and ORLAU ParaWalker. Orthoses with low lateral stiffness may not overcome the problem of adduction of the stance hip when attempting to have the swing leg clear the ground (see figure 3.21). High lateral stiffness at the stance hip can overcome this problem, and allow the supporting crutch to be moved outwards to increase its leverage against the body's center of mass. A stiffer HKAFO may allow more weight to be borne by the orthosis and less by the crutches. However, most people cannot afford the additional energy cost associated with greater HKAFO weight. Thus more sophisticated engineering techniques and materials need to be applied to develop HKAFOs which have increased lateral stiffness and are lightweight.

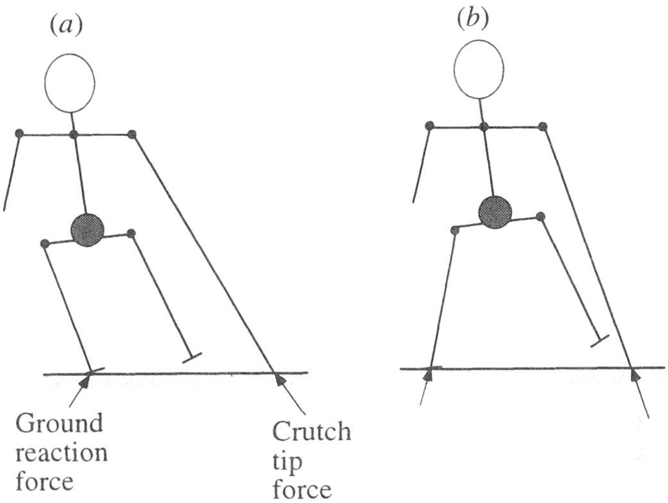

Figure 3.21. *Frontal plane illustration of reaction to temporary loss of balance. Illustration (a) shows the problem of stance hip collapse when attempting to clear the swing foot. Illustration (b) shows that the increased effective stance width, due to high lateral stiffness, prevents the fall. High lateral stiffness also prevents hip collapse. In (b) notice that the center of mass is not over the stance foot and remains close to the mid-line unlike in figure (a).*

The use of an AFO helps to stabilize the ankle and knee to allow functional ambulation for people with weakened hamstrings or quadriceps depending upon the amount of dorsi or plantor flexion incorporated into the ankle. An AFO can provide support and stabilization for people with ankle hyperreflexia, decreased voluntary control, and/or sensory deficits. An AFO generates stabilizing knee forces by shifting the body's center of mass location with respect to the feet. If an individual has sufficient functional lower limb musculature, but requires some support, ankle movement is provided. In this case the AFO serves to limit ankle motion. AFOs are commonly made of lightweight molded plastics. As such, they are lightweight and can have minimal effect on energy cost during low-speed ambulation. The gait pattern of ambulation with an AFO can be very close to normal, and then deviates as more support is required due to greater muscle weakness. Clinicians characterize walking velocity, step length, cadence, and stance time before prescribing an orthosis. There are two basic types of AFOs in use: the tone-inhibiting dynamic orthosis (TIAFO); and the static AFO. Many types of static AFOs are pre-fabricated. When walking velocity is held constant, the TIAFO yields longer step length and stance time with lower cadence.

The Vannini-Rizzoli stabilizing orthosis (VRSO) is an AFO that is similar in appearence to a high leather ski boot. The VRSO commonly runs the

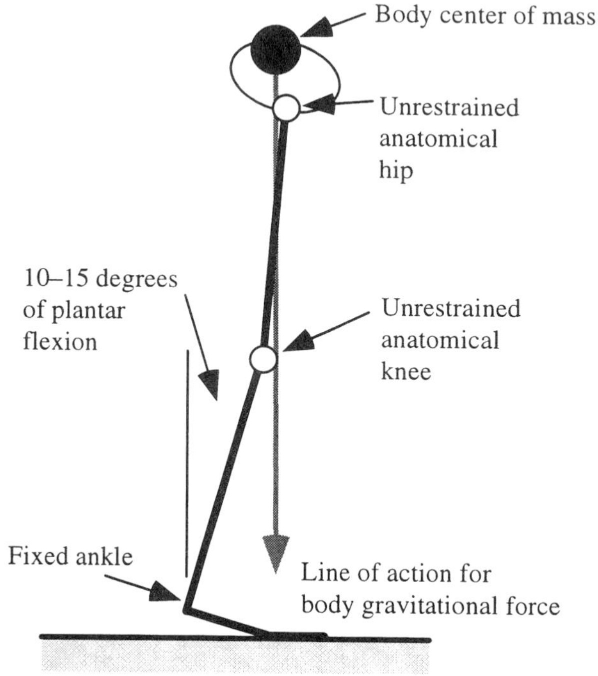

Figure 3.22. *Sagittal view stability diagram for VRSO during the dual-stance phase. The body gravitational force acts through the center of the foot to stabilize the knee in flexion. The trunk is used to hold the hip in flexion, the center of mass location can also help to stabilize the knee. Ten to fifteen degrees of ankle plantar flexion are required to attain the necessary joint orientations.*

entire length of the user's shank. The VRSO stabilizes the ankle and foot in approximately 10 to 15 degrees of plantar flexion. The fixed plantar flexion position is designed to stabilize the knee into extension. In an upright posture, the head is best held upright, and the hips and knees are best held in extension (see figure 3.22). This posture yields a quasi-stable errect position. The user initiates walking by shifting his/her trunk to one side (i.e., either left or right) which causes the unweighted foot to swing forward like a pendulum. This motion is reciprocated between contralateral limbs. Crutches or a walker are used to maintain balance. Effective gait is obtained by controlling pelvic rotation, tilt and lateral motion; knee flexion; and foot movement.

When ambulating with a VRSO, tilt of the pelvis is controlled by using the upper extremities to shift the trunk laterally. Forward rotation of the raised hip allows the unweighted leg to swing forward of the body's center of gravity. The stance leg's knee is locked in extension throughout this phase. Knee flexion is maintained by the fixed plantar flexion at the ankle. Flexion of the stance leg knee is undesirable, because it results in an unstable posture. The step length for VRSO users is about 50 mm, but can be longer for people with some lower-limb function. People with complete lower-extremity paralysis can effectively ambulate about 150 m before volitional exhaustion. Hence, the VRSO may be useful as an exercise device or for mobility within confined space; it is not a functional mobility device for most people with lower limb impairment. People with complete spinal cord injuries above T 10 (i.e., thoracic vertebrae 10) rarely ambulate functionally with an orthosis.

3.3. FUNCTIONAL NEUROMUSCULAR STIMULATION FOR MOVEMENT RESTORATION

Functional neuromuscular stimulation (FNS) is defined as activation of skeletal muscle in a purposeful manner. FNS has been applied in attempts to restore useful movement in upper and lower extremities of people with impairments to the central nervous system. When an action potential of a neural cell is excited an electrical stimulus propagates along the unimpaired and interconnected neural pathways to muscle fibers. Controlled neural activation can be elicited through the application of low levels of electrical current. Cardiac pacing, i.e., pace makers, has been the most successful example of functional electrical stimulation (FES). Moe and Post in 1962 expanded the application of FES to skeletal muscle as a functional orthosis. Some scientists and clinicians have promised brilliant results in very short time periods. Although many successful clinical applications and promising research results related to FNS have been documented in the scientific literature, overstatements and broken promises have made consumers and clinicians very skeptical.

FNS involves developing solutions to the complex problems of restoring human locomotion and manipulation. FNS requires knowledge of muscle and nerve physiology, electrical engineering, electrical safety, biomechanics, and control systems. Investigators from around the world have vigorously researched FNS since 1962. However, FNS has not reached widespread clinical application, or daily use by people with mobility and/or manipulation impairments. At the University and Medical Center in Ljubljana, Slovenia, peroneal nerve stimulators have been applied to over 2500 people with hemiplegia. This group has also provided over 100 people with spinal cord injuries with FNS systems that can allow them to independently ambulate with a simple four-channel reciprocal gait stimulator. Researchers at Case Western Reserve University, the Cleveland Metro Health Center, and Cleveland Veterans Affairs Medical Center have worked together to successfully improve functional grasp among 50 people with quadriplegia. They have successfully developed peroneal and implanted systems. The Cleveland research team has also been able to restore complex gait for about 30 people with paraplegia. Trials are ongoing to demonstrate that the technological advancement of FNS devices is sufficient to support wider clinical acceptance, and substantial commercialization.

FNS can be applied as a rehabilitation method that helps restore some aspects of limb function in cases where the central nervous system has been damaged and cannot convey voluntary movement commands to peripheral motor nerves. The best candidates for permanent FNS application appear to be people with incomplete paraplegia or moderately affected people with hemiplegia. Neuromuscular structures have been reported to produce the best motor responses when stimulated by a train of rectangular, charge-balanced electrical pulses with specific pulse width, amplitude, and repitition rate parameters. In FNS the pulse width varies from 100 μs to 500 μs at a frequency of about 30 Hz. The amplitude may vary from several volts to nearly 100 V, depending upon the distance between the active tissue and the electrodes, tissue excitability, and the stimulation electrode design.

There are three recognized FNS technologies used to excite body tissues: surface stimulators; percutaneous stimulators; and fully implantable stimulators. Surface stimulators use electrodes placed on the skin over the muscle or nerve to be stimulated. They have the least selectivity, but are simplest. Percutaneous stimulators use permanently implanted electrodes with wires chronically penetrating the skin which connect to an external pulse generator. An implantable stimulator is an electronic device contained in a biocompatible enclosure connected to electrodes permanently located near excitable tissue. A transmitting antenna on the skin surface delivers power and information to multiple stimulation sites. Implanted devices are placed where they are least intrusive and wires are guided to the stimulation electrodes (see figure 3.23). Deep muscles and nerve tissue can be excited with implanted electrodes.

A useful FNS system must provide a flexible triggering scheme, programmability or simulation parameters, small size, and ease of use by the patient. FNS systems must be very reliable and safe. Functional movement must be capable of processing relevant biomechanical parameters (e.g., joint angles, foot contact, and push-off) and developing appropriate control signals. Muscle recruitment and the ability to maintain force against fatigue are issues which affect limb performance. The contributions of stimulation patterns and voltages to knee stability have been studied. For efficient gait it is desirable to develop full knee extension during certain phases of the gait cycle. To obtain full knee extension with FNS continuous stimulation is required which results in rapidly developing fatigue. For FNS to restore lost motor function in people with central nervous system impairments, coordinated control over graded contractures of many muscles is necessary. For people with upper motorneuron lesions, selective stimulation of nerve fibers of peripheral nerves may be a means of enhancing motor function (see figure 3.24). Selective stimulation could allow tension control through

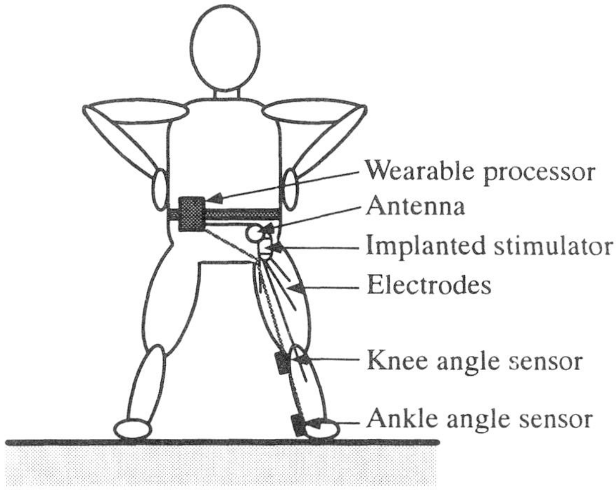

Figure 3.23. *Example configuration for an implantable multichannel FNS ambulation system.*

recruitment starting with the slowest to the fastest twitch motor units. This provides more physiologically based control. Slow twitch motor units provide less tension than fast twitch motor units, but do not fatigue as rapidly. The unimpaired motor system appears to recruit slow twitch fibers for activities which require low forces, and fast twitch fibers for activities requiring high speed and/or high force. Slow twitch fibers are recruited for frequently occurring activities which require low force. For example, FNS efficiency could be increased if slow twitch fibers could be recruited to maintain postural stability while standing or maintain limb position with the upper extremity. Fast twitch fibers could be used to facilitate movement. For example, fast twitch fibers would be recruited to initiate and generate a step or to lift an object. Once the task is completed, fast twitch fibers could be deactivated, and slow twitch fibers could return to maintaining posture of position. Selective recruitment could provide finer control as well as more efficient control. For example, lifting a fork to the mouth could begin with slow twitch fibers being used to manipulate fork and hand position for the transition to the mouth. Once the positioning is correct, fast twitch muscle fibers could be recruited to bring the fork to the mouth. When the fork gets close to the mouth slow twitch fibers can be recruited to provide fine motor control so that the person is not harmed. The concept of using slow twitch fibers prior to exciting fast twitch fibers is called 'size principle recruitment'.

Myoelectric activity is an important indicator of stimulated muscle performance. The electromyogram (EMG) provides information about recruitment during voluntary muscle contractions. Fine wire or needle electrodes can be used to record the activity of single motor units. When a voluntary muscle receives a stimulus, it responds with a brief contraction

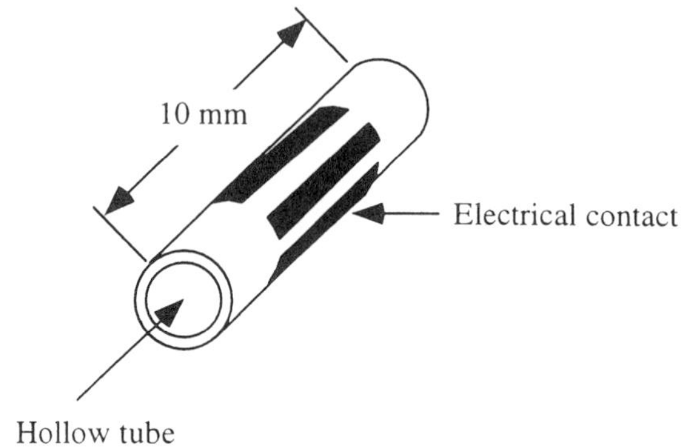

Figure 3.24. *Stimulation electrode nerve cuff for selective stimulation. Tube is placed around a motor nerve.*

called a 'twitch'. Motor units are characterized based upon their peak twitch force, time to peak force, and the half relaxation time (i.e., the time it takes for the twitch force to decay to one half its peak twitch force). The descriptions slow twitch and fast twitch muscle fibers comes from the identification of two categories of peak twitch force and time to peak force responses. Slow muscles are used for postural and position control. Slow muscles are nearly continuously active, making frequent small contractures. Fast muscles are electrically silent for long periods and then discharge short high-frequency bursts as the person makes occasional rapid or high-strength bursts.

Accurate external control of the musculo-skeletal system requires models of each muscle and tendon to be actuated. A substantial limitation of FNS locomotion and manipulation is that few musculo-tendon variables (e.g., muscle force, tendon force, muscle fiber length, rates of muscle contraction, and muscle fatigue) can be measured in humans during locomotion. Therefore two approaches have been taken to develop models of the human locomotor system for FNS. The musculo-tendon system can be treated as a black-box and a model can be developed using system identification techniques based upon input–output data (see figure 3.25). This approach uses a dynamometer targeted to measure parameters of direct importance to attaining the task at hand. For example, for FNS-controlled manipulation of a paralyzed upper limb a dynamometer can be used to measure forearm torque, grasp, and elbow angle with various loads and stimulation patterns. If the loads and stimulation patterns are varied over a wide range (i.e., sufficiently rich) then models of the muscle and limb response can be developed. Least squares and maximum likelihood techniques have been applied to develop linear and non-linear models. These systems can also be used to develop sufficient strength, through progressive exercise, to enable the performance of functional tasks.

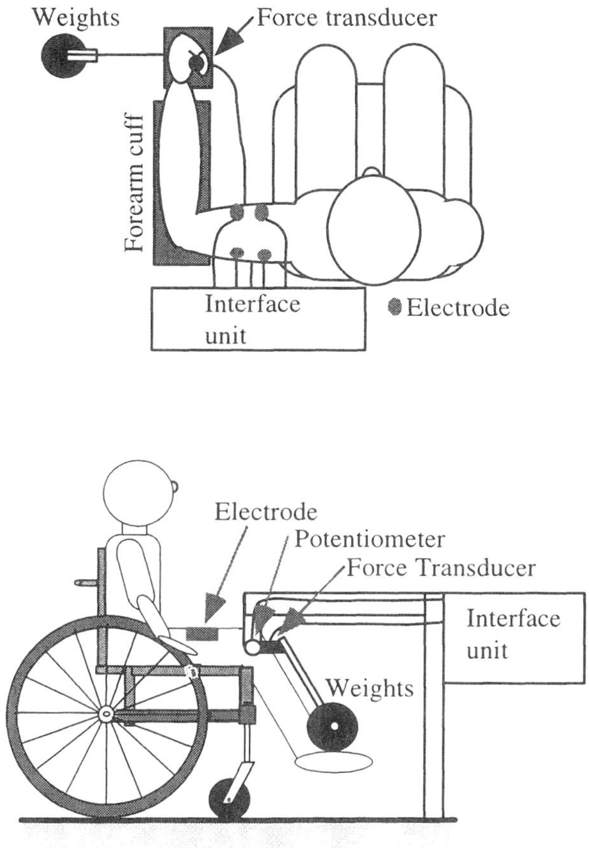

Figure 3.25. *Experimental set-ups for parametrization of upper extremity and lower extremity for FNS control. These systems can also be used to develop sufficient strength for manipulation or ambulation.*

Another approach is to use a forward dynamics or optimization method. The anatomical structure and location of muscles and tendons can be modeled (see figure 3.26). Models are either classified as subcellular models or mechanical models. Subcellular models are used to describe molecular and biochemical events associated with muscle contraction, and have not been applied to the development of FNS control algorithms. This is primarily because their complexity does not lend itself well to control algorithms. Mechanical models rely on the underlying anatomy, simple physiology, and input–output modeling of muscle fibers and tendons. An advantage of this approach is that it uses information about the musculo-skeletal system. Mechanical models can also rapidly become complex, therefore low-order differential equation models (e.g., first- or second-order) are used to represent each muscle and its tendons. Low-order lump parameter models rely on only a few basic properties of muscle fibers and tendons: the pination angle (i.e., the angle made between the muscle fibers and tendons), the orientation of the sarcomeres within a muscle fiber, the orientation of the fibers within the muscle, the passive and active mechanical properties of sarcomeres, the dynamics of excitation–contraction coupling, and the stress–strain properties of tendons. Muscles are considered to generate force proportional to activation level. The activation level is assumed to be related to the electrical activity, but the exact relation has eluded scientists. Several models have been developed for isometric contractions, but no general model relating EMG to muscle activation exists.

A classical anatomical method of specifying muscle functional roles is based upon the premise that a muscle moves the joint it spans in the direction of its applied moment. For example, the hamstring muscle crosses the back of the knee, and is believed to cause the knee to flex. Classical anatomy does not take into account muscle dynamics and the possibility of changing roles of muscles in response to dynamic limb positions and external loads. Muscles often cross more than one joint, and can vary their roles with differing anatomical orientations. The complexity of muscles, tendons, and functional anatomical relations has made clinical success with FNS difficult to obtain. To obtain functional ambulation or manipulation, numerous muscles must be coordinated and controlled. If the muscles are not properly stimulated a fall can occur (see figure 3.27). For example, premature ground strike or knee hyperextension due to prolonged swing can initiate a fall. A walker is often used to assist the person in maintaining balance.

The utilization of FNS to improve muscle performance and function has not been as universally promising as expected. Investigators have not been able to adequately address the biomechanical and metabolic demands of upright posture and ambulation for people with complete lower-limb paralysis. The presence of discomfort in people with incomplete lower-limb paralysis, when electrical stimulation is used to produce useful contractions, has also limited the clinical and in-home application of FNS. People with either complete or incomplete paralysis may eventually benefit from FNS, but a substantial number of problems persist despite the investigative efforts applied to this line of research.

FNS has been applied to restore hand function in people with upper-extremity paralysis. People with cervical spinal cord lesions have been the primary beneficiaries of such technology. The focus of upper-extremity FNS research has been to restore functional grasp and release. Upper-extremity FNS is not hindered by the problems of FNS ambultation. However, the tasks performed with the upper extremity necessitate fine control and complex motion patterns. FNS has been successfully applied to produce the two types of grasps identified by Keller *et al* as being most widely used by unimpaired individuals. Lateral prehension (e.g., key grip) and palmar prehension (e.g., three-jaw pinch) are obtained by establishing a stimulus map which translates

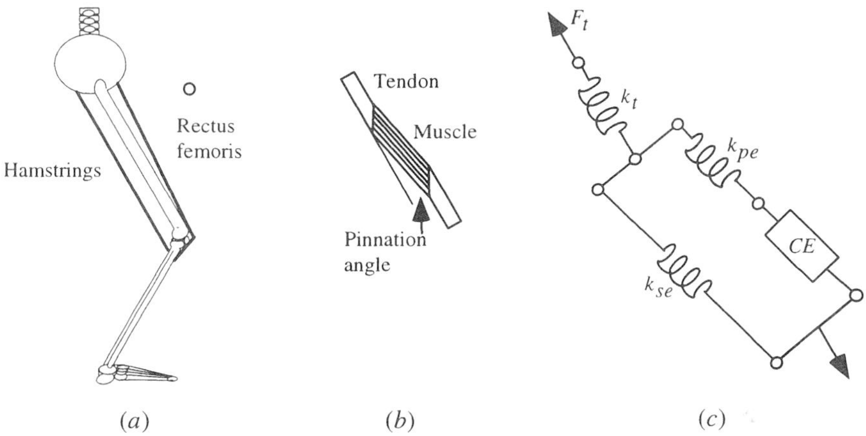

Figure 3.26. *Simple musculo-tendon-actuator model for developing functional neuromuscular stimulation ambulation control algorithms. The anatomical locations of the muscles to be stimulated are shown in (a). A simple biomechanical model of the rectus femoris muscle is shown in (b). A mechanical analog for the rectus femoris is shown in (c). In the mechanical model, F_t represents the tendon force, k_t the tendon spring constant, k_{pe} is the parallel elastic element constant, k_{se} is the series elastic element constant, and CE is the active contractal element. A forced differential equation can be used to represent the mechanical analog model.*

the control signal into a stimulus applied to each electrode. For lateral grasp and release, activation of thumb extensors, flexors, and adductors together with finger flexors and extensors is required. For palmar grasp and release activation of thumb abductors, finger extensors and finger flexors is required. Investigation is ongoing for controlling wrist extension to eliminate the need for a wrist splint during FNS grasping. Research on elbow extension control is likely to achieve voluntary arm placement by people with C4 lesions (i.e., lesions at the fourth cervical vertebrae) and to increase the functional workspace of people with C5/C6 lesions. This research could lead towards a solution to the 'hand to mouth' problem which would allow many people to feed themselves with greater independence.

3.4. BIOMECHANICS OF WHEELCHAIR PROPULSION

Mobility via wheelchair pushrim propulsion requires examination from a variety of perspectives. With over 20 million people worldwide relying on wheelchairs as their primary source of mobility there remains a need for properly matching each individual's characteristics (e.g., anthropometry, physiology, strength, range of motion, mobility goals) with the wheelchair

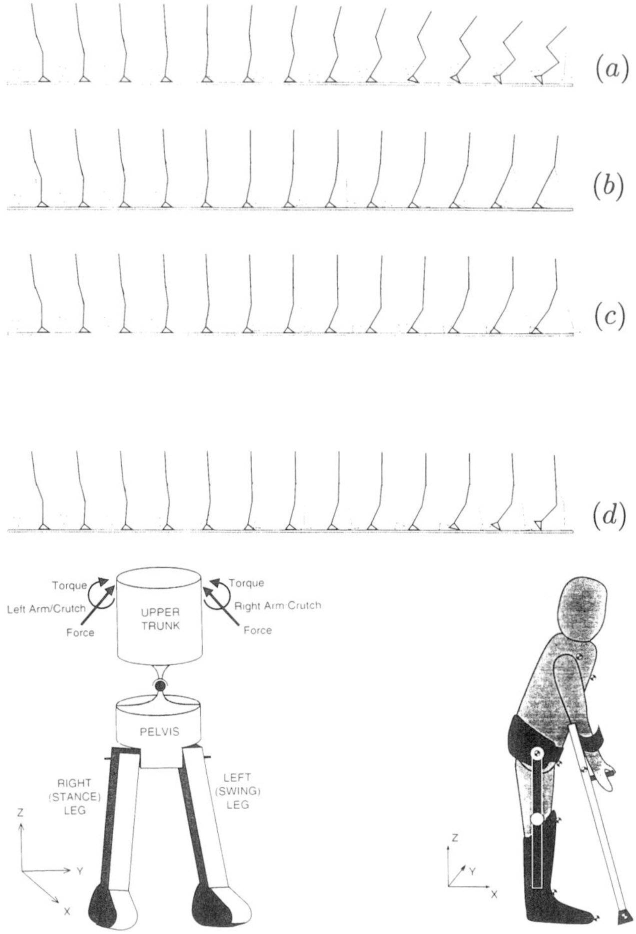

Figure 3.27. *Walking orthosis and four-segment model of a person with a spinal cord injury ambulating with FNS. Stick figure simulation sequence shows stance leg collapse (a), tripping because of premature ground contact (b), knee hyperextension at the end of swing due to a long step (c), and a normal step (d). (Figure reproduced with permission from Zajac F E, Schutte L, Tashman S, Yamaguchi G, Perkash I 1994 Functional Neuromuscular Stimulation (FNS) in Walking and Pedaling Rehabilitation Research and Development Center 1994 Report (Palo Alto, CA: Department of Veterans Affairs Medical Center).)*

characteristics (e.g., dimensions, mass, added features) to optimize comfort, safety, and performance.

The efficiency of wheelchairs depends on physiological capabilities of the user, the biomechanics of the stroke, and the design of the wheelchair and its appropriateness for the user. Some investigations have suggested that wheelchair propulsion demonstrates around 10% gross mechanical efficiency, which is less efficient than other forms of mobility (e.g., arm crank vehicles, lever drive systems, multi-speed pushrims). However, when performance is optimized, as in the case of racing wheelchairs, efficiency may reach 30%, which exceeds that of running. Gross mechanical efficiency is affected by wheelchair design, and the optimality of the interaction between the person and wheelchair.

Alternatives to pushrim wheelchair propulsion have not performed as well and have been less effective under a wide variety of circumstances. Conventional wheelchairs currently remain most widely available and the most commonly selected mode of mobility among people with lower limb impairments.

3.4.1. Wheelchair propulsion injury mechanisms

The most commonly reported site of musculo-skeletal injury in manual wheelchair users (MWU) is the shoulder. Surveys of MWUs show the prevalence of shoulder pain to be between 73% and 31%. Sie *et al* interviewed 103 patients with paraplegia; 36% reported shoulder pain. They found prevalence increased over time until 20 years post-injury, when it decreased slightly. Nichols *et al*, in a survey of MWUs, found a 51.4% prevalence of shoulder pain. They reported that the frequency of shoulder pain increased with time since onset of disability. Pentland and Twomey reported on pain complaints, strength and range of motion in 11 women with paraplegia and 11 healthy controls. This study found 73% of women with paraplegia less than 15 years out from their spinal cord injury (SCI) experience shoulder pain during activities. Gellman *et al* interviewed 84 subjects with paraplegia and found 100% of the subjects more than 15 years out from an SCI had shoulder pain as compared with 20% of those less than 15 years out from injury.

Bayley *et al* studied 94 veterans with complete paraplegia. Each veteran had a physical examination focusing on the upper extremity. Thirty-one patients reported a history of shoulder pain and 23 were found to have signs of impingement on examination. All 23 subjects with pain on examination had x-rays and arthrography which revealed rotator cuff tears in 65% and aseptic necrosis of the humeral head in 22%. No comment on association of time since injury and shoulder problems was made. Although this study includes objective evidence of shoulder pathology, testing was only performed on symptomatic subjects and thus subclinical pathology was likely to have been missed. Wylie and Chakera reviewed medical and surgical records of 51 individuals who were all over 20 years out from an SCI. They found radiographic shoulder abnormalities in 32% of the subjects. They also reported that patients with greater activity levels had fewer complaints of pain. The type of radiographic abnormality was not reported and shoulder films were not present on all subjects.

Although the shoulder is the most common site of musculo-skeletal injury in MWUs, elbow, wrist, and hand pain are also commonly reported. Sie *et al* reported elbow, wrist, and hand pain in 16%, 13%, and 11% respectively. This did not include patients with carpal tunnel syndrome (CTS). In addition, Sie *et al* further defined significant pain as that which required analgesia, occurred with two or more activities of daily living, or required cessation of activity. Using this definition the prevalence of all upper-extremity pain complaints was 20%, 5 years post-injury, and 46% from 15 to 19 years post-injury. Other studies have shown the prevalence of forearm, wrist, and hand pain to be between 55% and 8%. In all of the studies on upper-extremity pain the authors felt the pain was related to overuse of the arm during transfers or wheelchair propulsion and that additional work aimed at prevention strategies is needed.

The possible mechanism of injury to the rotator cuff has been extensively explored in the general population. Fu *et al* and Frieman *et al* provide two thorough reviews of this literature. The mechanism of injury to the rotator cuff has been divided into intrinsic and extrinsic factors. Intrinsic factors relate to the anatomy of the tendon itself, whereas extrinsic factor relate to surrounding structures. The most commonly cited intrinsic factor associated with rotator cuff disease is a critical zone for injury at the insertion of the supraspinatus tendon into the humeral head. This critical zone has been found to have decreased vascularity. In the Bayley *et al* study of shoulder pain in MWUs they found that interarticular pressure was more than twice arterial pressure when performing a transfer. They believed that this increased pressure further stressed the vasculature of the rotator cuff tendon.

The most commonly cited extrinsic factor is impingement of the rotator cuff by surrounding structures. Neer reported changes in the undersurface of the structures forming the coracoacromial arch which they related to impingement. They attributed 95% of all rotator cuff tears to impingement, and stated that this impingement took place with the arm abducted and forward flexed. Bigliani *et al* reported that specific shapes of the acromion correlated with an increased risk of tears of the rotator cuff. Of note, Wylie and Chakera reported that 18% of active wheelchair users had joint space narrowing in the shoulder. Although both of these theories of pathophysiology have come into question recently, it is likely that both the shape of the coracoacromial arch and the vascularity of the rotator cuff tendons contribute to pathology.

Any activity which forces the humeral head further into the glenohumeral joint can cause impingement under the acromioclavicular arch and thus inflammation. In the majority of the literature the humeral head is placed in close proximity to the acromioclavicular arch during overhead activities. In MWUs the vast majority of activity is not overhead, however; there are forces which tend to drive the humeral head up into the glenohumeral joint. These forces occur during transfer activities and during wheelchair propulsion when a downward force is necessary to create the friction against the pushrim.

Another extrinsic factor leading to impingement and rotator cuff tear (RCT) is instability of the glenohumeral joint. The instability is thought to relate to a combination of attenuation of supporting structures of the glenohumeral joint, such as the glenoid labrum, and to muscle imbalance. Muscle imbalance, caused by overuse, is thought to lead to abnormal biomechanics and thus injury. The most common disparity in strength associated with RCT is an imbalance between the internal and external rotators of the shoulder. Burnham and Steadward were able to demonstrate muscle imbalance in a group of wheelchair athletes and were able to correlate this imbalance to shoulder pain.

As with most conditions, RCT can be diagnosed with a thorough history and physical examination. Historically the patient will complain of pain with shoulder movement, commonly in the overhead position. Physical examination findings consist of pain with resisted internal or external rotation and pain at the end of the range of motion. The impingement test described by Neer involves forward flexion and internal rotation which if it causes pain is considered positive. Other findings on physical examination include pain on palpation of the subdeltoid bursa and biceps tendon. The most sensitive non-invasive test for RCT is magnetic resonance imaging (MRI). Iannotti *et al* compared MRI to operative findings in 91 patients who were operated on for the first time. MRI was found to be 100% sensitive and 95% specific for complete tears, and 93% specific and 87% sensitive for tendinitis. Many other smaller series have found similar figures in terms of sensitivity and specificity of MRI. While MRI is the non-invasive gold standard for rotator cuff examination, signal abnormalities have been reported in asymptomatic individuals primarily on proton density images. These investigations have not included surgical or arthoscopic confirmation for the absence of disease, but the inclusion of strict morphologic and signal criteria should virtually eliminate this as a source of bias.

The most common neurologic cause of upper-extremity pain in MWUs is CTS. The prevalence of CTS in this group is between 49 and 73%. In addition to CTS, ulnar nerve damage has also been reported by a number of investigators. Gellman *et al*, in a study of 77 individuals with T2 or below paraplegia, found 49% had signs and symptoms of CTS. Sie *et al* interviewed 103 subjects with paraplegia and found historical or physical examination evidence of CTS in 66%. Both of these large clinical series found the incidence of CTS increased with increased duration of paralysis. To add objective criteria to the diagnosis of CTS a number of investigators have performed nerve conduction studies (NCS) on this population. Aljure *et al* studied 47 patients with an SCI below the T2 level and found electrodiagnostic evidence of CTS in 63% and clinical evidence of CTS in

40%. This study also found an increased prevalence of CTS with duration of paralysis. Tun and Upton found that 50% of individuals with paraplegia below the T1 level had slowed motor conduction of either the ulnar or median nerves at the wrist. 44% of patients with electrical abnormalities were asymptomatic. Davidoff *et al* studied 31 patients with paraplegia below the T1 level and found electrodiagnostic evidence of CTS in 55% and symptoms of numbness or tingling in the hand in 74%. In both of these studies there was no association between duration of paralysis and nerve conduction abnormalities. Burnham and Steadward studied the incidence of nerve injury in a group of 28 wheelchair athletes. Exhaustive nerve conduction studies were performed and the prevalence of electrically diagnosed CTS was found to be 46%. It is important to note that by clinical criteria only 23% of the athletes had nerve damage. Duration of disability was found to correlate with some of the electrodiagnostic tests.

Stefaniwsky *et al* in 1980 studied 12 patients with various levels of SCI and found slowed conduction velocity in the ulnar nerve. Slowing of the ulnar nerve was seen in both the above-elbow to below-elbow segment and the below-elbow to wrist segment. In this small study there was no correlation between time since injury and ulnar nerve damage. In three of the studies described above, electrodiagnostic evidence of ulnar nerve injury was also documented. The prevalence of ulnar nerve injury in these studies varied between 15 and 40%. None of the studies commented on the association of duration of paralysis with ulnar nerve abnormalities.

From these studies it is apparent that CTS and ulnar nerve injuries are common problems in MWUs. Most of the studies found a greater prevalence of abnormalities on NCS then actual clinical symptoms. This may signify that subclinical nerve damage exists in a number of these individuals. Whether this subclinical damage goes on to become clinically important has not yet been determined. CTS is generally thought to be caused by compression of the median nerve within the carpal tunnel. Extremes of wrist flexion and extension have been shown to greatly increase the pressure within the carpal tunnel, more so in patients with CTS. Gellman *et al* studied patients with spinal cord injuries and found that the pressures in the carpal tunnel were higher in wrist flexion in this group than in a group of controls. Thickening of the flexor tendon sheaths secondary to repetitive motion, and contraction of the tendons within the carpal tunnel have also been implicated as a cause for compression of the median nerve. The majority of scientists who have investigated CTS in MWUs have implicated the repetitive activity of propelling the chair as a causative factor.

The symptoms of CTS include numbness and paresthesias in a median distribution in the hand. The pain is often described as stinging, burning, or aching. Symptoms typically worsen at night and are commonly worse with activities which involve pinching or grasping objects. Two clinical tests for CTS are the Phalen's and Tinel's signs. Unfortunately, the sensitivity and

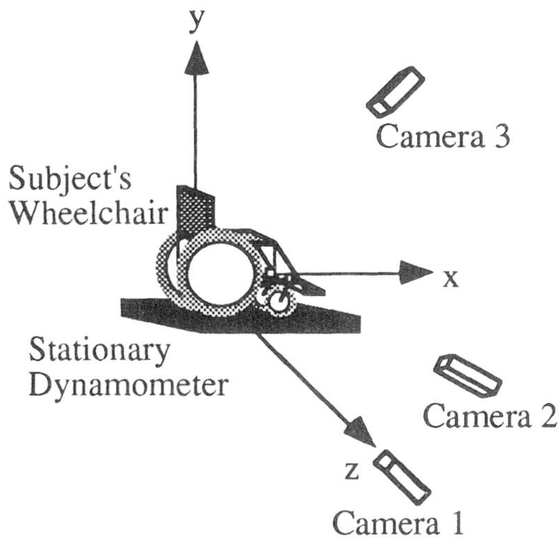

Figure 3.28. *Three-camera set-up on a stationary wheelchair dynamometer for recording 3D kinematics.*

specificity of these two tests are relatively poor. More sensitive tests for CTS are NCS. A number of studies have looked at the most appropriate NCS for CTS—Jablecki *et al* provides an excellent review which sorts through the literature and picks out those studies which meet rigid criteria. Two of the most sensitive NCS are median and ulnar mixed mid-palmars and radial and median sensory to the thumb. Jablecki *et al* presented three articles which met his criteria and evaluated the specificity and sensitivity of these two tests. Mid-palmar studies have been found to be 97% specific and between 70 and 85% sensitive in symptomatic hands. Radial and median sensory studies to the thumb have been found to be 99% specific and 60 to 70% sensitive in symptomatic hands. Each of these tests has an established protocol and an established upper limit of normal.

3.4.2. Kinematics of wheelchair propulsion

Biomechanical aspects of wheelchair propulsion are often examined by videotaping people while propelling at different speeds against various resistances (see figure 3.28). The videotape is digitized and analyzed with a computer to determine linear and angular positions, velocities, and accelerations. Kinematic analyses provide information about point of pushrim contact, phases of the propulsion-recovery cycle, and limb-joint motion patterns. Kinematic measures do not provide direct measures of the forces and torques applied to the pushrims.

Marker sites and sets: The selection of marker sites must coincide with areas where soft tissue is minimum. Further consideration should be given to the rigid-body models used during analysis of limb and joint motion.

Bony prominences on the body are best suited for marker placement. The choice of marker sets must reflect the model of the segment being developed and the underlying anatomy (see figure 3.29). For example, most wheelchair propulsion studies use a single marker to represent the hand despite the fact that the hand and wrist are composed of over 25 bones and joints. The appropriate number of markers and their orientation needs to be determined to obtain clinically relevant information.

Rigid body motions: Rigid-body constraints can also be used to minimize skin motion error and improve kinematic data. Anatomical configurations and limitations can be used to reduce soft tissue error. We propose to incorporate rigid-body constraints via a computer wire frame and a mathematical adjustment that ensures a constant spacing between two or more markers that are attached to the same rigid-segment model. With skin motion, the distance between markers will change. This change in marker location can be measured by comparison with the resting wire frame. The difference can be attributed to skin motion and digitizing error. The direction of motion can be weighted based upon orientation of local soft tissue underlying skeletal tissue. The actual limb orientation can be determined by using the most likely orientation based upon the wire-frame model and anatomical constraints (see figure 3.30). The wire-frame models will be constructed based upon anatomical and anthropometric literature and direct measurement from the subject.

The orientation of the markers defining the wire frame to the underlying skeletal structure can be measured statically in different orientations. This provides an estimate of marker motion relative to underlying skeletal structure motion. The screw axis can be used to determine the long-bone orientation. With the appropriate screw axis selected, the dynamic and static marker locations can be compared. When a particular marker or set of markers is consistently producing the same type of error, then there is the possibility of discovering the source of error and compensating for it. For example, if a particular motion is known to produce skin motion of known direction, then a goniometer can be used as a reference to compensate for this systematic error.

If the needs and abilities of the wheelchair user are not matched closely to the design and set-up of the wheelchair, the user may be subject to repetitive strain injury. The optimal interaction between the user and the wheelchair would provide the user a maximum degree of freedom in performing desired activities with minimal risk of injury. Kinetic, kinematic and multimodal (e.g., joints moments, joint forces) data may help to identify potentially injurious propulsion techniques and wheelchair set-ups.

Wheelchair propulsion is accomplished by the bilateral simultaneous repetitive motion of the upper extremities. The pushrim is grasped or struck and pushed downward and forward, in turn, rotating the wheels. Davis *et al* showed that the upper extremities during wheelchair propulsion go through

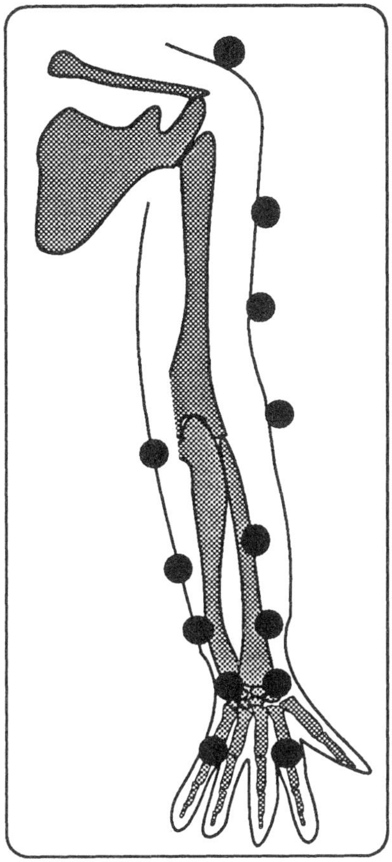

Figure 3.29. *Optical marker placement with respect to underlying anatomy for studying the kinematics of wheelchair propulsion.*

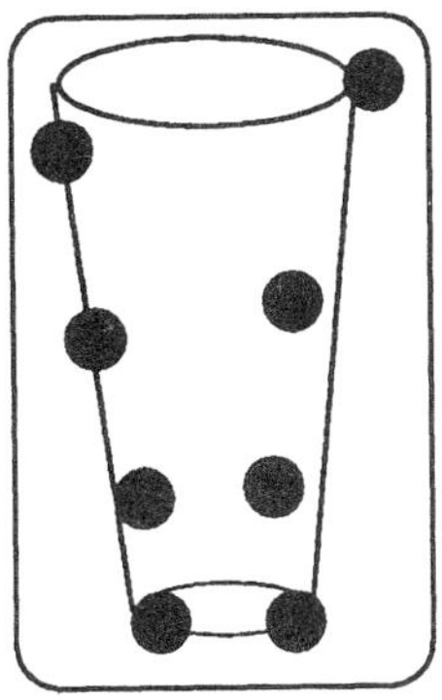

Figure 3.30. *Conical section model for estimating long-bone axis orientation during wheelchair propulsion.*

flexion and extension at this time. Conventional wheelchair propulsion can be divided into two general phases: push and recovery. The push or propulsion phase is defined as the time when the hand is in contact with the rim. The push phase is characterized by a force phase, the time during which force is being applied to the rim. The push phase usually begins with the hands gripping the top of the rim, or at a point just behind the top. The phase ends when the hands leave the rim, usually when the arm is extended. The period in which the hand is in contact with the rims is called the push time (PT). Van Der Woude *et al* defined PT as the time between the initial deflection from baseline level of the moment about the wheel axle to the lowest torque value of the descending curve. The recovery phase is the time when the hand is not in contact with the pushrim and is returning to the start of the push phase. The recovery phase can be defined as the nonpropulsive phase or the phase when no force is exerted to the pushrim. Recovery time (RT) is defined as the time equal to the stroke time (ST) minus the PT, where ST is the time which the user takes to complete one full stroke (RT+PT). RT is further defined as the time between the lowest axial torque value (usually baseline) and the next deflection from the baseline level on a torque curve.

Van Der Woude *et al* measured wheelchair propulsion technique parameters (cycle frequency, push time, recovery time and push angle) during two different workload strategies. The first strategy consisted of four consecutive workloads, which increased 0.28 $m\,s^{-1}$ every 3 minutes. The workload began at a slope of 0, 1, 2, or 3 degrees for tests 1 to 4, respectively. The second strategy consisted of a constant speed of 0.55, 0.83, 1.11, or 1.39 $m\,s^{-1}$, while the slope of the treadmill was increased 1 degree every 3 minutes for tests 5 to 8, respectively. The treadmill was set at 0 degree slope for the first workload. Each subject was allowed to used their own everyday wheelchair with 24 inch rear wheels and standard pushrims. The study was conducted on an Enraf Nonius, model 3446, motor-driven treadmill. The stroke time and push time were determined visually by means of 16 mm high-speed film. The recovery time was a function of mean stroke time minus mean push time. The push angle, the angular trajectory of the hand with respect to the wheel axis between initial hand contact to termination of hand contact, was measured visually from the film. Stroke time was also determined by visualization. The push time was found to decrease with increasing speed, while it remained fairly stable when the slope was increased. The recovery time continued to be relatively constant with increasing speed, but was found to decrease with increasing slope. The increase in treadmill slope and pushrim speed appeared not to influence push angle (PA), which essentially remained stable between 67 and 80 degrees. A constant PA supports the importance of force transmission (peak force) in power regulation. Therefore, an increase in the work-cycle, along with a constant PA, will produce an increase in peak force. Kinematic analyses illustrated that as pushrim velocity increased due to increasing stroke frequency, PT decreased, while RT did not show any significant change. Increasing stroke frequency and the amount of work per cycle, while decreasing ST, will cause an increase in mean power output (speed or slope).

Veeger *et al* measured the three-dimensional forces and torques on the pushrim. The experiment consisted of two different exercise tests at two levels of power output (0.25 and 0.50 $W\,kg^{-1}$). Kinematic analysis was provided by filming (90 frames s^{-1}) each subject from the sagittal view. The start, end and stroke angle were calculated from the kinematic data. Push cycle and recovery time was determined by the information acquired through the synchronization of force and kinematic data. The tests were performed at four different speeds (0.83, 1.11, 1.39 and 1.67 $m\,s^{-1}$) each for 3 minutes. The wheelchair was adjusted to each subject so that the elbow angle was 120 when the hand was at top dead center. The kinematic analysis indicated that the start angle started at 1.28 radians, while the end angle was at 2.63 radians from the horizontal. The stroke angle showed no significant change with increasing speeds. The PT and ST significantly decreased, while the RT showed no significant change with increasing speeds. The study concludes that the subjects did not decrease the time of recovery phase or increase stroke angle to increase pushrim speed.

Veeger *et al* measured kinematic features of wheelchair propulsion using five male wheelchair athletes. The test was performed on an Enraf Nonius motor driven treadmill at 0.56, 0.83, 1.11 and 1.39 $m\,s^{-1}$ for three minutes at each speed. The test was executed in two sessions, at a slope of two and three degrees. Expired gases were collected in Douglas bags during every third minute of each speed. During the same period, a DBM-55 Teledyne camera system was used to film the subject for three complete strokes. Anatomical markers were placed on the subject's acromion, lateral humeral epicondyle, radio-carpal joint and the distal end of the third metacarpal. Push time, stroke time and recovery time were estimated by the number of film frames and film speed. The push angle was calculated as the difference between the start angle (SA) and end angle (EA). A decrease in cycle time during increasing speeds was determined to be a function of decreasing push time, since there were no significant changes in RT. The increase in treadmill slope and pushrim speed did not influence PA. There was a significant decrease in ST and RT with increasing slope at the same speed, while PT showed no significant change. It was suggested that the subjects could improve their propulsion technique by extending the PA or increasing their stroke frequency to increase the overall percentage of PT.

Sanderson and Sommer investigated the kinematic characteristics of wheelchair propulsion. Three world-class paraplegic athletes volunteered for the study. The subjects pushed for 80 continuous minutes on a conventional pushrim style wheelchair at 60 to 65% of their $VO_{2\,max}$. A Locam, model 5001, was used to film the subjects in 20 minute intervals from the sagittal view. Anatomical landmarks were placed on the subjects' acromial processes, lateral condyle of the elbow and the styloid processes at the distal end of the ulna. Three consecutive strokes were digitized with a Bendix Datagrid Digitizer. The data revealed that there were no changes to the subjects' propulsion style during the 80 minute testing period, although there were significant differences found between subjects. The angular kinematic of the elbow and shoulder analysis showed no significant change in propulsion style during the entire testing period. When the hand was in contact with the pushrim, the push cycle was found to be very consistent. The push time was found to vary only 3% from the average PT throughout the 80 minute testing period. There were two distinct stroke styles found, circular and pumping. The two subjects utilizing the circular stroke technique followed the path of the pushrim. The subject who used the pumping technique had a short and abrupt stroking style that followed a small arc around the pushrim. It was concluded that the circular stroking style was more advantageous because the subject can prolong the propulsive phase, producing a greater impulse to the pushrim.

Cooper *et al* have developed a method, based upon Fourier transforms, to determine a wheelchair user's characteristic stroke pattern from wheelchair propulsion kinematic data. This method is incorporated into a clinical workstation to reduce wheelchair propulsion injuries. The method exploits the periodic nature of wheelchair propulsion kinematic data and fits the optimal minimum mean squared error Fourier series to the data. Data collected from several subjects were analyzed using this method. Fourier series representation provides a means for extracting relevant information from a potentially very large set of data corrupted by wide-spectrum noise. The results of Cooper *et al* show a double loop in the wrist contour data for experienced wheelchair users with paraplegia, which does not appear among inexperienced wheelchair users and wheelchair users with quadriplegia. Analysis of the elbow trajectory showed one or two loops in the sagittal plane contour plots for experienced wheelchair users and multiple loops for nonwheelchair users. The higher number of loops has been identified with varus-valgus motion at the elbow, which may lead to injury.

Kinematic data are often corrupted by noise. Noise may come from digitization error, shuttering, quantization error, obscuring of markers and a number of other sources. One of two assumptions is often made about the noise spectrum: either the noise is of higher frequency than the signal or it is white noise (i.e., constant power across the frequency spectrum). If the noise frequency is correctly assumed higher than the signal, then an appropriate low-pass filter will remove much of the noise. If the noise power is assumed white, then more sophisticated techniques are required. Researchers and clinicians often attempt to optimize the sampling frequency of the motion (kinematic) and force/torque (kinetic) data. This results in two different sampling rates. To calculate multimodal data (e.g., net joint moments, net joint forces), the two sampling rates must be synchronized. This has been accomplished using various spline techniques. However, spline methods only provide local synchronization, are heavily dependent upon boundary conditions, and do not guarantee stable results. Cooper *et al* have developed algorithms to exploit the periodicity of kinematic data to develop the optimal cosine series representation for smoothing (for noise reduction) and synchronization.

Wheelchair propulsion is typically characterized by symmetric use of the arms for propulsion. Simultaneous contractions of the arms and the concomittant impact with the pushrim cause a rectified sine-wave-type moment curve. Cooper *et al* have investigated asymmetrical wheelchair propulsion as a possible means of maintaing efficiency of movement by lowering the amplitude of the moment curve and in turn lengthening the propulsion phase. Their results showed an increase in energy cost with asynchronous wheelchair propulsion.

Efficiency of wheelchair propulsion may be reflected in the kinematics data. Cooper *et al* have studied the kinematics of racing wheelchair propulsion. Through a wheeling economy protocol, they have been able to correlate some kinematic variables with physiological and performance variables. They have found elbow height during follow through, amplitude of head and trunk motion and wrist trajectory all to be related to economy and performance. In order to effectively push a racing wheelchair, the upper body must be used to rapidly apply a large impulse to the pushrim. Applying an impulse to the pushrim causes a change in momentum of the racing wheelchair and rider. Wheelchair athletes must learn how to best use their body segments to apply as large an impulse as possible to the pushrim for the metabolic energy cost they can afford. They must also understand the factors which allow the impulse to the pushrim to be transmitted as efficiently as possible. Cooper and coworkers have divided the racing wheelchair into five phases. The stroke starts with the elbows in their highest position and finishes with the elbows coming back to the top. Each phase is defined in terms of its contribution to the total impulse generated with consideration for maximizing the impulse during each phase. The five phases are defined as follows: (1) drive forward and downward, (2) pushrim contact, (3) pushing through to the bottom of the pushrims, (4) push-off or follow through and (5) elbow drive to the top. Standard manual wheelchair propulsion can be divided in to six phases (see figure 3.31).

3.4.3. Kinetics of wheelchair propulsion

The ability of an individual to push a wheelchair efficiently and without injury is in large part related to the way in which the user applies force to the pushrim during propulsion. A number of factors influence the interaction between wheelchair user and chair, including level of spinal cord injury,

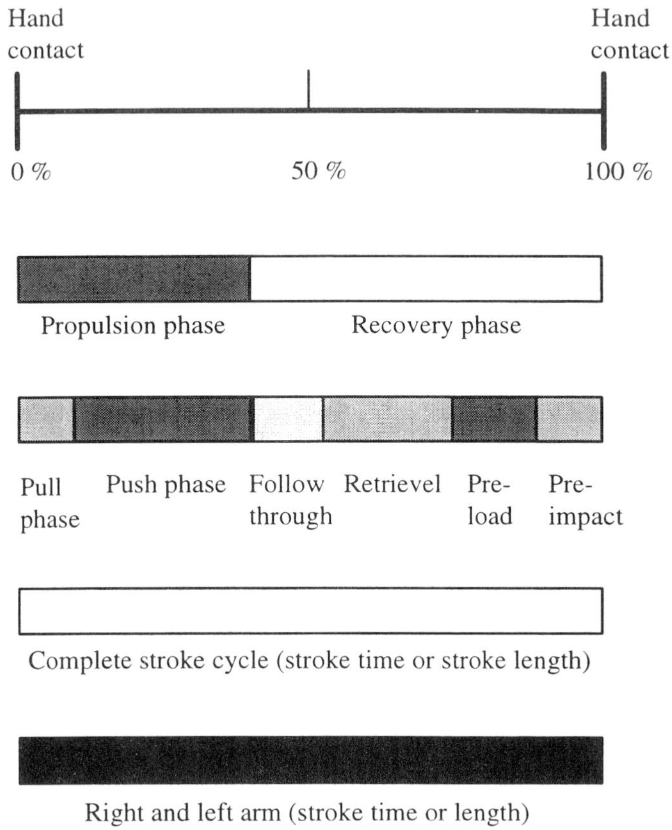

Figure 3.31. *Six phases of the standard manual wheelchair propulsion cycle. Critical phases can be identified, similar to those in walking, which can be used to identify and classify propulsion abnormalities and inefficiencies.*

design of the wheelchair, fit between user and chair, stroke mechanics, user fitness levels and history of upper-extremity injury. Understanding how forces generated by the individual are applied to the pushrim will provide insight into how these forces are related to optimizing efficiency, improving performance, identifying mechanisms of injuries (such as carpal tunnel syndrome, elbow tendonitis and rotator cuff injuries), developing injury prevention techniques and implementing changes in wheelchair design.

The complexity of developing a system for measuring pushrim forces is evidenced by the paucity of data in the literature on the kinetics of wheelchair propulsion. A number of researchers have attempted to develop a force-sensing system with varying degrees of success. The wheelchair kinetic data reported in the literature can be divided into three categories: (1) static force measurements, (2) external devices for measuring forces and torques and (3) measurement of force components at the pushrim—indirectly or directly.

Static force measurement: Brauer and Hertig used a system of springs to restrain a pushrim. Static torque was measured for wheelchair and nonwheelchair users at six different positions ranging from −10 to 40 degrees relative to the vertical. Males were found to generate more torque than females. Ranges of torque values were from 27.9 to 46.6 N m for males (wheelchair and nonwheelchair users) and 17.1 to 32.1 N m for females. The authors of this study concluded that the amount of torque produced by the user at the pushrim varies with frictional characteristics of the pushrim, grip location, handedness, grip strength and/or how well the wheelchair fit the anthropometric measurements of each subject. Brubaker *et al* measured static pushing and pulling forces for four grip positions using strain gaged beams mounted to the pushrims. A moveable set-up allowed various seat positions. The forces ranged between 500 and 750 N and varied considerably with rim and seat position.

External devices for measuring forces and torques: Tupling *et al* used a force plate to measure the force generated during the initiation of wheelchair propulsion for the grab and start techniques. They found that the grab start was more effective in initiating the movement and that the individual's strength determines their ability to generate an impulse. Samuelsson *et al* described a wheelchair ergometer which was a wheelchair with a gear attached to the hub connected by a chain and gear to an isokinetic dynamometer. This device allowed torque and power output to be calculated. Results of a pilot study showed that subjects could produce peak torque values in the range of 70 N m at 120 deg s^{-1}. Ruggles *et al* tested three different wheelchairs through a roller system connected to a Cybex Isokinetic Machine. Peak torque, angular displacement, work and angular impulse were compared. Differences among chairs were found and the authors suggest that wheelchair design and dimensions relative to the anthropometry of the user have a great influence on the mechanical characteristics of propulsion.

Measurement of force components at the pushrim: Niesing *et al* have given a description of a stationary ergometer designed for the analysis of various simulated wheelchair pushing conditions such as varying resistance, velocity and slope. This system allows for seat configuration changes, different pushrim sizes and adjustments in camber. Ergometer forces are measured by three-dimensional force transducers mounted on the support brackets. A torque transducer is mounted between the pushrim and the wheelchair axis. Although this ergometer allows for varying conditions of wheelchair propulsion to be studied, it does not allow the user to push their own wheelchair or any actual wheelchair and it can only be used in the laboratory. Van der Woude's laboratory has used this dynamometer for measuring the biomechanical and physiological characteristics of wheelchair propulsion on nonwheelchair and wheelchair users.

Veeger *et al* compared trained and untrained subjects on the previously described dynamometer during a 30-second sprint test. Of particular interest were the estimated pushrim forces. These were defined as F_{tot} (the resultant of the three orthogonal forces), F_m (the effective force on the pushrims measured by dividing the pushrim radius into the moment measured at the hub—in effect the tangential force), FEF (fraction of effective force determined as the ratio between the tangential force and the total force). F_m and F_{tot} were mean values determined over the duration of the push phase and F_{eff}, the effective force component, of F_{tot} was calculated based on the magnitudes of the three orthogonal forces, the position of the hand on the pushrim and the wheel camber. Differences in estimated force application existed between both groups which in part was attributed to the use of the elbow flexors. However, FEF was not significantly different. Large within-group variability was noted. It was concluded that nonwheelchair users should not be used for research directed toward optimization of wheelchair design, as has been shown in several previous studies.

In another study, Veeger *et al* tested nine nonwheelchair using subjects at different speeds of propulsion. They showed that the F_x (forward) forces were considerably smaller than the F_y (vertical), while F_z (medial/lateral) had a more consistent pattern with a large outward component at the end of the push phase. This resulted in a total force component which was directed predominantly downward. It was also found that FEF decreased significantly with speed and that the forces increased with external power output, while FEF parameters did not. The researchers concluded that the low FEF may be the result of non-optimal wheelchair dimensions, an ineffective propulsion technique, or the need for a large friction force between hand and rim. How alterations in these factors affected performance during other activities of daily living was not elucidated.

Roeleveld *et al* tested nine wheelchair athletes during a 30 s sprint test on the same system. Results showed that the fraction of effective force was low for these subjects and only slightly higher than what was found for less experienced subjects. This was attributed to the total force not being tangentially directed to the geometry of the wheelchair. Utilizing a similar protocol on the device, Dallmeijer *et al*, in addition to physiological measurements, compared forces applied to the pushrim by a group of spinal cord injured subjects during a 30 s sprint test. FEF was also found to be low in this study and, although a number of differences were found in performance between the high-lesion group (C4–C8) and the other lower-lesion groups, FEF was not different. None of these studies describes how the subjects were positioned in the dynamometer and whether seating position was varied to justify their comments about wheelchair geometry forcing the tangential force to be submaximal.

It is apparent from the literature that there are a number of shortcomings in information related to wheelchair pushrim forces. Cooper *et al* have developed a system which reliably and accurately measures pushrim forces in order to accomplish the goals of improving ride comfort, increasing

Figure 3.32. *Force and torque sensing SMART^{Wheel} for studying wheelchair propulsion dynamics.*

efficiency and reducing injuries. An early version of the present force-sensing pushrim, was reported by Cooper and Cheda to dynamically measure racing wheelchair pushrim forces and torques. A slotted disk was used to mount different size pushrims which mounted to the hub with three beams instrumented with strain gages. A later version of this force-measuring wheel, SMART^{Wheel}, allows measurement of pushrim forces in the plane of the wheel and the turning moment about the hub axis when mounted on everyday chairs (see figure 3.32 (photograph)).

Cooper's group has recently begun using the SMART^{Wheel} to study the kinetics of wheelchair propulsion. The SMART^{Wheel} has been fully described in the engineering literature and has been found to have a precision of 0.2 N and a resolution of 2 N. Correlation analysis showed the measured forces to be highly correlated with known forces ($r^2 = 1.0$, $p = 0.0001$). Accuracy was determined to be 4.1% for the worst case. Cooper has examined wheelchair propulsion as an optimal force and energy control problem. Cooper and coworkers have developed models of the upper limbs and how they must interact with the neuromotor control system to produce effective wheelchair propulsion. The modeling of wheelchair joint moments and forces had been limited to theoretical analysis until the development of the pushrim force-sensing technology (see figure 3.33).

Net joint moments and forces can be calculated from kinetic data acquired from the SMART^{Wheel} and kinematic data obtained through film or video analysis. The methods described earlier in this chapter were used to estimate

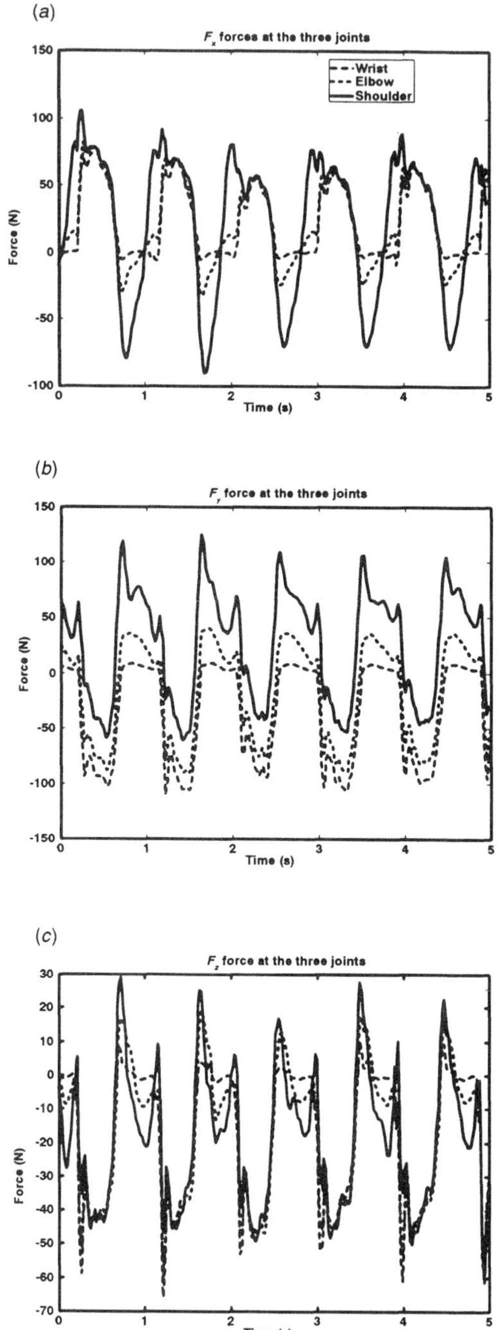

Figure 3.33. *Net joint moments and forces during manual wheelchair propulsion at* $5\,km\,h^{-1}$ *on a wheelchair dynamometer set to 10 W. (Continued opposite)*

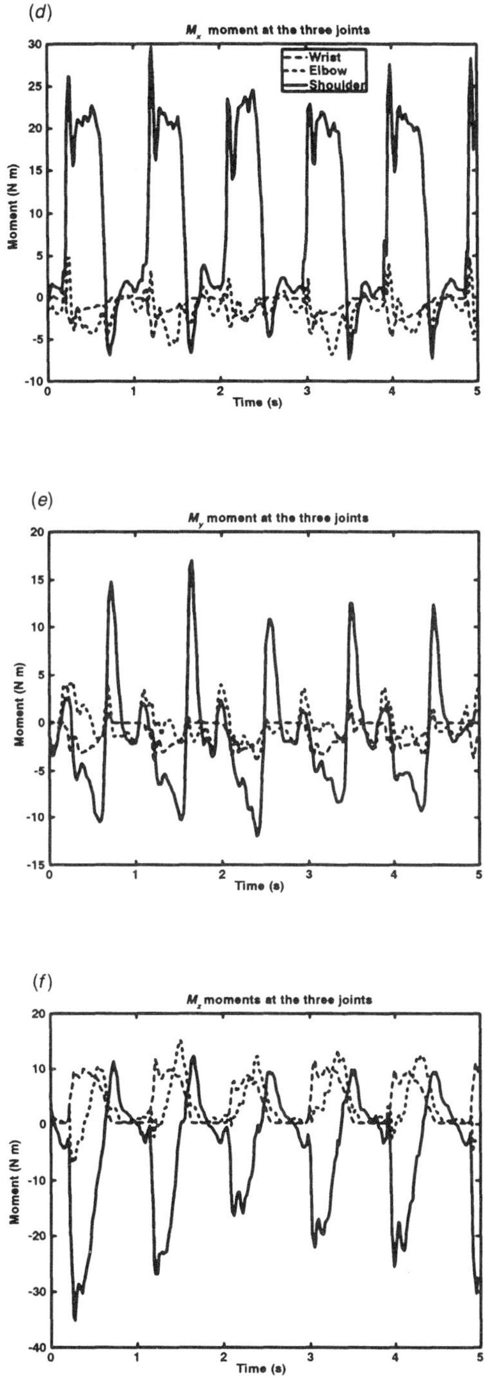

Figure 3.33. *Continued.*

the joint moments. Net joint moment and forces may be useful in improving propulsion efficiency and for understanding the underlying mechanisms related to neurological and orthopaedic injury to the upper extremities during wheelchair propulsion.

3.5. BIOMECHANICS OF SEATING

When seated the body is positioned such that its weight is supported mainly by the ischial tuberosities of the pelvis and the surrounding soft tissues. Depending on the wheelchair and posture, some proportion of the total body weight will be transferred to the seat, backrest, legrests and armrests of the wheelchair. Proper seating posture can provide stability to perform precise visual and motor control tasks. Proper seating can reduce the strain on muscles, tendons and joints. Proper seating posture can reduce hydrostatic pressure and improve venous return in the lower extremities.

The spine is particularly important to seating posture, as are the upper and lower extremities. Many people with disabilities develop or have developed spinal deformities. Curvature of the spine requires special consideration when designing seating support structures.

3.5.1. Seating and the spine

For the purpose of designing seating and postural support devices, the spine can be divided into three regions: the flexible cervical and lumbar spine and the rigid thoracic spine. The thoracic spine is sandwiched between the cervical and lumbar spine. The cervical spine is attached to the head which is assumed rigid, whereas the lumbar spine is connected to the sacrum which is rigidly affixed to the pelvis.

Sitting can be divided into three postures: anterior, median and posterior. The distinction is based upon the location of the person's center of mass which affects the distribution of body mass over the seat, backrest and legrests. The shape of the spine is also affected by posture. When in a relaxed median position the center of mass is directly above the ischial tuberosities and the lumbar spine is either straight or in slight kyphosis. Forward rotation results from forward rotation of the pelvis with a straight spine or by little rotation of the pelvis and large kyphosis. Contoured seating structures offer reduced pressure and shape deformation around the ischial tuberosities in a relaxed median position. This cushion design may reduce the risk of decubitis ulcers; however, the pelvis must remain free to rotate to prevent extreme kyphosis when the person moves to an anterior posture. In a posterior posture the person's center of mass is behind the ischial tuberosities and corresponds to rearward rotation of the pelvis and kyphosis of the spine. The design of the backrest has a substantial effect on posterior seating posture.

The postures of the cervical spine, shoulders and upper extremities are often dependent upon the assistive devices used, and the work layout. Hence, seating posture and task performance are interrelated. The posture of people with disabilities is dependent upon their anthropometry, disability etiology, level of impairment and sitting habits. The height, inclination and depth of the seat affect seating posture. The position, shape, height and angle of the backrest also influence seating posture, as do the length, angle and shape of the legrests. Other structures of the wheelchair or seating system (e.g., armrests) influence seating as well. Therefore when designing/selecting a seating system and wheelchair the tasks to be performed in the system must be considered as well as the physical attributes of the person. Also it is important that alterations in posture be permitted.

Radiographic studies have shown pelvic rotation and flattening of the lumbar spine lordosis while seated. A properly designed lower back support can mitigate flattening of lumbar lordosis. The shape of the lumbar spine during sitting depends on the rotation of the pelvis. When seated, the sacral end-plate become nearly horizontal and normal lumbar lordosis becomes flattened. Providing some sacral horizontal angle can help to provide a more neutral lumbar lordosis. Several suggestions have been made with respect to preventing flattening of the lumbar spine: a forward tilted seat, pelvic fixation, increased seat height, and increased backrest height. Lumbar or total back support (provided scapula movement is unrestricted) is most successful in providing healthy seating posture.

Certain postures place greater pressure on the disks of the lumbar spine. There is an increase in trunk load moment upon the lumbar disks when the pelvis is rotated backwards and the lumbar spine and torso are rotated forwards. There is deformation of the disks themselves with lumbar lordosis flattening. Inclination of the backrest angle can result in decreased disk pressure. An optimal angle of about 110 degrees has been published. Studies have also shown a decrease in disk pressure when lumbar support was added at the level of the fourth and fifth lumbar vertebrae. The use of armrests has also been shown to reduce disk pressure.

$$f_c = f_H \frac{a_H}{a_c} \qquad f_L = \frac{f_T b_T + f_H b_H}{b_L} \tag{3.29}$$

where $a_c =$ moment arm of the cervical spine extensor muscles
$a_H =$ moment arm of the head's weight
$b_H =$ moment arm of the head's weight
$b_L =$ moment arm of the lumber spine extensor muscles
$b_T =$ moment arm for the torso's weight
$f_h =$ weight of the head and neck
$f_t =$ weight of the torso
$f_c =$ force due to the cervical spine extensor muscles
$f_l =$ force due to the lumbar spine extensor muscles.

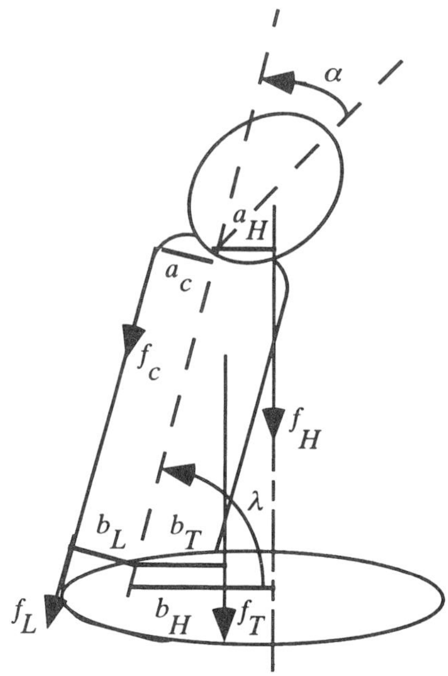

Figure 3.34. *Schematic diagram of forces acting on the cervical spine.*

Head and neck position is usually determined by visual requirements. Head and neck flexion are often related to visual tasks. For some people the head and neck provide an important control input. By altering the position of devices and monitors the amount of head and neck flexion can be controlled. This is a common problem among users of augmentative communication devices who have certain demands upon their arms for control of the device and other demands upon their head and neck for observing the display. From a simple biomechanical perspective, there are two moment equilibrium problems to be solved when balancing head and neck flexion. The cervical spine extension muscle force f_c and the lumbar spine extensor muscle force are resolved separately (see figure 3.34).

There is little variation in the values of the muscle moment arms with changes in posture, therefore the spine muscle forces are approximately proportional to the load moments. The magnitude of the load moments depend upon the angles α and λ shown in figure 3.34. From these relationships one can determine that even a slight forward lean posture can create large muscle forces. A flexed neck posture should be avoided when possible.

Bending and reaching have been shown to increase disk pressure due to large external moments imparted on the spine. This factor should be considered when integrating multiple assistive technologies with the wheelchair. For people who use wheelchairs for many hours during the day, a zero-shear reclining system should be used to permit changes in posture. The layout of other assistive devices should be such that lifting and moving is minimized. Operation of such devices should be performed with the person's joints and limbs in their near-neutral position.

3.5.2. The extremities while seated

It is not uncommon for therapists, rehabilitation engineers and users to integrate assistive devices with the wheelchair. Even simple devices (e.g., lap trays) may alter the working seating position. The placement of the work surface and devices relative to the person is important not only because they influence the spine but also because they influence the posture of and loads on the upper extremity as well. When performing work on a lap tray or counter top a shoulder abduction angle of less than 20 degrees and a flexion angle of less than 25 degrees should be attempted.

The ball-in-socket arrangement of the glenohumeral joint provides the arm with a large amount of mobility. This mobility is enhanced by six additional joints which comprise the trunk-arm complex or shoulder girdle. This extreme mobility sacrifices intrinsic stability. Stability is provided by the ligaments, joint capsules and musculature of the upper extremity. The head of the humerous is held in the glenoid socket by the scapula through the combined actions of the rotator cuff muscles and the pectoral muscles. Many muscle and skeletal structures are involved in movement of the shoulder. During many movements demand on one set of muscles is compromised at the expense of another set of muscles.

When integrating technologies or designing a workspace for the wheelchair user a particular concern should be to minimize the length and number of times the hands must be lifted above shoulder height. This type of activity may aggravate the stress induced on the shoulder by manual wheelchair propulsion and may lead to degenerative tendonitis in the biceps and supraspinatus muscles. In extended reach positions the shoulder joint is flexed and the elbow is extended. Load moments at the elbow and shoulder can become large relative to the flexor strength moments required at both joints. Use of an adjustable elbow support can reduce the moments on the shoulder and elbow.

Arm and forearm postures are not only dictated by the location of the hand, but also by hand orientation around the long axis of the forearm. If the hand is supinated, then the arm will normally be adducted and close to the torso. If operation of a device requires the hand to be prone, then the arm will normally be abducted and elevated. The relation between hand posture and arm orientation is important when designing or selecting devices to integrate with a wheelchair workstation.

Legrests provide critical support while mobile and while stationary. Proper leg support can help reduce the load on the buttocks and thighs. Legrests should be adjusted so that the feet sit squarely and firmly on the footrests. The weight of the lower legs and feet should be solely support by the legrests and not by the front of the thighs. Pressure to the thighs behind the knee can increase edema in the legs and place pressure on the sciatic nerve. Range of motion exercises and changes in seating posture can reduce edema in the legs. The seat base pan should be about 3 to 5 cm below the knee when loaded. The edge of the cushion or seat should be contoured to relieve pressure. About 5 to 10 cm of clearance should be provided between the edge of the seat pan and the center of the knees.

A seat angle of about 5 degrees is typical. Some wheelchair users prefer a larger seat angle because of the greater stability it provides. However, seat angles in excess of 15 degrees are not recommended. Excessive seat angle can result in small knee flexion angle which may transfer the weight of the trunk over a small area at the ischial tuberosities. The spine may also become flexed as the pelvic rotates backwards. In addition, the abdominal organs are compressed in this posture when one leans forwards.

3.6. BIOMECHANICS OF MANIPULATION

3.6.1. Assessment of upper-extremity orthoses

The team of physician, therapist and rehabilitation engineer or orthotist assesses the anatomy and neuromuscular function of the impaired limb. Based upon biomechanical and functional deficits, an orthosis is designed or specified. Upper-limb orthoses have been successfully applied to provide enhanced function for all joints of the upper extremity, from interphalangeal to glenohumeral. Most upper-extremity orthoses are used to enhance the function of the fingers, hand and wrist. Upper-extremity orthoses can either be static or dynamic. Static orthoses provide increased function by stabilizing, restricting motion and by positioning. Static orthoses are also used to enhance medical treatment by protecting weak muscle and connective tissue, inhibiting the development of contractures, to facilitate surgical repair and to support healing of soft tissue injuries and fractures. Dynamic orthoses are designed to enhance functional tasks and movement. Dynamic orthoses can redirect forces and moments from other muscles or joints to provide functional movement at other joints. They can also be used to provide functional movement by providing a substitute power source for weak or paralyzed muscles, or by allowing normal-strength muscles to move a joint or joints with the orthosis controlling the direction or alignment of the movement.

Static functional orthoses are often used to support the wrist during activities of work, self-care and leisure. Static functional orthoses should protect the joint while providing maximal fucntion of the splinted joint.

The wrist extensor orthosis (WEO) is one of the most widely prescribed functional orthoses. The WEO supports the user's wrist in extension while permitting flexion/extension of the metacarpophalangeal joints and thumb movement. The WEO should maintain a moderate distal transverse arch while conforming to the hand.

There are several types of WEOs available and all have their strengths and weaknesses. There are two widely used WEOs: the volar orthosis and the dorsal style orthosis. The volar orthosis is roughly rectangular in shape, with a large hole accommodating the thenar eminence (e.g., thumb hole). A radial connector bar can be incorporated into the volar orthosis. The connector bar passes from the palm across the web space to the dorsal radial aspect of the hand, ending at the second metacarpal head. A radial connector bar limits radial deviation and acts as a connector for a distal strap. The dorsal style orthosis uses a radial projection that curves across the palm to support the distal transverse arch of the hand.

Orthoses are manufactured from a variety of materials. Some materials can be molded directly over the person's body (e.g., low-temperature plastics that become malleable when placed in warm water) while others are formed over a positive mold. Polycaprolactone-based (e.g., Polyform) low-temperature thermoplastics are often used to make orthoses. A positive mold can be made by using bandage material soaked in plaster of Paris which is wrapped around the limb. The limb is positioned as desired with the completed orthosis. Angles can be verified with goniometers. Once the casts are removed, they must be sealed and reinforced with additional bandages soaked in plaster of Paris. Once the cast is completed, it is filled with plaster of Paris with a steel rod insert. The steel rod is used to ease handling of the completed positive mold. After the positive mold has hardened the remaining bandage is removed, rasps are used for smoothing, and a fine wire mesh is added to strengthen the positive mold. The type of the orthosis, the fabrication process and the patient's special needs (e.g., skin sensitivity, sensation, strength) are all interrelated. Complex dynamic upper-limb orthoses can be used to provide function to people with very little upper-limb motor function. The flexor hinge hand orthosis (i.e., tenodesis splint) and the balanced forearm orthosis are used by individuals with cervical-level spinal cord lesions to provide functional movement by redirecting some voluntary control. More complex orthoses require extensive training programs.

Although most upper-extremity orthoses are used to enhance function of the fingers, hand and wrist, little is understood about how different orthoses affect hand function during various common tasks. Greater understanding of the relationship between hand function and orthosis design becomes more pressing as the number of upper-extremity orthoses (UEOs) grows. For example, UEO are being more widely used for people with rheumatoid arthritis or direct wrist trauma. Hand function is primarily context driven. Clinical hand function assessments isolate function from context. This is accomplished by measuring hand strength (e.g., grip and pinch) and manual dexterity (e.g., finger dexterity, ability to grasp and release, right/left hand interaction) as means of approximating hand function during activities of daily living. Rehabilitation teams should also be concerned about reducing pain among many wrist–hand orthosis (WHO) users.

The Jebsen hand function test was developed to assess hand function while performing a variety of simple tasks. This test has been applied to evaluate manual dexterity with WHOs. The Jebsen test requires a wide range of unilateral hand functions and involves direct and tool-assisted manipulation of objects. The seven subsets of the Jebsen test are: writing a sentence; turning a 760 mm by 1270 mm note card; grasping, moving and releasing small objects; stacking checkers; using a spoon to place beans in a can; moving empty cans; and moving 450 g cans. Each of the seven tasks is timed and has specific criteria for task initiation and completion. There is a significant practice effect with these tests, therefore test–retest reliability must be considered. An unencumbered and unimpaired hand provides the fastest manual dexterity for all Jebsen subtests. Stabilizing orthoses tend to be more effective at improving simple grasp and release tasks with large objects. However, free-hand grasp is much stronger than any static orthosis. Therefore, clinicians must consider strength and manipulation issues during functional testing.

3.6.2. *Assessment of upper-extremity prostheses*

Upper-limb amputation accounts for about 15% of acquired limb loss in the United States. Under the age of ten, congenital skeletal deficiency is the most common cause of upper-limb loss. For people of ages 10 to 20, tumor is the most frequent cause of amputation for both lower and upper extremities. For people of 15 to 45, trauma, disease and tumor are all responsible for upper-limb loss.

Before the medical team recommends amputation, the individual should receive counseling regarding the need for amputation, level of amputation and functional expectation following amputation. Pain is often associated with amputation. Therefore, individuals should understand incisional pain, phantom pain, phantom limb sensation and pain control. As with all amputations, the residual limb length should be as long as possible. The surgeon should ensure that the distal end of the bone is covered with a full thickness of skin and soft tissue. However, wrist and elbow disarticulations are an exception to this rule. Many people obtain greater function with a partially mobile hand with some sensation than with a prosthetic terminal device. Once the limb has been amputated, range-of-motion and strengthening exercises are required. A provisional prosthesis can be fitted immediately post-surgery. This has been shown to improve acceptance, increase use of a prosthesis and maximize the likelihood of a return to preamputation levels of functional activity.

Table 3.1. *Advantages and disadvantages of various common upper-limb prostheses.*

Type	Advantages	Disadvantages
Cosmetic (passive)	Lightweight Good cosmesis Minimal harnessing	High cost if custom Least functional Stain easily
Body powered	Moderate cost Moderate weight Very durable	Largest body movement Most harnessing Poor cosmesis
Externally powered	Moderate to no harnessing Least operation movement Moderate cosmesis Good proximal function	Heavy Expensive High maintenance Limited sensory feedback
Hybrid		
cable elbow with electric terminal device (TD)	Reliable cable excursion to elbow Increased TD pinch	Electric TD weights forearm Good for elbow disarticulation
electric elbow with cable terminal device (TD)	Reliable cable excursion to TD Low effort to position Low maintenance	Poor cosmesis Low pinch force for voluntary opening TD

When prescribing or designing an upper-limb prosthesis, one must first decide whether the prosthesis is to be predominantly cosmetic or functional. Upper-limb prostheses are typically body powered (e.g., cable operated) or externally powered (e.g., electromechanical). In some cases combinations of body and externally powered devices are used. These are often called hybrid prostheses. Prosthetic design must take into account functional, vocational and recreational needs. Physical and cognitive abilities and financial resources also affect design options. Rehabilitation engineers must work with clinical professionals to specify and/or design the prosthesis construction including the controls, socket, harness, mechanical joint components, wrist unit and terminal devices. Central to the prosthetic design and selection process are the wishes of the person with the amputation. The rehabilitation team should have knowledge of a wide array of choices as well as the advantages and disadvantages of various prosthetic options, see

During the fitting and design process, the new amputee must receive instruction in the proper wearing, use, and care of the prostheses. Improper or insufficient training can result in reduced function and in some cases abandonment of the prostheses. Manual work and many outdoor activities are readily accomplished with body-powered components. Many people with amputations who work in office environments prefer passive cosmetic or myoelectric prostheses. Since nearly 90% of activities of daily living can be performed with one hand by the individual with a unilateral amputation, the physical impairment is often overestimated while the psychological aspects

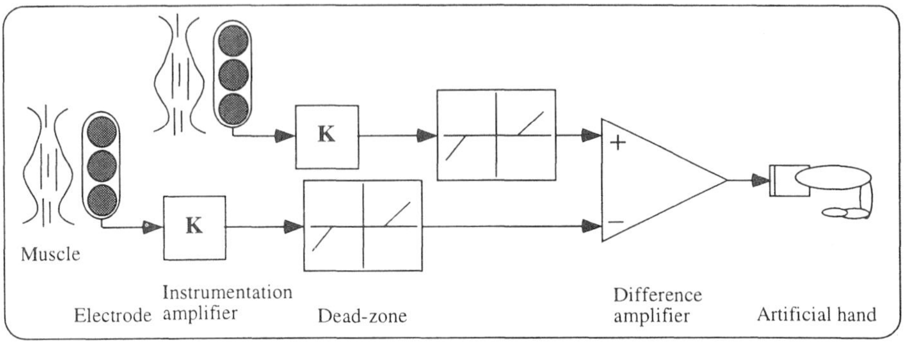

Figure 3.35. *Proportional myoelectric control of an electromechanical hand.*

are not given sufficient consideration. Simple passive prostheses to restore the appearence of the hand can be made using available materials and techniques. Molded and colored silicone rubber has been shown to work well for this application. A glove is made with several layers of skin-colored silicone rubber. The subtle effect of layering the low viscosity silicone achieves a life-like appearance. Most people with bilateral upper-limb amputations should initially be fitted with body-powered components. People with unilateral, proximal arm amputations may benefit from the use of an electric elbow and hand. For many people with hand amputations, aesthetic considerations are more important than the need for prehension.

The design and fitting of upper-extremity prostheses requires substantial effort and creativity and their significance has been increased by the development of myoelectrically controlled prosthetic arms. Although a prosthetic hand only serves as a modest substitute for the missing hand considerable energy has been expended to achieve the current level of success. With a myoelectric prosthesis, control of the electromechanical hand is accomplished with muscular electrical potentials generated within the residual limb. These potentials are received at the surface of the residual limb through electrodes. The signal is then amplified and transmitted to a microprocessor to produce control signals. An important factor in a successful fitting is a reliable and consistent interface between the prosthesis, the electrodes and the residual limb. A prosthesis must fit properly. A myoelectric prosthesis requires a particularly close connection to attain useful control. Users must learn to integrate the prosthesis into their own body patterns.

Myoelectric control is accomplished by interpreting surface potentials generated by muscular potentials inside the residual limb (see figure 3.35). When the body performs voluntary movements, control signals are transmitted from the central nervous system in the form of short tension pulses via nerve pathways to innervated muscle. This action generates an electrochemical reaction at the junction of the nerve fiber to the muscle, which produces contraction of the muscle fiber. Contractions of multiple muscle fibers can be measured as electrical potentials in the microvolt range on the skin surface. Conceptually, a myoelectric prosthesis receives control signals to create motion much in the same way the intact limb would. The person with the amputation must have sufficient strength in each of several muscles to be able to use myoelectric control. For example, finger extensors and flexors are required to attain grasp and release. In simple myoelectric systems relays or transistors are used to control the current through a motor. If the muscle signals indicate a change from flexion to extension the battery voltage across the motor is reversed which causes a corresponding reversal in motor current and direction. Myoelectric hands can offer an opening width of 100 mm, an average gripping speed of 85 mm s^{-1} and obtain a grip force of 80 N. The grip force is commonly limited via a slip clutch. The speed of closure and grip force are coupled. The speed of closure remains at maximum until a grip force of 15 N is exceeded, at which time speed is decreased. A difficult problem in electric prostheses is multichannel controls and feedback systems. All biological processes form feedback and feedforward loops. For the wearer of a prosthetic limb, the visual system provides the primary feedback system. Artificial feedback systems are being developed, but have so far not shown significant improvement in control.

The rehabilitation engineer and clinical staff must see the person at regular intervals to assure proper fit and maintenance of the upper-extremity prosthesis. Regular adjustment and maintenance can increase function and prevent problems developing with the residual limb (e.g., neurotrauma, joint contractures).

FURTHER READING AND REFERENCES

Aljure J, Eltorai I, Bradley W E, Lin J E and Johnson B 1985 Carpal tunnel syndrome in paraplegic patients *Paraplegia* **23** 182–6

Allin J and Inbar G 1986 FNS control schemes for the upper limb *IEEE Trans. Biomed. Eng.* **BME-33** 818–24

Allin J and Inbar G 1986 FNS parameter selection and upper limb characterization *IEEE Trans. Biomed. Eng.* **BME-33** 809–17

An K N and Chao E Y 1984 Kinematic analysis of human movement *Ann. Biomed. Eng.* **BME-12** 585–97

Andersson G B J and Oertengren R 1974 Lumbar disk pressure and myoelectric back muscle activity during sitting. III. Studies on a wheelchair. *Scand. J. Rehab. Med.* **3** 122–7

Andersson G B J, Murphy R W, Oertengren R and Nachemson A L 1979 The influence of backrest inclination and lumbar support on the lumbar lordosis in sitting *Spine* **4** 52–8

Armstrong T J and Chaffin D B 1979 Some biomechanical aspects of the carpal tunnel *J. Biomech.* **12** 567–70

Asato K T, Cooper R A, Robertson R N and Ster J F 1993 SMART^{Wheels}: Development and testing of a system for measuring manual wheelchair propulsion dynamics *IEEE Trans. Biomed. Eng.* **BME-40** 1320–4

Ashley R K, Vallier G T and Skinner S R 1992 Gait analysis in pediatric lower extremity amputees *Orthop. Rev.* **21** 745–9

Axelson P 1993 Wheelchair comparison *Sports 'N Spokes* **18** 34–40

Bajd T and Jaeger R (ed) 1994 *Basic and Applied Myology: FES for Movement Restoration* **4** 127–202

Baldini F D, Cooper R A, Jones D, Robertson R N and Widman L 1992 Economical aspects of wheelchair propulsion *Med. Sci. Sports Med.* **24** S32

Baldini F D, Skinner B F and Cooper R A 1990 The economy of synchronous (syn) and asynchronous (asyn) wheelchair propulsion *American College of Sports Medicine South West Conf.*

Balmaseda M T, Koozekanani S H, Fatehi M T, Gordon C, Dreyfuss P H and Tanbonliong E C 1988 Ground reaction forces, center of pressure and duration of stance with and without and ankle-foot orthosis *Arch. Phys. Med. Rehab.* **69** 1009–12

Banta J V, Bell K J, Muik E A and Fezio J 1991 ParaWalker: energy cost of walking *Eur. J. Pediat. Surg.* **1** (suppl. 1)

Barter J T 1957 Estimation of the mass of body-segments *Report (WADC TR 57-260) Wright-Patterson Air Force Base, OH*

Bayley J C, Cochran T P and Sledge C B 1987 The weight bearing shoulder. The impingement syndrome in paraplegics *J. Bone Joint Surg.* **69** 676–8

Beasley R W 1987 Hand and finger prostheses *J. Hand Surgery* A **12** 144-7
Beasley R W and De Bese G M 1986 Upper limb amputations and prostheses *Orthop. Clin. North Am.* **17** **3** 395–405
Bernotas L A, Crago P E and Chizeck H J 1987 Adaptive control of electrically stimulated muscle *IEEE Trans. Biomed. Eng.* **BME-34** 140–7
Bigliani L U, Ticker J B, Flatow E L, Soslowsky L J and Mow V C 1991 The relationship of acromial architecture to rotator cuff disease *Clin. Sports Med.* **10** 823–38
Bloomquist L E 1986 Injuries to athletes with physical disabilities: prevention implications *Phys. Sportsmed.* **14** 96–105
Bowker P, Messenger N, Oglive C and Rowley D 1992 The energetics of paraplegic walking *J. Biomed. Eng.* **BME-14** 344–50
Brauer R L and Hertig B A 1981 Torque generation on wheelchair handrims *Proc. 1981 Biomech. Symp. ASME/ASCE Mechanics Conf. Colorado* pp 113–6
Braune W and Fischer O 1889 Über den Schwerpunkt des menschlichen Korpers mit rucksicht auf die Ausrustung des deutschen Infanteristen *Abh. D. math-phys. K. Sachs. Gesselsch. der Wiss.* **26** 561–672 W M Krogman and F E Johnston (ed) *Human Mechanics—Four Monographs Abridged* Wright-Patterson Air Force Base, OH (AMRL-TDR-63-123)
Brienza D M, Chung K-C, Brubaker C E and Kwiatkowski R J 1993 Design of a computer-controlled seating surface for research applications *IEEE Trans. Rehab. Eng.* **BME-1** 63–6
Broer M R and Zernicke R F 1979 *Efficiency of Human Movement* (Philadelphia, PA: Saunders)
Brown D D, Knowlton R G, Hamill J, Schneider T L and Hetzler R K 1990 Physiological and biomechanical differences between wheelchair-dependent and able-bodied subjects during wheelchair ergometry *Eur. J. Appl. Physiol.* **60** 179–82
Brubaker C E 1986 Wheelchair prescription: An analysis of factors that affect mobility and performance *J. Rehab. Res. Dev.* **23** 19–26
Brubaker C E, Ross S and McLaurin C A 1982 Effect of seat position on handrim force *Proc. 5th Ann. Conf. Rehab. Eng., (Houston, TX)* p 111
Burnham R S and Steadward R D 1994 Upper extremity peripheral nerve entrapments among wheelchair athletes: prevalence, location and risk factors *Arch. Phys. Med. Rehab.* **75** 519–24
Butler P B and Major R E 1987 The ParaWalker—a rational approach to the provision of reciprocal ambulation for paraplegic patients *Physiotherapy* **73** 393–7
Cambell G S, Gow D and Hooper G 1992 Low cost cosmetic hand prosthesis *J. Hand Surgery* B**17** 201–3
Carlson J D and Trombly C A 1983 The effect of wrist mobilization on performing of the Jebsen Hand Function Test *Am. J. Occupational Therapy* **37** 167–75
Carroll G J 1987 Comparison of median and radial nerve sensory latencies in the electrophysiological diagnosis of carpal tunnel syndrome *Electroencephalog. Clin. Neurophysiol.* **68** 101–6
Caspar R M 1971 On the use of gamma ray images for determination of human body-segment parameters *Quantitative Imagery in Biomedical Sciences (Houston, TX, 1971)*
Challis J H and Kerwin D G 1992 Calculating upper limb inertial parameters *J. Sports Sci.* **10** 275–83
Chukwuduziem O U, Levangie P K and Nelson D L 1985 Effects of selected assistive devices on normal distance gait characteristics *Phys. Therapy* **65** 1188–91
Clauser C E, McConville J T and Young J W 1969 Weight, volume and center of mass of segments of the human body *Report (AMRL-TR-69-70) Wright-Patterson Air Force Base, OH*
Contini R 1972 Body segment parameters, part II *Artificial Limbs* **16** 1-19

Contini R, Drillis R and Bluestein M 1963 Determination of body-segment parameters *Human Factors* **5** 493–504

Cooper R A 1989 The racing wheelchair *SOMA Eng. Human Body, ASME* **3** 28–34

Cooper R A 1989 Simulating wheelchair racing *Proc. 12th Ann. RESNA Conf. (New Orleans, June 26–30, 1989)* pp 450–1

Cooper R A 1990 A systems approach to the modeling of racing wheelchair propulsion *J. Rehab. Res. Dev.* **27** 151–62

Cooper R A 1990 A force/energy optimization model for wheelchair athletics *IEEE Trans. Syst. Man Cybern.* **20** 444–9

Cooper R A 1990 A systems approach to the modeling of racing wheelchair propulsion *J. Rehab. Res. Dev.* **27** 151–62

Cooper R A 1990 An exploratory study of racing wheelchair propulsion dynamics *Adapted Phys. Activity Q.* **7** 74–85

Cooper R A 1990 Wheelchair racing sports science: a review *J. Rehab. Res. Dev.* **27** 295–312

Cooper R A 1991 System identification of human performance models *IEEE Trans. Syst. Man Cybern.* **21** 244–52

Cooper R A 1991 High tech wheelchairs gain the competetive edge *IEEE Eng. Med. Biol. Mag.* **10** 49–55

Cooper R A 1991 A comparison of pulmonary functions of wheelchair racers in their racing and standard chairs *Proc. 14th Ann. RESNA Conf. (Kansas City, MO, 1991)* 245–7

Cooper R A 1992 Contributions of selected anthropometric and metabolic parameters to 10 K performance—a preliminary study *J. Rehab. Res. Dev.* **29** 29–34

Cooper R A 1994 Blood lactate response of wheelchair racers to maximal exercise *Proc. 17th Ann. RESNA Conf. (Nashville, TN, 1994)* pp.269–71

Cooper R A, Asato K T, Robertson R N and Ster J F 1992 2-dimensional kinetic analysis of manual wheelchair propulsion with an improved SMART^{Wheel} *Proc. 14th IEEE/EMBS Int. Conf. (Paris, 1992)* vol 14(4) pp 1544–5

Cooper R A, Baldini F D, Langbein W E, Robertson R N, Bennett P and Monical S 1993 Prediction of pulmonary function in wheelchair users *Paraplegia* **31** 560–70

Cooper R A and Bedi J F 1990 Gross mechanical efficiency of trained wheelchair racers *Proc. 12th Ann. Int. Conf. IEEE/EMBS (Pennsylvania, PA, 1990)* vol 12 pp 2311–2

Cooper R A and Bedi J F 1992 Function of class and disability groups in the results of Elite wheelchair road racers *Palaestra* **8** 36–41

Cooper R A, Brienza D M and Brubaker C E 1994 Wheelchairs and Seating *Current Opinion Orthop.* **5** 101–7

Cooper R A and Cheda A 1989 Measurement of racing wheelchair propulsion torque *Proc. 11th Annual Int. Conf. IEEE/EMBS (Seattle, Washington, 1989)*

Cooper M, Cooper R A, Faria I and Ster J F 1991 A comparison of torque production of clip-in pedals and conventional toe clips and cleats *Proc. 15th Ann. Am. Soc. Biomechanics Meeting (Tempe, AZ, 1991)* pp 212–3

Cooper R A, Horvath S M, Bedi J F, Drechsler-Parks D M and Williams R E 1992 Maximal exercise responses of paraplegic wheelchair road racers *Paraplegia* **30** 573–81

Cooper R A, Robertson R N and VanSickle D P 1993 Smoothing of wheelchair propulsion kinematic data *Proc. 15th Ann. IEEE/EMBS Int. Conf. (San Diego, CA, 1993)* pp 1296–7

Cooper R A, Robertson R N and VanSickle D P 1993 A recursive back propagation algorithm for computing net muscle moments and net joint forces *Proc. 16th Ann. RESNA Conf. (Las Vegas, NV, 1993)* pp 277–9

Cooper R A, Robertson R N, VanSickle D P, Langbein W E and Fehr L 1993 Wheelchair injury clincal workstation: method for determining a characteristic stroke, *Proc. 15th Ann. IEEE/EMBS Int. Conf. (San Diego, CA, 1993)* pp 1294–5

Cooper R A and Ster J F 1992 An investigation of retinal detachment and neck exercise *Proc. NACOB II: The Second North American Congress on Biomechanics (Chicago, IL, 1992)* pp 435–6

Curtis K A, McClanahan S, Hall K M, Dillon D and Brown K F 1986 Health, vocational and functional status in spinal cord injured athletes and non-athletes *Arch. Phys. Med. Rehab.* **67** 862–5

Dallmeijer A J, Kappe Y J, Veeger H E J, Janssen T W J and van der Woude L H V 1994 Anaerobic power output and propulsion technique in spinal cord injured subjects during wheelchair ergometry *J. Rehab. Res. Dev.* **31** 120–8

Davidoff G, Werner R and Waring W 1991 Compressive mononeuropathies of the upper extremity in chronic paraplegia. *Paraplegia* **29** 17–24

Davis R, Ferrara M and Byrnes D 1988 The competitive wheelchair stroke *NSCA J.* **10** 4–10

Delgrosso I and Biollat M A 1991 Carpal tunnel syndrome: role of occupation *Int. Arch. Occup. Environ. Health* **63** 267–70

Dempster W T 1955 Space requirements of the seated operator *Report (WADC TR 55–159) Wright-Patterson Air Force Base, OH*

Dempster W T 1955 The anthropometry of body action *Ann. NY Acad. Sci.* **63** 559–85

Douglas R, Larson P, D'Ambrosia R and McCall R E 1983 The LSU reciprocating gait orthosis *Orthopaedics* **6** 834–9

Drillis R and Contini R 1966 Body segment parameters *Report No. 1163–03, Office of Vocational Rehabilitation, Department Health, Education and Welfare, New York*

Drillis R, Contini R and Bluestein M 1964 Body segment parameters—A survey of measurement techniques *Artificial Limbs* **8** 44–66

Edelstein J E and Berger N 1993 Performance comparison among children fitted with myoelectric and body powered hands *Arch. Phys. Med. Rehab.* **74** 376–80

Ensminger G J, Cooper R A and Robertson R N 1995 SMART^{Wheeler} A clinical workstation interface for the analysis of wheelchair propulsion *Proc. 16th Ann. IEEE/EMBS Int. Conf. (Baltimore, MD, 1994)* vol 16 pp 494–5

Esquenazi A, Leonard J A, Meier R H, Hicks J E, Fisher S V and Nelson V S 1989 Prosthetics orthotics and assistive devices. 3. Prosthetics *Arch. Phys. Med. Rehab.* **70** S-206–S-209

Ferrara M S, Buckley W E, McCann B C, Limbird T J, Powell J W and Robl R 1992 The injury experience of the competitive athlete with a disability: prevention implications *Med. Sci. Sports. Exerc.* **24** 184–8

Fess E E, Gettle K S and Strickland J W 1981 *Hand Splinting: Principles and Methods* (St. Louis: Mosby)

Fischer O 1906 *Theoretische Grundlagen für eine Mechanik der Lebenden Korper mit Speziellen Anwendungen auf den Menschen sowie auf einige Bewegungsvorgange an Maschinen* (Leipzig: Teubner)

Flood-Joy M and Mathiowetz V 1987 Grip-strength measurement: a comparison of three Jamar dynamometers *Occup. Ther. J. Res.* **7** 235–43

Frieman B G, Albert T J and Fenlin J M Jr 1994 Rotator cuff disease: a review of diagnosis, pathophysiology and current trends in treatment *Arch. Phys. Med. Rehab.* **5** 604–9

Fu F H, Harner C D and Klein A H 1991 Shoulder impingement syndrome. A critical review *Clin. Orthop.* 162–73

Fujikawa K 1963 The center of gravity in the parts of the human body *Okijimas Folia Anat. Japan* **39** 117–25

Gaines R F, Zomlefer M R and Zhao W 1984 Armstroke patterns of spinal cord injured wheelchair users *Arch. Phys. Med. Rehab.* **65** 616

Gehlsen G M, Davis R W and Bahamonde R 1990. Intermittent velocity and wheelchair performance characteristics *Adapted Phys. Act. Q.* **7** 219–30

Gelberman R H, Hergenroeder P T, Hargens A R, Lundborg G N and Akeson W H 1981 The carpal tunnel syndrome. A study of carpal canal pressures *J. Bone Joint Surg.* **63** 380–3

Gellman H, Chandler D R, Petrasek J, Sie I, Adkins R and Waters R L 1988 Carpal tunnel syndrome in paraplegic patients *J. Bone Joint Surg.* **70** 517–9

Gellman H, Sie I and Waters R L 1988 Late complications of the weight-bearing upper extremity in the paraplegic patient *Clin. Orthop.* **233** 132–5

Gibbons D T, O'Rain M D and Philippe-Auguste S 1987 An above-elbow prosthesis employing programmed linkages *IEEE Trans. Biomed. Eng.* **BME-34** 493–8

Glattly H W 1964 A statistical study of 12 000 new amputees. *South Med. J.* **57** 1373–8

Glaser R M, Foley D M, Laubach L L, Sawka M N and Suryaprasad A G 1979 Metabolic and cardiopulmonary responses to wheelchair and bicycle ergometry *J. Appl. Physiol.: Respir. Environ. Exercise Physiol.* **46** 1066–70

Granat M H, Ferguson A C, Andrews B J and Delargy M 1993 The role of functional electrical stimulation in the rehabilitation of patients with incomplete spinal cord injury—observed benefits during gait studies *Paraplegia* **31** 207–15

Hagberg M, Morgenstern H and Kelsh M 1992 Impact of occupations and job tasks on the prevalence of carpal tunnel syndrome *Scand. J. Work Environ. Health* **18** 337–45

Hanavan E P 1964 A mathematical model of the human body *Report (AMRL-TR-64–102) Wright-Patterson Air Force Base, OH*

Hanavan E P 1966 A personalized mathematical model of the human body *J. Spacecraft* **3** 446–8

Hardy D C, Vogler J B and White R H 1986 The shoulder impingement syndrome: prevalence of radiographic findings *Am. J. Roentgenol.* **147** 557–61

Harless E 1962 The static moments of the component masses of the human body *Report (FTD-TT-61–295) Wright-Patterson Air Force Base, OH*

Hatze H 1975 A new method for the simultaneous measurement of the moment of inertia, the damping coefficient and the location of the center of mass of the body in situ *Eur. J. Appl. Physiol.* **34** 217–226

Hay J G 1973 The center of gravity of the human body *Kinesiology IV* (Washington, D C: AAHPER)

Hay, J G 1974 Moment of inertia of the human body *Kinesiology IV* (Washington, D C: AAHPER)

Hertzberg H T E, Daniels G S and Churchill E 1954 Anthropometry of flying personnel—1950 *WADC Technical Report 52–321 Wright-Patterson Air Force Base, OH*

Hjeltnes N and Lannem A 1990 Functional neuromuscular stimulation in four patients with complete paraplegia *Paraplegia* **28** 235–43

Huang C T, Jackson J R, Moore N B, Fine P R, Kuhlemeier K V, Traugh G H and Saunders P T 1979 Amputation: energy cost of ambulation *Arch. Phys. Med. Rehab.* **60** 18–24

Hughes C J, Weimar W H, Sheth P N and Brubaker C E 1992 Biomechanics of wheelchair propulsion as a function of seat position and user-to-chair interface *Arch. Phys. Med. Rehab.* **73** 263–9

Iannotti J P, Zlatkin M B, Esterhai J L, Kressel H Y, Dalinka M K and Spindler K P 1991 Magnetic resonance imaging of the shoulder: sensitivity, specificity and predictive value *J. Bone Joint Surg.* **73** 17–29

Ishida A and Mayzaki S 1987 Maximum likelihood identification of a posture control system *IEEE Trans. Biomed. Eng.* **BME-34** 1–5

Jablecki C K, Andary M T, So Y T, Wilkins D E and Williams F H 1993 Literature review of the usefulness of nerve conduction studies and electromyography for the evaluation of patients with carpal tunnel syndrome. AAEM Quality Assurance Committee *Muscle Nerve* **16** 1392–414

Jackson D A and Clifford J C 1989 Electrodiagnosis of mild carpal tunnel syndrome *Arch. Phys. Med. Rehab.* **70** 199–204

Jansen E, Vittas D, Hellberg S and Hansen J 1982 Normal gait of young and old men and women: ground reaction force measurement on a treadmill *Acta Orthopaedia Scandanavia* **53** 193–6

Jacobsen S C, Knutti D F, Johnson R T and Sears H H 1982 Development of the Utah artificial arm *IEEE Trans Biomed Eng.* **BME-29** 249–56

Jarvis S and Rolfe H 1982 The use of an inertial dynamometer to explore the design of children's wheelchairs *Scand. J. Rehab. Med.* **14** 167–76

Jebsen R H, Taylor N, Trieschmann R B, Trotter M J and Howard L A 1969 An objective and standardized test of hand function *Arch. Phys. Med.* **50** 311–9

Jendrzejczyk D J 1985 Flexible socket systems *Clin. Prosthet. Orthot.* **9** 27–31

Jones D, Baldini F D, Cooper R A and Monical S 1992 Ten kilometer performance and predicted velocity at VO_2 peak among well-trained wheelchair racers *Sports Med. Training Rehab.* **3** 233

Kaplan P A, Bryans K C, Davick J P, Otte M, Stinson W W and Dussault R G 1992 MR imaging of the normal shoulder: variants and pitfalls *Radiology* **184** 519–24

Keegan J J 1953 Alterations of the lumbar spine related to posture and seating *J. Bone Joint Surg.* **35-A** 589

Keller A D, Taylor C L and Zahm V 1947 Studies to determine the functional requirements for hand and arm prosthesis. Final report National Academy of Sciences Contract No VAM-21223

Kendall D 1960 Etiology, diagnosis and paraesthesiae in the hands *Br. Med. J.* **2** 1633–40

Kimura J 1979 The carpal tunnel syndrome: localization of conduction abnormalities within the distal segment of the median nerve *Brain* **102** 619–35

Kraft G H and Detels P E 1972 Position of function of the wrist *Arch. Phys. Med. Rehab.* **53** 272–75

Law H T 1981 Engineering of upper limb prostheses *Orthop. Clin. North Am.* **12** 929–51

Lehman J F and Warren C G 1976 Restraining forces in various designs of knee ankle orthoses: their placement and effect on the anatomical knee *Arch. Phys. Med. Rehab.* **57** 430–7

Lehman J F, Warren C G, Hertling D, McGee M, Simons B C and Dralle A 1976 Craig-Scott orthosis: a biomechanical and functional evaluation *Arch. Phys. Med. Rehab.* **57** 438–42

Leonard J A, Esquenazi A, Fisher S V, Hicks J E, Meier R H and Nelson V S 1989 Prosthetics, orthotics and assistive devices. 1. General concepts *Arch. Phys. Med. Rehab.* **70** S-195–S-201

Liou J T, Wilson A J, Totty W G and Brown J J 1993 The normal shoulder: common variations that simulate pathologic conditions at MR imaging *Radiology* **186** 435–41

Liu Y K, Laborde J M and Van Buskirk W C 1971 Inertial properties of a segmented cadaver trunk: their implications in acceleration injuries *Aerospace Med.* **42** 650–7

Loslever P and Ranaivosoa A 1993 Biomechanical and epidemiological investigation of carpal tunnel syndrome at workplaces with high risk factors *Ergonomics* **36** 537–55

Lundborg G, Gelberman R H, Minteer-Convery M, Lee Y F and Hargens A R 1982 Median nerve compression in the carpal tunnel–functional response to experimentally induced controlled pressure *J. Hand Surg.* **7** 252–9

Masse L C, Lamontagne M and O'Riain M D 1992 Biomechanical analysis of wheelchair propulsion for various seating positions *J. Rehab. Res.* **29** 12–28

McLaurin, C A and Brubaker, C E 1991 Biomechanics and the wheelchair *Prosthet. Orthot. Int.* **15** 24–37

Meeh C 1985 Volummessungen des menschlichen Korpers und seiner einzelnen Theile in den verschiedenen Altersstufen *Z. Biol.* **31** 125–47

Miller D I and Morrison W E 1975 Prediction of segmental parameters using the Hanavan human body model *Med. Sci. Sports* **7** 207–12

Miller D I and Nelson R C 1973 *Biomechanics of Sport—A Research Approach* (Philadelphia, PA: Lea & Febiger)

Mizrahi J, Braun Z, Najenson T and Graupe D 1985 Quantitative weightbearing and gait evaluation of paraplegics using functional electrical stimulation *Med. Biol. Eng. Comput.* **23** 101–07

Moe and Post 1962 *The Lancet* **212** 285

Mori M and Yamamoto T 1959 Die Massenanteile der einzelnen Korperabschnitte der Japaner *Acta Anat.* **37** 385–8

Murray M, Mollinger L, Sepic S, Gardner G and Linder M 1983 Gait patterns in above-knee amputee patients: hydraulic swing control versus constant-friction knee components *Arch. Phys. Med. Rehab.* **64** 339–45

Naeder M 1990 The artificial substitution of missing hands with myoelectric prostheses *Clin. Orthop. Related Res.* **258** 9–17

Neer C S 1972 Anterior acromioplasty for the chronic impingement syndrome in the shoulder: a preliminary report *J. Bone Joint Surg.* **54** 41–50

Nene A V and Jennings S J 1992 Physiological cost index of paraplegic locomotion using the ORLAU ParaWlaker *Paraplegia* **30** 246–52

Nene A V and Major R E 1987 Dynamics of reciprocal gait of adult paraplegics using the ParaWlaker (hip guidance orthosis) *Prosthet. Orthot. Int.* **11** 124–7

Nichols P J, Norman P A and Ennis J R 1979 Wheelchair user's shoulder? *Scand. J. Rehab. Med.* **11** 29–32

Niesing R, Eijskoot F, Kranse R, Den Ouden A H, Storm J, Veeger H E J, Van Der Woude L H V and Snijders C J 1990 Computer-controlled wheelchair ergometer *Med. Biol. Eng. Comput.* **28** 329–38

Nilsen R, Nygaard P and Bjorholt P G 1985 Complications that may occur in those with spinal cord injuries who participate in sport *Paraplegia* **23** 152–8

Nuber G W 1988 Biomechanics of the foot and ankle during gait *Clin. Sports Med.* **7** 1–13

Oderkerk B J and Inbar G F 1991 Walking cycle recording and analysis for FNS-assisted paraplegic walking *Med. Biol. Eng. Comput.* **29** 79–83

Ostrosky K, VanSwearigen J, Buffett R and Gee Z 1994 A comparison of gait characteristics in young and old subject *Phys. Ther.* **74** 637–44

Ozaki J, Fujimoto S, Nakagawa Y, Masuhara K and Tamai S 1988 Tears of the rotator cuff of the shoulder associated with pathological changes in the acromion. A study in cadavera *J. Bone Joint Surg.* **70** 1224–30

Pentland W E and Twomey L T 1991 The weight-bearing upper extremity in women with long term paraplegia *Paraplegia* **29** 521–30

Pettit D J, Robertson R N and Cooper R A 1992 Development of a dynamometer for measuring dynamic hand function *Proc. NACOB II: The Second North American Congress on Biomechanics (Chicago, IL, 1992)* pp 449–50

Phalen G S 1970 Reflections on 21 years' experience with the carpal-tunnel syndrome *JAMA* **212** 1365–7

Phillips L and Nicosia A 1990 An overview...with reflections past and present of a consumer *J. Rehab. Res. Dev.-Clinical Suppl. No. 2: Choosing a Wheelchair System* 1–7

Pierson W R 1957 Non-topographic photogrammetry as a research technique *FIEP Bull.* **27** 48–449
Pierson W R 1959 The validity of sterophotogrammetry in volume determination *Photogram. Eng.* **BME-25** 83–5
Pierson W R 1961 Monophotogrammetric determination of body volume *Ergonomics* **4** 213–7
Pierson W R 1961 Photogrammetric determination of surface area *Photogram. Eng.* **BME-27** 99–102
Pierson W R 1963 Photogrammetric technique for the estimation of surface area and volume *Ann. NY Acad. Sci.* **110** 109–12
Plagenhoef S 1971 *Patterns of Human Motion—A Cinematographic Analysis* (Englewood Cliffs, NJ: Prentice-Hall)
Pope C, Wilkerson J and Ridgway M 1992 Wheelchair racing: A second look *Palaestra* summer 21–35
Prochazka A 1993 Comparison of natural and artificial control of movement *IEEE Trans. Rehab. Eng.* **BME-1** 7–17
Rathbun J B and Macnab I 1970 The microvascular pattern of the rotator cuff *J. Bone Joint Surg. Br.* **170** 540–53
Reswick J B 1986 Evaluation of the Seattle Foot *J. Rehab. Res. Dev.* **23** 77–94
Ridgway M, Pope C and Wilkerson J 1988 A kinematic analysis of 800 meter wheelchair racing techniques *Adapted Phys. Act. Q.* **5** 96–107
Robertson R N and Cooper R A 1993 Kinetic characteristics of wheelchair propulsion utilizing the Smart^{Wheel} *Proc. 17th Ann. Meet. Am. Soc. Biomechanics (Iowa City, IA, 1993)* pp 202–3
Robertson R N, Cooper R A, Ensminger G J and Stewart K J 1994 Upper extremity net joint forces and moments during wheelchair propulsion *Proc. 17th Ann. RESNA Conf. (Nashville, TN, 1994)* pp 155–7
Rodgers M M, Gayle G W, Figoni S F, Kobayashi M, Glaser R M and Gupta S C 1992 Kinematic and kinetic responses to wheelchair propulsion during fatigue in SCI individuals: A pilot study *Proc. of the 15th Ann. RESNA Conf. (Toronto, 1992)* pp 202–3
Rodgers M M, Gayle G W, Figoni S F, Kobayashi M, Lieh J and Glaser R M 1994 Biomechanics of wheelchair propulsion during fatigue *Arch. Phys. Med. Rehab.* **75** 85–93
Rodgers M M, Tummarakota S, Lieh J and Schrag D R 1992 Three-dimensional dynamic analysis of joint reaction forces and moments during wheelchair propulsion *Proc. of NACOB II (Chicago, IL, 1992)* pp 457–8
Roeleveld K, Lute E M C, Veeger H E J, Gwinn T and van der Woude L H V 1994 Power output and technique of wheelchair athletes *Adapted Phys. Act. Q.* **11** 71–85
Rose G K 1979 The principles and practice of hip guidance articulations *Prosthet. Orthot. Int.* **3** 37–43
Rose G K and Henshaw J T 1972 A swivel walker for paraplegics: medical and technical considerations *Biomed. Eng.* **BME-7** 420–5
Rosenrot P, Wall J C and Charteris J 1980 The relationship between velocity, stride time, support time, and swing time during normal walking *J. Human Movement Sci.* **6** 323–5
Rossi J 1988 Concepts and current trends in hand splinting *Occup. Ther. Health Care* **4** 53–68
Rovick J S and Childress D S 1988 Pendular model of paraplegic swing-through crutch ambulation *J. Rehab. Res. Dev.* **25** 1–16
Ruggles D L, Cahalan T and AN K 1994 Biomechanics of wheelchair propulsion by able-bodied subjects *Arch. Phys. Med. Rehab.* **75** 540–4

Samuelsson K, Larsson H and Tropp H 1989 Wheelchair ergometer with a device for isokinetic torque measurement *Scand. J. Rehab. Med.* **21** 205–7

Sanderson D J and Sommer III H J 1985 Kinematic features of wheelchair propulsion *J. Biomech.* **18** 423–9

Sankarankutty M, Stallard J and Rose G K 1979 The relative efficiency of 'swing through' gait on axillary, elbow and Canadian crutches compared to normal walking *J. Biomed. Eng.* **BME-1** 55–7

Santschi W R, DuBois J and Omoto C 1963 Moments of inertia and centers of gravity of the living human body *Report (AMRL-TDR-63-36) Wright-Patterson Air Force Base, OH*

Sears H and Shaperman J 1991 Proportional myoelectric hand control: an evaluation *Am. J. Phys. Med. Rehab.* **70** 20–8

Seeger L L, Gold R H, Bassett L W, Ellman H 1988 Shoulder impingement syndrome: MR findings in 53 shoulders *Am. J. Roentgenol.* **150** 343–7

Sie I, Waters R L, Adkins R and Gellman H 1992 Upper extremity pain in the post rehabilitation spinal cord patient *Arch. Phys. Med. Rehab.* **73** 44–8

Silcox D H, Rooks M D, Vogel R R and Fleming L L 1993 Myoelectric prostheses *J. Bone Joint Surg.* **75-A** 1781–9

Silverstein B A, Fine L J and Armstrong T J 1987 Occupational factors and carpal tunnel syndrome *Am. J. Ind. Med.* **11** 343–58

Skinner H B and Effeney D J 1985 Gait analysis in amputees *Am. J. Phys. Med.* **64** 82–9

Snell F 1994 Concepts of externally powered upper limb prostheses *J. Arkansas Med. Soc.* **90** 436–7

Solomonow M, Baratta R, Shoji H and D'Ambrosia R 1986 The myoelectric signal of electrically stimulated muscle during recruitment: an inherent feedback parameter for a closed loop control scheme *IEEE Trans. Biomed. Eng.* **BME-33** 735–44

Spivak C D 1915 Methods of weighing parts of the human body *JAMA* **65** 1707–8

Stallard J, Dounis E, Major R E and Rose G K 1980 One leg swing through gait using two crutches, an analysis of ground reaction forces and gait phases *Acta Orthop. Scand.* **51** 71–7

Stefaniwsky L, Bilowit D S and Prasad S S 1980 Reduced motor conduction velocity of the ulnar nerve in spinal cord injured patients *Paraplegia* **18** 21–4

Stein R B, Belanger M, Wheeler G, Wieler M, Popovic D B, Prochazka A and Davis L A 1993 Electrical systems for improving locomotion after incomplete spinal cord injury: an assessment *Arch. Phys. Med. Rehab.* **74** 954–9

Stern E B 1991 Wrist extensor orthoses: dexterity and grip strength across four styles *Am. J. Occup. Ther.* **45** 42–9

Stewart K J, Cooper R A and Robertson R N 1994 The analytical inadequecies of treadmill mounted force platforms *Proc. 17th Ann. RESNA Conf. (Nashville, TN, 1994)* pp 167–9

Stewart K J, Cooper R A and Robertson R N 1995 A method for comparison of contiguous ground reaction force data and single stance ground reaction force data *Proc. Ann. RESNA Conf.* pp 186–8

Strauss M G, Maloney J, Ngo F and Phillips M 1991 Measurement of dynamic forces during manual wheelchair propulsion *Proc. Ann. Meet. Am. Soc. Biomechanics (Tempe, AZ, 1991)* 210–1

Strauss M G, Moeinzadeh M H, Schneller J and Trimble J 1989 The development of an instrumented wheel to determine the handrim forces during wheelchair propulsion *Proc. Ann. Winter Meet. ASME (San Francisco, CA, 1989)* pp 53–4

Swearingen, J J 1962 *Determinations of centers of gravity of man (ASTIA AD 10 410)* Federal Aviation Administration, OK

Troy B S, Cooper R A, Robertson R N and Grey T L 1994 An analysis of working postures of manual wheelchair users in the working environment *Proc. 17th Ann. RESNA Conf. (Nashville, TN, 1994)* pp 483–5

Tun C G and Upton J 1988 The paraplegic hand: electrodiagnostic studies and clinical findings *J. Hand Surg.* **13** 716–9

Tupling S J, Davis G M, Pierrynowski M R and Shephard R J 1986 Arm strength and impulse generation: Initiation of wheelchair movement by the physically disabled *Ergonomics* **29** 303–11

Van Der Woude L H V, Hendrich K M M, Veeger H E J, Van Ingen Schenau G J, Rozendal R H, De Groot G and Hollander A P 1988 Manual wheelchair propulsion: effects of power output on physiology and technique *Med. Sci. Sport Exercise* **20** 70–8

Van Der Woude L H V, Veeger H E J and Rozendal R H 1989 Ergonomics of wheelchair design: A prerequisite of optimum wheeling conditions *Adapted Phys. Act. Q.* **6** 109–32

Van Der Woude L H V, Veeger H E J and Rozendal R H 1989 Propulsion technique in hand rim wheelchair ambulation *J. Med. Eng. Technol.* **13** 136–141

Van Der Woude L H V, Veeger H E J, Rozendal R H and Sargeant A J 1989 Optimum cycle frequencies in hand-rim wheelchair propulsion *Eur. J. Appl. Physiol.* **58** 625–32

Van Der Woude L H V, Veeger H E J, Rozendal R H and Sargeant T J 1989 Seat height in handrim wheelchair propulsion *J. Rehab. Res.* **26** 31–50

Veeger H E J, Lute E M C, Roeleveld K and Van Der Woude L H V 1992 Differences in performance between trained and untrained subjects during a 30-s sprint test in a wheelchair ergometer *Eur. J. Appl. Physiol.* **64** 158–164

Veeger H E J, Van Der Woude L H V and Rozendal R H 1989 The effect of rear wheel camber in manual wheelchair propulsion *J. Rehab. Res.* **26** 37–46

Veeger H E J, Van Der Woude L H V and Rozendal R H 1989 Wheelchair propulsion technique at different speeds *Scand. J. Rehab. Med.* **21** 197–203

Veeger H E J, Van Der Woude L H V and Rozendal R H 1991 Within-cycle characteristics of the wheelchair push in sprinting on a wheelchair ergometer *Medicine and Science in Sports and Exercise* **23** 264–271

Veeger H E J, Van Der Woude L H V and Rozendal R H 1992 Effect of handrim velocity on mechanical efficiency in wheelchair propulsion *Med. Sci. Sports Exercise* **24** 100–7

Veeger H E J, Van Der Woude L H V and Rozendal R H 1992 A computerized wheelchair ergometer: Result of a comparison study *Scand. J. Rehab. Med.* **24** 17–23

Vosse A J, Cooper R A and Dhaliwal B 1990 Computer control of a wheelchair dynamometer *Proc. RESNA 13th Ann. Conf. (Washington, DC, 1990)* pp 59–60

Walton J S 1970 A template for locating segmental centers of gravity *Res. Q. Amer. Assoc. Health Phys. Ed.* **41** 615–8

Wang Y T, Deutsch H, Hedrick B and Morse M 1991 Relationship between kinematic factors and muscle activity during wheelchair propulsion *Proc. Ann. Int. Conf. IEEE Eng. Med. Biol. Soc.* **13** pp 1841–42

Watanabe K T, Cooper R A and Ster J F 1991 A device for studying wheelchair propulsion dynamics *Proc. IEEE-EMBS 13th Int. Conf. (Orlando, FL, 1991)* 1817–8

Watanabe K T, Cooper R A, Vosse A J, Baldini F D and Robertson R N 1992 Training practices of athletes who participated in the national wheelchair athletic association training camps *Adapted Phys. Act. Q.* **9** 249–60

Waterland J C and Shambes G M 1970 Biplane center of gravity procedures *Percept. Motor Skills* **30** 511–4

Waters R L, Perry J, Antonelli D and Hislop H 1976 Energy cost of walking of amputees: the influence of level of amputation *J. Bone Joint Surg.* **58A** 42–6

Waters R L, Yakura J S, Adkins R and Barnes G 1989 Determinants of gait performance following spinal cord injury *Arch. Phys. Med. Rehab.* **70** 811–8

Weidenhielm L, Olsson E, Broström L-Å, Börjesson-Hederström M and Mattsson E 1993 Improvement in gait one year after surgery for knee osteoarthrosis: a comparison between high tibial osteotomy and prosthetic replacement in a prospective random study *Scandinavian J. Rehab. Med.* **25** 25–31

Weinback A P 1938 Contour maps, center of gravity, moment of inertia and surface area of the human body *Human Biol.* **10** 356–71

Werner C O, Elmqvist D and Ohlin P 1983 Pressure and nerve lesion in the carpal tunnel *Acta Orthop. Scand.* **54** 312–6

Whitsett C E 1963 Some dynamic response characteristics of weightless man *Report (AMRL-TDR-63-18) Wright-Patterson Air Force Base, OH*

Williams M and Lissner H R 1962 *Biomechanics of Human Motion* (Philadelphia, PA: Saunders)

Winter D A 1990 *Biomechanics and Motor Control of Human Movement* (New York, NY: Wiley)

Wu Y and Krick H 1987 Removable rigid dressing for below-knee amputees *Clin. Prosthet. Orthot.* **11** 33–44

Wylie E J and Chakera T M 1988 Degenerative joint abnormalities in patients with paraplegia of duration greater than 20 years *Paraplegia* **26** 101–106

Zernicke R F, Hoy M G and Whiting W C 1985 Ground reaction forces and center of pressure patterns in the gait of children with amputation: preliminary report *Arch. Phys. Med. Rehab.* **66** 736–41

Zlatkin M B, Reicher M A, Kellerhouse L E, McDade W, Vetter L and Resnick D 1988 The painful shoulder: MR imaging of the glenohumeral joint *J. Comput. Assisted Tomography* **12** 995–1001

Zook D E 1932 The physical growth of boys: A study by means of water displacement *Am. J. Disabled Children* **43** 1347–432

Zuniga E, Leavitt L, Calvert J, Canzoneri J and Peterson C 1972 Gait patterns in above-knee amputees *Arch. Phys. Med. Rehab.* **53** 373–82

CHAPTER 4

UNIVERSAL DESIGN AND ACCESSIBILITY

4.1. BARRIER-FREE DESIGN

The objective of barrier-free design is to integrate the needs of all potential users into the design of a structure or system. Barrier-free design does not mean that structures and facilities are built to accommodate people with disabilities, because this implies that other users are of secondary importance. In barrier-free design all users' needs are considered. With proper design, public facilities will accommodate a large portion of the population with minimal inconvenience and maximal safety.

Everyone benefits from barrier-free design. Elderly and people with disabilities are not the only people who are affected by architectural barriers. Approximately 56% of the people in a typical community could directly benefit from barrier-free design. Bicyclists, shopping cart users, parents, people recovering from accident/illness can all benefit from design which integrates people's needs. Figure 4.1 illustrates the distribution of a typical population which would benefit from barrier-free design.

The majority of people in a community benefit from barrier-free design, and at little or no inconvenience to other people in the community. The cost of barrier-free design is also no greater than standard design. Barrier-free design only requires a rethinking of how people use structures and facilities. If inclusion and integration become primary foci of public facility design, then a greater number of people will be accommodated at little or no additional cost.

Eliminating architectural barriers is good business. Business opportunity increases with the number of potential customers. Designs that account for human needs and allow for pedestrian traffic flow patterns of all people can also be of benefit to business as there would be fewer pedestrian accidents. Barrier-free design can lead to lower insurance premiums. By incorporating the needs of potential users into the design of the structure and facilities

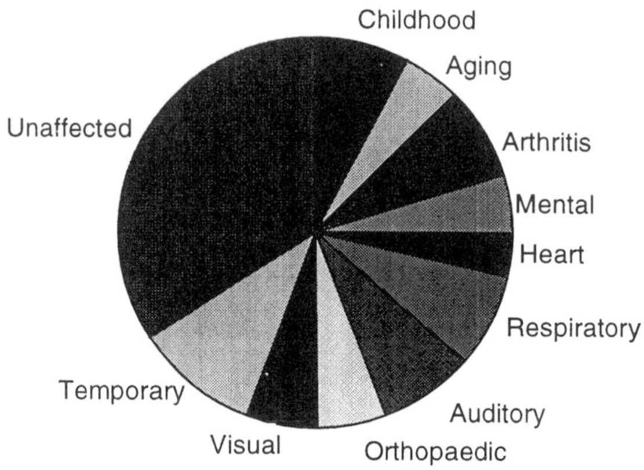

Figure 4.1. *Distribution of people by impairment in a typical population.*

the owner's liability will be reduced, the risk of injury due to fire will be reduced, and injuries to workers can be reduced.

Several state and federal regulations mandate reasonable accommodation of people with disabilities. It is, generally, more costly to make modifications to an existing structure or facility, than it is to design for inclusion. However, there are tax incentives and some federal funds available for making structures, and facilities accessible. Some accommodation requires only simple changes in fixtures (e.g., door handles, bathroom fixtures, telephones). Other accommodation requires minor architectural modifications (e.g., ramps, widening of entrances, enlarging bathrooms).

Designing or modifying structures and facilities to accommodate most people can create favorable public relations. Facilities open and usable by everyone increase the number of people who will use the facilities, and better serve the public interest.

4.2. ELEMENTAL RESOURCE MODEL

Whether designing a building for universal access or specifying rehabilitation technology to provide mobility within the building and to manipulate the environment, an interface between the user and environment is required. Each person has a finite set of human resources or abilities. The resources must be matched with tasks. When task demands exceed resources, then devices or systems are required to increase resources or reduce demands. Generalized systems performance theory (GSPT) provides a framework for assessing and graphically illustrating resources and demands. Kondraske have developed the elemental resource model (ERM) using GSPT. Their model can be used to conceptualize the rehabilitation process. In rehabilitation, the systems of interest are humans, assistive devices, the environment, and their interaction.

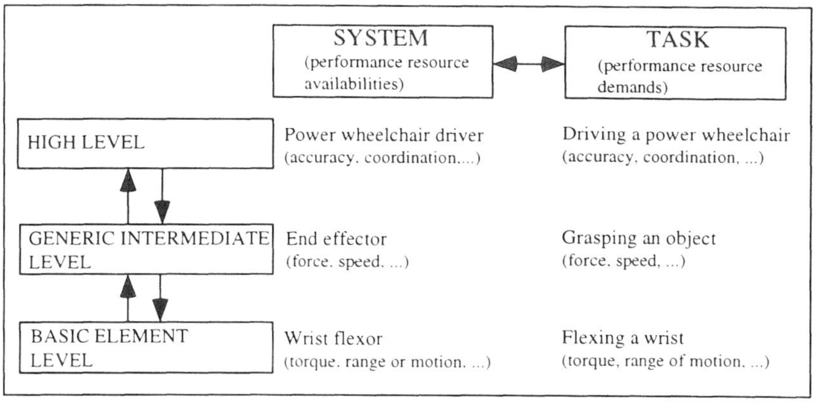

Figure 4.2. *The environmental resource model consists of multiple hierarchical categories. Performance resources at the 'basic elemental level' are dictated by finite sets of human subsystems (i.e., musculoskeletal, cardiovascular) and their respective performance dimensions. Higher levels are created by configuring and combining basic elements.*

Most knowledge that exists related to performance and its quantification has evolved from a desire to understand a specific situation. Performance is influenced by numerous factors which may involve complex relationships. Performance is the process of attaining the best of something, and not just the process of describing behavior. The objectives of GSPT are to: provide a common conceptual definition and measurement of critical aspects of an arbitrary system's performance; provide a common conceptual basis for facilitating system–task interface assessments and decision making; identify cause-and-effect principles to explain system operation. The GSPT includes a consistent use of a resource to model all aspects of a system's performance. The GSPT applies the non-linear, threshold, effect associated with resource economic mathematics. A task which has a resource demand cannot be performed by a system which does not have the necessary resource allocations. Once minimum resources to perform the task are available, either naturally or artificially, the quality of task performance can be assessed.

The elemental resource model is derived by applying the concept of monadology and GSPT to a human system (figure 4.2). Monadology is the concept of using the combination of a finite set of 'basic elements' to describe complex systems. The alphabet is an example of a monadology used to describe language. Performance resources at the basic element level are finite in number and reflect a finite set of human subsystems and their performance dimensions. Collective systems form higher levels, and higher levels can be combined in a similar fashion.

The person is modeled as a pool of elemental performance resources ($R_{A_{ij}}(t)$). These are shown grouped in four different domains: life sustaining; environmental interface; central processing; and information (figure 4.3).

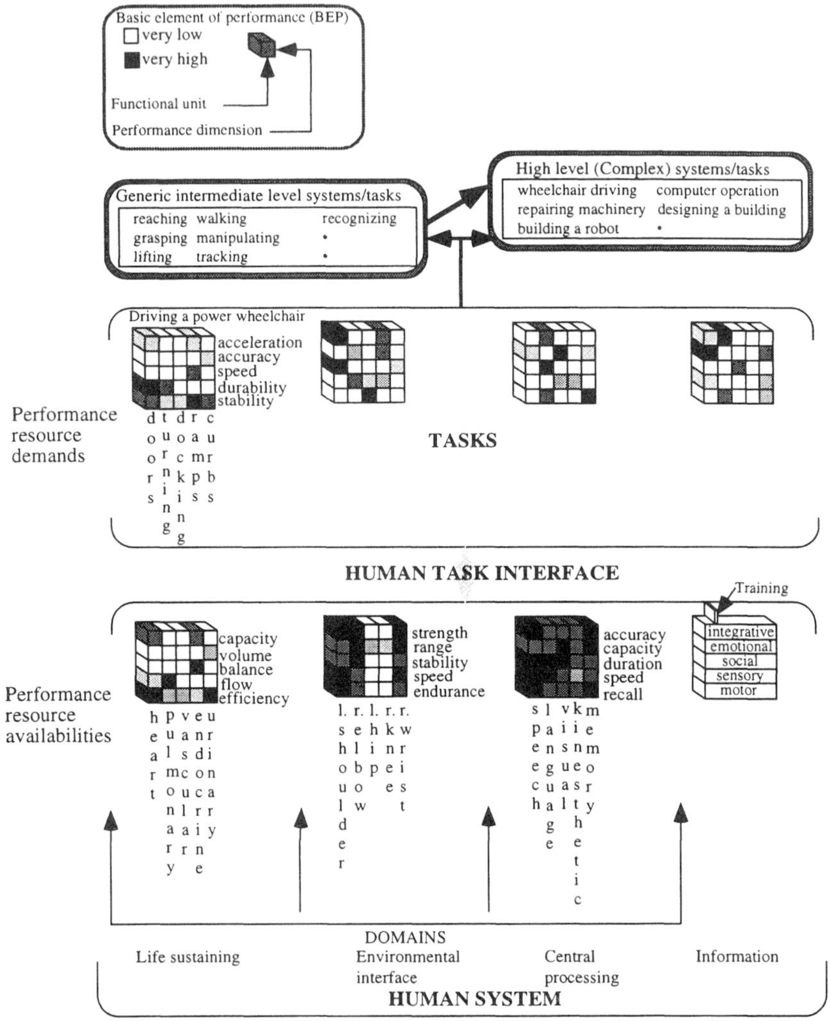

Figure 4.3. *Summary of the major components of the elemental resource human performance model. Tasks are built from basic elements and allocated resources. This model emphasizes human-task interfacing based upon resource economics.*

Other domains can be defined, and may simplify the resource demand-availability mapping for other tasks. Choosing the appropriate model can simplify the human–task interface mapping. Within domains, physical subsystems referred to as functional units can be identified. GSPT is applied to each subsystem to yield dimensions of performance. Performance dimensions must be quanitifiable and are generally presented in a positive manner. For example, the inability to perform a task must be traced back to the lack of some resource, or the ability to perform a task well must be related to an abundance of one or more performance dimensions. Often human performance assessment focuses on inhibiting factors. The ERM focuses on resources. A single basic element of performance (BEP) is defined by specifying a basic functional element and a dimension of performance. Physical domains are represented by tangible and, in some cases, intangible resources. Domains like the information domain may use a different structure. Cognitive functional units are placed under central processing. However, the information available for cognitive processing is placed as a separate domain. This permits a clearer representation of complex functions. This structure also permits abstract concepts such as happiness and motivation to be accommodated.

Any given task can be defined in terms of a unique set of resource demands ($R_{D_{ij}}(t)$) which are imposed on the BEPs and information resources. Shadings in figure 4.3 imply the amount of demand for each component of the task. Likelihood of achieving performance goals for a given task component is governed by economic principles which require that $R_{A_{ij}}(t)|Q \geqslant R_{D_{ij}}(t)$ for all i and j (e.g., $R_{A_{11}} \geqslant R_{D_{11}} \cap R_{A_{12}} \geqslant R_{D_{12}} \cap R_{A_{13}} \geqslant R_{D_{13}} \ldots$). Q represents a unique condition, in terms of a set of values R_i along other identical dimensions under which a specific R_{A_i} is measured (e.g., $Q_{1,k} = \{R_{1,k}, \cdots, R_{p,k}\}$ for all $p \neq i$). The subscript 'k' distinguishes one of several conditions under which a given availability is measured. All task demands when translated to the individual subsystems must be within the performance resource availability envelope. Adequacy of a given resource associated with a task is a necessary, but not a sufficient condition for success.

The ERM is rather new and has not been widely applied, but shows promise for conceptualizing the requirements for the design and application of universal access and rehabilitation technology. In order to apply the ERM, substantial knowledge of the person, technology, and environment are required in order to describe each basic element. However, with the appropriate framework, knowledge can be transferred from one expert to another or from one problem to another. This could lead to a better approach to matching the resources available to people with disabilities to rehabilitation technology and to the development of universal access.

4.3. FACTORS AFFECTING BARRIER-FREE DESIGN

4.3.1. Human needs

Several factors affect barrier-free design: the needs and abilities of the users, the needs of the community, and the resources available. All of these factors must be weighed, and the optimal structure or facility given these

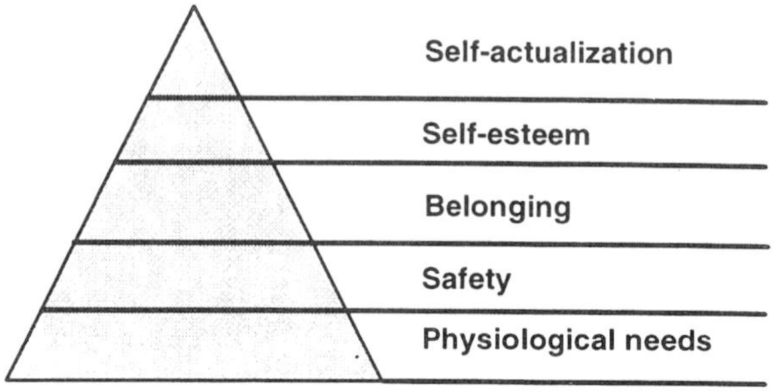

Figure 4.4. *Hierarchy of human needs.*

constraints can be designed. Determining the needs of the users, and how to address those needs is often one of the most difficult challenges facing many architects and engineers.

Architects and engineers need to know several things about the building/facility users. Building/facility users are all people affected by the building or facility. This includes people who will work in, use, pass by, live in the area of, or pay for the building or facility. Direct and indirect users must be considered. Many people who never enter the building or facility are affected by its presence. The needs of people displaced by the building or facility must also be considered.

A hierarchy of human needs (Maslow) must be considered when designing a structure or facility. The five basic human needs (figure 4.4) form a basis for integrating people into designs. The facility or structure serves to meet all basic physiological needs (e.g., protection from the elements, toilet facilities). People must also feel and be safe in the facility or structure.

Safety incorporates structural safety, emergency preparedness, and physical security. The risk of injury, and crime must be minimized by design. People have varying levels of ability to function in any environment, and special consideration must be given when people are in an unfamiliar environment or during an emergency. Design should consider the needs of all people with respect to information processing, traffic flow, and personal safety. For example, some people have limited use of their legs and arms, and have difficulty communicating verbally. Such people must have access to information, especially safety information.

People require a sense of belonging. Barriers to integration of all people into the mainstream of society must be removed. When people are confronted with architectural barriers, a basic human need is threatened. People can contribute more to society, and can achieve higher personal satisfaction if they have a sense of belonging. Buildings and facilities need to be designed so that people can participate in all activities (e.g., social activities, personal communication, professional activities). For example, a corporation may design a conference room with modern telecommunications equipment, and not consider the needs of hearing impaired or visually impaired employees. These employees would be excluded from the group despite having physical access to the conference room.

Promoting positive self-esteem benefits building users and society. A facility which permits all people to perform their business in an efficient manner will be successful. All people should be able to perform their desired tasks at their chosen speed with minimal disruption of other people. The ability to contribute to the group, and to perform to one's ability promotes self-esteem.

Self-actualization is essentially being able to achieve one's own goals based upon an internal sense of pride and fulfillment. The design of facilities can promote self-actualization. For people to achieve their personal goals, the facilities must be in place to permit them access to the means required to achieve their goals. Architectural and attitudinal barriers can prevent people from achieving their fullest potential. For example, if buildings are designed so that wheelchair users must constantly go through back corridors and use service elevators to accomplish their work it will be difficult for them to achieve their fullest potential. If a wheelchair user must ask for assistance to ascend stairs to perform a job or personal function, s/he will likely have difficulty achieving her/his fullest potential.

All people are not alike. There are differences among occupants. There is no such thing as the 'average' building or facility user. Architects and engineers must consider a number of factors when designing facilities and structures. People of various ages and experiences have different abilities. Children and senior citizens do not have the same needs or abilities. The age of potential users must be considered in the design selection of fixtures, information systems, and traffic flow patterns. There are gender differences to be considered when designing facilities and structures. The most obvious may be that often separate toilet facilities are expected for men and women. However, men and women sometimes have different perspectives on how facilities and structures should be designed. Considerations should be made for child-care facilities and making them accessible to all people when appropriate.

It is of the utmost importance to maintain the health and well being of all building/facility users. Buildings/facilities must be designed to promote the physical and mental well being of all users. Pre-existing health conditions of some users must also be considered. Simple accommodation may include parking spaces near entrances and exists, and elevators.

Education of the users must also be considered when designing a building or facility. Technology plays an important role in many modern facilities. People with different levels of education may have varying levels of familiarity with technology. Engineers and architects must consider the knowledge required by the user to properly access different aspects of the facility. For example, a highly sophisticated information retrieval system may be of little use if many people do not know how to use it.

Economic status must also be considered in designing facilities. People of different economic statures sometimes have other priorities. People with lesser financial means may not be able to afford parking fees to use public facilities, or they may require public transportation systems to access certain facilities.

People vary in social status and that affects their expectations, motivations, and attitudes. A wealthy person may evaluate a facility based upon its creature comforts, whereas a person of lesser means may be more interested in proximity to public transportation.

The nature of employment or business will affect the design of the facility or structure. A medical office may require a clean, clinical appearance, whereas an automotive garage may require space for automobiles and tools. Some equipment requires special construction of the building. If the structures are designed with all people in mind, then more people will have access to the specialized equipment and facilities.

Ethnic and religious heritage may affect the design and/or use of a facility or structure. In some cultures it is common practice to have multiple generations of a single family share the same home. In other cultures the single family dwelling is more common. Architects and engineers must recognize cultural practices when designing buildings and facilities. Different religious practices dictate architectural design. In some religions the proximity of the toilet facilities to eating areas are specified. In addition, religious structures require a multitude of special considerations. However, access by all people must be considered when designing religious buildings, just as with other buildings.

4.3.2. Motivator–hygiene model applied to barrier-free design

Motivator–hygiene theory proposes that people have two basic sets of needs: hygiene needs and motivator needs. The hygiene needs relate to a person's environment, and serve primarily to prevent dissatisfaction. The motivator needs relate to the ability to achieve and experience psychological growth. The motivators lead to satisfaction because of an inherent need for growth and self-actualization. If these factors are not met, no satisfaction will occur.

When applied to barrier-free design, the motivator–hygiene model suggests a two-step process. The environment must be designed so that the hygiene needs of all people are met. These needs are centered around the avoidance of loss of life, hunger, pain, sexual deprivation, and other primary drives. Hygiene needs parallel the physiological and safety needs of Maslow's hierarchy model. Most institutions satisfy sanitary and environmental protection needs. Lavatories must be accessible, functional, and clearly identifiable. Emergency information must be clearly described and displayed in terms clearly communicated to a variety of people (e.g., symbolically, in braille, audibly). Signs should be easily located even in panic situations. Loudspeakers can be used to announce information for visually impaired people, whereas clearly displayed signage helps hearing-impaired people. Hygiene needs of wheelchair users may be helped with automatic doors, enlarged lavatories, and accessible sinks and soap dispensers. Most accessibility standards focus on meeting hygiene needs. For example, minimum ramp widths and maximum ramp gradients are set to ensure the safety of wheelchair users.

Meeting hygiene needs only leads to no dissatisfaction. Motivator needs must be addressed to bring about satisfaction. Motivator needs relate to psychological growth through: being creative, individuality, independence, autonomy, and exposure to new concepts. People with disabilities strive to lead self-directed lives in an environment of their choosing with variable degrees of independence. Motivator needs are similar to belonging, self-esteem, and self-actualization in Maslow's model. A simple implementation of a motivator is to provide access for people with disabilities at the primary entrance to the building. If traffic flow through a building promotes interaction and cross-communication between people with disabilities and other building occupants it will motivate people. Using microphones, augmentative communication devices, and translators for hearing-impaired persons in meeting facilities can act as motivators, and bring about greater productivity. Computer-generated displays, adjustable writing surfaces, and laser pointers can assist with interaction and communication. For example, some aquariums use 'touch pools' where children can touch starfish and other small sea creatures. These exhibits promote understanding and participation. Providing access to people with disabilities to interact with the sea creatures and to experience new sensations helps to build psychological growth. Professionals needs to use their imagination to develop means for people with disabilities to use their abilities and senses to experience the environment rather than to be spectators.

4.3.3. Designing with a team

A team approach works best for including integration into design. People of various professions and life experiences bring different perspectives to the design process. The variety helps to ensure that a number of important factors are considered in the final design. For architectural projects a number of people should be involved. The owners of the building must be involved in the design, because they are often funding the project, and it is their needs which form the core of the design. End users can contribute information about how they envision using the finished facility. An architect, or team of architects, must convert the different demands and desires, and convert them into designs. A construction engineer must determine if the design is structurally sound and the appropriate construction materials have been used. The contractor should be involved because s/he is going to be performing the actual construction. A rehabilitation engineer should be part of the team to ensure that the facility is usable by all people.

Architects must at times act as social scientists. Architecture can affect social interaction. Architects must attempt to understand how their designs will affect the interaction between people. The goal of the architect is to complete a building which is functional, yet permits the expression of creativity and individuality.

Space usage has significant impact on social interaction. Peoples' needs and perceptions must be considered during the design phase. People often desire their own space, over which they have control. People can be territorial. Crowding can cause discomfort and conflict. People must be given sufficient space to accommodate their activities. Privacy is also important to people. If the building is to accommodate living space, then individual tenants require some degree of privacy. If the building is used for office space, then individual employees also require some privacy.

The design team may accommodate the physical needs of all potential users, and yet the building will not appeal to people. The layout and design of the building must also be intuitively appealing. The design team must consider distance perception in their plans. A building may not be appealing if it appears cramped or if distances between various facilities appear great. A related issue is the spatial mapping of the building. The design team must consider how the space will be used, and by whom. For example it is best to put commonly used facilities adjacent to each other. A mistake sometimes made is to alternate men's and women's toilet facilities on adjacent floors (e.g., men's toilets on odd numbered floors, and women's toilets on even numbered floors) of office buildings. This may require a large portion of the office to move between floors each day.

Architecture can also have social impact. The design team must consider the location of the structure, the history of the region, and acceptable aesthetics for the region. It is not uncommon in some regions for buildings to be required to be designed with a specific type of exterior appearance (e.g., Spanish style in some Californian cities).

Some questions cannot be answered based solely on education and experience. Sometimes the design team, or architects must conduct research to determine how people will respond to different structure designs. Studies can be conducted with models or with partitions in an existing structure. Much can be learned by studying how people use existing structures of various designs.

4.3.4. Design factors

Human factors must be considered along with a number of other factors when designing a building. Long-term use of the building must be considered.

Buildings may undergo several transformations during their lifetimes. The design team should anticipate some of the changes a building may go through. All projects are initiated with limited resources. Both natural (e.g., space, layout) and financial resources must be considered. Whenever possible the design should complement the existing surroundings. As in all design cases, architectural designs are often faced with conflicting priorities. When faced with conflicting priorities, the design team must use its collective wisdom to make the best compromise within the law.

There is an explosion of information about how to incorporate human factors into design. All members of the design team must be aware of current ideas in their subject. However, there must come a time in the design when it becomes finalized, and construction begins. Designs should be flexible enough to accommodate new information and technological advances in the future. Design team members should question traditional values related to design, and give importance to quality of life issues for all potential users of the facility.

4.4. INTERIOR SPACE DESIGN

The interior of a building requires special consideration with regards to human factors, and barrier-free design. Interior space can be divided into two areas: fixed feature and semi-fixed feature space. Fixed feature space is typified by specialized rooms. The user has only limited freedom to modify the room's primary features (e.g., bathrooms). Semi-fixed feature space is one where limited environmental modification is possible (i.e., arranging furniture in a living room). From the design team's perspective, fixed feature space requires greater consideration for barrier-free design.

The design team must consider fixtures as well as the interior layout. The fixtures (e.g., sink, bathtub, mirrors, toilet, door handles) must be usable by a wide variety of people with different abilities. Typically, a 95 percentile value is used (i.e., 95% of all people could use the fixture). The layout of the fixtures must be such that people have ready access to them. Careful consideration during fixture selection, and interior design layout facilitates barrier-free design.

Doorways require special consideration. Door widths must be large enough for wheelchair users to enter, typically 36 inches (900 mm). The door handles must be easily operated by a large number of people, and the door's swing weight (i.e., the force required to open the door) must be low enough for most people to operate the door independently.

Stairs may present a tremendous risk for injury. Stairs require markings for people with visual impairments. There is some confusion in US law: if an elevator is available the stairs may not have to meet access codes for people with visual impairments. Stairs must be designed with sufficient footing and with landings to allow people to recover and safely pass. Doorways must only open to landings, and there must be sufficient space for people to pass safely while a door opens. Each floor of the building must be marked in braille and raised lettering.

Windows require some special consideration. The location of windows partially determines their accessibility and affects the atmosphere of the room. Windows also contribute to the layout of the room. Some windows are easier to open and close than others. Opening/closing mechanisms must have handles or lock mechanisms which can be identified, reached, and grasped by most people (i.e., 95th percentile of all people). The force/torque required to operate the opening/closing mechanism must be low enough for most people to operate it easily. Windows should be placed within the visual perspective of range of the people using the space, and should be located to minimize the risk of somebody attempting egress through them.

4.5. DESIGN FOR PEOPLE WITH DISABILITIES

Needs vary greatly for various types of disabilities. Therefore, a functional understanding of how various disabilities affect design must be considered. For example, a blind person may feel more comfortable in smaller spaces where most elements are within reach, while a wheelchair user may prefer open spaces where it is easier to maneuver. In addition, a floor textured to help guide blind people may impede wheelchair users, but both groups function well with hard surfaced floors. Often, compromises must be made to accommodate the greatest number of people or to best accommodate the most prevalent users of the facility.

Environmental competence is a function of environmental design. Proper environmental design facilitates ready access by nearly all people, and makes people feel comfortable. The abilities and experiences of the user must be considered. Physical, mental, psychological, and sensory abilities must be considered. Integration not segregation of persons with disabilities must be the primary goal of environmental design.

People with disabilities must not only be able to use or have access to facilities, but they must be able to do so in a nearly normal fashion. In order to design facilities for use by all people, the design team must consider a number of factors.

If an environment is not designed well, then people can become disoriented. For example, it is not uncommon for courtyards to include planter boxes. Sometimes these planter boxes are round, and placed periodically throughout the courtyard. Without a clearly defined guide-strip people with visual impairments may become disoriented. A person who is blind may enter the courtyard, and orient him/herself with one of the round planter boxes. It may take two or three laps around a planter box to realize there is a problem. This process may continue from planter box to planter box until the person finds a better landmark. Some people have difficulty understanding signage if it is not properly placed or thought out.

Without proper planning, tasks that may be thought to be simple may require uncommon acts of agility, stamina, and reaction time for some people. The agility and stamina of people decreases, and reaction time may increase with age and some forms of disability. For example, an older person in a wheelchair may not have the agility, stamina, and reaction time required to cross a busy parking lot. Therefore, parking spaces and public transportation stations need to be provided close to services which would be frequented by people with severe agility, stamina, and reaction-time limitations.

Size and maturity also affect environmental design. People who are much larger or smaller than usual must be considered during environmental design. People who are very tall may have difficulty reaching items which are placed low, whereas people who are much shorter than usual may have difficulty reaching items which are placed high. In some stores, items which are most frequently used are placed on the middle shelves to be readily accessible to most people. The weight of the items must also be considered, as the difference between the carriage height and shelve height should be minimized. The maturity of an individual is also important. Many stores are displayed to attract customers' attention to certain products. This marketing technique is often targeted at adults but may be extremely confusing for young children.

People have varying degrees of flexibility (e.g., elderly people, people with arthritis). Some people may be unable to reach products located on lower shelves. Other people may have difficulty opening and closing drawers or cabinets. It is also difficult for some people to sit in some seats of a vehicle. Design teams must consider what requirements their designs place upon the end user.

Some people have difficulty with incontinence. Therefore, toilet facilities must be made available, accessible, and clearly marked. Larger, and private toilet facilities are more accommodating for many people with incontinence problems. For larger facilities, it may be advisable to offer changing rooms. Many of these accommodations benefit parents of small children as well as some people with disabilities.

The enabler is a simple model to help design teams incorporate the needs of various people with disabilities into environmental designs (figure 4.5).

4.5.1. Accessible design guidelines

To create facilities which are accessible to people with disabilities, the design team must have an understanding of their unique capabilities and limitations as well as thorough knowledge of the needs of other occupants

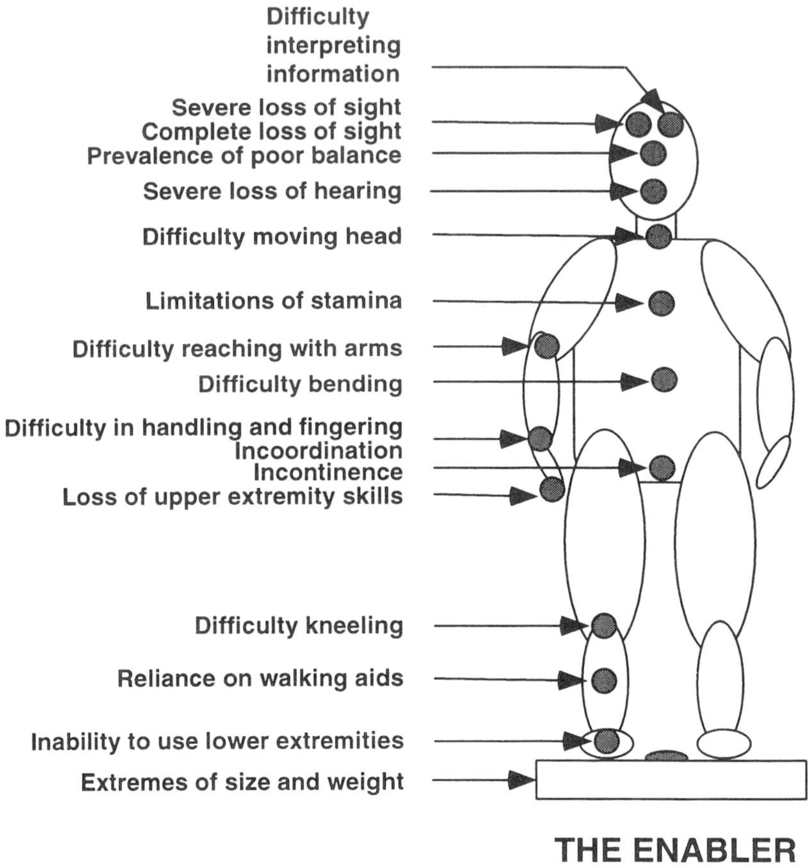

Figure 4.5. *The enabler—a model for accessible design.*

or users. At various times and in different ways, many people are affected by temporary or permanent disabilities. Efforts towards regaining and recapturing independence should be encouraged. Before a project can begin, the needs of the users must be defined.

The design team, including people with disabilities, must meet and develop written descriptions of the needs of people with disabilities, and the goals for the project. The design team must determine the affects of mobility, sensory, and cognitive disabilities. Do not fall into the trap of solving problems as they are articulated. Take the time to record all of the users' needs, and then methodically approach all of the problems once they have been articulated. This will help to ensure optimal design. Accessible design projects require a team approach. Few people are familiar with mobility and sensory limitations, and also know about sources of suitable, high-quality adaptive products and equipment. Most contractors and architects are under-equipped with working knowledge and experience in these areas.

Architects can help to define the project's scope and goals. They will verify that the design meets municipal requirements. Architects will prepare preliminary and final plans, specifications, and equipment lists. Architects will develop specification packages for contractors, and secure municipal authority approval. They can assist with monitoring and supervising the contractor's work. A primary consideration that many people often overlook is that the client can only control any two of the basic project elements: quality, scope, and cost. The architect and contractor will always control the remaining element.

The *US Fair Housing Amendments Law* of 1988 provides guidance for accessible design. The law contains two primary components: guidelines for accessibility, and the scope of accessibility mandated. The *Fair Housing Amendments Law* requires that buildings have 'accessible routes', 'adaptable elements', and 'usable space'. This has been interpreted to mean that people with disabilities must have reasonable access to the building. People must be able to use the various elements (e.g., elevators, bathrooms) within the building, and there must be sufficient space usable by people with disabilities within the building. The *Fair Housing Amendments Act* requires that all ground floor units in walk-up apartments, and all living units in elevator buildings be accessible/adaptable. All public and common-use areas must be readily accessible.

The 1986 *American National Standards Institute (ANSI) A117.1 Standard* is more specialized when compared with the *Fair Housing Amendments Law* in the area of public and private multi-family housing. However, *ANSI A117.1* covers a much broader spectrum of accessibility issues, and it provides engineering design guidelines.

4.5.2. Bathrooms and toilet facilities

Bathrooms and toilet facilities are one of the more important specialized design areas for people with disabilities. These facilities must be designed to accommodate the space requirements of wheelchair users, without being exceedingly expensive. Bathrooms in residential settings must meet different requirements than toilet facilities in public buildings. In both settings, the layout and equipment must be such that most people can use them independently. However, space must be available if an assistant is required. Equipment should accommodate people with mobility, and sensory impairments. Fixtures and door handles should be of the paddle type. The force required to use toilet equipment (e.g., doors, sinks, towel dispensers) should not exceed ten pounds. Latches must be usable by people with limited hand function and coordination.

In a residential setting, a bathroom implies that a bathtub and/or shower must be incorporated into the design. This requires additional consideration.

The bathroom must incorporate sufficient maneuvering space for the user to easily access all equipment. Typically, 45 to 55 square feet (4 to 5 square meters) of floor space can accommodate most people. Wheelchair users must have access to the tub/shower in a position to make transferring easier. The tub/shower controls must be accessible from a wheelchair before transferring. Ideally, the usable floor space could have an area where the wheelchair user can complete a full revolution (i.e., the floor space dimensions should be greater than the length of the distance from the rearmost point on one side of the wheelchair to the forward-most point on the opposite side of the wheelchair). Figure 4.6 provides an example of an accessible residential bathroom.

Public toilet rooms must meet additional requirements. The right-hand side of figure 4.6 illustrates a possible layout for a toilet stall. Toilet stalls must be large enough for wheelchair users to maneuver within the stall and to make a side transfer to the toilet. This means that a typical stall width of five feet is required. The toilet itself should be elevated to between 17 and 19 inches (430–480 mm). The kneespace beneath the sink should be 30 inches (760 mm) wide and 20 inches (500 mm) deep. The height of the sink should be between 27 and 34 inches (690–860 mm). The mirror height should be no more than 36 inches from the floor. Grab bars must also be provided around the toilet. A 1.25 to 1.5 inch (30–40 mm) diameter tube should be used as the handle on grab bars. The grab bars should be offset about 1.5 inches (40 mm) from the wall. Grab bars should be located from 33 to 36 inches (840–900 mm) off the floor. In the United States, *American National Standards Institute A117.1* standards for buildings and facilities to provide accessibility and usability for people with physical disabilities should be consulted for exact design specifications.

4.5.3. Home kitchen design

None of the functional spaces in a home is more complex to design for people with disabilities than the kitchen. The process of food preparation and clean-up involves a number of interrelated tasks requiring an array of appliances. Designing a kitchen that accommodates people with disabilities requires careful planning in advance of any construction or renovation.

Each person has different capabilities and limitations as well as his/her own preferences and desires. Therefore, it is imperative to first establish the precise requirements that should be met in the kitchen design. It is important to determine whether the kitchen will be used by 'able-bodied' people as well as people with disabilities. The future resale of the home is also an important consideration. Some people may not be capable of or interested in performing certain activities; therefore these tasks may be given a low priority.

To be efficient and effective, a kitchen must incorporate three basic activity centers: the stove and oven; the refrigerator and freezer; and the sink,

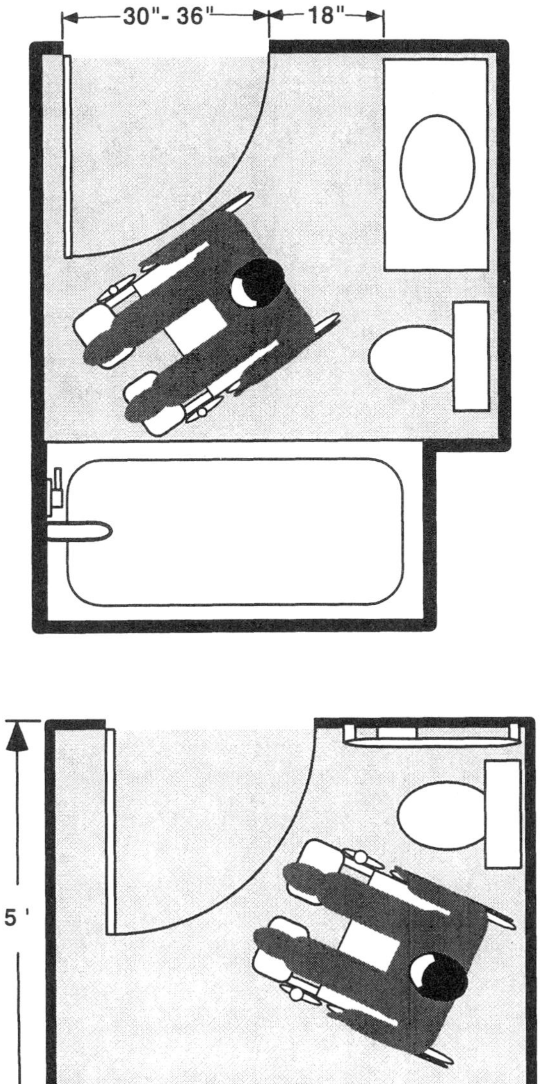

Figure 4.6. *Schematic diagram of accessible bathroom and toilet facilities.*

disposal and dishwasher. These activity centers form what has been deemed the work triangle. Optimally, these centers would be arranged to follow the natural sequence of food preparation and clean-up. Food preparation and clean-up patterns can be accommodated with many different layouts. Three common layouts are U-shaped, L-shaped, and galley kitchens, as shown in figure 4.7. Of these plans the U-shaped kitchen often provides

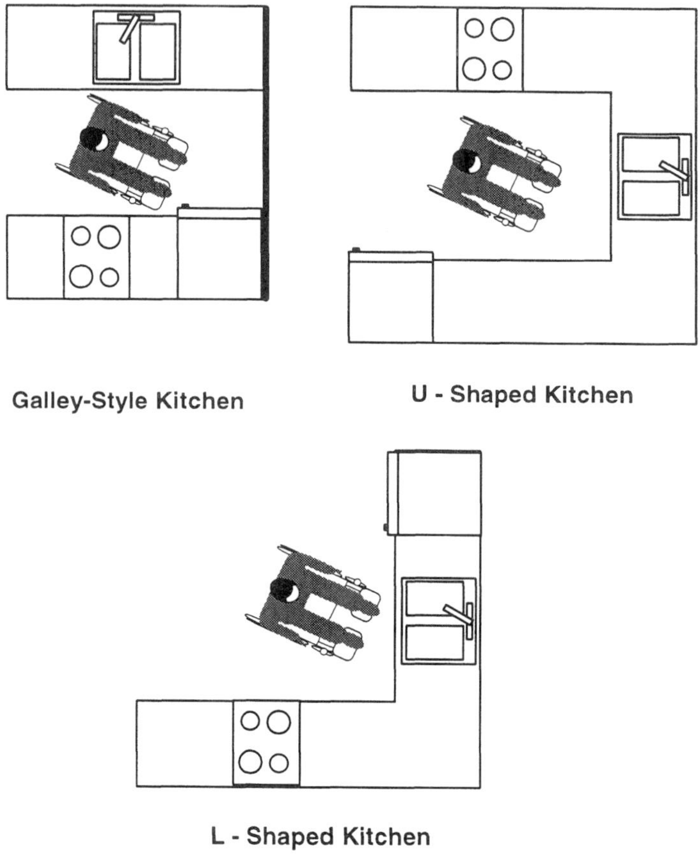

Figure 4.7. *Examples of kitchen arrangements.*

greatest efficiency because the counter-top has greater surface area and is more continuous.

Adequate counter space adjacent to each of the work centers is very important. Many people with disabilities must set an item aside as they orient themselves for the next task. The refrigerator/freezer should have at least 18 inches (500 mm) of counter space adjacent to its opening side. The sink and dishwasher area should provide between two and three feet of counter space on both sides of the sink. The stove/oven should have two feet of counter space on both sides.

A minimum of ten linear feet of counter top should be provided in the kitchen. The counter top should be as continuous as possible, and the work center layout should be designed to minimize disruptions in work flow. Standard counter heights are 36–37 inches (900–925 mm). Tasks which require precise action are best performed with higher counter tops, whereas, tasks requiring more physical strength are easier to perform on lower counter tops. Using a counter top height of 34 inches (850 mm) improves use by wheelchair users and smaller people. Another approach is to lower sections of the counter top; however, this can make the kitchen awkward to use. Electrically operated, adjustable-height counter tops are available which permit optimal use by ambulatory people and wheelchair users.

Adequate storage space at each work center is essential for kitchen task efficiency. The vertical reach of most wheelchair users extends from 9 inches (200 mm) above the floor to 54 inches (140 mm) above the floor. A horizontal reach of 24 inches (600 mm) is considered maximum. Cabinets can be equipped with sliding shelves or deep drawers. Lazy-Susan corner units can be used to maximize use of storage space. Wall and full-height cabinets can be equipped with fold-out racks. Many items can be purchased off-the-shelf, avoiding the need for expensive custom cabinetry.

The sink is often considered the center of the kitchen. To make the sink accessible, the design must provide adequate clear floor space, functional counter-top height, and provisions for knee space. The space beneath the sink should be 30 inches (760 mm) wide, 20 inches (500 mm) deep, and as high as necessary for the user with a minimum of 27 inches (700 mm). Some sink manufacturers place the drain near the back, which provides additional knee space. The disposal unit can be offset to one side or the other. A single lever-type handle that controls both water temperature and flow rate is recommended. A hand-held spray hose, and a laboratory-type spigot can make it much easier to fill pots and bottles. The dishwasher should be located approximately 24 inches (600 mm) from the edge of the sink.

Knee-space requirements for the cook top are the same as those for the sink. A cooktop unit instead of conventional stove allows for knee-space below the burners. This arrangement makes it easier for people in wheelchairs to position themselves to place pots and pans. A smooth ceramic-top cooking unit allows utensils to be easily slid and maneuvered, minimizing the risk of spilling.

A wall oven can be positioned to allow either direct or side access. The oven should be located near the end of a heat resistant counter with knee space. A side-hinged oven is easier to operate; however, the user must be protected from spills.

A vertical side-by-side refrigerator/freezer provides greatest access with both refrigerator and freezer storage at similar reach levels. The doors should be permitted to swing to 180 degrees when open.

Accessible kitchens can be designed at reasonable cost with some creativity, and careful attention to stock items. The designer must have a clear understanding of the needs of all users. 4.6.4

4.5.4. Parking spaces

Many people with disabilities require additional space to enter and exit a personal vehicle. Some people require more time to perform various activities than others, and some people do not have the ability to travel great distances. For these and other reasons, specialized parking spaces were designed for people with a medical disability as defined by US Federal law (*PL100-641*). Public law *100-641* establishes a uniform system of handicap parking. The law defines 'qualifying persons' with disabilities, provides guidelines for issuing special license plates and removable placards, establishes toll-free parking policies, and centralizes the authority to the state government to issue plates and placards.

The law recommends that states include in their statutes a section on reciprocity, to ensure that states, cities, and counties recognize all special license plates and removable placards. Advocacy groups support ensuring equal access to parking via parking policies that would permit parking in metered spaces without time limitations and cost.

Some states have legislated a policy pertaining to gasoline stations. In many states, motorists with a proper handicap license plate or placard may purchase gas at the self-service price, while having a full-service attendant dispense the gasoline.

Many parking spaces are based upon the ANSI standard for space dimensions (*A117.1, section 4.6.2*) which calls for a parking space to be a minimum of eight feet (2.4 m) wide, with an adjacent aisle a minimum of five feet (1.5 m) wide. Advocacy groups have proposed the universal handicapped parking space (figure 4.8). This space has larger dimensions and is safer for drivers and passengers with disabilities. It requires a parking area of 11 feet (3.4 m) and an access aisle of five feet (1.5 m). Many vans have side entry with a lift, which exceeds the width of a standard-size space. The parking space should be 18 feet (5.5 m) deep.

Parking spaces are to be located near commonly used entrances. Parking spaces are designed for people with disabilities to safely enter and exit a vehicle, and to safely enter the adjacent buildings. The *Americans with Disabilities Act Title III* specifies guidelines for parking spaces. In small parking lots (1–25 total spaces) one accessible parking space must be van accessible. In medium parking lots (101–150 total spaces) five accessible spaces are required, and one must be van accessible. For parking lots with 151–1000, five plus 2% of the spaces should be accessible and one in every eight of these should be van accessible. Parking lots with 1001 or more spaces should have 2% (20 spaces accessible) plus 1% of the spaces over 1000. One in every eight of the accessible spaces must be van accessible. Signs reserving each space must display the international access symbol.

4.5.5. Doors and doorways

Exterior doors are often the first potential obstacle encountered by people with disabilities when visiting a building. Many factors affect the opening and closing force required at exterior doors (e.g., wind, door weight,

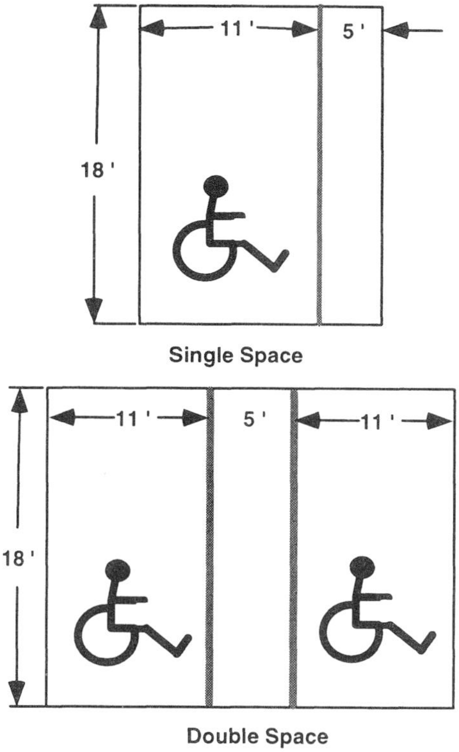

Figure 4.8. *Universal handicapped parking space.*

ventilation systems). These factors can make it difficult for people with disabilities to operate exterior doors. Automated door openers can make access easier for people with disabilities. Automated exterior doors use various operating principles (e.g., balanced doors, revolving doors). Many mechanisms can be used to activate doors, some systems are automated (e.g., pressure sensors, motion sensors) others require the user to activate the door (e.g., button, push-plate). Switches for manual doors should be located about 42 inches (1065 mm) above the finished floor. In some instances, exterior entrances may use standard manually operated hinged doors, in which case, the maximum opening force should not exceed 8.5 pounds (38 N) force (*ANSI A117.1*). Exterior doors should have a minimum clearance width of 36 inches (900 mm). Paddle-type door handles should be used with manual doors. This type of handle is easier to grasp and operate. The door threshold should not exceed 0.5 inch (1 cm) in height.

Doors installed within individual dwellings are subject to different standards than those located in public spaces. The minimum requirement for a residence entrance door is a 32 inch (800 mm) clear width (US Department of Housing and Urban Development (HUD)). The width of doors is also an important determinant of accessibility. For interior hallways a 34 inch (860 mm) doorway should be used off a 36 inch (900 mm) interior corridor (HUD), a 32 inch (800 mm) doorway may be usable if the corridor is widened to 48 inches (1200 mm). There is always a trade-off between corridor width and doorway width. It is also helpful to allow for 18 inches (500 mm) of clearance after the doorway. This makes maneuvering easier for wheelchair users and for those who walk with a walker, crutches, or canes.

4.5.6. Ramps

Ramps and curb-cuts are commonly employed to provide access to some facilities by wheelchair users, and for people who have difficulty walking. The slope of a ramp or curb-cut may not exceed 8.33% (i.e., one foot rise for a twelve foot run). Cross slopes shall not exceed 2% (i.e., one foot rise for 50 feet run). Ramps and passages must provide a minimum of 36 inches of clear width (915 mm). Ramps and passages less than 60 inches wide (1525 mm) must provide passing space at reasonable intervals. Passing space must be at least 60 inches by 60 inches (1525 mm by 1525 mm) in clear width. If the height of the ramp exceeds one foot a handrail should be used. The handrail should be at a height of 27 inches (690 mm) above the surface of the ramp. Ramps which must traverse longer distances should incorporate a landing every 30 feet (9 m). The landing should be at least five feet (1.5 m) long.

The surface should be firm, stable and slip resistant. Gratings and differences in level between segments (e.g., boards, cement footings) should be minimized. Vertical differences must not exceed 0.25 inch (6.5 mm), and horizontal distances (i.e., gaps between adjacent boards or plates) must not exceed 0.5 inch (13 mm).

4.5.7. Selected safety issues

In some places objects may be mounted on or near walls adjacent to passage ways. If not properly placed these objects can be hazards. Objects projecting from walls (e.g., signs, telephones, canopies) which have their leading edges between 27 inches and 80 inches (685–2030 mm) above the floor or sidewalk surface must not protrude more than four inches into the room or walkway.

Gratings for drainage may present a hazard to people who use wheelchairs or who have difficulty walking. Gratings located in walking surfaces must have spaces no greater than 0.5 inch (13 mm) wide in one direction. Gratings with elongated openings must be placed so that the long dimension is perpendicular to the dominant direction of travel.

Crossing streets can be difficult for many people. Safe interaction of people with automobile traffic requires careful planning. Curb ramps and other sloped areas at marked crossings must be wholly contained within the markings, flared sides may be excluded. Curb ramps and sloped areas must be protected or located to prevent their obstruction by parked vehicles. The edges of curb ramps must be well defined.

At controlled crossings people must make their presence known to the crossing controls. The controls must be accessible and operable by a wide variety of people. Buttons for operating traffic lights should be raised, and be a minimum of 0.75 inch (19 mm) in the smallest dimension. The force required to activate the switch must be less than 5 pounds force (22.2 N). The controls should be located on a level area as close as possible to the curb or ramp. Pedestrian-activated controls should not be lower than 24 inches (600 mm) or greater than 48 inches (1200 mm) above the sidewalk. Sidewalks should be delineated with contrast markings or materials (e.g., paint, brick, tile). A raised marking about 0.25 inch (6.5 mm) high is helpful for people with visual impairments.

Where provided, controls and operating mechanisms at roadside emergency communications systems call boxes should be operable with one hand, and not require tight grasping, pinching, or twisting of the wrist. The highest part required to be operated by the user must not exceed 54 inches (1370 mm) above the graded surface at the call box. The force required to operate any of the controls must not exceed 5 pounds force (22.2 N). Emergency communication systems must provide both visual and auditory acknowledgment of call receipt, and must not require voice communication.

Inadequate hand-rails can present a hazard to people with mobility impairments. Hand-rails must be used when the walking surface is at one foot or greater above the ground. Handrails should be about 27 inches (685 mm) high from the surface of the walkway. Hand-rails with a 1.5 to 2 inch (37–50 mm) diameter work well for most people. Designers must be aware that hand-rails must be sufficiently strong to withstand several times a 95th percentile persons body weight. Hand-rails on stairs or ramps should extend 18 inches (500 mm) beyond the landing.

4.6. ACCESSIBLE TRANSPORTATION

4.6.1. Bus and highway transportation

Accessible bus and paratransit services improve the quality of life for people with disabilities. Properly designed accessible transportation systems can assist people who might not otherwise have the means to be employed or to perform many daily activities independently (e.g., shopping, attending school). Public transportation can reduce dependency on public assistance. Many people have no other viable means of transportation.

The *Urban Mass Transportation Act (UMTA)* of 1970, the *Federal Aid Highway Act (FAHA)* amended in 1975, and the *Americans with Disabilities*

Act (ADA) of 1990 are the three major sources of federal legislation to improve the access of mass transit. People with disabilities must have the same right to use mass transportation as other individuals. Accommodations must be designed and implemented to ensure ready access to public transit. UMTA and FAHA had two primary weaknesses: 'special efforts' to ensure that people with disabilities had access to public transit were not clearly defined, thus allowing inaction; and the Department of Transportation (DOT) did not use UMTA and FAHA to set standards by which transit authorities must comply. Section 504 of the *Rehabilitation Act* of 1973 has been used more effectively to secure the rights of people with disabilities to use public transportation.

Since its issuance in 1977 by the Department of Health, Education, and Welfare (HEW), each federal department and agency has been required to implement regulations to comply with the intent of section 504. DOT's 1976 regulations as well as court decisions applying them attempted to define the special efforts necessary to comply with section 504, and formulated what DOT considered appropriate transit programs. The results was the creation of separate bus systems for people with disabilities (paratransit). In 1979 after HEW had published its section 504 regulation guidelines, DOT interpreted these guidelines as requiring all mainline fixed-route buses purchased with federal funds to be accessible. The new DOT regulations required buses purchased after July 1979 to be lift equipped. Two court cases invalidated the 1979 DOT regulations concerning accessibility, Southeastern Community College v. Davis, and American Public Transit Association v. Lewis decided that although transit authorities had some obligation to provide accessible transportation for people with disabilities, they were not necessarily required to make extensive modifications to such programs. Guided by these court decisions, DOT developed interim regulations published in 1981 which required 'special efforts' to be made, and in so doing established separate transportation programs.

DOT's interim regulations did not include the specificity required for planning by regional transit authorities. In 1982, section 317(c) of the *Surface Transportation Act* was passed which mandated DOT to publish permanent regulations. These DOT regulations were completed in 1986. The 1986 regulations give regional transit authorities the flexibility to choose one of three options: a special service system (paratransit), an accessible fixed-route bus system, or a mixed system of special services and accessible buses. The regulations also mandate participation by people with disabilities in all aspects of planning, implementation, and modification of a particular system. Regional transit authorities were required to meet six service criteria:

- All people with disabilities who are physically unable to use public transportation must be eligible to use special services.
- Special service transportation must be provided within 24 hours of request.
- There may be no restrictions or prohibitions based upon trip purpose.

- Fares must be comparable to those charged to the general public.
- Special services must operate the same hours and days as services for the general public.
- Special service transportation must be available through the same service area as that available to the general public.

Unfortunately, DOT interpreted congressional intent to include a limit requiring no more than 3% of a regional transit authority's operating budget to be used to implement accessible transportation systems regardless of whether the six criteria were met. In 1990, the *Americans with Disabilities Act* was indicated. The ADA removed the 3% limit, and required regional transit authorities to develop plans for meeting the six criteria above. ADA implies that mainstream fixed-route public transportation should also become accessible.

Mass transit should not restrict the ability of people with disabilities to participate in employment, educational, and cultural opportunities. All new buses purchased should be equipped with lifts and restraints necessary to accommodate people with mobility and sensory impairments. New and remodeled transit facilities should be designed to provide access for people with disabilities. It may not be practical to accommodate some individuals with disabilities using fixed-route public transportation. Therefore a multimodal system that provides both accessible paratransit services and an accessible fixed-route system may accommodate the most people equitably.

Accessible bus transportation relies on wheelchair lifts. Technical improvements, preventative maintenance, and operator–mechanic familiarity with lift systems can significantly increase their reliability. A lift is simply a small elevating platform by which people with mobility impairments can board buses, vans, and commuter rail systems. There are two classes of lift mechanisms: active and passive. Active lifts, used on many paratransit systems, block the doorway or entrance in which they are installed when stowed. Passive lifts, used on many fixed route buses, allow unimpeded entrance through the doorway in which they are installed when stowed. A passive lift often consists of a series of steps that, when activated by the driver, fold out to form an elevating platform. Once the wheelchair passenger is securely on the lift platform, a safety barrier is raised to prevent the wheelchair/user from rolling off. Some power wheelchairs can override some safety barriers. The platform is elevated and the user rolls onto the bus. Disembarking follows this same procedure in reverse.

The complexity of passive lifts makes them susceptible to technical difficulties. Some passive lifts have demonstrated reliability problems in cold weather. The sand and salt used in cold climates accelerates wear of components which are not well lubricated or which are not protected. The hydraulic oil used in lift pumps must be selected for lift operation in the ambient temperature of the region. Using lighter weight oils during winter months can minimize the risk of the oil freezing. Ferrous surfaces should be plated with protective coatings to reduce corrosion or these parts should be designed using stainless steel. Shielding devices help to reduce or prevent build-up of sand and salt. Frequent preventative maintenance and lubrication can significantly increase the lifetime of lifts. UMTA guidelines specify that when a lift is used and maintained properly it should have a useful life equal to that of the vehicle. The practice of cycling lifts can increase reliability and operator familiarity. Cycling is defined as operating the lift though a complete cycle. Many transit authorities require the driver to cycle the lift before taking the bus out.

A major consideration is the safety of all passengers throughout the entire trip. This is especially true for riders who use wheelchairs. Special securement devices are required to provide adequate safety for all passengers once a wheelchair user boards a bus. There are a variety of wheelchair types (e.g., scooters, power wheelchairs, ultralights) that must be accommodated by securement mechanisms, and wheelchairs continue to evolve, which makes the task of designing securement mechanisms difficult. Most often new wheelchair designs show no consideration for securement mechanism designs. Securement mechanism designers must regularly update their designs to accommodate new wheelchairs. UMTA has accepted the position that securement devices serve to secure the wheelchair with the same level of security as regular passenger seats.

Securement devices must be quick and easy to use by most people with disabilities. They must be able to accommodate a variety of manual and powered wheelchairs. The wheelchair must be secured facing forwards using at least two points, preferably symmetrical points. The securement mechanism must retract when not in use. Wheelchair securement mechanisms must restrain the loaded wheelchair in a test which simulates a 30 mile per hour (50 $km\,h^{-1}$) impact at 20 gravities to be considered engineered well. The test has been abbreviated the 20/30 test, which has been accepted by the US DOT. Many securement mechanisms incorporate occupant restraints using a seat belt which is mounted to the floor of the vehicle, and a shoulder belt which is mounted to the vehicle body. Automatic securement mechanisms are available which use a docking device permanently attached to the wheelchair and a receiver mounted on the floor of the vehicle. The docking device is captured by the receiver, locking in the wheelchair until an electric motor is activated to release it. Because of the differences among people with disabilities, securement mechanisms cannot guarantee safety. Osteoporosis, muscle paralysis, and other manifestations of disability may cause a wheelchair user to become injured in an accident even when using a securement device.

Sometimes people who are elderly or who have a disability may need to use paratransit. Paratransit vans are typically modified to be small buses. Paratransit vans often include fiberglass tops reinforced with a steel roll cage which must be certified for roll-over protection. The roll cage is attached to the vehicle body and frame during construction. The fiberglass tops are required to achieve six feet four inches of internal headroom for ambulatory passengers. The interior of paratransit vans are easily washable, with vinyl sidewalls, and a heavy-duty ribbed rubber floor. Seating consists of bench seats, and folding jump seats. The floor or walls are equipped with tie-down or securement mechanisms that use belts and mechanical clamps. Paratransit wheelchair lifts are usually semi-automatic attendant operated and not designed to be used independently. Having the drivers operate the lifts gives the owners experienced lift operators and lowers their liability.

4.6.2. Rail travel

Rail transportation offers an alternative to other more dominant modes of transportation. Increased traffic congestion and the lower ecological impact have drawn some people's attention toward rail travel. Some areas are planning new rail systems to relieve traffic problems, and some rail lines once used exclusively for freight transportation are being converted to passenger travel. People with disabilities need to be included in all aspects of the planning and implementation of rail transportation. Many rail systems were not originally designed with access for people with disabilities in mind. However, some creative design can accommodate most people.

The National Railroad Passenger Corporation, also known as Amtrak, is the US's intercity passenger rail service. Amtrak serves approximately 500 communities in 44 states and Canada. Since beginning its operation in 1971, this national rail passenger service has made considerable effort to improve the accessibility of its trains and stations. As Amtrak has built new stations, full accessibility has been provided. They have made renovated stations partially if not fully accessible. Many stations incorporate raised platforms which give wheelchair users roll-on–roll-off access.

Trains typically incorporate at least one car with a wheelchair accessible bathroom, and wheelchair storage space. Each Superliner sleeping car has on its lower level, a fully accessible sleeping compartment, and restrooms. Access between cars is via stairs. However, a car attendant is available for food and beverage service at the passenger's seat or room. Viewliner sleeping cars have fully accessible bedrooms complete with restroom and shower.

ADA is the most important piece of legislation to affect rail travel. The ADA prohibits the purchase or lease of a new intercity rail vehicle, commuter rail vehicle, rapid rail vehicle, or light rail vehicle to be used for public transportation which is not readily accessible and usable by people with disabilities. Rail transportation providers must make a good-faith effort when purchasing used or remanufactured rail vehicles to purchase vehicles which are readily accessible and usable by people with disabilities. The ADA requires all new and remodeled facilities to be accessible. Existing facilities must become accessible over a specified time period.

4.6.3. Air carrier transportation

Air transport has had a profound affect on society in this century. People with disabilities desire the same level of access to air travel as other people. The 1979 *Section 504* requirements, the *Air Carrier Access Act*, and the *Americans with Disabilities Act Titles II and III* provide guidelines for accessible air carrier transportation. The courts have ruled that compensatory and emotional distress damages are necessary and appropriate under the *Air Carrier Access Act*. The *Air Carrier Access Act* requires widebody aircraft built after 1990 to have accessible lavatories, removable armrests on at least half of the aisle seats, and must have priority space for storing a wheelchair in the cabin. Aircraft with an accessible lavatory must have an onboard wheelchair. Carriers must give wheelchairs priority storage in baggage compartments, and must accept battery powered wheelchairs; when necessary the carrier must package the batteries at no cost to the passenger. Three primary areas have provided challenges for people with disabilities and the air carriers: access to small commuter aircraft, access to lavatories on full-size carriers, and airport accessibility.

The primary challenge facing small-commuter-aircraft accessibility is the ability to get into the aircraft. There are manufacturers of lift devices suitable for small commuter aircraft which do not use a level entry. Some air carriers prefer to continue using on-board wheelchairs and airline personnel to facilitate boarding by passengers with disabilities.

Lavatories on full-size carriers are often too small to accommodate people who use wheelchairs, even on-board wheelchairs, especially those people who require assistance. Air carriers fear that accommodating the lavatory needs of people with disabilities will reduce the number of seats, and will be costly. Solutions which provide additional lavatory space include placing two lavatories adjacent to one another with a folding partition and double doors which can be opened to provide more space, another alternative is to use a movable wall between lavatories to provide additional maneuvering space for an onboard wheelchair.

People with disabilities need to be treated with dignity. Service personnel who work with travelers who have a disability must learn to speak directly to the person, maintain eye contact, and ask if assistance is needed. Personnel must treat the traveler's assistive technology with great care.

4.7. ACCESS LEGISLATION

Many laws and regulations reflect past failures. People with disabilities have worked tirelessly to improve physical access to various aspects of society and there is a long legal history related to accessibility.

4.7.1. US accessibility legislation

The *Smith–Fess Vocational Rehabilitation Act* of 1920 followed World War I, and was the first act to include civilians with disabilities in federal vocational rehabilitation plans. The *Social Security Act* of 1935 provided unemployment compensation for people who were unable to work due to disability. After World War II, the *Vocational Rehabilitation Act* amendment (PL78-113) was established to provide job training for people with disabilities. The federal government would pay 100% of the cost for veterans, and would pay 50% of the cost for civilians. People with mental impairments were included for the first time with the deletion of the term 'physically' from the law. *Vocational Rehabilitation Act* amendment (PL83-565) moved rehabilitation under the new department of HEW. The amendment included research grants, demonstration grants, and workshops.

The *Civil Rights Act* of 1964 began to relieve discrimination in public education and accommodations based upon race, color, religion, and national origin. During the Vietnam conflict, the *Vocational Rehabilitation Act* amendment of 1965 (PL89-333) was authorized. The primary goal remained employment. However, section 15 authorized the 'National Commission on Architectural Barriers'. In that same year, the *Elementary and Secondary Education Act* (PL89-10) was initiated. This act funded various programs to support 'Educationally Deprived Children.'

In 1968, the National Committee on Architectural Barriers released its two-year report entitled *Design for All Americans*. The report stated that of 2975 architects surveyed 709 replied of which only 251 were aware of ANSI 117.1. None of the four major codes established referred to barriers. Manufacturers and suppliers of building supplies were unaware of standards. Transportation was identified as a major problem, but no program to improve transportation was proposed. The *Architectural Barrier Act (ABA)* (PL90-480) was passed that same year. The ABA required federally funded buildings to be accessible, and suggested states enact similar legislation (34 already had). The law demanded revisions of all building codes, and established ANSI 117.1 as a model.

One of the most far-reaching and fundamental pieces of legislation to address issues important to people with disabilities is the *Rehabilitation Act* of 1973 (PL93-112). This law shifted the emphasis from vocation to quality of life. This law was enacted by the US Congress after two vetoes (10/72, 3/73) by then President Nixon. Section 504 was arguably the most important component of the piece of legislation. However, section 504 was not enacted until 28 April 1977, when HEW Secretary Califano reacted to calls for action by many people with disabilities and their supporters. Sit-ins were staged, with wide public and political support, at HEW offices in Washington, DC and San Francisco (for 24 days). On 3 June 1977 the regulations became effective.

People with disabilities were often at a disadvantage when fighting for their rights, because fewer than 50% of people with disabilities received an appropriate education. The *Bill of Rights for Handicapped Children* (PL94-142) was passed in 1975. This law provided for education in the least restrictive environment, which started 'mainstreaming.' It also provided the impetus for large-scale busing of children with disabilities.

In 1981, the US Architectural and Transportation Barrier Compliance Board adopted the final format for its 'Minimum Guidelines and Requirements for Accessible Design' as published in the 1/27/82 Federal Register. These guidelines set minimum standards for General Services Administration (GSA), Department of Defense (DOD), Department of Housing and Urban Development (HUD), and the US Postal Service (USPS).

Many people with disabilities continued to have difficulty in gaining access to new assistive technology. States were often unaware of technology to improve the integration of people with disabilities into mainstream society. In 1987, the *Assistive Technology Related Assistance Act* (PL100-407) was signed into law. This law provides funds to states on a competitive basis to develop programs to increase the availability of assistive technology.

The most wide-sweeping piece of legislation ever enacted in the US related to people with disabilities is the *Americans with Disabilities Act* of 1990. The ADA prohibits discrimination against people with disabilities in employment (public and private), all public services, public accommodations, transportation, and telecommunications.

4.7.2. European accessibility legislation

Many European countries have legislation governing accessibility standards which are comparable to US standards. Sweden has substantial legislation relating to people with disabilities. The Fokus Society, an independent consumer-oriented group, was established in Sweden to promote the belief that people with disabilities must be as self-sufficient as possible. The Fokus Society helps people with disabilities to find and modify apartments. Sweden has required public buildings to be accessible since 1966, and places of employment since 1971. Accessibility legislation is integrated into the Swedish National Building Code. Swedish building requirements apply to new construction of government buildings and facilities, places of work, and publically financed dwellings. Application of requirements to existing dwellings is left to local authorities. The Decree on Construction in Finland specifies that 'buildings and facilities intended for the use of the public shall be designed so that they are suited to disabled people'. These regulations apply to private buildings which offer administrative, legal, social or religious services, as well as banks, accommodations, and restaurants. Compliance with accessibility regulations is required before a building permit is issued. Norway's building regulation of 1985 is very comprehensive and addresses new and renovated buildings, including transport-related buildings. Access to buildings, internal areas, lavatories, telephones, coat racks, buttons, switches, and signs are regulated.

France has accessibility standards similar to those of the US, but specifies fewer specific dimensions in accessibility guidelines. This helps architects to make accommodations appropriate for the particular architectural structure. French regulations apply to all new construction of facilities that serve the public, even if they are privately owned. Existing national buildings, community buildings, and public buildings that are administrative, scientific or cultural in purpose must be accessible. The accessibility guidelines of Belgium are similar to those of the US Americans with Disabilities Act Accessibility Guidelines (ADAAG). Building permits for new construction and significant alterations, based upon added value, are only issued after accessibility standards are met. These standards include reserved parking places, access routes to buildings, doors, public telephones, lavatories, and corridors. Special dispensation is possible if justified by local or technical circumstances.

Britain, Greece, Ireland, and Italy have legislation and guidelines for new construction. Britain also has extensive guidelines for accessibility of existing buildings. British regulations for existing buildings provide guidelines with few specific details. This permits flexibility specific to particular architectural constraints. The regulations in Ireland apply to all new public buildings, including railway stations and air terminals. In Greece regulations apply to new buildings, including interior design and signage.

In response to United Nations and European Economic Community initiatives, Holland, Germany, and Switzerland have all recently enacted national accessibility legislation. These countries previously treated accessibility as a regional or local issue. German codes focus on personal and public dwellings as opposed to public facilities. These codes give specific dimensions for ramps, hallways, elevators, and toilets. Germany has also integrated accessible parking, lowered curbstones, and improved wheelchair accessibility to building codes for many public and private buildings and shops. In Switzerland new construction and substantial renovation must take measures in favor of people with motor handicaps.

4.7.3. American accessibility legislation (Canada and Mexico)

The National Building Code of Canada entitled *Barrier Free Access* was completed in 1989. These standards apply to all new buildings except houses and residential facilities. The content and specifications are similar to US regulations, with modifications for local considerations. Canadian codes provide guidelines for ramps, lavatories, doors, and corridors.

Mexico has recently begun to develop accessibility standards. In 1990, a group was formed in Mexico City to work toward the elimination of 'physical barriers in public buildings and public transport'. The Mexican Institute of Social Security wrote guidelines for access. These guidelines were tested in two pilot studies beginning in 1991. The standard of the Ministry of Health is the only legislation applicable across the country. Although the standard is very specific, it is very limited in scope in that it only applies to new or remodeled health-care facilities. In 1993, a project was initiated to make the city of Oaxaca accessible using standards prepared by the Mexican Institute of Social Security. Mexico has very limited resources to develop and implement accessibility legislation. The lack of national accessibility requirements means that barrier elimination in Mexico is largely based on good will.

4.7.4. Asia and the Pacific accessibility legislation (Australia and Japan)

The building code of Australia sets accessibility requirements which apply to all buildings except nonhabitable outbuildings and certain references. Health-care facilities must be accessible in all areas.

Japan has taken a somewhat unique perspective with regards to accessibility legislation. Accessibility is a matter of local and regional pride and honor more so than a legislative issue. Society in Japan favors 'sequestered protection' for people with disabilities. Therefore, Japan has no accessibility standards which are established by law on a national level. However, assistive devices and accessible faculties are readily found in Japan. Builders and contractors make buildings accessible of their own volition and at their own expense. Accessibility is integrated into design because it projects an image of quality and responsibility. Despite an apparent lack of accessibility legislation, Japan has made itself largely accessible because individual and local pride have fostered competition for accessibility.

4.7.5. South African accessibility legislation

South Africa has accessibility standards which are centered upon wheelchair users. All clinics and health centers must comply with accessibility standards. Other buildings must comply unless they are all on one level with level access. South African accessibility standards address parking, elevators, lavatories, signage, ramps, and clearance (doors and hallways).

FURTHER READING AND REFERENCES

Abeyratne R I 1991 Mobility laws in the international transportation by air of the elderly and disabled—a legal dilemma for the airlines? *Air Law* **16** 155–66

ADA: public accommodations and commercial facilities 1991 *Paraplegia News* **45** 21–3

Adams T, Sauer B and Reger S I 1992 Kinematics of the wheelchair seated body in crash simulation *Proc. RESNA Int. '92 (Toronto, Ont., 1991)* pp 360–2

Americans with Disabilities Act Accessibility Guidelines 1992 *Federal Register* **57** 60613–74

Andersen R W 1992 Architectural barriers legislation and the range of human ability: of civil rights, missed opportunities, and building codes *Willametee Law Rev.* **28** 525–45

Anfield R N 1992 Americans with Disabilities Act of 1990: A primer of title i provisions for occupational health care professionals *J. Occup. Med.* **34** 503–9

Angerman B and Dreschler H 1991 Results of statutory rehabilitation measures in Germany *Int. J. Rehab. Res.* **14** 195–202

Architectural and Transportation Barriers Compliance Board 1992 Americans with Disabilities Act Accessibility Guidelines; state and local government facilities; proposed rulemaking *Federal Register* **57** Monday 21 December 60613–74

Armstrong M J 1993 Disability self-help organizations in the developing world: a case study from Malaysia *Int. J. Rehab. Res.* **16** 185–94

Barrier Free Environments, Inc 1993 UFAS *Retrofit Guide, Accessibility Modifications for Existing Buildings* (New York: Van Nostrand–Reinhold)

Bayes K and Franklin S 1971 *Designing for the Handicapped* (London: Godwin)

Beasley K 1989 Federal laws to mainstream accessibility *Paraplegia News* **43** 29–30

Beasley K A and Davies T D 1991 Kitchen design for the wheelchair user *Paraplegia News* **45** 37–40

Becker K P and Greenberg R A 1985 *Educational Rehabilitation of the Handicapped in the German Democratic Republic and in the United States of America: An Overview* (New York: Pergamon)

Berube B 1981 Barrier-free design: making the environment accessible to the disabled *Can. Med. Assoc. J.* **124** 68–78

Bolte B 1991 The misconception of political correctness *Mainstream* **15** 41–2

Branson G D 1991 *The Complete Guide to Barrier Free Housing* (Charlottesville, VA: Betterway)

California Governor's Committee for Employment of the Disabled, 1991 *Language Guide on Disability*

California State Accessibility Standards, Title 24, California Code of Regulations (CAR); California Building Code

Couch R H 1993 Rehabilitation innovations in Central America *Int. J. Rehab. Res.* **16** 13–22

Daley R and Page L 1991 Handicap parking: problems and policies *Paraplegia News* **45** 52–6

Davidson C 1991 Affordable accessible housing *Paraplegia News* **45** 41–2

Davies T D 1990 ANSI A1171 Undergoing changes *Paraplegia News* **44** 29

Davies T D 1991 Fair housing: final guidelines *Paraplegia News* **45** 48–9

Davies T D and Beasley K A 1988 *Design for Hospitality: Planning for Accessible Hotels and Motels* (New York: Nichols)

Davies T D and Beasley K A 1990 Housing alternatives for the future *Paraplegia News* **44** 24–32

Davies T D and Beasley K A 1992 *Fair Housing Design Guide for Accessibility* (Washington, DC: Paralyzed Veterans of America)

DeJong G and Lifchez R 1983 Physical disability and public policy *Scientific American* **248**

Department of International and Social Affairs 1986 *Disability: Situation, Strategies and Policies, United Nations Decade of Disabled Persons, 1983–1992* (New York: United Nations)

Eccleston J E and Gaal R P 1990 Design of an inexpensive, portable, remote-controlled door opener *Proc. RESNA 13th Ann. Conf. (Washington, DC, 1990)* pp 429–430

Editors 1992 Understanding access *The Disability Rag* **13** 1–12

Enderle J D 1992 *Engineering Senior Design Projects to Aid the Disabled* (Washington, DC: National Science Foundation)

Eriksson J, Bjerken A, Ekerup M and Johansson G 1993 Simulation of physical disabilities in computer designs of environments *Proc. 2nd Eur. Conf. Adv. Rehabilitation Technol. (Stockholm, 1993)* pp 40.3

Eur. Conf. of Ministers of Transport (ECMT) 1990 Transport for Disabled People: A Review of Provisions and Standards for Journey Planning and Pedestrian Access (Paris, France: OECD Publications Service)

Evan T and Associates 1992 *Americans with Disabilities Act: Facilities Compliance Workbook* (New York: Wiley)

Ficke R C 1992 *Digest of Data on Persons with Disabilities* (Washington, DC: National Institute on Disability and Rehabilitation Research)

Fitzpatrick T K 1963 *Rehabilitation Programs and Facilities in Western Europe* (Washington, DC: US Department of Health, Education and Welfare)

Gostin L O and Beyer H A 1993 *Implementing the Americans with Disabilities Act: Rights and Responsibilities of All Americans* (Baltimore, MD: Brookes)

Greer N R 1987 The state of the art of design for accessibility *Architecture* January

Hahn H 1985 Disability policy and the problem of discrimination *Am. Behav. Sci.* **28** 293–318

Hahn H 1985 Toward a politics of disability: definitions, disciplines and politics *Soc. Sci. J.* **22** 87–105

Hahn H 1988 The politics of physical differences: disability and discrimination *J. Soc. Issues* **44** 39–47

Herzberg F 1966 *Work and the Nature of Man* (New York: The World Publishing Company)

Herzberg F 1968 One more time: How do you motivate employees? *Harvard Business Rev.* January–February 53–62

Ifukube T, Sasaki T and Peng C 1991 A blind mobility aid model after echolocation of bats *IEEE Trans. Biomed. Eng.* **38** 461–5

Ison T G 1993 Disability and the law of employment *Int. J. Rehab. Res.* **16** 91–6

Johansson G and Eriksson J 1993 Computer-aided planning of adaption of homes for disabled persons *Proc. 2nd Eur. Conf. Adv. Rehabilitation Technol. (Stockholm, 1993)* pp 33.2

Jurgen R K 1991 Smart cars and highways go global *IEEE Spectrum* **28** 26–36
Kearney D S 1992 *The ADA: Compliance and Costs* (Kinsington, MA: Means)
Knowler D 1991 Adapting your home: getting started *Paraplegia News* **45** 45–6
Kondraske G V 1987 Human performance: science or art *Proc. 13th Northeast Bioengineering Conf.* pp 44–7
Kondraske G V 1988 Experimental evaluation of an elemental resource model for human performance *Proc. 10th IEEE-EMBS Conf. (New Orleans, LA, 1988)* pp 1612–3
Kondraske G V 1990 A PC-based performance measurement laboratory system *J. Clin. Eng.* **BME-55** 467–77
Livneh H 1982 On the origins of negative attitudes toward people with disabilities *Rehab. Lit.* **43** 338–47
Malassigne P and Amerson T L 1992 A survey on the accessibility of bathroom fixtures *Proc. IEEE-EMBS 14th Int. Conf. (Paris, 1992)* vol 14 No 4 pp 1513–4
Malassigne P and Amerson T L 1992 A survey on the usage of bathroom fixtures by disabled people *Proc. RESNA Int. '92 (Toronto, Ont, 1992)* pp 273–5
Martin S S, Moakley T J, Berger H R, Calcaterra R M, Weisman J J, Dempsey G, McDermott S and Capozzi D M 1989 Rail accessibility today *Paraplegia News* **43** 21–34
Mayagoitia R E 1993 Achieving disabled access in the city of Oaxaca, Mexico *Proc. IEEE-EMBS 15th Int. Conf. (San Diego, CA, 1993)* vol 15 No 3 pp 1316–7
Mayagoitia R E 1991 Establishing of Access Standards for the Mexican Institute of Social Security *Proc. IEEE-EMBS 13th Int. Conf. (Orlando, Florida, 1991)* vol 13 No 4 pp 1825–6
McCloskey M 1991 Air carrier access act: court decision *Paraplegia News* **45** 26
Morris T and Trimble J 1992 *Virtual Barrier-Free Design* (Hines, IL: Rehabilitation Research and Development Center)
Nesmith L 1987 Designing for 'special populations' *Architecture* January 63–4
Park L D 1975 Barriers to normality for the handicapped adult in the United States *Rehabilitation Literature* **36** pp 108–11
Peters A 1985 When the cheering stops *Disability Rag* March-April
Philippen D P 1990 West German standards to be transferred as a new general barrier-free building code *Rehabilitation Research and Development Progress Reports* US Department of Veterans Affairs vol 28 No 1 pp 222–3
Philippen D P 1993 Barrier-free accessible technology (Smart House) and tele-service technology is depending upon a barrier-free environment *Proc. 2nd Eur. Conf. Adv. Rehabilitation Technol. (Stockholm, 1993)* pp 40.1
Philippen D P and Marx L 1990 Applicable barrier-free concepts adaptable to modern technology developments *Rehabilitation Research and Development Progress Reports* US Department of Veterans Affairs vol 28 No 1 pp 206
Philippen D P and Shields 1990 Development of methods of egress of the disabled and elderly from private or public buildings *Rehabilitation Research and Development Progress Reports* US Department of Veterans Affairs vol 28 No 1 pp 224–5
Poulson D F and Richardson S J 1993 Issues in the uptake of adaptable smarter home technology *Proc. 2nd Eur. Conf. Adv. Rehabilitation Technol. (Stockholm, 1993)* p 14.1
Raffa M F 1985 Removing Architectural Barriers: the Architectural Barriers Act of 1968 *Mental Phys. Disability Law Rep.* **9** 304–8
Rigger R 1989 A barrier-free house: part 1 *Paraplegia News* **43** 39–42
Rigger R 1989 A barrier-free house: part 2 *Paraplegia News* **43** 21–4
Robbins S 1989 More care for all fares *Paraplegia News* **43** 11–3
Robbins S 1992 Tearing down barriers *Paraplegia News* **46** 13–8
Romano S 1992 Access to the screen *Paraplegia News* **46** 50–1
Rothenburg R S 1991 Keys to finding an accessible apartment *Paraplegia News* **45** 43–4

Rovig S M and Goodwin D M 1990 Modifying office chairs for people with special needs *Proc. RESNA 13th Annual Conf. (Washington, DC, 1990)* pp 397–8
Shown D 1991 Evaluation of Rehabilitation Services in Nigeria *Int. J. Rehab. Res.* **14** 182–4
Sommer R 1974 *Tight Spaces: Hard Architecture and How to Humanize It* (Englewood Cliffs, NJ: Prentice-Hall)
Sorensen R J 1979 *Design for Accessibility* (New York: McGraw-Hill)
Steinborn A and McDermott S 1988 Mass Transportation *Paraplegia News* **42** 23–39
Suarz de Balcazar Y and Fawcett S B 1990 Handicap parking violations *Paraplegia News* **44** 34–5
Texas Young Lawyers 1990 *The Americans with Disabilities Act: An Overview* (Austin, TX: Texas Young Lawyers Association)
Transportation Technology Development for California 1990 *Program and Policy Review* (Sacramento, CA: California Department of Transportation, Office of New Technology and Research Management)
Thompson D L, Thomas K R and Fernandez M S 1994 The Americans with Disabilities Act: Social Policy and Worldwide Implications for Practice *Int. J. Rehab. Res.* **17** 109–21
United States of America 101st Congress 1990 *Public Law* 101–366 26 July
US Architectural and Transportation Barriers Compliance Board 1991 Americans with Disabilities Act (ADA): accessibility guidelines for buildings and facilities *Federal Register* **54** 144
Walter F 1971 *Four Architectural Movement Studies for the Wheelchair and Ambulant Disabled* (London: Disabled Living Foundation)
Walton M L and Cottrell E H 1989 Technical paper on accessibility codes and standards: a comparison of domestic and selected foreign standards and codes for accessible facilities *US Architectural and Transportation Barriers Compliance Board Task 2 Final Report*
Wolfson R I 1991 Technology: catching up with workplace needs *Paraplegia News* **45** 32–4

CHAPTER 5

PERSONAL TRANSPORTATION

5.1. INTRODUCTION

Special adaptive equipment needs increase with the degree of impairment and desired degree of independence in areas such as personal care, mobility, leisure, personal transportation, and employment. Disability negatively affects the financial status of many people. Adaptive driving equipment can become expensive depending upon the adaptations required. People are also concerned that they receive the proper equipment for them to safely operate their vehicle. Most people with disabilities purchase vehicles and adaptive driving equipment with personal funds. Personal vehicles and adaptive equipment also require maintenance.

Access and egress equipment have the greatest maintenance requirements. Other devices such as hand controls, steering equipment, securement mechanisms, and interior controls require less maintenance. Most users of adaptive driving equipment are satisfied with the performance of such equipment. Most frequent equipment problems are minor and are repaired by consumers themselves.

For many people with disabilities, their rehabilitation includes an evaluation to determine driving potential with the use of assistive driving equipment. The driving evaluation process usually begins with an assessment of the individual's potential for safely operating a motor vehicle. Driver training programs are currently unregulated, and there is no nationwide policy for quality assurance. The quality of the assessment and training program is dependent upon the education, training, and experience of the evaluator, and the availability of assistive driving equipment.

Physical and driving characteristics of the driver being evaluated are important factors related to an individual's potential to safely operate a motor vehicle. Cognitive and visual tests should be used in evaluating driving potential, as well as physical tests. Physical functional abilities such as range of motion, manual muscle strength, sensation, grip strength, pinch strength, fine motor dexterity, and hand–eye coordination all may be related to driving potential. Driving characteristics must also be evaluated when determining an individual's potential to safely operate a motor vehicle. Throttle force, brake force, steering force, brake reaction time, and steering reaction time are all factors which influence an individual's driving potential.

Driving evaluators rank driving characteristics of greater importance than physical functional abilities. However, they measure physical characteristics more frequently than driving characteristics. This may be because of the additional complexity and equipment required to measure driving characteristics. Most driving evaluators indicate that the ability to drive is best measured in a vehicle during actual driving situations, and hence they do not rely heavily on clinical tests.

Assistive driving equipment vendors and vehicle modifiers are also involved in the process of ensuring consumer safety and satisfaction. The knowledge, experience, and methods these vendors bring to the assessment and implementation process are very important to providing appropriate personal transportation. Some US states and other countries regulate vehicle modifications and assistive driving technology. The National Mobility Equipment Dealers Association (NMEDA) has begun the development of a certification process for assistive transportation technology. The Society of Automotive Engineers (SAE) is developing quality assurance standards to improve the practice of vehicle modification, hand-control design, power-assist vehicle control device design, wheelchair lift design, securement system design, and occupant restraint design.

Vehicle modifications are often paid for by third-party providers (e.g., US Department of Veterans Affairs, State Department of Vocational Rehabilitation, and Workers Compensation). Many vendors are equipped to handle most common van modifications (e.g., lift installation, rewire controls, raise roofs, and drop floors). Fewer vendors are qualified to install zero/reduced effort power vehicle controls.

The varied functional abilities of the consumer are reflected by the amount of custom fabrication and modification of equipment. Vendors report that occupant restraint systems often require modification. However, original equipment manufacturers (OEM) of automobiles do not design their restraint systems to operate with wheelchairs. Nonetheless, occupant restraint systems must be modified or designed to accommodate each individual's needs. Modification or alteration of OEM restraints may compromise their integrity. Lifts, external controls, and six-way power seats have been reported to require the most maintenance. Lifts, external controls, and raised roofs have been reported to be the simplest to install.

Many states rely on independent vehicle modification vendor and equipment manufacturers for recommendations on adaptive driving devices. Most vendors are unable to access the medical appropriateness of specific conversion options, or if the client is cognitively able to process the complex tasks that driving demands. Rehabilitation engineers have been called upon to assist in the development and implementation of systematic procedures for procuring appropriate adaptations for drivers with disabilities.

The independent operation of a motor vehicle can potentially be one of the most hazardous situations that a rehabilitation engineer is faced with. Inappropriate identification and prescription of specific driving devices can jeopardize the safety of the driver and the public.

Before driving for the first time with a disability, and the identification and prescription of an adaptive driving system, each individual should complete a thorough driver evaluation. After having evaluated a person's driving potential and requirements, a rehabilitation engineer should assist with the identification and prescription of adaptive driving equipment and vehicle modifications.

Driver evaluation programs should address driving potential, assistive technology and training required, vehicular modifications required, and OEM options and availability. Neurological and physical ability to operate a motor vehicle must be assessed. The prospective driver must be evaluated in a similar vehicle while performing various driving maneuvers. The results of the dynamic driving test are used to indicate the need for further driver training. Many people require multiple hours of traffic exposure with a trained instructor to develop competency with control interfaces and traffic safety. In some instances vehicle modifications are necessary. Evaluations should include determining transfer potential, appropriate primary and secondary control interfaces, and vehicle access enhancements. OEM options should be selected to minimize conversion requirements. The rehabilitation engineer can facilitate the proper assessment of driving potential of a person with a disability, and recommend the appropriate vehicle modifications and adaptive driving technology.

5.2. SELECTING A VEHICLE

It is often a difficult task to find an automobile which meets the specific needs of a particular wheelchair user; no automobile meets the needs of all wheelchair users. Consumers with disabilities are concerned about cost, maintainability, and handling as are most consumers. However, automotive consumers with disabilities are also concerned about ease of entry, stowage space for the wheelchair, and seat positioning. Reduced size, increasingly sloping windshields, lower roofs, and higher sills of new cars make selecting a new vehicle difficult for wheelchair users. In addition, there are fewer two-door and full-size sedans. The ability to load the wheelchair into a vehicle is essential. Some individuals with sufficient strength and the suitable vehicle are able to stow their wheelchairs inside the vehicle without the use of assistive devices. Many people must rely on an external loading device.

Table 5.1. *Typical ranges of accessibility dimensions for sedans (units are inches).*

Wheelbase	93–108
Door height	33–47
Door width	41–47
Headroom	36–39
Max space behind seat	9–19
Min space behind seat	4–9
Seat-to-ground distance	18–22
Width of door opening	38–51

The selection of the appropriate vehicle should be based upon the client's physical abilities and social needs.

Many people with paraplegia or quadriplegia have had to develop new ways of transferring into their vehicles. Many use passenger-side transfers in the smaller cars, as opposed to driver-side transfers when cars were larger. Lightweight wheelchairs may assist people in stowing their wheelchair in their vehicle. Chairs where the rear wheels can be removed, become much smaller and lighter, making it easier to stow them on the passenger seat or behind the driver's seat.

Another approach some people have used to overcome the problems associated with a smaller car is to use a car-top wheelchair carrying device. These devices lift the wheelchair to the top of the car, fold it, and stow it. They have been designed to work with four-door sedans, light trucks, and compact automobiles.

There are several critical dimensions to an automobile when determining wheelchair accessibility (figure 5.1). The wheelbase of the automobile is often used by auto manufacturers to determine vehicle size (e.g., full-size, mid-size, compact). Rehabilitation engineers should become familiar with various vehicle designs, and their critical accessibility dimensions. After observing various people with disabilities transferring to and from different vehicles, one can gain an appreciation of the accessibility of various vehicles. Typical ranges for passenger vehicles are presented in table 5.1.

The possibility of designing an accessible vehicle has been explored. Many people could benefit from vehicles designed to accommodate the needs of a larger portion of the population, and many people who do not require accessible features may prefer them. Accessible vehicles should be designed to accommodate entry and operation of the vehicle by people with paraplegia, quadriplegia, hemiplegia, amputations, ambulation impairments and without disability. People who are obese or short or tall or pregnant must also be accommodated. The vehicle must be accessible by manual- and power-wheelchair users. The vehicle should be maintenance accessible, and easily maintained. It should have accessible storage, be secure from

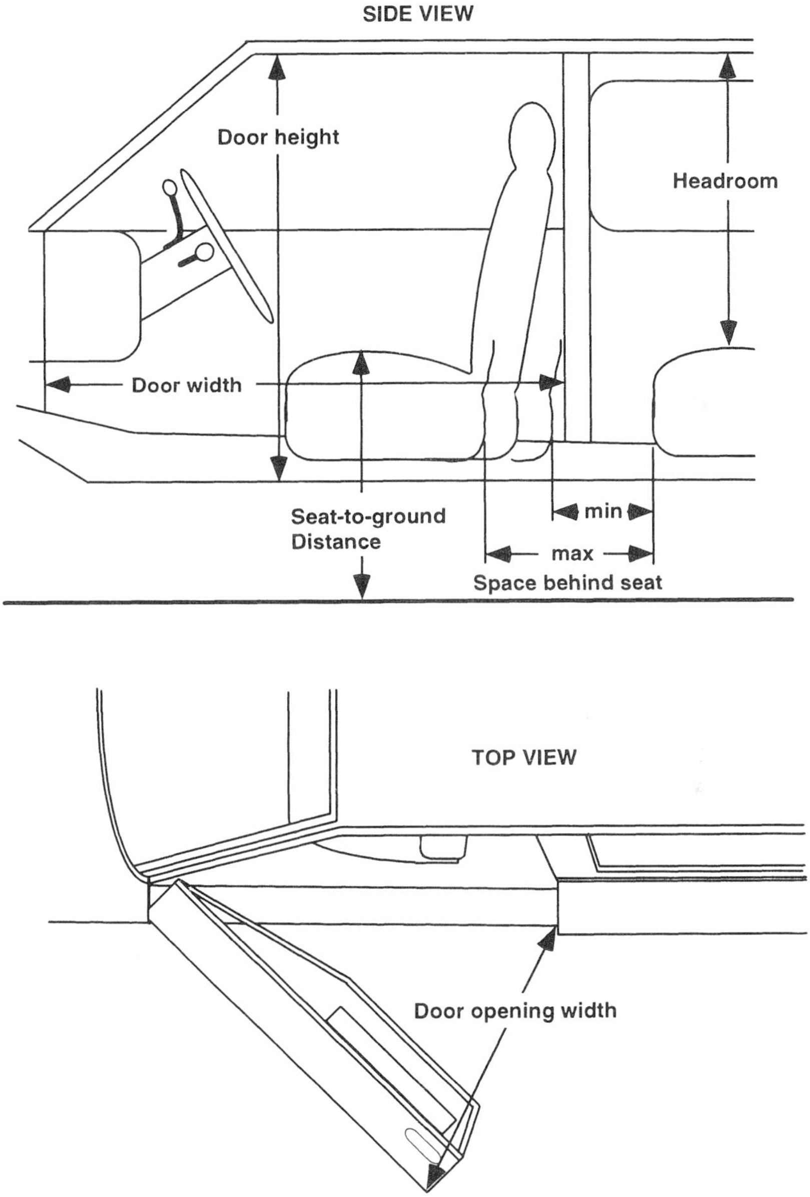

Figure 5.1. *Critical dimensions for determining accessibility of a personal automobile.*

vandalism and theft, and provide clear visibility for driver and passengers. The vehicle styling should be attractive, and be identifiable as an accessible vehicle if desired.

An accessible vehicle may have the appearance of a mini-van with an electronically activated sliding door and automatic drop access floor section (i.e., a built-in lift). Spring-loaded rear seat bases (e.g., stadium- or theater-type seats) would aid wheelchair entry. Window and panel dimensions should conceal the wheelchair when the vehicle is in operation. Accessible vehicles would minimize the expensive modifications required for some people, and would make driving more pleasant and affordable for many others.

5.3. LIFT MECHANISMS

Many wheelchair users who cannot transfer into a passenger vehicle seat, or prefer a larger vehicle, drive vans equipped with wheelchair lifts. There are several types of wheelchair lift devices for vans. The most common are the platform lifts and crane lifts. Platform lifts may use a lifting track, a parallelogram lifting linkage, or a rotary lift. Lift devices are either electromechanically powered, or electrohydraulically powered. Platform lifts have a large flat surface about the size of the wheelchair which ascends and descends from the van's floor level to the ground (figure 5.2). The platform often folds into the side doorway of the van (figure 5.3). Crane lifts, also called swing-out lifts, have a platform which elevates and folds or rotates into the van. Lifts may either be semi-automatic or automatic. Semi-automatic lifts are operated at least partially by hand. In many cases semi-automatic lifts require the user to initiate various stages (e.g., unlocking door, door opening, lowering lift) of the lifting process. Automatic lifts are designed to perform all lift functions. They usually have an outside key-operated control box, or an interior radio-controlled control box. Many lifts fold for storage. When the loading/unloading process is completed, the lift platform is folded parallel the van's sidewalls. Folding may be hand operated or powered.

Lifts may use a variety of power sources. Some lifts use electrohydraulic actuators to lift and fold, with valves and gravity used to lower the lift. Many lifts may be mounted to side or rear doors of full-size vans and side doors of mini-vans. Crane lifts may swing out from a post at the front or rear of the side door. Interlocking mechanisms are available with some lifts to prevent the lift from being operated while the door is closed. This helps to prevent possible vehicle damage. Lifts are generally constructed of steel and aluminum. The platform and other parts which must be lifted are often aluminum.

The Society of Automotive Engineers (SAE) has developed guidelines for the testing of wheelchair lift devices for entry and exit from a personal vehicle. Testing of lift devices is important for consumers, prescribers, and manufacturers. The SAE/ISO-proposed tests establish minimum performance

Figure 5.2. *Fully automatic rear-door van lift mechanism for wheelchair with rider. (Photograph courtesy The Braun Corporation.)*

criteria, and only devices which satisfy all of the tests are considered to be in compliance. The standards are intended to set an acceptable level of reliability and performance for van lifts.

The SAE/ISO procedures consist of four series of tests which include receiving inspection, visual inspection of the installed lift, specification and disclosure requirements, and a series of stress tests. The receiving inspection includes a review and inventory of documents and parts before installation. The visual inspection test reviews the requirements for installation, examination for compliance with design standards, and verification of conformance with dimension standards. Inspectors also look for possible failure points, electrical hazards, lift and vehicle load distribution, vandal protection, and maintenance requirements. The specification tests require

Figure 5.3. *Semi-automatic side-door van lift mechanism for wheelchair with rider. (Photograph courtesy The Braun Corporation.)*

examination of control switches, electrical power requirements, platform tilt angles, finish surface, maximum platform acceleration/deceleration, and safety features. Many lifts include important safety features like threshold warnings, wheelchair retaining devices, and manual back-up operating systems. The stress tests include, in the order performed, an accelerated life-cycle test, a static load test, and a static ultimate load test.

Safety is an important aspect of wheelchair lift device design. Thousands of people rely on their wheelchair lift for access to their vehicle, and they must be able to rely on its safe operation. Occupants must be protected from potential hazards. Lifts must be equipped with safeguards. Lifts should not have unprotected sharp edges or projections. Shear and pinch points must be eliminated or covered from the user. People come in close contact with lift mechanisms, and they are often installed in expensive vehicles, therefore dirty, oily, or greasy surfaces must be isolated from the user, and exposed surfaces of the vehicle.

Wheelchairs must sit on lift mechanisms, and people often stand on them. Platform surfaces must be designed to be non-slip or coated with a non-slip surface to provide adequate tire traction, and shoe traction. Wheelchair retaining devices must also be incorporated to prevent the wheelchair and occupant from rolling off the lift and to keep the lift from moving if not properly interlocked. The platform angle deflection under load must be kept to a minimum to prevent the rider and wheelchair from rolling off the platform, and to prevent unsafe operation. Lift mechanisms and vehicle openings must be integrated so that a wheel or caster cannot slip between the vehicle opening and the lift platform.

The electrical systems of wheelchairs must also be safe and reliable. Electrical hazards which may cause harm to the user must be prevented. The user must be isolated from short circuits in the electrical system. Lifts should be designed to avoid electrical fires and heating. Electrical systems may fail for a variety of reasons, therefore, methods to allow manual operation of the lift for egress must be possible with minimal assistance in case of malfunction or power loss.

5.4. WHEELCHAIR SECUREMENT SYSTEMS

Securement systems are used to temporarily attach wheelchairs to vehicles during transport. Securement systems must be designed with the safety of the wheelchair rider, and the other passengers in the vehicle, in mind. Many power wheelchair users can operate a motor vehicle from their wheelchair, but are unable to transfer into a vehicle seat. Public transportation systems often transport people in wheelchairs, and there must be a means of safely securing the wheelchair. The type of motor vehicle affects the crash pulse, and severity of the collision. Auto safety standards have reduced the number of US automobile accident fatalities despite an increase in the number of vehicles. Engineers, researchers and scientists have been working since 1970 to develop standards for wheelchair securement systems to realize the same benefits. US standards for wheelchair securement devices for privately operated vehicles are the responsibility of the Society of Automotive Engineers (SAE) and International Standards Organization (ISO). Securement standards specify the crash pulse profile, severity of test collision, and a reusable surrogate wheelchair. Securement mechanism, tie-down, pretension, and crash dummy, Hybrid III, positioning are important as well.

The crash pulse determines the severity of the collision of the test sled, and hence real-world. A crash pulse maps the deceleration profile in terms of multiples of the Earth's gravity, of the test system mass (figure 5.4). Accelerometers mounted to the sled are used to record the crash pulse. Securement systems are tested with a surrogate wheelchair at 30 miles per hour (48 +2/-0 kilometers per hour) with a 20g deceleration. Wheelchairs must be safely restrained when experiencing an impact of this magnitude and no part of the wheelchair shall protrude into the occupant space where it might cause injury. Tests are performed with a flat structurally rigid platform with the Wheelchair Tie-Down and Occupant

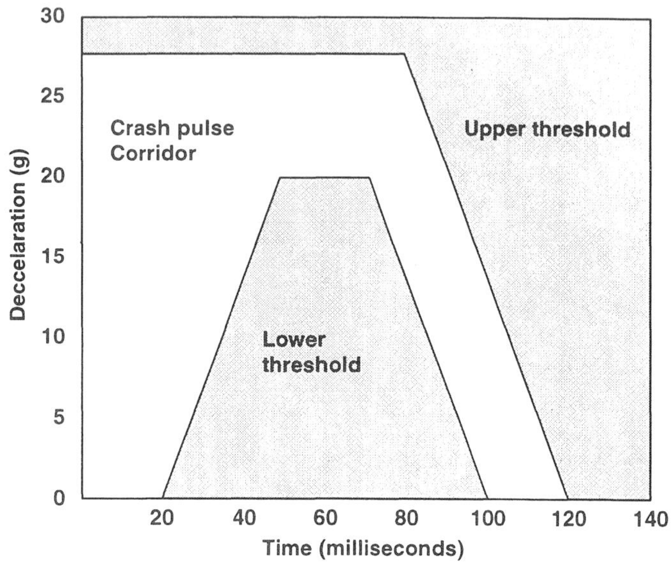

Figure 5.4. *Securement system test crash pulse corridor in terms of deceleration in g's and time in ms.*

Restraint System (WTORS) installed as per the manufacturer's specification. An anthropomorphic test dummy with a minimum total mass of 73.5 kg is used. Preferred anthropomorphic wheelchair test dummies are the Hybrid II, Hybrid III, OPAT, and TNO 10. Dummies must be calibrated at intervals not exceeding 50 tests and prior to the first test following maintenance. A surrogate wheelchair which complies to ISO or SAE specifications is used during testing.

The safety of the wheelchair tie-down and occupant restraint systems (WTORS) is determined by the excursion of the dummy's head and knee relative to the test trolley, which is measured using high-speed photography or video recording. A minimum frame rate of 500 frames per second and a background grid or calibrated displacement measurement system are required. Images must be corrected for parallax error. The excursion of the dummy's head is defined as the distance between the resting position of the head mid-line and the most forward point on the head at peak deflection. The knee excursion is the maximum distance moved by the knee-joint marker. Wheelchair excursions are measured with respect to a marker located on the wheelchair which coincides with the hip joint of the dummy in the resting position. The maximal horizontal excursion of the wheelchair's main axle should be measured for design and evaluation purposes. All excursion measurements are made with respect to the test platform (sled). WTORSs are evaluated on whether the dummy remained in the wheelchair, the wheelchair remained in the test frame, and changes in orientation of the wheelchair during and after the test. Test engineers must be careful to note any separated or fragmented parts as well as changes to any or all adjusters. The functional status of release mechanisms and securing buckles must be assessed post-impact.

WTORS are important for transport of people in motor vehicles while in their wheelchairs. WTORS consist of several components and subsystems. The anchor point is where the center line of the restraint belt passes into the anchor fitting. The final point of attachment on the vehicle is called the anchorage. A common manually operated four-point tie-down system is depicted in figure 5.5. Some tie-down systems are automatic. These systems may be activated when the wheelchairs rolls into position. Vehicle floor modifications may simplify restraining the wheelchair. A dropped floor is used to lower a certain portion of the vehicle floor, whereas a flat floor is used to eliminate uneven surfaces in the vehicle floor. A power-operated platform (power pan) is sometimes used to allow raising and lowering portions of the vehicle floor with respect to the vehicle body. These modifications may facilitate access, improve vision, improve control positioning, and increase headroom. Vehicles often require reinforcement at anchor points for tie-down systems. Some tie-down systems place the wheelchair passenger facing forwards, others sideways, and still others facing rearwards. Many tie-down systems are based upon the four-point tie-down. A four-point tie-down uses four belts attached to the wheelchair at four distinct points and with the opposite ends of the belts attached to four separate points on the vehicle. Wheechair channels may be made by lowering portions of the vehicle's floor to align the wheels of the wheelchair.

Wheelchair tie-downs must be designed so that wheelchairs can be attached and released from the vehicle without the use of tools. Wheelchair tie-down mechanisms must capture the wheelchair in such a way that the wheelchair remains restrained in the event of wheelchair component deformation or if one or more of the tires deflate. Tie-down mechanisms must not deform any part of the wheelchair during normal operation and use. All metal parts should be made corrosion resistant. All parts should be smoothly finished (that is, free from sharp edges, burrs, and irregularities) which may present a hazard or may cause discomfort to vehicle passengers. Components should be designed to minimize damage to passenger's clothing. Designers should also consider emergency egress in the design of their tie-down systems. If air bags are used in the vehicle, their effectiveness must not be reduced by the tie-down or wheelchair occupant restraint system.

If the tie-down system is designed to have passengers in wheelchair ride facing rearwards, head and back supports must be used to protect the occupant during vehicle frontal impact. If the wheelchair rider faces forwards a head support is only required if a backrestraint attached to the vehicle is used instead of the wheelchair backrest. Head and backrest designs must provide users with adequate protection during an impact. Manufacturers

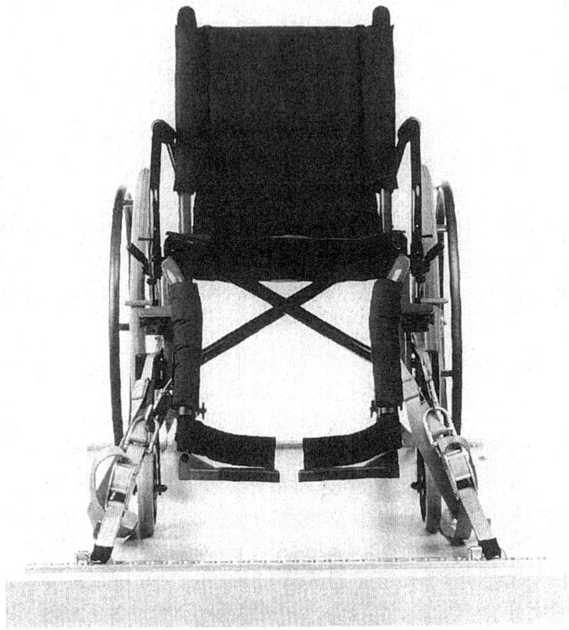

Figure 5.5. *Manually operated four-point wheelchair tie-down system for a motor vehicle*

and designers must clearly identify the types of wheelchairs which are inappropriate for use with their device. WTORSs should incorporate either a means of removing slack or limiting movement of the wheelchair to not more than 25 mm in the horizontal plane when a force equal to the weight of the wheelchair is applied through the rider's resting hip joint position.

5.5. PASSENGER RESTRAINT SYSTEMS

Proper use of lap and shoulder belts is critical to protecting passengers in automobile seats. A similar level of crash protection is required for individuals who remain in their wheelchairs during transportation. Wheelchairs require special consideration when designing passenger restraints. Wheelchairs are flexible, higher than a standard automobile seat, and not fixed to the vehicle. The passenger is restrained using a harness of at least one belt to provide pelvic restraint and two shoulder or torso belts that restrain both shoulders. A head support may also be used to prevent rearward motion of the head during impact or rebound. A three-point restraint is the combination of a lap belt and a shoulder belt (e.g., pelvic torso restraint, lap-sash restraint, lap-shoulder restraint).

5.5.1. Basic laws and concepts of vehicle motion

Newton's laws of motion along with the concepts of work, energy, and the law of the conservation of energy provide a basis for vehicle analysis. Newton's first law states that a body at rest will remain at rest while a body in motion will remain in motion unless acted upon by an external force. A force is the cause or agent that alters the motion of an object and represents as a single entity the cumulative effect of millions of molecular interactions that occur between two bodies. There are two types of forces which can act on objects: body forces which can act from a distance without contact (e.g., gravity, magnetism); and traction forces which are applied to the surface of bodies. Forces are vector quantities represented by magnitude and direction.

Newton's second law defines the relationship between the applied force and a body's motion as a function of duration of force application, change in the object's velocity, and the object's mass. This law states that if force (F) is applied to a body for some period of time (Δt), the body will experience a resultant change in velocity (Δv) equal to

$$\Delta v = \frac{F \Delta t}{M} \tag{5.1}$$

where M is the mass of the body. A more familiar form of this equation is

$$F = Ma \qquad a = \frac{\Delta v}{\Delta t} \cong \frac{\mathrm{d}v}{\mathrm{d}t} \tag{5.2}$$

where a is the acceleration of the body. Newton's third law states that for every force there is an equal and opposite force or reaction force. The concepts of energy and work must also be understood to analyze the motion of vehicles. Energy is the capacity for doing work and overcoming resistance. Work is the transference of energy from one body to another. If an object is displaced along the direction of an applied force, then that force does work on the body. The work done on the body is equal to the magnitude of the force (F) times the distance over which it works (d)

$$W = Fd = \int f(t)d(t)\,\mathrm{d}t = \text{work.} \tag{5.3}$$

There are two types of energy associated with a body: kinetic energy and potential energy. Potential energy is stored within the body. Kinetic energy is associated with the motion of the body

$$KE = \frac{Mv^2}{2} \tag{5.4}$$

where again M is the mass of the body and v is the magnitude of its velocity. The law of the conservation of energy states that the total amount of energy within a controlled volume remains constant unless energy is either transmitted into or out of the control volume. These concepts can be applied to study vehicle dynamics. Newton's first law is in effect when acquiring velocity to move from one place to the next, when stopping, and when parked. Newton's second law states that the velocity acquired by each passenger is proportional to the product of the external forces, the duration for which they are applied, and inversely proportional to the mass of each occupant. The second law implies that to accomplish a rapid change in velocity, a large force must be applied. Vehicles and passengers are deformable objects; however, the capacity to deform does not invalidate the laws of motion.

The concept of deformability does have an affect on energy calculations. For example, if two identical cars were traveling at 50 $km\,h^{-1}$ along the same direction of the same roadway, but decreasing the velocity to zero by different processes: one by hitting a solid wall and the other by applying its brakes (figure 5.6). In both vehicles, the product of the force by the time duration will be identical. The braking car will experience a small force over a long time, while the crashing car will experience a large force over a very short period of time. The braking car translates with respect to the ground and performs work which removes energy from the car. The barrier does not move and therefore does not remove any energy from the car. The kinetic energy of the vehicle and passengers must be released. This energy is absorbed in the crushing of the vehicle and in the passenger restraints.

The occupants of a vehicle must also dissipate the kinetic energy that they have obtained during both braking and crashing. During braking the passengers should experience a large number of small forces acting against their bodies. These forces come from seat friction, foot-to-floorboard contact, and hands to steering wheel contact. During crashing the forces will be higher and must be applied safely. The unrestrained occupant behaves much like the vehicle in the previous example (figure 5.7). In a crash, the vehicle stops abruptly and the passenger, when unrestrained, continues to move forward through the passenger compartment until impacting an interior surface. Although most interior surfaces are padded or contoured to reduce impact injuries, substantial injury and death can result from even low-speed accidents. During impact with an interior surface, the occupant must absorb the total amount of kinetic energy accumulated due to the speed prior to the vehicle impact. This transformed kinetic energy is what results in injury. The introduction of an occupant restraint or air bag significantly alters the sequence of events that the occupant experiences.

Air bags and seat belts have found wide acceptance and application as vehicle restraints. Common belt systems apply their restraint forces to both the torso and pelvis to cause the needed change in acceleration while air-bag systems apply their restraint forces to the upper torso and head. With a belt system that rests on the surface of the body two conditions must be met for the system to be effective: the belt must have a tensile load within it; and the belt must change direction while in contact with the body. A change in direction is required to apply a restraining force (figure 5.8). As the belt

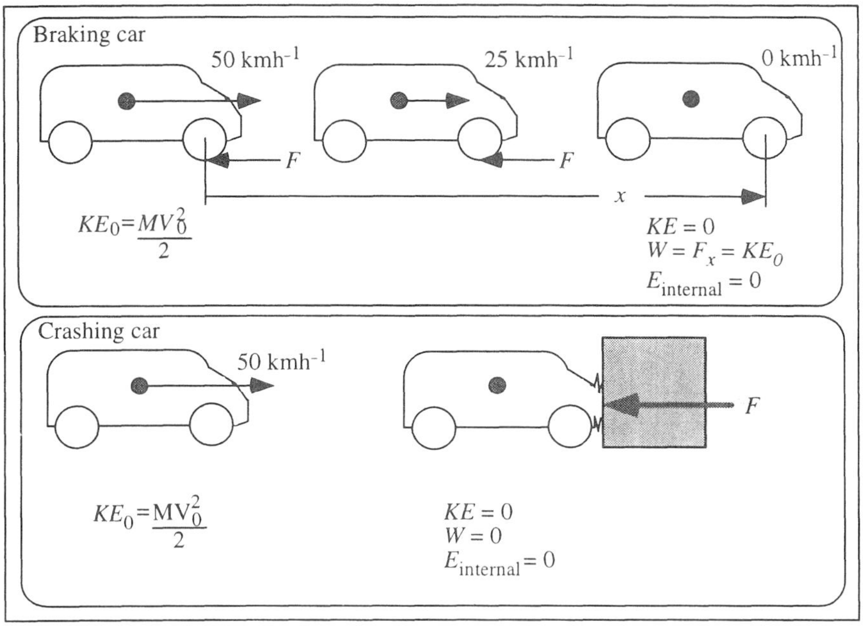

Figure 5.6. *Comparison of work and energy absorbed by an automobile during braking and crashing. During braking, energy is dissipated through performing negative work, while crashing the energy must be absorbed by the internal structure of the vehicle.*

changes direction, the forces applied to the body increase. Some portions of the body are better suited to withstanding external forces. Typically, boney regions can withstand higher loads with less damage than soft tissue. Therefore, the belt must also be guided to change directions so that is rides over the tissue which is most tolerant to the increasing forces. Conversely if the belt is applying loads over areas which cannot tolerate high force, then the belt must be redirected as the force increases.

The passenger seated in a wheelchair must be restrained safely during transport. This requires development and analysis of passenger and wheelchair restraints. The loading points and dynamic behavior of the chair in a crash must be considered. Wheelchairs are designed differently than standard automobile seats, and must be evaluated for their transport suitability.

5.5.2. Restraints for the passenger seated in a wheelchair

The relationship between injury criteria and the mechanics of restraint systems are important to ensure the safety of wheelchair users in motor vehicles. Hip and head deflection are often-used criteria for determining potential injury. The independence of wheelchair users and their ability

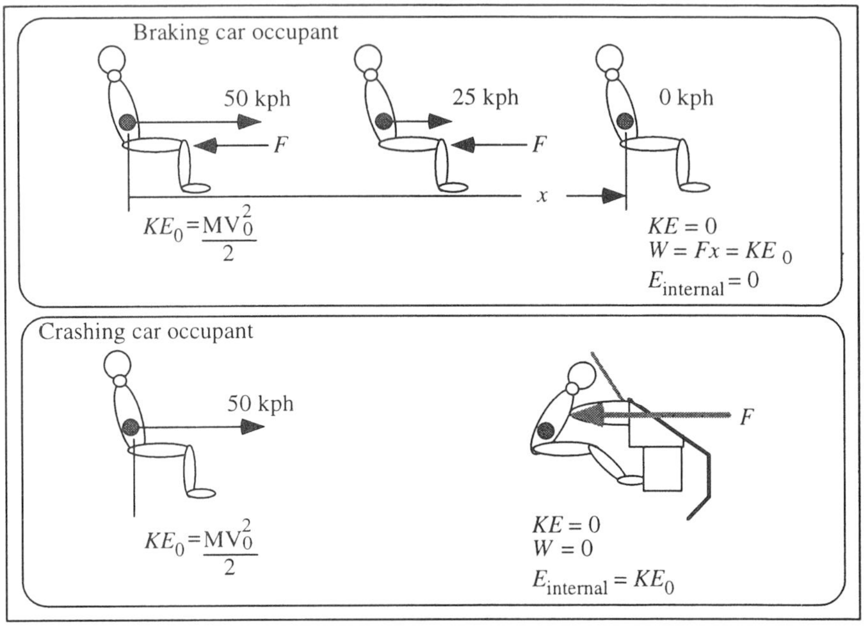

Figure 5.7. *Comparison of work done and energy absorbed by occupant during braking and during a crash. During a crash without restraints, the passenger's body impacts with the interior of the vehicle.*

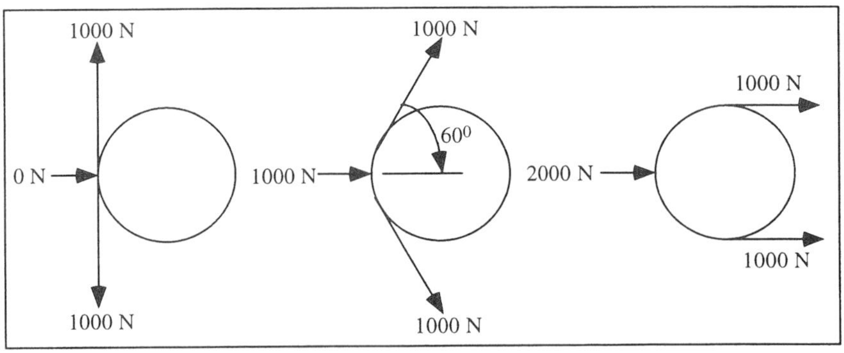

Figure 5.8. *Applied belt force as a function of belt wrap angle. The belt must change direction while in contact with the body to apply force. Body tissue under the belt must be suitable for tolerating belt force.*

to maintain employment is heavily dependent upon their ability to obtain adequate transportation. The automotive industry has invested considerable effort in research and development to protect vehicle passengers. Research is not nearly as extensive for the passenger who remains seated in a wheelchair while traveling. Many wheelchair and occupant restraint systems copy the

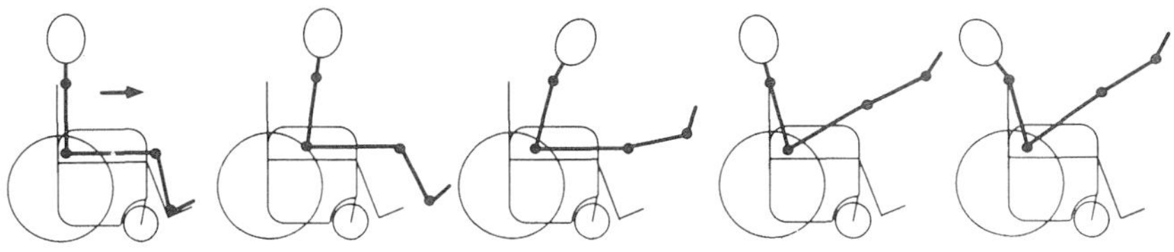

Figure 5.9. *Wheelchair impact kinematics.*

designs used for standard automobile seats. This type of design may not be appropriate.

Crash tests, performed by Adams and Reger (1993), have shown that for 10 or 20g impacts of 100 ms duration people may sustain injuries despite being restrained. When shoulder belts mounted 60 inches above the floor were used to restrain a 50th percentile male dummy it was found that the torso was well controlled, and head and chest excursions were limited. When shoulder belts were anchored 36 inches above the floor they were ineffective in controlling torso movement. Kinematic results and head injury criteria (HIC) can be used to estimate the extent of injury sustained by a human passenger. HIC of 1000 or greater indicate a serious or fatal head injury. Generally, a HIC value approaching or exceeding 1000 is indicative of head impact with some portion of the vehicle interior. The open space typically surrounding a wheelchair user in a public bus precludes impact with the bus's interior. High HIC values may occur when the torso is effectively restrained and there is a high degree of neck flexion. If the chin strikes the chest then there may be an impact great enough to cause head injury. In 1972, the US National Highway Traffic Safety Administration proposed the HIC to assess head injury potential in automobile crash test dummies; the HIC are based upon the resultant translational acceleration. The development of the HIC are described in *SAE Information Report*, J885-APR84 (1984).

If the torso of the passenger is left unrestrained, then there is a risk of impact with the vehicle interior, lumbar fracture from excessive hip flexion, and abdominal injury due to high lap-belt forces. Restraining the torso without compensating for impact dynamics may result in excessive neck flexion and head and/or neck injury. Safe restraint systems must accommodate the large variation in wheelchairs and seating systems used by the public.

Adams *et al* (1992) used a Hyge Impact Simulator to study the kinematics of 10g and 20g impacts over 100 ms. Crash kinematic data were collected for a Hybrid III 50th percentile dummy seated in a wheelchair and restrained with either a lap and/or shoulder belt. Power and manual wheelchairs were tested, as were scooters. They were able to identify four phases, defined by the movement of the hip, neck, back, and head (figure 5.9).

During the linear translation phase, 80 to 100 ms after impact, the dummy moves forward 8 to 11 inches with joint angles remaining constant.

Table 5.2. *Driving situations for evaluation of passenger restraints.*

Normal driving
Non-impact evasive maneuvers
Low-speed impact
High-speed impact

The forward rotation phase is represented by reversal of hip motion, and forward rotation of the trunk, head, thighs, and shins. Maximum forward displacement of the head, chest, and legs occurs during this phase. A properly fitting shoulder belt can minimize excursion of the trunk during this phase. An inclined footrest, such as commonly used on scooters, can reduce leg rotation. Individuals are most likely to incur head and neck injuries during this phase. During the rebound phase, the dummy moves towards its initial sitting position. The movement progresses slowly until the dummy impacts with the backrest. The extended rotation completes the crash kinematics. During this final phase, the dummy continues to move after contacting the seat and backrest which in some instances may cause the neck to hyperextend. These kinematic phases are independent of the type of wheelchair tested.

Many wheelchair users require postural supports to ensure functional stability. The ability of postural supports to provide protection to their users in dynamic transportation environments is important. The US Federal Government regulates transportation of occupants in a passenger vehicle. Federal Motor Vehicle Safety Standards (FMVSSs) must be met before a vehicle is eligible for sale. Regulations exist regarding occupant restraints to protect vehicle occupants during a crash. These regulations do not apply to occupants who remain seated in their wheelchair.

Anterior chest supports and pelvic restraints may be attached to a wheelchair, and remain secured around the user during transport. The presence of these postural supports may increase the individual safety or may increase their risk of injury. Several situations need to studied when evaluating the efficacy of passenger restraint systems (table 5.2).

Modifications of FMVSS regulations regarding static testing of passenger restraints for estimating dynamic crash worthiness can be applied to postural support systems. If these modifications are followed then a pelvic restraint must withstand 5000 pounds force (22.2 kN) and must not extend more than 7 inches (18 cm). A pelvic/torso restraint must withstand a combined load of 6000 pounds force (26.7 kN), or 5000 pounds force (22.3 kN) as the pelvic component and 3000 pounds force (13.4 kN) as the torso portion. The pelvic/torso restraint must not extend more than 10 inches (25 cm).

Postural support mechanisms and restraints experience failures in several locations. Pelvic restraints may fail at the wheelchair attachment points.

Overlapping the webbing ends substantially increases the static strength of the attachment. Most commonly available postural support systems perform adequately during normal driving conditions and evasive maneuvers, but require substantial redesign or modification for 30 mile per hour 20g crash situations. Wheelchair-mounted postural supports which are suitable as occupant restraints would be beneficial to people who require both.

Whiplash constitutes the most prevalent form of injury to vehicle occupants struck from the rear. As the trunk is accelerated forwards through contact with the backrests, the head snaps back and the neck becomes hyperextended. This hyperextension is often followed by hyperflexion as the head rebounds. This hyperflexion is accentuated if the vehicle hits an object as it moves forwards. Research on automotive seat headrests has clearly demonstrated their effectiveness in reducing neck injuries. Some wheelchair users use headrests as part of their seating and postural support system.

Merz and Patrick (1971) recommended that hyperextension of the neck be limited to 80 degrees, and if possible 60 degrees to minimize risk of injury. More recently Foret-Bruno *et al* (1991) studied the frequency and severity of injuries to occupants in rear impact collisions and found that headrests were effective in reducing the risk of cervical injury by 30% for both men and women. They also conducted dynamic tests using Hybrid III dummies to study the influence of horizontal head–headrest distance, headrest stiffness, backrest elasticity, and impact speed on neck measurement loads for automobile seats. The Hybrid III test results of Foret-Bruno *et al* were in agreement with the collision data and showed that the effectiveness of the headrest is dependent on the length of time the head remains in contact with the headrest. As the distance between the head and headrest increases, the shearing force at the neck and the rotation of the head relative to the thorax increase.

US Federal Motor Vehicle Safety Standard 202 specifies requirements for head restraints to reduce the frequency and severity of neck injuries in collisions. Head restraints must satisfy either the dynamic test or the static and dimension tests. The dynamic test requires that head restraints limit angular displacement of the head to 45 degrees with respect to the torso under the influence of an 8g rearward deceleration applied to the seat support structure.

The static test procedure uses a headform to apply a rearward load 2.5 inches (6.4 cm) below the top of the headrest perpendicular to the torso centerline. A 3300 inch pound (373 N m) torque about the seating surface reference point is applied. This corresponds to a headrest load of about 132 pounds (587 N). The horizontal displacement of the headform must not exceed four inches (10 cm) from the torso centerline. The headrest must withstand a static load of 200 pounds force (890 N) or until seat failure. The width of the headrest must be 6.75 inches (17 cm) for a bucket seat.

Many wheelchair headrests contain common system components such as attachment hardware, vertical adjusters, anterior–posterior adjusters, and headrest pads. Wheelchair headrests should be able to withstand 587 N, and dissipate a minimum energy of 29 J for a 9 cm head displacement. Headrests must be able to withstand 890 N without plastic deformation. Many wheelchair headrests can meet these criteria with minimum vertical adjustment. The bending load is concentrated at the point of attachment of the vertical adjustment and the backrest, and this is where most failures occur. Headrests exhibit three common modes of failure: plastic deformation of the vertical adjuster; plastic deformation or failure of the attachment receptacle; and yielding of the backrest. Headrests also show bending or deformation at the bend or angle of the vertical adjuster. The effectiveness of the headrest during rear impact is increased if the head is in contact with the pad at impact. This is the case for many wheelchair users whose headrests are adjusted to support the head in a neutral position.

Restraints should be designed to apply forces to skeletal areas of the occupant. Pelvic restraints should rest low over the pelvic bone and extend downwards and rearwards to the vehicle anchor points. Torso belts should lie across the clavicle, sternum and ribs extending upwards and rearwards to vehicle anchor points. Three-point belts should be designed so that the junction of the shoulder belt and lap belt is near the passenger's hip. Fasteners and straps commonly used for postural support of wheelchair seated people should not be relied upon for occupant restraint and protection in a crash unless they meet with all requirements in the SAE and ISO standards. Webbing should be protected from sharp or rough edges to prevent premature wear.

5.6. AUTOMOBILE HAND-CONTROLS

Hand-controls allow many people the freedom to travel wherever and whenever they please. Hand-controls are made for sedans and vans. Hand-controls are available for automatic and manual transmission vehicles. However, hand-controls for manual transmission automobiles must be custom made. There are also portable hand-controls, and long-term hand-controls. Portable hand-controls are designed to easily attach to most common automobiles with a minimal number of tools. This facilitates using rental or loaner vehicles. Hand-controls are available for either left- or right-hand control.

5.6.1. Manual hand-controls

Many hand-controls are attached to the steering column. Some vehicles have shrouds over the steering column which must be removed or modified before

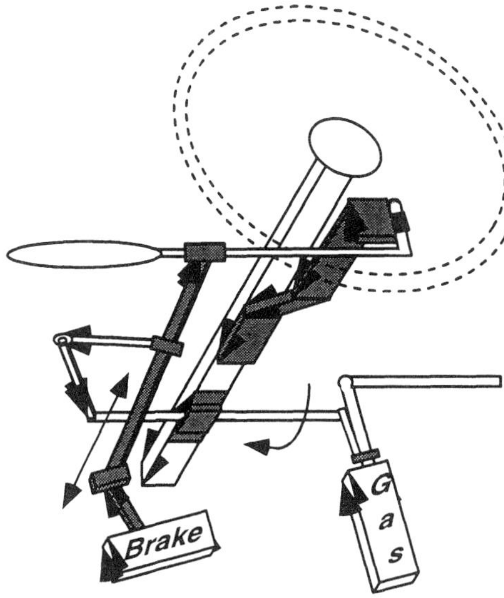

Figure 5.10. *Left-hand control using push for brake and pull for gas.*

installing hand-controls. Hand-controls either clamp to the steering column or are attached to a bracket which is bolted to the steering column or dash, typically where the steering column bolts to the dash. Installation of the hand-control should not interfere with driver safety features (e.g., air bags, collapsible steering columns). Some instances of air-bag inflation have been recorded during installation which may be hazardous to the installer. Some hand-controls are not compatible with air bags. The push rods of the hand-control either clamp directly to the pedals or the levers connected to them.

Most systems activate the brakes by having the driver push forwards on a lever with a hand grip. This allows the driver to push against the back of the seat, creating substantial force, and braces the driver in the event of a collision. The throttle, gas pedal, is operated in a number of ways. Some systems use a twist knob or motorcycle-type throttle. Other systems actuate the throttle by pulling on the brake throttle lever (figure 5.10). Another method is to rotate the throttle–brake lever downwards (i.e., pull the lever towards the thigh at a right angle to the brake application to operate the throttle) (figure 5.11). Using a push-actuated brake and twist-activated throttle allows the driver to simultaneously control throttle and brake.

Manual transmission hand-controls have been designed which allow control of the throttle, brake, and clutch. The clutch and brake require substantial force to activate. Manual hand-controls must be designed to withstand the fatigue stresses induced by repetitively operating the clutch. Manual transmission automobiles can be operated by using a push-activated

Figure 2.2. *Fishbone diagram used to identify factors related to a problem area. Here the example problem of consumer abandonment of assistive technology in a service delivery program is examined.*

brake, and a twist-activated clutch similar to that used on automatic transmission vehicles to operate brake and throttle. A motorcycle-type throttle is mounted on the handle of the brake–clutch activation lever. This system permits simultaneous operation of brake, clutch, and throttle. However, the steering hand must be removed from the steering wheel to operate the shift lever. This may result in unsafe operating conditions.

Many types of handle are used on hand-controls. People sometimes drive for long periods, and their hands may fatigue if the appropriate handle is not installed. Users have different preferences and therefore some experimentation is appropriate. Controls use rounds knobs, oblong knobs, handgrips, forks, and loops. People with limited hand function should use automobiles with power steering, and a fork or loop hand-control. The hand is placed in the fork or loop which is slightly larger than the thickness of the hand.

Adaptations are often made to the steering wheel to improve the ability to perform steering maneuvers. Many people use a rotating steering knob which allows the driver to turn the wheel full circle with one hand. This aids parking and maneuvering in close quarters. People with limited hand function use a fork or loop type handle mounted on a rotating steering knob. Typically, the prongs of the fork or the loop extends upwards from the steering wheel. The hand is placed in the fork from above or the fingers are threaded through the loop reducing the grip force required to steer the vehicle.

It is common to have the same vehicle driven by several people which may require the vehicle to be safely operated with hand-controls and the OEM foot controls. Many hand-controls permit operation of the vehicle via foot controls by operating in parallel with the OEM controls. Care must be taken to ensure that the lever and brackets of the hand-controls do not restrict the driving motions of foot-control drivers.

5.6.2. Automatic (fly-by-wire) hand-controls

Many people have the motor control necessary to operate a motor vehicle, but they do not have the strength required to operate manual hand-controls. Automatic or fly-by-wire hand-controls use external actuators (e.g., air motors, servo mechanisms, hydraulic motors) to reduce the force required to operate various vehicle primary controls. Some primary controls can be purchased from the factory with power assist. Power steering, power brakes, six-way power seats, and power adjustable steering columns can be purchased as factory options on many vehicles.

Power steering employs either a hydraulic actuator or servo actuator to reduce the force required to turn the steering wheel. Power steering systems sense the force applied to the steering wheel by the driver, and the actuator is activated to assist turning the wheels in the desired direction. The sensitivity of the power steering system determines the amount of force required by the operator to turn the vehicle. Power brakes operate on the same principle. Force applied to the brake pedal is sensed by the power braking system, which then amplifies the braking force. Automatic braking systems (ABS) are used on some vehicles to limit the brake pad force. This is because vehicle control and braking are reduced when the wheels begin to slide. ABS minimize the vehicle stopping distance, and provide the driver with greater control of the vehicle by sensing the motions of the wheels and holding them at the brink of skidding.

Six-way power seats are used to provide greater postural support and positioning than standard automotive seats. They can be controlled by a few switches to move fore–aft, incline–recline, and up–down. This allows the user to position the seat for easy entry and exit, and for optimal driving comfort. Power adjustable steering columns also make vehicles more accessible. By using a few buttons the steering column can be tilted upwards or downwards allowing positioning for entry/exit into the vehicle and for optimal driving control.

Custom devices are available for people who require more than the OEM options for power assistance. Microprocessor and electronic technology have dramatically changed how motor vehicles are designed. Many functions of an automobile are controlled electronically or with electromechanical–electrohydraulic controls. This change in vehicle design has made a wide variety of options available for people who require advanced vehicle controls.

Modifications can be made to the electrohydraulic control system used for power braking to provide near zero-effort braking. This can be accomplished by adding a larger actuator to the OEM power braking control system or by adding an electromechanical, electrohydraulic, or electropneumatic actuator to the OEM braking system. The actuator can receive either a mechanical input (e.g., lever, push-rod) or electronic input (i.e., signal proportional to some input device). Zero-braking control systems permit the brakes to be operated with a joystick or slide bar (i.e., linear potentiometer, linear encoder). Many automobiles use electronic fuel injection. Electronic fuel injection systems convert the position of the accelerator pedal to a serial digital signal which is used by a microcontroller to inject the optimal fuel–air mixture into the automobile at the proper time during the piston stroke. The electronic signal for the accelerator position can be provided by another control device (e.g., joystick, slide-bar).

Power steering is a common option on many vehicles. Near zero-effort steering can be accomplished by adding an electromechanical, electrohydraulic, or electropneumatic actuator to the power steering control system. The actuator can be driven by an alternative control device. Some vehicles use an absolute encoder or potentiometer attached to the steering wheel shaft to provide an input signal to the power steering control system.

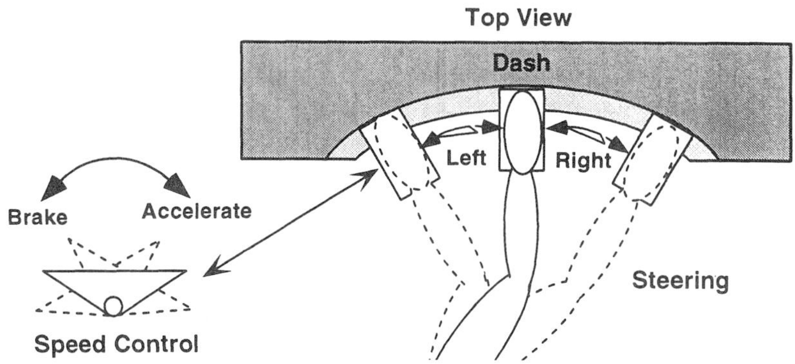

Figure 5.12. *Pivoting fly-by-wire hand-control system.*

A pivoting hand-control system provides operation of motor vehicles for many people (figure 5.12). A pivoting hand-control operates by having the driver pivot one hand forwards (with fingers) for braking, back for acceleration (with palm of hand), steering is controlled by sliding the hand-control left and right along a curved tray (dash) by pivoting the arm and shoulder. The resistance of the controls must be adjustable to each individual's abilities. Tremor damping can also be included for some individuals. Secondary controls (e.g., turn signals, warning signals, lights) can be operated by the other hand. Some functions can be placed on the sliding handle if necessary.

5.7. CONTROL OF SECONDARY FUNCTIONS

The placement and variety of controls for secondary automobile functions such as turn signals, environmental controls (e.g., heater, air conditioner), auxiliary devices (e.g., radio, door locks, dome lights) can impose barriers to driving a motor vehicle for people with mobility impairments. Greater control of secondary automobile functions can be achieved by using a local interface unit which is accessed via an appropriate control strategy. Commercial systems exist which allow primary functions controls (e.g., steering, braking) to be adapted to incorporate secondary functions. Customized secondary control systems have also been developed. Strategies developed for alternative computer access can be applied to secondary vehicle controls using a microcontroller and interface hardware.

Modular dash systems can allow the user flexibility to select and position controls and accessories for ready access. Dash components can be deleted, added, or moved as necessary. By finding the proper placement of secondary controls many people's needs can be accommodated.

Other people require more specialized secondary control systems. Systems may use scanning-based strategies to provide control over the approximately 50 secondary functions of an automobile. Scanning is an effective method only if the user has the ability to customize the scanning sequence and strategy. Secondary controls may be operated through switches mounted to a central panel. The panel is placed within the user's functional reach space. Switches must be chosen which allow control of the secondary system and which can be easily operated by the user. Another strategy is to combine the use of switches and scanning. The combination of switches and scanning allows frequently used secondary controls to be simply operated by a switch, whereas less frequently used functions are scanned hierarchically. The combination also reduces the number of switches, making operation of secondary controls less confusing.

In some cases it may be best to use a selection or scanning screen and annunciator or status screen for secondary controls. The annunciator screen displays the status of each of the secondary controls, or in some cases only active controls are displayed. Typically icon-based status symbols are most appropriate for the driving environment. Modularity and flexibility of the central interface units of secondary control systems are very important. Systems must accommodate a variety of input devices (e.g., joystick, switch, voice) and output devices (e.g., digital switches, relays). Standardized interface hardware and protocols should be used whenever possible.

FURTHER READING AND REFERENCES

Adams T C and Reger S I 1993 Factors affecting wheelchair occupant injury in crash simulation *Proc. 16th Ann. RESNA Conf. (Las Vegas, NV, 1993)* pp 80–2

Adams T C, Sauer B and Reger S I 1992 Kinematics of the wheelchair seated body in crash simulation *Proc. RESNA Int.'92 (Toronto, Ont, 1992)* pp 360–2

Babirad J, Mendelson L and Flis M 1990 Appropriate development of vehicle modification protocol for maximizing traffic safety of disabled drivers *Proc. RESNA 13th Ann. Conf. (Washington, DC, 1990)* pp 15–6

Barclay H W, Torrible B and Kohar R 1992 *Transportation for the Severely Disabled* Advanced Mobility Systems Corporation, 621 Justus Drive, Kingston, Ontario, Canada, K7M 4H5

Beck K 1990 Wheelchair lifts for vans *Paraplegia News* **44** 34–5

Brubaker C E, Digges, Sprigle S H and Thacker J G (ed) 1991 *Transportation I: A State of the Art Conf. on Personal Transportation for Persons with Disabilities* (Charlottesville, VA: RESNA)

Cardoso F L and Thacker J G 1993 Performance testing of wheelchair lifts for personal licensed vehicles *Proc. 16th Annual RESNA Conf. (Las Vegas, NV, 1993)* pp 74–6

EEs' tools and toys 1993 Human interfaces in a hurry *IEEE Spectrum* **30** 49

Falanga M, Chang J H and Baeckstroem M 1985 Automobile design for physically disabled and able-bodied populations *Proc. 8th Ann. RESNA Conf. (Memphis, TN, 1985)* pp 15–7

Falanga M 1984 Independent transportation for the physically handicapped *Automotive Eng.* June/July 39–41

Farr S, Mifsud M, Nauman S and Young M 1991 The development of a multi-function vehicular control interface unit for quadriplegic drivers *Proc. 14th Ann. RESNA Conf. (Kansas City, MO, 1991)* pp 409–10

Fisher W E, Seeger B R and Svensson N L 1987 Development of an Australian standard for wheelchair occupant restraint assemblies for motor vehicles *J. Rehab. Res. Dev.* **24** 23–34

Foret-Bruno J Y, Dauvilliers F and Tarriere C 1991 Influence of the seat and headrest stiffness on the risk of cervical injuries in rear impact *Proc. 13th ESV Conf. (Paris, 1991)*

Gianutsos R, Campbell A, Beattie A and Mandriota F 1992 The driving advisement system: a computer-augmented quasi-simulation of the cognitive prerequisites for resumption of driving after brain injury *J. Assistive Technol.* **4** 77–86

Hinkenkemper J A M, Hekstra A C and Veenbaas R 1993 The development of simulator tests to assess the crashworthiness of car adaptions *Proc. 2nd Eur. Conf. Adv. Rehabilitation Technol. (Stockholm, 1993)* p 3.2

Karg P E and Sprigle S H 1993 Determining the transportation safety of chest supports and pelvic restraints *Proc. 16th Annual RESNA Conf. (Las Vegas, NV, 1993)* (National Science Foundation) pp 83–5

LeGuellec P and Mollendorf J C 1991 Wheelchair tie-down mechanism system: securely fastens a wheelchair in a van *NSF 1991 Engineering Senior Design Projects to Aid the Disabled* pp 114–5

Metz H J and Patrick L M 1967 Investigation of the kinematics and kinetics of whiplash *Proc. 11th Stapp Conf.* SAE paper #670919

Metz H J and Patrick L M 1971 Strength and response of the human neck. SAE paper #710855

Morris B, Sprigle S H and Karg P E 1992 Assessment of transportation technology: Survey of consumers *Proc. RESNA Int.'92 (Toronto, Ont., 1992)* pp 349–50

Nicole C, Ross T and Richardson S 1993 Identification and grouping of requirements for drivers with special needs *Proc. 2nd Eur. Conf. Adv. Rehabilitation Technol. (Stockholm, 1993)* pp 3.1

Pilkey W D, Kang W and Kitis L 1993 Crash simulations for wheelchair-occupant systems in transport *Proc. RESNA 16th Ann. Conf. (Las Vegas, NV, 1993)* pp 280–2

Robbins S 1990 Hand controls *Paraplegia News* **44** 41–3

Roush L and Koppa R 1992 A survey of activation importance of individual secondary controls in modified vehicles *J. Assistive Technol.* **4** 66–9

Shaw G, Lapidot A, Scavnicky M, Haxel B, Bolton J and Klopp G 1993 Testing procedures for wheelchair securement system standards *Proc. 16th Ann. RESNA Conf. (Las Vegas, NV, 1993)* pp 77–9

Smart E D, Bluj A, Dezern J, Pearson W and Pendleton W 1990 A portable driver measurement system *Proc. RESNA 13th Ann. Conf. (Washington, DC, 1990)* pp 13–4

Sprigle H S, Morris B O and Karg P E 1992 Assessment of transportation technology: survey of driver evaluators. *Proc. RESNA Int.'92 (Toronto, Ont, 1992)* pp 351–3

Sprigle S H, Morris B O and Karg P E 1992 Assessment of transportation technology: survey of equipment vendors. *Proc. RESNA Int.'92 (Toronto, Ont., 1992)* pp 354–6

Thacker J G and Cardoso F L 1992 Performance testing of wheelchair lifts for personal licensed vehicles *Proc. 15th RESNA Int. Conf. (Toronto, Ont., 1992)*

Veterans Administration Standard Design and Test Criteria for Safety and Quality of Automatic Wheelchair Lift Systems for Passenger Motor Vehicles VAPC-A-7708-3, Veterans Administration Prosthetic Assessment and Information Center, 104 South Gay Street, Baltimore, MD, 21202

Yoganandan N, Skrade D, Pintar F A, Reinartz J and Sances A 1991 Thoracic deformation contours in frontal impact *Proc. 35th Stapp Car Crash Conf. (San Diego, CA, 1991)* pp 47–64

CHAPTER 6

WHEELCHAIR SAFETY, STANDARDS AND TESTING

6.1. INTRODUCTION

Standards have been developed for numerous products due to the concern of consumers, manufacturers and government for safety and the availability of quality products. Some industries regulate themselves, while others are regulated by state and/or federal agencies. Recently, there has been movement towards international standards because of a growing global economy.

Wheelchair standards have been developing since the mid-1960s. The US government took an interest in developing quality and performance standards for purchasing wheelchairs. The US Veterans Administration, now the US Department of Veterans Affairs, which is one of the world's largest purchasers of durable medical equipment took the US lead in standards development. Early wheelchair standards were based upon simple tests to determine static strength, reliability, controllability and energy consumption. Wheelchair standards have continued to evolve ever since.

The American National Standards Institute (ANSI) and the International Standards Organization (ISO) have been working with a multidisciplinary committee of RESNA on the development of wheelchair standards.

Standards consist of two primary components: tests and normative values. Tests and normative values are required to objectively evaluate products.

6.2. STANDARD TESTS

If a product is to bc ccrtificd by a national or international organization, it must undergo and pass a series of standardized tests. A variety of tests are used to evaluate the safety and durability of devices. Tests are designed so that similar results can be achieved at any recognized test facility. The requirements for testing are quite rigorous to ensure repeatability.

Manual and power wheelchairs are tested. The tests vary slightly depending on which type of product is being tested. The tests for all wheelchairs can be grouped into three categories: (1) stability tests (static and dynamic), (2) strength tests (static, impact and fatigue) and (3) energy consumption tests.

Tests consist of various processes and circumstances to which the wheelchair is subjected, to determine numerical values describing the wheelchair's function. These values are of little practical use without information about how similar wheelchairs perform on the same tests and during actual use.

6.3. NORMATIVE VALUES

Once a wheelchair has been tested, the performance of the wheelchair must be evaluated against normative values for similar products. These values may be derived by manufacturers through testing several models of the same wheelchair or by testing a product line. Normative values for use by consumers, purchasers and manufacturers must be developed by testing several wheelchairs from a number of different manufacturers. Normative values must be developed for products designed for use in similar situations.

Wheelchairs are broadly divided into two categories: manual and power. Within these two categories there is further division. Manual wheelchairs are divided into active duty lightweight, active duty normal and depot or institutional wheelchairs. Active duty lightweight wheelchairs are designed for demanding and physical users, these include ultralight wheelchairs and sports wheelchairs. Active duty normal wheelchairs are designed for moderate daily use. Depot or hospital wheelchairs are primarily designed for temporary, short-term or indoor use.

Power wheelchairs are also divided into three groups: active indoor and outdoor use (class 3), active indoor with some outdoor use (class 2) and primarily indoor use (class 1). Standardized tests vary somewhat for the different classes of power wheelchairs.

Normative values should differ for the various types of wheelchairs, although all wheelchairs should attain some minimum level on each test. The development of normative values follows the development of standardized tests. Normative values are determined by applying the standardized tests to current products. Several groups throughout the world are working to gather enough information so that normative values may be developed for different wheelchairs with the various standardized tests.

6.4. STATIC STABILITY

A simple measure of how stable and secure a wheelchair is during normal activities of daily living is to determine static stability parameters.

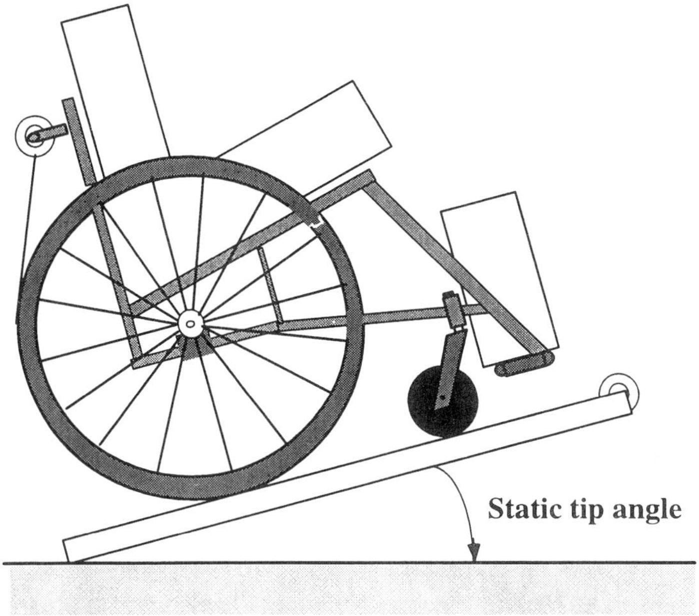

Figure 6.1. *Static tip angle is measured by slowly raising a platform with the wheelchair loaded with the appropriate sized dummy. When both uphill wheels lift off the platform the angle is recorded.*

Wheelchairs are tested on a simple tilting platform (figure 6.1). A wheelchair with a person or appropriate test dummy is rolled onto a tilting platform. Once in place, the incline of the platform is slowly increased until the wheelchair tips or slides. If the wheelchair slides, then the wheelchair is restrained by either placing a block behind the rear wheels (40 mm high) or by using a strap attached to the platform which wraps around the rear wheels and attaches to the wheelchair backrest. The latter method does not significantly alter the axis of rotation, whereas the former does. The appropriately loaded wheelchair is tilted facing up the incline until an ordinary piece of paper can just pass under the front wheels without them turning. The tip angle is recorded to the nearest degree. The test is performed with the rear wheels locked and unlocked.

The tip angle is also measured in a similar manner with the loaded chair facing down the tilt platform and facing to the side of the tilt platform. These tests help to determine the static stability of the wheelchair while the person performs various activities of daily living. Some typical values are given in table 6.1. The static tip angle for eight manual wheelchairs using three load cases (55 kg person with paraplegia, 100 kg ambulatory person and a 100 kg ISO fatigue dummy) were measured. Measurements were made with three different restraints: (1) the wheelchair's wheel locks, (2) a block

Table 6.1. *Static tip angle, in degrees, with loaded chair facing up the inclined plane, M=mean, SD=standard deviation.*

	Strap		Brake		Block	
	M	SD	M	SD	M	SD
55 kg person with paraplegia	10.47	2.181	11.38	1.784	20.08	3.104
100 kg ambulatory person	11.47	1.993	11.39	1.957	18.66	3.685
100 kg ISO	7.31	2.590	8.30	2.796	14.38	4.991
dummy	7.31	2.590	8.30	2.796	14.38	4.991

behind the rear wheels and (3) a strap fixed to the rising edge of the platform whose other end, after being wrapped around the rear wheels, was attached to the backrest of the wheelchair. Only backward tip angle was measured and the axle was in the furthest most rearward position in all cases. Static tip angle was measured when a standard piece of paper could pass under the front wheels without turning them. The angle was measured using a machinist protractor and level. All of the wheelchairs used in this study had seat widths in excess of 42 cm, hence a 100 kg dummy was appropriate.

A three-factor analysis of variance, ANOVA, was performed on the data (load device, chair type and test condition). Scheffe *post-hoc* tests were performed to determine factors related to significant results. A significance level of $p < 0.05$ was set, *a priori*.

There were no significant differences ($p > 0.05$) in tip angle found between the types of wheelchairs, i.e., depot versus active. There were significant differences ($p = 0.0095$) found between the type of load device. The data for the 55 kg person and the 100 kg person were significantly different ($p < 0.05$) from that of the ISO dummy, with the dummy being the more conservative measurement. There were no significant differences ($p > 0.05$) found between the data for the two people.

There were significant differences ($p = 0.0001$) found between test conditions. Both the tests using the wheelchair's wheel locks and the straps differed significantly ($p < 0.05$) from the blocks. However, no significant difference ($p > 0.05$) was found between the results using the wheelchair's wheel locks or the strap method. Results of using the three test conditions were significantly different for each of the load devices, ISO dummy ($p = 0.0026$), 55 kg person ($p = 0.0001$) and the 100 kg person ($p = 0.0001$).

The results indicate that there is no difference between the static tip angle when measured using the wheelchair's wheel locks or when using a flexible strap around the rear wheels. The current test procedure using the block behind the rear wheels, when the wheel locks slip, biases the results in favor of wheelchairs which require the block. The block nearly doubles the static tip angle. The standard should be changed to use the strap or flexible membrane method instead of the block.

The static tip angles measured with the ISO dummy were consistently lower than they were for the human subjects. The wheelchair static tip angle using the dummy was about 75% of that for the human subjects. Because the dummy yields more conservative estimates of a wheelchair's static tip angle, it should be used when making these measurements rather than a person.

The depot and active wheelchairs with the rear axle in the furthermost back position yielded similar results for static tip angle. This suggests that manufacturers of active duty wheelchairs design their wheelchairs to incorporate the more conservative rear axle position of depot wheelchairs and then allow greater forward adjustment as the user becomes more skilled.

6.5. A GEOMETRIC APPROACH TO STATIC STABILITY

The geometric approach to stability analysis is primarily a qualitative method of examining the static stability of a vehicle. This approach should assist the reader who is not mathematically inclined in developing a more intuitive sense for the notion of stability of complex systems.

When a wheelchair is motionless on a flat surface, there are as many points of contact with the surface as there are wheels. If adjacent contact points are connected with lines, the footprint is constructed. The footprint of a three-wheeled chair is a triangle, whereas the footprint of a four-wheeled chair is a rectangle.

For a wheelchair to be statically stable (i.e., when the chair and rider are not moving) the center of gravity of the rider/wheelchair system must remain within the footprint. If the rider/wheelchair system is tilted on a platform, then the perpendicular projections of the points of contact of the wheels with the platform onto the floor form the new footprint. One might imagine a rider/wheelchair system sitting on a glass platform tilted with respect to the floor, a carpenter's square is placed on the floor directly below the point of contact of each wheel with the glass platform and a corresponding point on the floor. The points on the floor will define the new stability footprint. If the platform were tilted 90 degrees with respect to the floor, the resulting footprint would be a line (i.e., this is similar to balancing a bicycle while stopped). The point is that as the rider/wheelchair system are tilted the footprint becomes smaller and the system becomes more apt to fall. The tilting of the platform is analogous to a hill. Thus when on a hill the rider/wheelchair system is more apt to fall. This is because the footprint (i.e., the region in which the center of gravity must remain) is reduced.

The degree of static stability is related to the area of the footprint. All other factors being equal, which is seldom the case, the larger the area

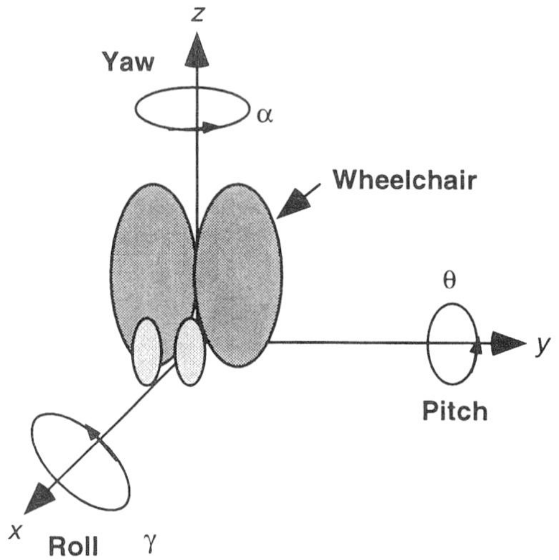

Figure 6.2. *Coordinate system for wheelchair stability analysis.*

of the footprint the greater the stability. One might interpret this to mean that if the area of the triangular footprint of a three-wheeled chair were made the same as the area of the rectangular footprint of a four-wheeled chair the chairs would be equally stable. This may not be true. If all of the other rider/wheelchair dimensions were equal and the center of gravity were located in the geometric center of each respective footprint, the chairs would not have the same degree of stability in all directions. This is because the distance of the center of gravity from each edge of the respective footprints cannot be equal for three and four wheeled chairs as one is a triangle while the other is a rectangle.

6.6. STABILITY WITH ROAD CROWN AND INCLINATION

The effects of road crown and road inclination can be examined using coordinate transformations from the wheelchair coordinate frame to the world coordinate frame. The coordinate definitions are given in figure 6.2. The influence of gravity on the stability of the racing wheelchair is only dependent upon the wheelchair's orientation, hence the coordinate transformations are purely rotations. Road crown and inclination alters the projected height of the wheelchair/pilot system center of gravity and also the projected width of the wheelbase.

The coordinate transformation matrices are defined below.

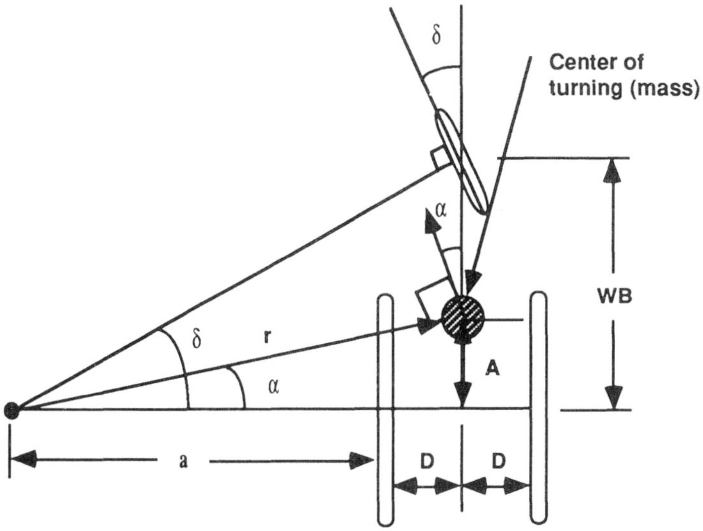

Figure 6.3. *Top view schematic diagram of turning geometry for stability analysis of steered wheelchairs and scooters.*

$$R_{x,\gamma} = \begin{bmatrix} 1 & 0 & 0 \\ 0 & c_\gamma & -s_\gamma \\ 0 & s_\gamma & c_\gamma \end{bmatrix} \qquad R_{z,\alpha} = \begin{bmatrix} c_\alpha & -s_\alpha & 0 \\ s_\alpha & c_\alpha & 0 \\ 0 & 0 & 1 \end{bmatrix}$$
$$R_{y,\theta} = \begin{bmatrix} c_\theta & 0 & s_\theta \\ 0 & 1 & 0 \\ -s_\theta & 0 & c_\theta \end{bmatrix}. \tag{6.1}$$

Angle γ defines yaw (road crown), angle θ defines pitch (incline), angle α defines roll (the angle that the chair turns from the center-line of the road). For this case, the order of the transformations is pitch, roll, yaw. Hence, the above matrices are multiplied and transposed (to determine the inverse) in the specified order and the resulting relationship is given by

$$R^{\text{World}}_{\text{Chair}} = R^0_1 = \begin{bmatrix} c_\alpha c_\theta & s_\alpha c_\theta & -s_\theta \\ -c_\gamma s_\alpha + c_\alpha s_\gamma s_\theta & c_\gamma c_\alpha + s_\alpha s_\theta s_\gamma & s_\gamma c_\theta \\ s_\gamma s_\alpha + c_\alpha s_\theta c_\gamma & -s_\gamma c_\alpha + s_\alpha s_\theta c_\gamma & c_\gamma c_\theta \end{bmatrix}. \tag{6.2}$$

The projections, Lp and Dp, of the height of the center of gravity (L) of the wheelchair rider system and the distance (D) from the point of contact with the road of the outermost wheels to the center of gravity onto the road surface must be determined (figure 6.3).

The origins of the world coordinate system ($x_0 y_0 z_0$) and chair coordinate system ($x_1 y_1 z_1$) are located directly below the center of gravity in the plane of the bottom of the wheels. The vectors that define L and D in the chair coordinate system are given by

$$\boldsymbol{L}_1 = \begin{bmatrix} 0 \\ 0 \\ L \end{bmatrix} \quad \text{and} \quad \boldsymbol{D}_1 = \begin{bmatrix} 0 \\ D \\ 0 \end{bmatrix}. \tag{6.3}$$

The definitions for L and D in the world coordinate system with respect to the chair coordinate system are given

$$\boldsymbol{L}_0 = R_1^0 \boldsymbol{L}_1 = \begin{bmatrix} l_x \\ l_y \\ l_z \end{bmatrix} \quad \text{and} \quad \boldsymbol{D}_0 = R_1^0 \boldsymbol{D}_1 = \begin{bmatrix} d_x \\ d_y \\ d_z \end{bmatrix}. \tag{6.4}$$

The projections of L and D onto the $x_0 y_0$-plane are the l_x, l_y and d_x, d_y components of $\boldsymbol{L}_0$ and $\boldsymbol{D}_0$

$$\boldsymbol{L}_p = \begin{bmatrix} l_x \\ l_y \end{bmatrix} \quad \text{and} \quad \boldsymbol{D}_p = \begin{bmatrix} d_x \\ d_y \end{bmatrix}. \tag{6.5}$$

The vectors describing L and D in the world coordinate frame can be calculated using equations (6.2) and (6.3)

$$\begin{bmatrix} l_x \\ l_y \\ l_z \end{bmatrix} = \begin{bmatrix} -s_\theta \\ s_\gamma c_\theta \\ c_\gamma c_\theta \end{bmatrix} L \quad \text{and} \quad \begin{bmatrix} d_x \\ d_y \\ d_z \end{bmatrix} = \begin{bmatrix} s_\alpha c_\theta \\ c_\gamma c_\alpha + s_\alpha s_\theta s_\gamma \\ s_\alpha s_\theta c_\gamma - s_\gamma c_\alpha \end{bmatrix} D. \tag{6.6}$$

The projections of L and D onto the $x_0 y_0$-plane are given by l_x, l_y and d_x, d_y of equation (6.6)

$$\begin{bmatrix} l_x \\ l_y \end{bmatrix} = \begin{bmatrix} -s_\theta \\ s_\gamma c_\theta \end{bmatrix} L \quad \text{and} \quad \begin{bmatrix} d_x \\ d_y \end{bmatrix} = \begin{bmatrix} s_\alpha c_\theta \\ c_\gamma c_\alpha + s_\alpha s_\theta s_\gamma \end{bmatrix} D. \tag{6.7}$$

6.6.1. *Fixed dynamic analysis of racing wheelchair roll stability*

The matrix approach derived in this chapter presents a general fixed dynamic approach to wheelchair stability analysis. The component of the projection of the vector $\boldsymbol{D}$ (rotated) in the world coordinate frame which is orthogonal to z_0 is required. By selecting the appropriate transformation matrix, this is simply the magnitude of the projection of $\boldsymbol{D}$ onto the $x_0 y_0$-plane which is given by

$$D_p = D\sqrt{(s_\alpha c_\theta)^2 + \left(c_\gamma c_\alpha + s_\alpha s_\theta s_\gamma\right)^2}. \tag{6.8}$$

Equation (6.8) can be represented in more compact form as a dot product

$$D_p = D \cos \tilde{\beta} \quad \text{where} \quad \tilde{\beta} = \cos^{-1}\left(\sqrt{(s_\alpha c_\theta)^2 + \left(c_\gamma c_\alpha + s_\alpha s_\theta s_\gamma\right)^2}\right). \tag{6.9}$$

To determine the roll stability of a wheelchair about its outermost wheels, the component of the vector (rotated) defining the height of the center of gravity in the chair frame which is orthogonal to the z_0 axis and parallel to the vector $\boldsymbol{D}$ in the chair coordinate frame is required. This can readily be determined by using the fact that $\boldsymbol{D}$ and $\boldsymbol{L}$ are orthogonal in the chair coordinate frame. Hence

$$L_p = L \sin \tilde{\beta} = L \sin \left[\cos^{-1} \left(\sqrt{(s_\alpha c_\theta)^2 + \left(c_\gamma c_\alpha + s_\alpha s_\theta s_\gamma\right)^2} \right) \right]. \quad (6.10)$$

The equations for the critical velocity for three- and four-wheeled wheelchair cases are given by

$$v_3 = \left[\frac{rg}{L} \left(\frac{D_p(WB_3 - A) - L_p WB_3}{WB_3} \right) \right]^{\frac{1}{2}} \quad (6.11)$$

$$v_4 = \left[\frac{rg}{L} (D_p - L_p) \right]^{\frac{1}{2}}. \quad (6.12)$$

In equations (6.11) and (6.12), r is the radius of the turn, g is the acceleration due to gravity, v is the forward velocity of the wheelchair/rider system, WB_3 is the wheelbase of the three-wheeled chair and A is the distance from the center of gravity along the center-line of the wheelchair to a line connecting the rear axles. On an inclined road with crown, equations (6.8) and (6.10) are used directly in the critical velocity equations. Other situations are special cases of this case

$$\tilde{\beta} = \cos^{-1} \left(\sqrt{(s_\alpha c_\theta)^2 + \left(c_\gamma c_\alpha + s_\alpha s_\theta s_\gamma\right)^2} \right). \quad (6.9)$$

When these values are substituted into equations (6.11) and (6.12) for the critical velocities, the results are

$$v_{4c} = \left[\frac{rg}{L} (D_p - L_p) \right]^{\frac{1}{2}} \quad (6.13)$$

where

$$L_p = L \sin \tilde{\beta} \quad \text{and} \quad D_p = D \cos \tilde{\beta}$$

$$v_{3c} = \left[\frac{rg(D_{p3} - L_{p3})\sqrt{D^2 + WB_3^2}}{LWB_3} \right]^{\frac{1}{2}} \quad (6.14)$$

where

$$L_{p3} = (LWB_3 \sin \tilde{\beta}) / \sqrt{D^2 + WB_3^2}$$

$$D_{p3} = (D(WB_3 - A) \cos \tilde{\beta}) / \sqrt{D^2 + WB_3^2}$$

therefore

$$v_{3c} = v_{4c}\sqrt{\frac{D_p(WB_3 - A) - L_pWB_3}{(D_p - L_p)WB_3}}$$

$$WB_3 \gg A \quad \Rightarrow \quad v_{3c} \approx v_{4c}.$$

The relative fixed dynamic stability of three- and four-wheeled pilot/wheelchair systems follows directly from equation (6.14). In this most general case, for fixed dynamic analysis, equivalence is achieved by lengthening the wheelbase of the three-wheeled pilot/wheelchair system ($WB_3 \gg A$) or by moving the center of gravity closer to the rear wheels. However, as A approaches zero, yaw stability is reduced, making it difficult to control the direction of the chair.

6.7. IMPACT STRENGTH TESTS

Impact strength tests are used to determine the strength of the wheelchair and its components under conditions that simulate typical usage. All wheelchairs undergo some occasional mechanical stress due to normal use. The user may hit a bump or curb, which should not, under normal circumstances, cause the chair to fail. People may be lifted while seated in their wheelchair or they may be inadvertently dropped into their wheelchair or with the wheelchair. While these occurrences may not happen daily, they are important in determining the required static strength of the wheelchair and its components.

There are two basic types of tests performed to measure static strength: (1) static stress tests and (2) impact tests. Static stress testing is intended to determine whether the wheelchair can withstand the minimal static load prescribed. The minimal static load is representative of the static strength required of the wheelchair and its components under load levels that may occur during actual use. Impact tests are used to evaluate whether the wheelchair and its components are capable of withstanding occasional impact loading as may occur when bumping a curb or when being dropped.

The static strength of components is also important. The wheelchair's armrests, footrests, wheelie-bars and push-handles all may experience significant stress. The premature failure of any component may negatively affect the overall performance of the wheelchair. Therefore, adequate strength without excess weight must be ensured. Components of wheelchairs may be dropped, pulled on, pushed against, or otherwise stressed during use.

6.7.1. *Static stress tests*

Various components of the wheelchair are tested to determine if they exceed standards for minimum static strength. Armrests, footrests, wheelie bars and push-handles must all exceed static strength tests. Each of these components must be tested with loads to simulate them being pulled upon or pushed against. The components or their attachments must withstand the specified load without yielding in order to pass. Sometimes it is desirable to test some components until failure to determine the component's yield strength. These tests may be performed in any convenient order.

6.7.2. Impact tests

Wheelchairs experience a number of different types of impacts during their useful life cycle. These impacts are simulated during testing in order to evaluate whether the wheelchair and its components meet or exceed minimum standards. The wheelchair seat, backrest, and various components are tested.

The seat is tested by dropping a soccer ball filled with lead shot onto the seat from a specified distance (figure 6.4). This simulates someone being dropped onto the wheelchair seat. The backrest is tested by swinging a weight pendulum into the backrest from a prespecified height. The seat and backrest are visibly inspected after each test and any damage is noted. No damage that will inhibit normal use of the chair is permitted. The pushrims of the propulsion wheels must withstand impact from a pendulum dropped from a specified height. This is similar to the impacts that may be experienced by hitting doorways, walls, or other wheelchairs.

People must lift wheelchairs to place them in vehicles, to stow them away or for a number of other reasons. Inevitably, wheelchairs will be dropped at some time during their lives. The wheelchair must be capable of withstanding such impacts. During testing, the unladen wheelchair must be dropped so that the front wheels hit first and again so that the rear wheels hit first. Components are also tested to determine if they can withstand being dropped. The footrests and armrests must be dropped from a specified height and not have their function inhibited. The loaded chair is also tested. This test can be performed in two ways. The chair loaded with the appropriate fatigue test dummy (50, 75 or 100 kg) is dropped at a specified height or the height to yield a failure is determined.

The entire wheelchair may be rolled into a curb or other obstacle. Wheelchairs must be tested to determine their ability to withstand rolling impact. Therefore a loaded dummy is rolled into a rigid curb at $1\ \mathrm{m\,s^{-1}}$. The chair is angled to hit one of the front casters first. International Standards Organization (ISO) standard ISO-7176-08 specifies the procedures and test equipment specific to static, impact and fatigue testing of wheelchairs. A section of this standard describes the pushrim impact pendulum. During normal wheelchair use it is not uncommon for the pushrims of the wheelchair to experience impacts due to hitting stationary objects (e.g., doorways, walls) while the user propels the wheelchair. The same is true for casters and footrests. Impact tests are designed to ensure that a wheelchair and its

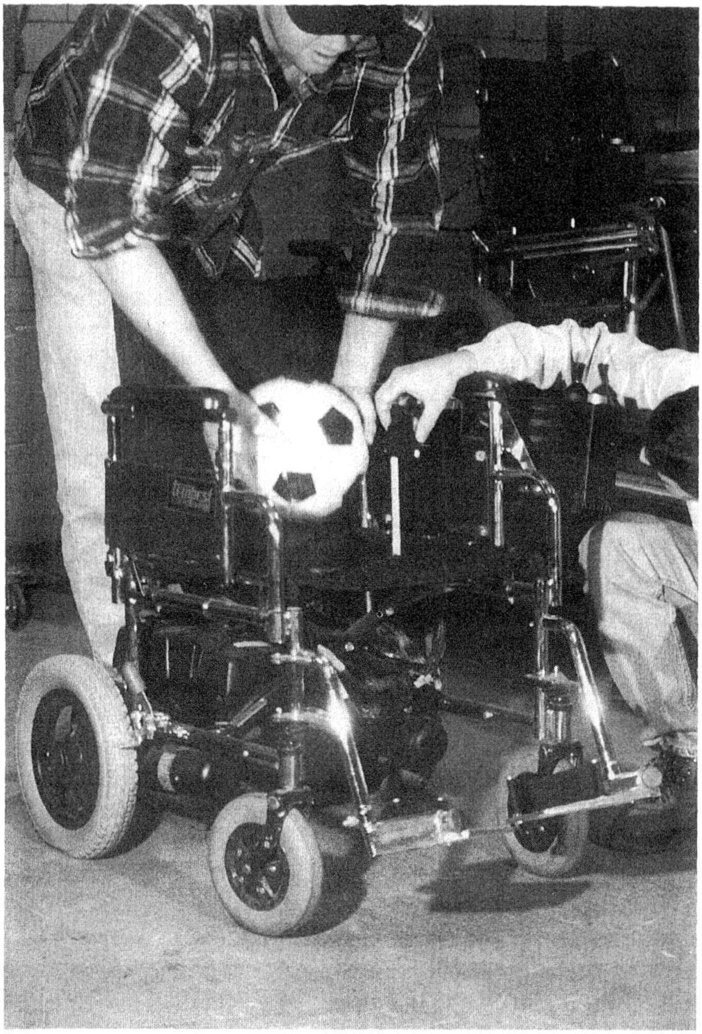

Figure 6.4. *Test laboratory scientists performing a seat impact test with a 25 kg lead-shot-filled soccer ball. This test evaluates the strength of the seating surface and seat material. (Photograph by R A Cooper.)*

components can withstand reasonably large impacts infrequently. Pushrim impacts are tested by using a pendulum designed for that purpose, whereas casters and footrests are tested by impacting a standard curb at 1 $m\,s^{-1}$. Although all three tests have similar purposes, two different pieces of test apparatus are used. In addition, there is some concern that the curb impact test may not be reliable. Use of the pushrim impact test pendulum for all impact tests may simplify the test procedures and improve test reliability.

Figure 6.5. *Caster impact test to determine the strength of the caster assembly. A 10 kg pendulum is allowed to swing freely into the wheelchair caster. The impact is aligned so that the wheel is hit squarely at the pendulum's center of percussion. The centerline of the caster is aligned at 45 degrees to the wheelchair centerline prior to testing. (Photograph by R A Cooper.)*

In addition, the number of personnel required to conduct the test can be reduced.

Pendulum impact performed using an ISO pushrim impact test pendulum must be at the center of percussion. The pendulum must aligned to hit the caster (figure 6.5) or footrest (figure 6.6) squarely. The angle of the pendulum should be measured to the nearest degree. The wheelchairs may need to be elevated from the floor so that the pendulum can strike the casters properly. Wheelchairs should be struck at 45 degrees with respect to the frame, but in line with the right caster. Wheelchairs should be struck on the right footrest in the sagittal plane and in the frontal plane. Each wheelchair should complete three trials at each angle. Strikes can be performed in order of increasing angle. The rear axles of each wheelchair should be placed in their most statically stable position. The dummy should be seated properly in the chair as per ISO 7176-07. The types of chairs, their footrest type and caster type must be recorded.

(a)

(b)

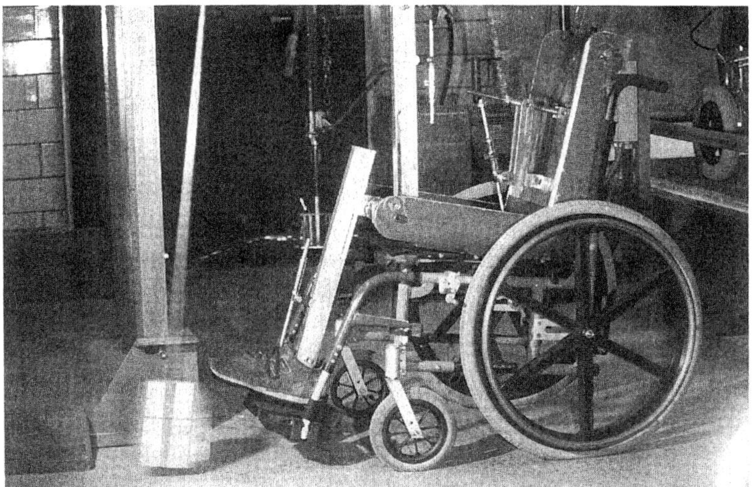

Figure 6.6. *(a) Footrest side impact test. The 10 kg wheelchair test pendulum is aligned to hit the front outside edge of the footrest at 90 degrees to the wheelchair's center line. (b) Footrest front impact test. The 10 kg wheelchair test pendulum is aligned to hit the outside edge of the footrest in line with the wheelchair's centerline. (Photograph by R A Cooper.)*

Test results indicate that the standard deviation in maximal resultant acceleration is about 25% between trials for each wheelchair using the curb impact method, whereas the standard deviation is about 10% using the pendulum method. This indicates that the pendulum method may be more repeatable. Differences in impact strength between wheelchairs can be detected by the pendulum impact tests, which indicates that these tests may be useful in evaluating wheelchair quality. Wheelchairs have experienced failures during the pendulum impact tests with a 60 degree release angle. The curb impact and pendulum impact tests are different. Video analysis has shown that there is substantial movement of the dummy upon impacting a curb. Often, there is little or no movement when the chair is impacted by the pendulum. However, the strength of the footrest and casters can be evaluated using either method. The pendulum impact tests are simpler to perform and require fewer personnel. This will make them appealing to test laboratories.

The curb impact test is not very stress inducing for the wheelchair. Failures are rarely seen with the curb impact test. However, wheelchair users report such problems. The pendulum impact tests shows consistently higher accelerations. Of course, the speed at impact with the curb could be increased. An angle of about 60 degrees is recommended for caster impacts and 45 degrees for footrest impacts when using the pendulum. Footrests should have a lower impact strength because these components are easily replaced and their failure may prevent damage to the wheelchair and user in some instances. It is important that people apply the impacts in a consistent manner. The center of percussion of the ISO Impact Pendulum must be used as the contact point for all tests.

6.7.3. *Forward impact stability*

Wheelchair users often cross door thresholds or hit small obstacles (e.g., sidewalk cracks, small stones) which may block or stall the front casters. When the front casters are stopped or substantially slowed, a moment is imparted on the wheelchair and rider which may cause the rider/wheelchair to flip forwards. The component of the wheelchair user's velocity orthogonal to the door threshold or other obstacle is used in the analysis of forward impact stability. It is assumed that both front casters hit the obstacle at nearly the same time. When a wheelchair/rider system hits an obstacle it undergoes an impulse at the point of contact of the front casters and the obstacle. The wheelchair may then rotate about the axles of the front casters. Taking moments about the axles of the front caster, the moment of the impulsive force is zero and angular momentum is conserved (figure 6.7).

The equation for impulsive motion is given by

$$I_0\dot{\theta} = Mvy \tag{6.15}$$

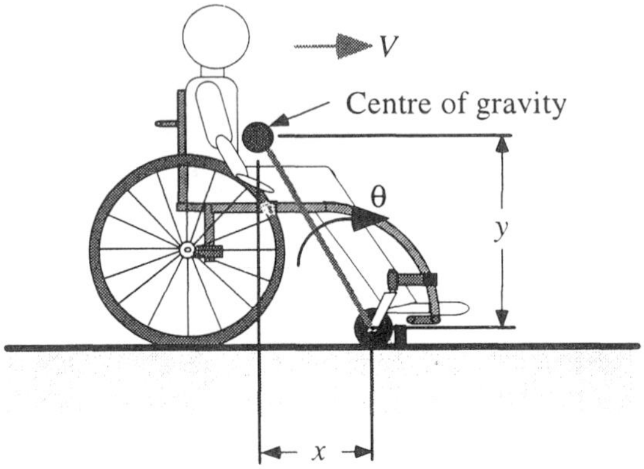

Figure 6.7. *Simple sagittal-plane forward-roll stability model of a wheelchair and rider.*

where M is the total mass of wheelchair and rider, I_0 is the moment of inertia of the total mass about the axles of the front casters, v is the forward velocity of the wheelchair/rider when they first strike the obstacle, y is the height of center of gravity above the axles of the front wheels, and θ is the angular velocity of the wheelchair and rider.

For the wheelchair and rider to overturn, the center of gravity must be raised above the front axles of the wheelchair

$$\Delta h = \left(\sqrt{x^2 + y^2} - y\right) \tag{6.16}$$

where x is the horizontal location of the center of gravity from the axles of the front wheels.

If the wheelchair/rider are to overturn, the kinetic energy remaining after the impulse must be sufficient to lift the center of gravity for the system through the distance given above. This condition is given by

$$\tfrac{1}{2} I_0 \dot{\theta}^2 \geqslant Mg\Delta h = Mg\left(\sqrt{x^2 + y^2} - y\right) \tag{6.17}$$

where g is the acceleration due to gravity.

Equations (6.15) and (6.17) can be combined to yield an equation that defines the critical velocity for flipping a wheelchair and rider about the front axles

$$v^2 \geqslant \frac{2gI_0}{My}\left(\sqrt{1 + \frac{x^2}{y^2}} - 1\right). \tag{6.18}$$

This means that for any particular wheelchair and rider, a minimum forward velocity required for overturning can be calculated. While it is realized that a more comprehensive model could be developed to describe this type of overturning, this model gives a good general estimate. The assumptions made during the calculations in the regression analysis combined with this method will probably result in errors of less than 10%.

6.8. FATIGUE STRENGTH TESTS

Fatigue testing is used to determine the durability of wheelchairs and their components under a large number of low-level stresses that can have a cumulative effect during the life of the wheelchair, typically three to five years. During fatigue strength testing, wheelchairs must be equipped for normal use, and be thoroughly inspected prior to and after testing. Multiple wheelchairs may be tested; however, all must undergo the complete fatigue testing. If wheelchairs have adjustable components, the test engineer must consider the effects of the full range of adjustability. Additional testing is at the discretion of the test engineer.

6.8.1. Double-drum fatigue testing

Failure can occur in wheelchairs and components due to large numbers of repeated stresses as may occur when riding over small obstacles (e.g., door thresholds, cobblestones, etc) or over uneven terrain (i.e., those that impart twisting moments).

The ISO double-drum test is designed to test and/or evaluate the durability of a manual wheelchair under dynamic loading (figure 6.8). The test requires that the wheelchair be placed upon two rollers, the front wheels on one roller and the rear wheels on the other roller. Prespecified straps of steel are bolted to the rollers to simulate bumps and the rollers are driven at a speed equivalent to 1 $m\,s^{-1}$. The test is run until a major structural failure occurs, or until the number of cycles specified by the manufacturers are exceeded. The greater the number of cycles the wheelchair is able to withstand the more durable it is presumed to be. A wheelchair must withstand 200 000 double-drum cycles to be in compliance with ISO standards.

Wheelchairs come in a variety of different designs and are constructed from a number of materials. The ISO double-drum test only examines the occurrence of catastrophic failures due to bumps (impulses) at a fixed speed. Different wheelchair designs may react differently to different types of disturbances and potential failures may go undetected. A double-drum tester must be capable of accommodating a wide variety of sizes and types of wheelchairs.

The wheelchair sits atop of four rollers; the rollers for the front wheels share a common shaft and the rollers for the rear wheels share a common shaft. The front wheels rotate 5% faster than the rear wheels so that over a complete testing cycle the wheelchair will not always receive the bump or undulation at the same time. This helps to excite more modes of the wheelchair and to minimize creating flat spots on the wheels. Each wheel rides on a separate roller that can turn either eccentrically or concentrically about its axle. This hub design allows the amplitude of the undulations to be changed quickly and easily. Using four separate rollers, one for each wheel,

Figure 6.8. *Wheelchair double-drum tester for measuring durability of manual and powered wheelchairs and scooters, when exposed to large numbers of small bumps. (Photograph by R A Cooper.)*

ensures that the magnitude of the undulations remains constant during testing if the wheelchair moves laterally.

Tests indicate that certain frame designs experience greater stress during different test scenarios than others. Rigid-frame wheelchairs, a design common to sport and ultralight models, are more susceptible to bumps (impulses) than are folding models. This is because the folding models are often flexible enough to conform to the bump. However, rigid-framed wheelchairs are not as affected by undulations as are folding wheelchair frames. Again the folding wheelchairs tend to follow the undulations by twisting and bending, whereas the rigid frames ride over the undulations. Tests indicate that rigid-framed wheelchairs may perform better on tests involving undulations than bumps, whereas the opposite is true for many folding-wheelchair designs.

6.8.2. *Curb drop testing*

It is not uncommon for wheelchair users to propel their wheelchairs off curbs, which when repeated can result in structural failure. This test was designed to simulate conditions which occur when a wheelchair and occupant free-fall from a specified height in a normal attitude (i.e., they must start the fall horizontally).

Any apparatus which provides free-fall for the wheelchair and dummy to a solid flat surface is acceptable. The wheelchair must start its fall horizontally and be free to rotate. The wheels must rotate to prevent creating flat spots and unrealistic loading of the wheelchair.

There are two methods currently employed to test for this standard: (1) the wheelchair is placed upon a large metal plate which is lifted and then accelerated away from the wheelchair, and (2) the wheelchair is lifted by cables attached to the dummy at one end and a carriage at the other and the chair is lifted as the carriage rises and then drops as the carriage is propelled downwards. The latter method requires a minimum number of constraints on the wheelchair (figure 6.9). The wheelchair is loaded with an appropriate fatigue test dummy (50, 75, or 100 kg) for 1/30 of the number of cycles specified for the double-drum tester, 6666 for compliance with ISO standards. Upon completion of the testing the rolling resistance is measured at about 0.5 $m\,s^{-1}$.

6.8.3. *Fatigue test results*

Results from fatigue testing of several wheelchairs on ISO double-drum and curb drop testers using a 100 kg dummy show the need for testing and for applying this information to wheelchair design (tables 6.2 and 6.3).

6.8.4. *Fatigue testing of common depot and rehabilitation wheelchairs*

Twelve manual wheelchairs (six identical depot wheelchairs and six identical rehabilitation wheelchairs) commonly purchased by third-party providers were tested using ANSI/ISO double-drum and curb drop testers. All of the wheelchairs were folding models, and all were tested to failure. The depot wheelchairs were placed on the double drum tester for 10 000 cycles and then moved to the curb drop tester for 350 cycles. This process was repeated in sets of 10 000 and 350 until the wheelchair broke or was permanently deformed. The rehabilitation wheelchairs were placed on the double drum tester for 100 000 cycles and 3500 cycles for the curb drop tester. Previous experience with these types of chairs led us to choose the number of cycles in each set so that both types of chairs would experience about the same

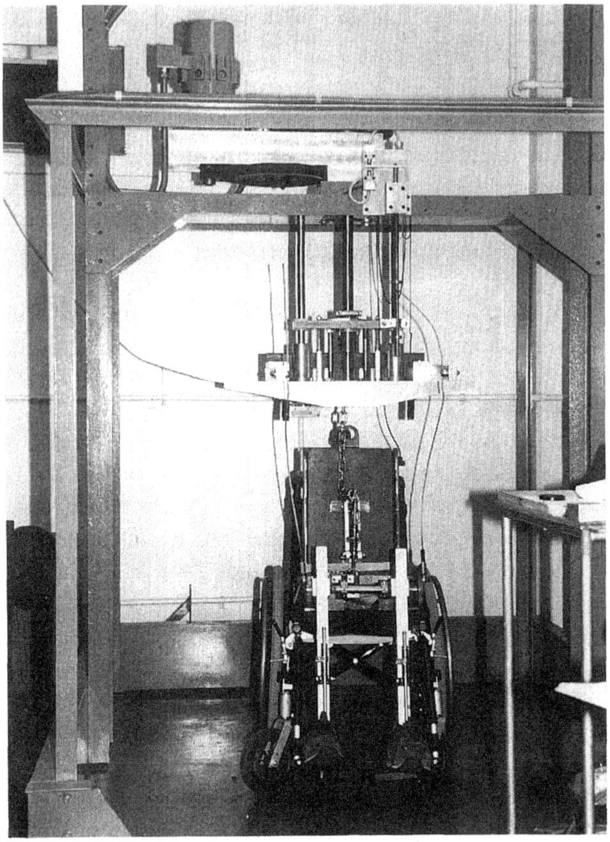

Figure 6.9. *Wheelchair curb drop tester for measuring the durability of manual and powered wheelchairs and scooters. (Photograph by R A Cooper.)*

percentage of double-drum and curb drop equivalent cycles during their lifetime. The results of the testing are presented in table 6.4.

The standard dummy was a version of the ANSI/ISO wheelchair test dummy called out in WC-11 of these standards. The contoured dummy was developed by Cooper *et al*. The most interesting result of this study is the fact that t-tests revealed that rehabilitation wheelchairs lasted significantly ($p < 0.05$) longer than depot wheelchairs. Perhaps more important to the prescribers and third-party payers, rehabilitation wheelchairs cost approximately twice as much as depot wheelchairs yet last more than ten times as long on average ($647\,961/44\,435 = 14.6$). This indicates that it would be more cost effective to purchase rehabilitation wheelchairs and not depot wheelchairs as is common with current practice. An additional rehabilitation wheelchair with pneumatic eight inch front casters was tested; all of the other wheelchairs had eight inch solid casters and this wheelchair

Table 6.2. *Fatigue test results for manual wheelchairs tested with a double-drum tester.*

Cycles	Comments
CHAIR 1	
20 133	Front caster alignment was adjusted. Casters had moved in excess of 15 degrees.
20 186	Left front caster spindle failed. Right front caster fell off upon removal from machine.
CHAIR 2	
129 688	No apparent damage to wheelchair or components.
CHAIR 3	
35 000	Cross (X) brace noticeably bent. The member extending from the upper right to the lower left of the chair was distorted by about one inch from one end to the other. The other cross brace was distorted about 0.25 inch.
42 600	Catastrophic failure occurred. The frame distortion led to overloading the right front caster. The right front caster subsequently failed. Both the rim and the tire were destroyed in this failure. The frame in the area of the right front caster was also severely distorted. The frame cracked at the intersection of the backrest vertical and horizontal member. The crack was presumably caused by a combination of the oscillating load from the dummy and the previous distortion of the frame in other areas. The cross members which make up the X-brace were further bent, the worst case being 1.5 inches of deflection from end to end. The vertical support member of the footrests was bent about 0.5 inch from top to bottom.
CHAIR 4	
30 991	The pushrim on the right wheel became detached from the wheels due to vibration. All of the screws which hold the pushrim in place had disengaged. The left wheel pushrim was in a similar condition, but the screws had only loosened. The left side of the frame had numerous cracks which included the complete severing of one section of the frame used in the folding mechanism. The right side of the frame also exhibited several cracks. The backrests where they intersect the vertical frame supports failed. The lower side frame members failed where they intersect with the X-brace members. The failures caused complete separation on the left side. On the right side the backrest failure caused complete separation. The X-brace members of the frame were found to be bent. The deviation was approximately 1.5 inches from end to end. Some of the deflection was initiated during static impact testing, but was amplified during fatigue testing.
CHAIR 5	
50 260	Left caster bearings failed. This caused the left front fork to pull out of the caster housing. The chair then rotated forward on the double-drum tester. Further frame damage was caused by forward rotation.
CHAIR 6	
38 682	Indications are that the X-brace on the right side failed as did the seat back. These failures led to numerous other structural failures.

Table 6.2. *continued.*

Cycles	Comments
CHAIR 7	
60 605	Rear pair of seat screws broke and were replaced at 13 624, 22 482 and 32 797 cycles. Both footrests failed. Obvious structural failure to footrest. Frame cracked at footrest attachment points.
CHAIR 8	
100 000	Bearings in the front caster housings were damaged. Swivel resistance was drastically increased and one or more bearings were flattened to yield an indexing effect.
CHAIR 9	
100 000	Bearings in the front caster housings were damaged. Swivel resistance was drastically increased and one or more bearings were flattened to yield an indexing effect. Caster housings no longer perpendicular to rolling surface (right displaced 4 degrees, left displaced 1 degree).

Table 6.3. *Fatigue test results for manual wheelchairs tested with a curb drop tester.*

Cycles	Comments
CHAIR 2	
554	Right front caster came loose and was adjusted.
1864	Left rear wheel not turning freely.
3238	Rear wheels show signs of splaying.
4232	Test completed.
CHAIR 8	
3500	Significant splaying developed in rear wheels, about 0.75 inch from top of armrests to lowest point on frame. Approximately 0.5 inch of toe-out developed in rear wheels. There was a weld fracture on the right hinge of the folding mechanism (fracture was between top of the hinge member and the main frame).
CHAIR 9	
3500	Significant splaying developed in rear wheels, about 0.75 inch from top of armrests to lowest point on frame. Approximately 0.5 inch of toe-out developed in rear wheels. Beam attached to footrest was bent down about 0.5 inch.

lasted over 2 000 000 total cycles without failure. Examination of table 6.4 reveals that the rehabilitation wheelchairs tended to experience component failures (figure 6.10) whereas the depot wheelchairs tended to experience frame failures. ISO testing indicates that design features can be related to fatigue test results and durability. However, rider comfort and wheelchair function need to be considered as well as wheelchair durability when suggesting design changes.

Table 6.4. *Results of comparison study for rehabilitation and depot wheelchair durability.*

Classification	Dummy type	Double drum	Curb drop	Total cycles[a]	Failure
Rehabilitation	Standard	400 000	10 890	726 700	Footrest
Rehabilitation	Contour	300 000	9060	571 800	X-brace
Rehabilitation	Standard	610 000	21 060	1 241 800	Caster Spindle
Rehabilitation	Contour	100 000	2291	168 730	Footrest
Rehabilitation	Standard	400 000	10 500	715 000	Caster Spindle
Rehabilitation	Contour	253 733	7000	463 733	Caster Spindle
Depot	Contour	13 175	350	23 675	Caster mount
Depot	Standard	59 785	1 750	112 285	Side frame
Depot	Contour	19 676	350	30 176	Side frame
Depot	Standard	20 002	350	30 502	Side frame
Depot	Contour	40 001	1 050	71 501	Caster mount
Depot	Standard	15 129	350	25 629	Caster mount

[a] Total cycles = (double-drum tester revolutions) + 30 (curb drop tester drops).

6.9. FINITE-ELEMENT MODELING APPLIED TO WHEELCHAIR DESIGN/TESTING

Finite-element modeling (FEM) can be used to optimize the design of and to analyze complex structures. FEM can also be used to analyze structures made of composite materials. FEM has been applied to wheelchair design to a limited extent. Previous work has shown the value of FEM to wheelchair design and optimization. However, there is a lack of data concerning actual road loads experienced by users and simulated road loads from ISO fatigue testing. Models have been developed for power-wheelchair frames, depot-wheelchair frames and for racing wheelchair frames. These frames have been loaded with rider loads (estimated static loads applied to the seating surfaces with the ground fixed) and road loads (estimated static loads applied to the wheels with the seating surface fixed). In every case, the applied loads were statically determined with some safety factor to account for dynamic loading. Although this approach is reasonably accurate if standard materials such as steel and aluminum are used, they are only very rough approximations. FEM modeling can be used to estimate reliability of wheelchair system components using statistical models. However, without actual road load data the probability distribution must be assumed; generally it is assumed normal and the reliability of the structure can be estimated from the intersection of the fatigue and failure probability curves. S–N curves are useful for estimating reliability of systems, but they are no substitute for testing.

(a)

(b)

Figure 6.10. *(a) Fatigue fracture of a manual wheelchair front caster stem. This wheelchair went nearly 600 000 double-drum cycles before this failure occurred. (b) Fatigue failure of a front caster. Insufficient composite filler material was used leaving the caster too brittle and of insufficient strength. This component failed after approximately 30 000 double-drum cycles. Both of these items could be simply replaced and the chair used further.*

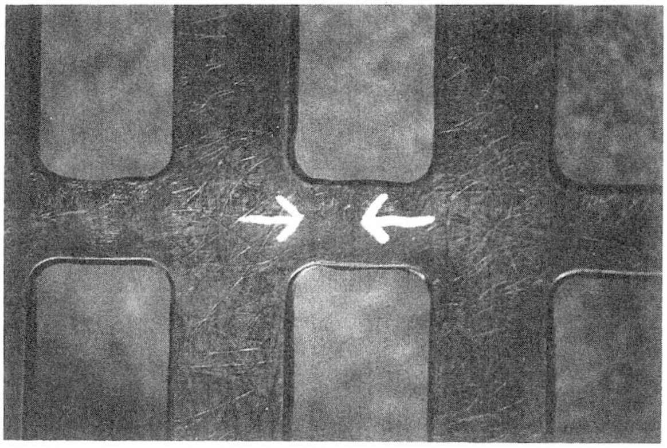

Figure 6.11. *Manual wheelchair seat base made from fiberglass embedded in a plastic matrix. Fatigue fracture developed after about 15 000 double-drum cycles due to insufficient distribution of fiberglass particles in the failure region. Seat base was redesigned with higher fiberglass content and stronger geometry. (Photograph by R A Cooper.)*

Some data have been collected on wheelchair frame fatigue, but none were suitable for integrating with FEM data. Fatigue testing is a very important part of the ISO and RESNA wheelchair standards. Fatigue testing gives engineers information about the life of the wheelchair and about weak points (figure 6.11). However, it is difficult to apply these results to improving quality without using tools like FEM. FEM allows changes in load conditions and design parameters to be analyzed efficiently. FEM combined with road load data has the potential to improve quality without causing a great increase in cost. Fatigue testing has been successfully used in the design of wheelchairs and is applied by some of the larger manufacturers. However, fatigue testing can be costly and making changes in wheelchair design based upon fatigue testing without knowledge of external loads, is analogous to shooting in the dark. Fatigue testing without measurement tells what broke, but does not give a clear indication why. FEM combined with road loads during activities of daily living and fatigue testing can give engineers the tools required to minimize wheelchair failures.

6.10. TEST DUMMIES

There are four dummies used for wheelchair testing (25, 50, 75 and 100 kg). Dummies can be used for all tests where the wheelchair is to be loaded. The ISO standard is based upon simple and inexpensive construction with mass distribution similar to a human. The dummies described in the ISO/RESNA standards are fragile, very difficult to handle and do not work with most ultra-light wheelchairs. Test dummies can be developed which meet ISO/RESNA standards (at least intentions) but are more durable, safer to handle and can be used with ultralight wheelchairs with narrow footrests as well as classical designs.

The ISO/RESNA dummy should be similar in anthropometry to a human wheelchair rider and should be much more durable than any wheelchair being tested. The dummy should not fail, especially during dynamic testing, before the wheelchair. The plywood suggested for the body in the standard is too weak to withstand more than a few fatigue tests. The feet of the dummy should be able to rest properly on the footrest(s) of a wide variety of wheelchairs, including those with tapered front ends. The dummy should be safe to handle. The dummy must to be capable of withstanding 10 million cycles in a wheelchair on an ISO/RESNA double-drum tester (this is approximately equivalent to 100 tests). The leg length and foot width were to be adjustable to accommodate a wider variety of wheelchair designs.

The ISO/RESNA standards call for dummies to be simple to construct and inexpensive. Although the initial investment for aluminum is higher than for plywood, it is a better material because the long-term cost should be lower with the extended product life. Aluminum is lightweight, as is plywood and therefore adequate freedom remains to adjust mass distribution. Steel plates can be used for additional mass. Steel is durable, readily available and easy to work with.

The width between the lower legs should be adjustable (suggested range is 200–250 mm) to accommodate a variety of footrest widths. The length of the lower legs should also be adjustable (suggested range is 300–500 mm).

Typically, wheelchairs are tested for between 100–200 thousand cycles on an ISO/RESNA double-drum tester and 3500–7000 cycles on an ISO/RESNA curb drop tester. These tests can be particularly punishing to the dummy. High-density foam is glued to the back, seat and feet to protect wheelchairs from the dummy's frame.

The standards do not specify how the mass is to be distributed within each component (i.e., the torso). This may cause undue stresses in some areas of the wheelchair. If the dummies are to be used for vibration analysis (e.g., ride comfort), then dummies which model human vibrational properties need to be developed. It has been observed that a patient transfer hoist works well when transporting test dummies.

6.11. POWER WHEELCHAIR RANGE TESTING

Energy consumption is measured over a distance of 1500 m while driving around a tennis court. The rolling surface is level in this case. The direction of rotation (i.e., direction was changed from clockwise to counter clockwise) is changed at 750 m. Wheelchairs are always driven in the forward direction. The wheelchair is warmed-up for 750 m at near maximal speed prior to data

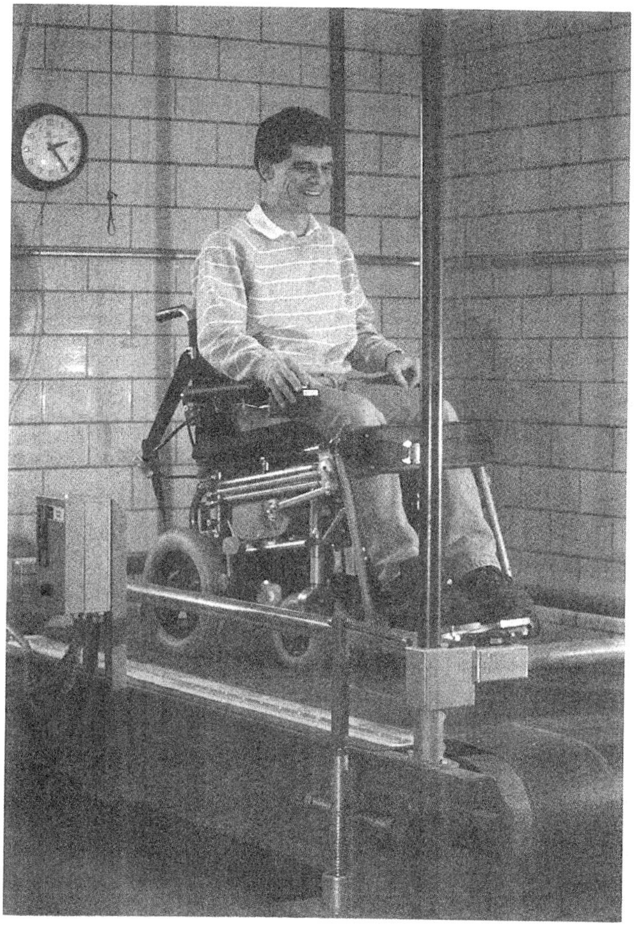

Figure 6.12. *Test scientist conducting a power-wheelchair range test on a motor driven treadmill. This test is used to estimate the distance which can be covered on a single charge with a freshly charged battery. (Photograph by R A Cooper.)*

collection. Hence each wheelchair is driven a distance of 2250 m. The time required to complete the 1500 m is recorded and later used to calculate maximum speed of each wheelchair while performing each experiment. A course is laid out around the perimeter of the tennis court and the distance measured with a steel tape. The pilot is instructed to follow the course as closely as possible. Some laboratories perform this test on a motor driven treadmill (figure 6.12).

Tests were performed on seven different wheelchairs loaded with an ISO 100 kg test dummy (table 6.5). A specially designed circuit is used to measure battery voltage and load current. Data were collected at a suitable

Table 6.5. *Examples of power-wheelchair range test results with 100 kg rider.*

	Current (A)	Power (W)	Energy (kJ)	Speed ($m\,s^{-1}$)
Chair 1	19.1	466	304	2.3
Chair 2	6.6	155	139	1.7
Chair 3	8.8	198	126	1.8
Chair 4	8.4	205	184	1.7
Chair 5	19.8	545	327	2.5
Chair 6	12.2	301	200	2.3
Chair 7	6.0	146	101	2.0

sampling rate (e.g., 20 Hz per channel). New batteries are to be installed prior to testing, and all batteries must be fully charged before each experiment. The chairs are configured as per the manufacturer's specifications and tires are properly inflated. Experienced test pilots (technicians) are used for the tennis court tests. The types of chairs and their battery type are provided in the test report.

The range of each chair is estimated using the measured data and equation (6.19) below. The 1 $m\,s^{-1}$ range data are susceptible to greater variability because data were collected for only ten seconds during steady state (table 6.6). Wet cell batteries provide longer range of operation than gel cell batteries in every instance. This is because the amp-hour rating of a wet cell battery is consistently higher than that of a gel cell battery

$$\text{range} = \frac{\text{nominal battery capacity(amp-hour)} \times \text{speed traveled (kph)}}{\text{amperes consumed (amp)}}\ \text{km}. \tag{6.19}$$

The predicted range for the tennis court test at maximum speed ranges from a low of 23.6 km to a high of 57.7 km. The range of the power wheelchair can be improved by the use of wet lead acid batteries in place of gel lead acid batteries. However, wet batteries often require greater maintenance and care during transport. No alternative batteries were tested. All of the manufacturers specified lead acid batteries for their wheelchairs. The range at 1 $m\,s^{-1}$ was typically greater than it was at maximum speed. This information may be useful to consumers, who when warned of a low battery could extend their range by reducing speed. Current draw on an incline will be greater than the values indicated here. Some wheelchairs incorporate regenerative braking which allows some of the energy expended while going up an incline to be regained through charging the batteries while driving down an incline. Range will also vary with driving habits. The tennis court test has an excess of 40 turns which require the pilot to slow the chair and then accelerate out of the turn.

Table 6.6. *Estimated range in km with 100 kg load at maximum speed and at approximately 1 m s^{-1}.*

	Range at 1 m s^{-1}	Range at full speed
Gel cells		
Chair 1	43.0	30.3
Chair 2	76.6	44.5
Chair 3	27.3	23.6
Chair 4	46.7	35.0
Chair 5	68.8	31.8
Chair 6	75.3	47.5
Chair 7	40.4	38.4
Wet cells		
Chair 1	52.2	36.8
Chair 2	87.8	51.0
Chair 3	41.0	35.4
Chair 4	53.5	40.1
Chair 5	83.5	38.6
Chair 6	91.4	57.7
Chair 7	60.6	57.6

6.12. POWER WHEELCHAIR CONTROLLER PERFORMANCE

Power wheelchairs generally have a number of adjustable performance features that are dealer and/or user adjustable. Traditionally, powered wheelchairs have had a high/low switch for indoor and outdoor use. More recently wheelchairs have incorporated programmable controllers which can be tuned to meet an individual's mobility needs (figure 6.13). There is considerable discussion among the rehabilitation engineering community as to who should be able to tune a power-wheelchair controller. Several features can be programmed into microcontrolled power wheelchairs. Maximum speed can be set within the wheelchair's capabilities. With some wheelchairs adjusting maximum speed also reduces maximum torque. However, for many this does not present a problem. Forward and lateral acceleration can be set to determine how rapidly the user can obtain maximum forward speed and how fast the user can turn the wheelchair. Forward deceleration can be adjusted to determine how quickly the wheelchair will come to a stop once the joystick is placed in the neutral position or reversed. Disconnecting the power to the wheelchair often results in more severe braking.

The forces and motor control required to operate input devices is important to safe and effective operation of a power wheelchair. Many wheelchair users have functional limitations which limit their ability to operate a joystick and/or other input devices. The force required to operate a device must reflect the abilities of the user. Some power-wheelchair users lack sufficient

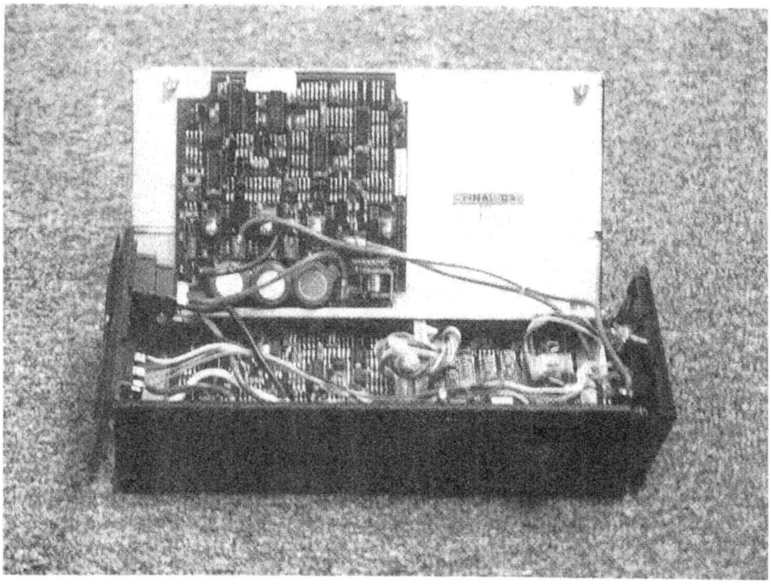

Figure 6.13. *Power-wheelchair programmable controller with DC servo amplifier. Tests are performed to determine the safety of controllers in various internal-fault scenarios. (Photograph by R A Cooper.)*

fine motor control to operate some analog input devices (e.g., joystick). The sensitivity of the controls must be matched to the user's motor control ability. Rehabilitation engineers should measure and report the forces required to actuate various operational control switches and input devices. A therapist or physician can provide assistance in determining the amount of force a user can exert for purposes of actuating control switches and input devices.

Extremes in temperature as experienced when gong from room temperature to a cold outside temperature should not adversely affect the performance of the power wheelchair (figure 6.14). Power wheelchairs should also operate safely when exposed to rain and moisture. Rain performance is evaluated by subjecting the wheelchair to a standardized shower. The wheelchair is driven immediately after the shower and one hour later. Any variation in driving behavior is noted. During climatic testing, the chair is exposed to $-25\,^{\circ}\mathrm{C}$ and $+50\,^{\circ}\mathrm{C}$ for three hours and then driven before returning to room temperature. To measure the effects of shipping and storage, the wheelchair is exposed to a temperature of $-40\,^{\circ}\mathrm{C}$ for three hours and then to a temperature of $+65\,^{\circ}\mathrm{C}$ for three hours. The chair is allowed to sit a room temperature for 24 hours and is then driven. Deviations in normal driving behavior are recorded. Climatic testing can be helpful wher designing or selecting a power wheelchair for local climate conditions.

Figure 6.14. *Power-wheelchair users must sometimes leave warm environments for cold environments. Simply leaving a warm building on a cold winter's day can be a substantial shock for a power wheelchair. Climatic testing is used to help ensure the wheelchair will function properly during activities performed by consumers. (Photograph by R A Cooper.)*

6.13. DESIGNING FOR SAFE OPERATION

Powered wheelchairs are not always driven on flat, level surfaces. People drive up and down hills, around turns and over curb-cuts. The dynamic stability of the power wheelchair is tested while driving up and down a ramp at full speed and at full braking for both cases. The power wheelchair is also turned while going up and down hill (figure 6.15). Power wheelchairs must be able to climb obstacles which are encountered during normal driving. A simple test for obstacle climbing ability is to drive the powered wheelchair onto an ever-increasing stack of wooden sheets (figure 6.16). The height of the maximum obstacle which can be safely overcome gives a good measure of the power wheelchair's obstacle climbing ability. The driver should be able to maintain safe control over the wheelchair through all of these maneuvers. For some wheelchairs, anti-tip casters may contact the road surface during the dynamic stability maneuvers. The rehabilitation engineer, clinician and consumer must determine if that is appropriate for the individual.

Wheelchairs must meet minimum requirements for flammability. ISO and

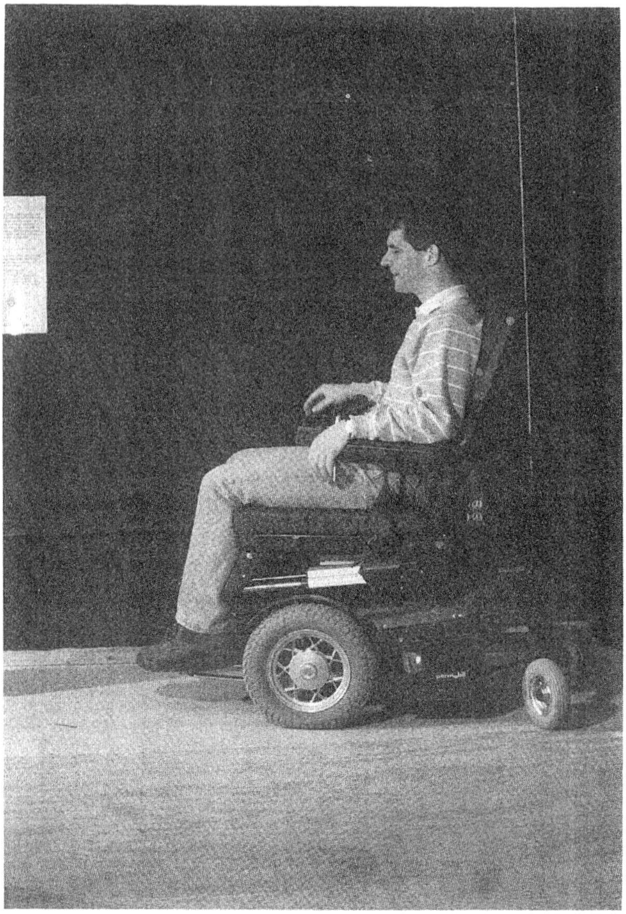

Figure 6.15. *Dynamic stability testing of a power wheelchair. Power wheelchairs are driven up and down slopes and along side slopes. Test technicians drive power wheelchairs on a ramp at various angles and record its behavior through a battery of tests. (Photograph by R A Cooper.)*

ANSI/RESNA Standards for flammability of fabric and mattress materials are used to test upholstered parts of wheelchairs. These tests provide the wheelchair user some assurance of safety when exposed to a smoldering cigarette or open flame.

Power-wheelchair batteries must be charged regularly and must perform safely (figure 6.17). The charger or wheelchair should not become damaged if connected improperly or improper connection should be prohibited. Different batteries require different charging characteristics and chargers should be matched to the appropriate batteries. Battery chargers must pass a comprehensive set of test procedures. The test procedures are different

Figure 6.16. *Power-wheelchair users encounter a variety of obstacles while driving. The ability to climb obstacles is important to some consumers. The obstacle-climbing ability test determines the maximum height of an obstacle that the power wheelchair can safely conquer (?). (Photograph by R A Cooper.)*

depending upon whether the charger is attached to the wheelchair or is a stand-alone unit. Battery chargers must also meet the power requirements of the local country's power grid. Some chargers can operate on either 220 or 110 V.

Wheelchair standards only represent minimum performance criteria and do not guarantee safe operation and design. Engineers and designers must ensure that their designs are based on sound principles, valid assumptions and thorough testing. It is also critical that the wheelchairs, their components and their accessories are properly prescribed. Inappropriately prescribed technology can destroy even the best design efforts. Consumers should be used in testing only after products have completed testing in the laboratory or factory. Some flexibility is required in the design of wheelchairs. One wheelchair design is not appropriate for every person or all activities. Many people may be served by the same style wheelchair if options are available which allow customization. For example, frame and upholstery materials could be sturdier for people of extreme weight. Pinch points and catch points must be avoided. Control must be designed and positioned for safe operation. Inappropriately designed controls can cause the wheelchair to respond undesirably and possibly cause injury. Manufacturers need to test

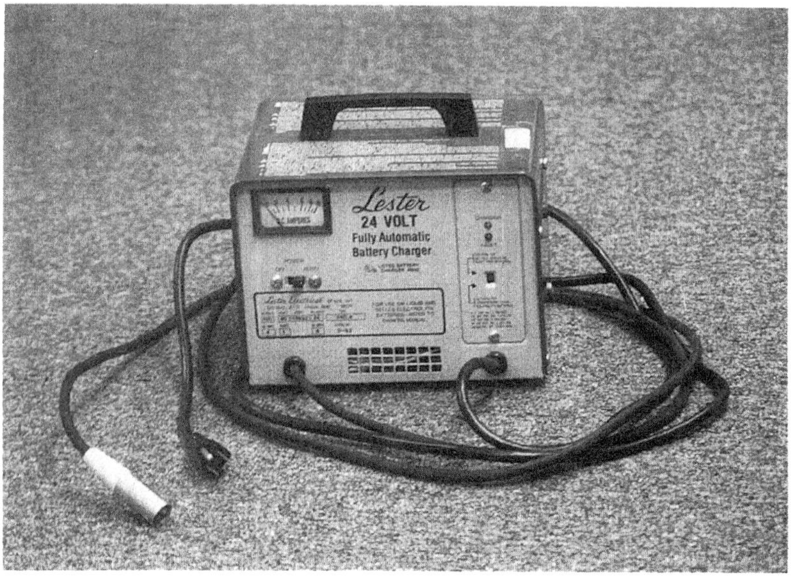

Figure 6.17. *Powered wheelchairs are mostly battery powered and the batteries require charging on a regular basis. Battery chargers must be safe and reliable for power wheelchairs. A series of standards has been developed for power-wheelchair and scooter-battery chargers. (Photograph by R A Cooper.)*

existing products periodically to ensure quality manufacture. Consumers need to keep manufacturers and regulatory agencies aware of failures they experience.

FURTHER READING AND REFERENCES

Axelson P, Minkel J and Chesney D 1994 *A Guide to Wheelchair Selection: How to Use the ANSI/RESNA Wheelchair Standards to Buy a Wheelchair* (Washington, DC: Paralyzed Veterans of America)

Axelson P and Wood Z 1992 Wheelchair standards and you *Paraplegia News* **46** 54–5

Baldwin J D, Dee L S and Thacker J G 1992 Dynamic structural response of a cross-tube wheelchair frame *Proc. RESNA Int. '92 (Toronto, Ont., 1992)* pp 622–3

Baldwin J D and Thacker J G 1993 Stress response of wheelchair frames to front caster impact *Proc. RESNA 16th Ann. Conf. (Las Vegas, NV, 1993)* pp 321–3

Baldwin J D, Thacker J G, Baber T T and Aylor J H 1990 *Simulation of a Random Fatigue Process Applied to Wheelchair Structures* (Washington, DC: RESNA) pp 63–4

Cooper R A 1993 Stability of a wheelchair controlled by a human pilot *IEEE Trans. Rehab. Eng.* **BME-1**

Cooper R A and MacLeish M S 1992 Racing wheelchair roll stability while performing turning maneuvers—a simple model *J. Rehab. Res. Dev.* **29** 23–30

Cooper R A, Myren C, Ster J F, VanSickle D P, Stewart K J, Reifman G and Heil T A 1993 Design of an anthropomorphic ISO-RESNA/ANSI wheelchair test dummy *Proc. 16th Ann. RESNA Conf. (Las Vegas, NV, 1993)* pp 283–5

Cooper R A, Robertson R N, VanSickle D P, Stewart K J and Albright S 1994 Wheelchair impact response to ISO test pendulum and ISO standard curb *IEEE Trans. Rehab. Eng.* **2** 240–6

Cooper R A, Ster J S and Heil T 1991 Development of a new ISO wheelchair two-drum tester *Proc. 13th Ann. IEEE/EMBS Int. Conf. (Orlando, FL, 1991)* vol 13 pp 1867–8

Cooper R A, Ster J S, Myren C and Pettit D 1992 An improved design of a 100 kilogram ISO/RESNA wheelchair test dummy *Proc. 15th Ann. RESNA Conf. (Toronto, Ont., 1992)* pp 210–2

Cooper R A, Stewart K J and VanSickle D P 1995 Evaluation of methods for determining rearward static stability of manual wheelchairs *J. Rehab. Res. Dev.*

Cooper R A and Watanabe K T 1991 A derivation of turning geometry and turning dynamics for racing wheelchairs *Proc. 14th Ann. RESNA Conf. (Kansas City, MO, 1991)* pp 240–2

Hekstra A 1993 Simulation models for test evaluation and product development of wheelchairs *Proc. 2nd Eur. Conf. Advancement Rehabilitation* p 3.3

Hekstra A C 1991 Human factors in wheelchair testing *Proc. of Workshop 'Ergonomics of Manual Wheelchair Propulsion: State of the Art' (Vrije University, Amsterdam)* pp 25–35

Johnson I and Andrich R 1991 *Eur. Rep. Wheelchairs Testing* Commission of the European Communities COMAC BME

Jones I S 1976 *The Effect of Vehicle Characteristics on Road Accidents* (Elmsford, NY: Pergamon)

Kauzlarich J J and Thacker J G 1985 Rear caster wheelchair directional instability *Proc. RESNA 8th Ann. Conf. (Memphis, TN, 1985)* pp 78–80

Majaess G G, Kirby R L and Ackroyd-Stolarz S A 1991 A new method to assess the dynamic rear stability of occupied wheelchairs. *Proc. RESNA 14th Ann. Conf. (Kansas City, MO, 1991)* pp 356–7

Mauger-Cote D, Audet J and Nolet J 1993 Wheelchair emergency repair services *Proc. 2nd Eur. Conf. Advancement Rehabilitation* p 5.3

McLaurin C A and Axelson P 1990 Wheelchair standards: an overview *J. Rehab. Res. Dev.—Clin. Suppl.* No 2 100–3

Pettit D J, Cooper R A and Bennett P 1992 Design and evaluation of a simple, inexpensive, ultralight wheelchair *Proc. 14th Ann. IEEE/EMBS Int. Conf. (Paris, 1992)* vol 14 No 4 pp 1515–6

RESNA Standard 1989 *Wheelchair Standards* (Washington, DC: RESNA)

Roebroeck M E, Van Der Woude L H V and Rozendal R H 1991 *Methodology of Consumer Evaluation of Hand Propelled Wheelchairs* Commission of the European Communities COMAC BME

Schnoll L 1991 *The ISO 9000 Series: Worldwide Standards for Quality. Medical Design and Material* February pp 36–40

Tam E W C, Chiu E Y M and Evans J H 1992 Using ISO standards for manual wheelchair testing: the Hong Kong experience *Proc. RESNA Int. '92 (Toronto, Ont., 1992)* pp 625–6

Thacker J G, Todd B A and Disher T D 1985 Stress analysis of wheelchair frames *Proc. RESNA 8th Ann. Conf. (Memphis, TN, 1985)* pp 84–6

Thacker J G, Gorman S F and Todd B A 1992 Evaluation of rider comfort with front caster suspension *Proc. RESNA 8th Ann. Conf. (Memphis, TN, 1992)* pp 87–9

VanSickle D P and Cooper R A 1993 Demonstration of a methodology for wheelchair acceleration analysis *Proc. 15th Ann. IEEE/EMBS Int. Conf. (San Diego, CA, 1993)*

VanSickle D P, Cooper R A and Robertson R N 1993 A 2-dimensional wheelchair dynamic load history using accelerometers *Proc. 16th Ann. RESNA Conf. (Las Vegas, NV, 1993)* pp 324–6

van Woerden J A 1993 M3S: A general purpose interface for the rehabilitation environment *Proc. 2nd Eur. Conf. Advancement Rehabilitation Technol. (Stockholm, 1993)* p 22.1

CHAPTER 7

MANUAL WHEELCHAIR DESIGN

7.1. INTRODUCTION

People have been injured or contracted disabling conditions for centuries. Those people with disabilities who survived for an extended period of time were transported on hammocks slung between poles which were carried by others. This was the preferred means of transportation of the upper class and thus carried no stigma. Later the wheelbarrow was developed and soon became a common mode of transportation for people with disabilities. During the dark ages of history people with disabilities were looked upon as outcasts from society. During the renaissance, the French court popularized the first wheelchairs. Wheelchairs were overstuffed arm chairs with wheels placed underneath them. This enabled movement, with assistance, indoors. Later the wooden wheelchair with wicker matting was developed. This type of chair remained the standard until the 1930s. Franklin D Roosevelt was not satisfied with the wooden wheelchair and had many common metal kitchen chairs modified with wheels. In the 1930s a young mining engineer named Everest experienced an accident that left his mobility impaired. He worked with a fellow engineer Jennings to develop steel wheelchairs. Within a few years, they formed a company Everest & Jennings to manufacture wheelchairs. Following World War II, medical advances saved the lives of many spinal cord injured veterans, who would have otherwise died. Veterans medical centers issued these veterans steel-framed wheelchairs with 18 inch seat widths. These wheelchairs were designed to provide the veteran some mobility within the hospital and home, and not to optimize ergonomic variables. Wheelchairs remained virtually unchanged until the disability rights movement began in the 1970s.

Some innovative users began to modify their wheelchairs (with saws and welders) and from there some began their own companies. Hence the ultralight wheelchair was born. Some of these companies still survive and have indeed captured a substantial portion of the wheelchair market. Of course the giants of wheelchair manufacturing have also taken note and produce their own ultralight wheelchairs.

Due to the development of this new class of manual wheelchair, the ultralight wheelchair, there has come some confusion. Ultralight wheelchairs require more knowledge to specify properly than do standard wheelchairs. This is because more variables can be specified for ultralight wheelchairs. Some of this difficulty can be overcome by specifying an ultralight wheelchair that allows adjustment of many key features. This permits some experimentation once the chair arrives. Otherwise a lighter and more efficient chair can be specified by not having the features adjustable if the user and the dealer know what is desired. Some compromise between having every feature adjustable and none adjustable can be developed.

7.2. CLASSES OF MANUAL WHEELCHAIRS

Manual wheelchairs have developed rapidly in recent years—fifty years ago there was only one style of wheelchair and it came in one color: chrome (figure 7.1). Now, there are numerous types of wheelchairs to choose from and they come in a wide range of colors. Wheelchairs have moved from being chairs with wheels designed to provide some minimal mobility to advanced orthoses designed to meet the mobility demands of the user. The proper selection and design of a wheelchair depends upon the abilities of the user and on the intended use. Thus specialized wheelchairs have been and continue to be developed to yield better performance for the user.

7.2.1. Depot (institutional) wheelchair

The depot or institutional wheelchair is essentially the same wheelchair that was issued to veterans in the 1940s. The wheelchairs may be a bit lighter, but the basic frame is unchanged. Depot wheelchairs are intended for institutional use where several people may use the same wheelchair. These wheelchairs are typically used in airports, hospitals, and nursing care facilities. Generally, they are inappropriate for active people who use wheelchairs for personal mobility (figure 7.2). Depot wheelchairs are designed to be inexpensive, accommodate large variations in body size, to be low-maintenance, and to be attendant-propelled. Hence, they are generally heavier and their performance is limited.

Many depot wheelchairs are manufactured in developing countries. The inexpensive labor makes the manufacture of inexpensive depot wheelchairs possible. A typical depot wheelchair will have swing-away footrests, removable armrests, a single cross-brace frame and solid tires. The swing-away footrests add weight to the wheelchair; however, they make transferring into and out of the wheelchair easier. Armrests provide some comfort and stability to the depot wheelchair user and they can aid in keeping clothing off the wheels. Depot chairs typically fold to reduce required storage area

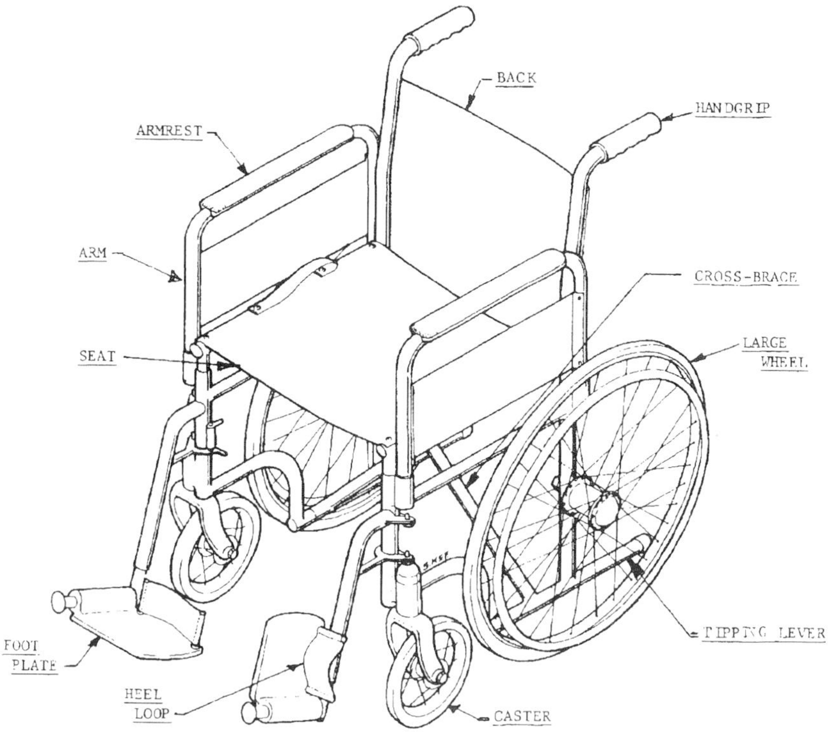

Figure 7.1. *Basic common wheelchair with folding cross-brace frame, 24 inch rear wheels, 8 inch front casters, and cloth seat and backrest upholstery. (Figure reproduced with permission from Wheelchairs: A Prescription Guide A Bennett Wilson, Jr, Demos, New York, NY, 1992.)*

and so that the chair will fit into the automobile of the borrower. Solid tires are used to reduce maintenance. However, solid tires typically dramatically reduce ride comfort. Pneumatic tires are recommended for outdoor usage. There is typically very little if anything that can be adjusted to fit the user on a depot chair. Depot chairs are available in various seat widths, seat depths and backrest heights. These dimensions must be specified to the manufacturer or distributor.

7.2.2. Amputee wheelchair

People with lower-limb amputations typically have a different center of gravity location than do people who have their lower limbs. When seated in a wheelchair the center of gravity of the person with lower limb amputations may be close enough to the rear axles of the wheelchair to require some modification to the wheelchair axle position. The amputee wheelchair came about because wheelchairs were originally designed for people who were

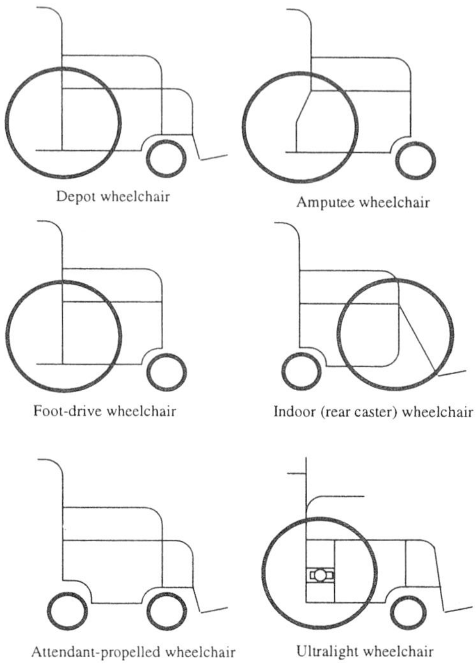

Figure 7.2. *Side view schematic showing differences between: depot chair, amputee chair, foot-drive chair, indoor chair, attendant-propelled chair, and ultralight chair.*

anatomically intact. Thus the center of gravity of the wheelchair and an amputee was too close to the rear axles. This caused the wheelchair to have too great a tendency to do a 'wheelie'. Hence, amputee wheelchairs had the rear axle positions extended rearwards on the frame. The frame of an amputee wheelchair has to be designed to accommodate the offset between the back of the seat and the rear axle housings. Not all people with lower-limb amputations require the more stable axle mountings and the abilities of the user must be evaluated before the wheelchair is specified or designed.

7.2.3. Hemiplegic (one-arm drive) wheelchair

People who have brain related motor impairments often require specialized wheelchairs which permit the optimum application of their motor abilities. Typically a one-arm drive wheelchair consists of a linkage connecting the rear wheels (figure 7.3). This allows the user to push upon the pushrim of one wheel and to propel both wheels. To effectively turn the wheelchair, the user must have the ability to disengage the drive mechanism and to propel each rear wheel independently. One-arm drive wheelchairs are modified versions of standard wheelchairs. Most types of wheelchairs can be modified

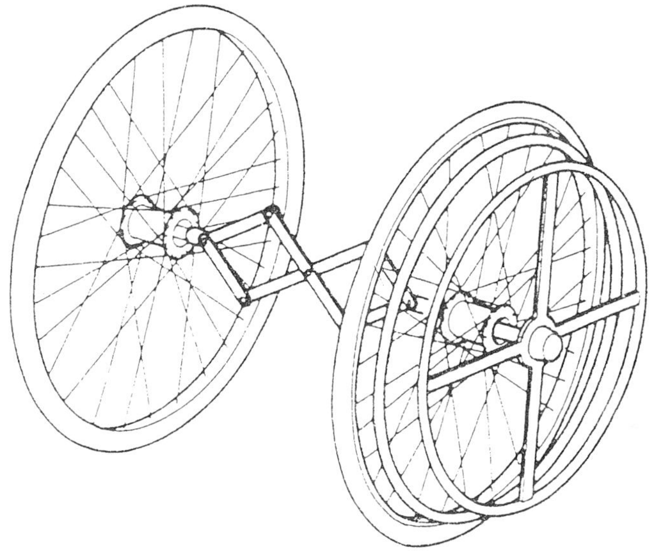

Figure 7.3. *Mechanism for a hemiplegic (one-arm-drive) wheelchair. (Figure reproduced with permission from Wheelchairs: A Prescription Guide 2nd edition, A Bennett Wilson, Jr, Demos, New York, NY, 1992.)*

to accommodate one-arm drive mechanisms. The proper fit of the wheelchair to the user is most critical with one-arm drive wheelchairs, because of the user's reduced physical strength. The wheelchair must fit such that the user can get maximal leverage for pushing the wheelchair and to control the wheelchair.

7.2.4. Foot-drive wheelchair

Some people have weakness of the upper and lower extremities and can gain maximal benefit from wheelchair propulsion by combining the use of their arms and legs or by using their legs. The design and selection of a foot-drive wheelchair depends greatly upon how the user can take greatest advantage of their motor abilities. The strength of the user's legs must be determined to decide whether it is best to pull or to push with the legs. When pushing the user moves with his/her back to the direction of motion. The design of the chair is affected by whether the user pushes or pulls. Typically if the user pulls the chair along with his/her leg(s) then a wheelchair with the footrests removed is effective.

If the person pushes the wheelchair, then modifications to the design of standard wheelchairs are required. It may be best to design the chair so that the seat faces opposite the direction it would face in a typical wheelchair. The casters should lead the rear wheels for the most common direction of travel.

This will help to reduce the possibility of the user flipping over when hitting an obstacle and will make the chair more directionally stable. The wheelchair should be set-up so that the user/wheelchair center of gravity is well within the footprint of the wheelchair. This may require some consideration of where the larger wheels are mounted.

7.2.5. *Indoor (rear caster) wheelchair*

Some wheelchairs are designed to meet the special requirements of indoor use. Indoors spaces are more limited and one is often required to get close to furnishings and fixtures to use them properly. Wheelchairs have been and can be designed to be used indoors. Commonly, indoor wheelchairs use rear casters because of the maneuvrability of these designs. However, rear caster designs make the wheelchair less stable in lateral directions. Indoor wheelchairs typically have shorter wheelbases. This makes them less stable and they are not recommended for outdoor use.

Some indoor wheelchairs have the ability to elevate the user to the standing position. Most manual wheelchairs cannot be propelled in the stand-up setting in order to ensure safe and stable operation. Standing gives the user the ability to reach cabinet and counter space which would otherwise be inaccessible. Some wheelchairs are made to operate in the standing position. These can be used to work at counter tops or for using chalk boards. Standing has the additional advantage of providing therapeutic benefits, i.e., hemodynamic improvements and amelioration of osteoporosis.

7.2.6. *Attendant-propelled wheelchair*

Not all wheelchairs are propelled by the person sitting in the wheelchair. In many hospitals and long-term care facilities wheelchairs are propelled by attendants. The design of these wheelchairs requires special consideration. If the wheelchair is to be solely propelled by attendants with no assistance from the rider, then there may be no need for the larger drive wheels.

Attendant-propelled wheelchair designs must consider the rider and the attendant as users. The rider must be transported safely and comfortably. The attendant must be able to operate and easily maneuver safely with minimum physical strain. The design of the wheelchair must be such that the attendant can assist the rider with transfers and maneuver around the chair in restricted spaces. The push handles should be such that the attendant experiences no undue stress to the hands, arms, and back. Consideration must be given for how the chair will be maneuvered over obstacles. The primary consideration is that the wheelchair has two users: the rider and the attendant.

7.2.7. *Ultralight wheelchair*

Active users often prefer highly maneuverable and responsive wheelchairs which fit their physical and psychosocial character. The ultralight wheelchair evolved from the desire of wheelchair users to develop functional ergonomic designs for wheelchairs. Ultralight wheelchairs are made of such materials as aluminum, alloy steel, titanium, or composites. The design of ultralight wheelchairs allows a number of features to be customized by the user or to be specified for manufacture.

Wheelbase, wheel size and type, axle position, camber, wheel alignment, caster type and size, seat angle and a number of other variables affect the performance of wheelchairs. Many of these characteristics can be adjusted or specified with ultralight wheelchairs. Ultralight wheelchairs are designed for active users and are made more durable with less weight than a typical wheelchair. There are a wide variety of ultralight wheelchair designs. The most common features of all are the low weight, the high-quality of materials used in their construction and their functional design. Many people can benefit from ultralight wheelchair designs.

7.2.8. Sports wheelchair

Sports wheelchairs are an outgrowth of the popular wheelchair sports movement. The desire to achieve better performances has lead wheelchair users, inventors and manufacturers to constantly develop specialized wheelchairs for sports. There is no real typical sports wheelchair, as the design depends heavily on the sport. Basketball and tennis wheelchairs are often thought to typify sports wheelchair design (figure 7.4). However, racing, field events or shooting wheelchairs have little in common with the former. Sports wheelchairs are designed to meet the demands of a particular sports. Some people may use an ultralight wheelchair to compete in sports, however, serious competition often mandates a custom sports wheelchair.

7.3. FRAME DESIGN

Several factors must be considered when designing a wheelchair frame: what are the intended uses; what are the abilities of the user; what are the resources available; and what are the existing products available. These factors determine if and how the frame will be designed and built. Successful designs of sports wheelchairs can only be accomplished with continuous input from and interaction with wheelchair users. The durability, aesthetics, function, ride comfort and cost of the frame are dependent on the materials for construction, frame geometry, and fabrication methods. One of the issues that makes wheelchair design more complicated is the fact that the user is dependent upon wheeled mobility everyday, nearly all day.

Presently all common wheelchair frames center around tubular construction. In other words wheelchairs are generally made out of some lightweight tubing. The tubing can either be welded together or bolted

Figure 7.4. *Aluminum three-wheeled court sports ultralight wheelchair with plastic foot-plate and tubular feet guards. This wheelchair also has radical camber. (Photograph courtesy of TOP END by action.)*

together using lugs. There are two basic frame types, folding and rigid and two common frame styles, the box frame (figure 7.5) and the cantilever frame (figure 7.6).

7.3.1. Frame styles

The box frame is named such because of its rectangular shape and because of the tubes outlining the edges of the 'box' (figure 7.7). Box frames can be very strong and very durable. A cantilever frame is so named because the front and rear wheels, when viewing the chair from the side, appear to be connected by only one tube; this is similar to having the front wheels attached to a cantilever beam fixed at the rear wheels (figure 7.8). Both frame types require cross-bracing to provide adequate strength and stiffness.

These frame types stem from slightly different design philosophies. The box frame provides great strength and rigidity. Thus the wheels are mounted to a fairly rigid framework. If designed and constructed properly the frame only deflects minimally during normal loading and most of the suspension is

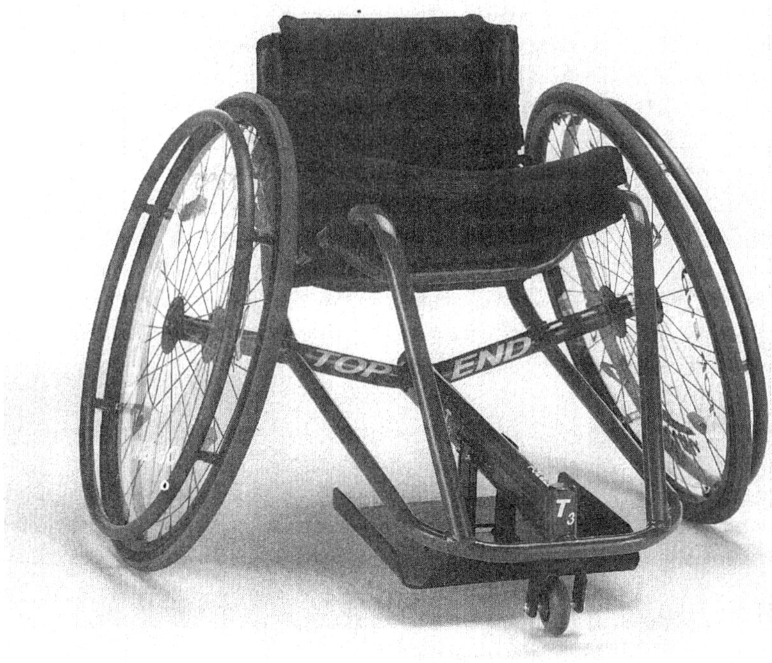

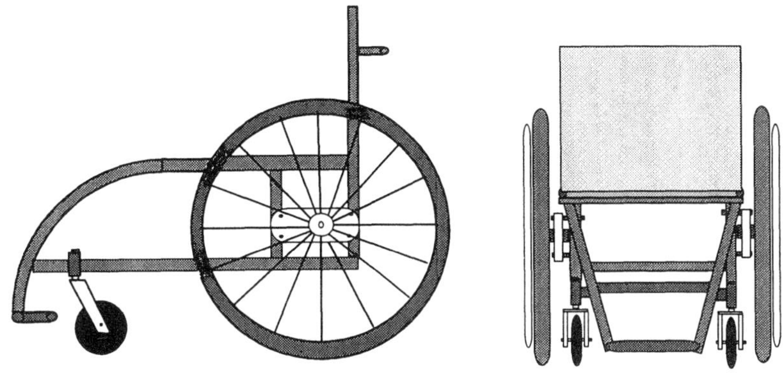

Figure 7.5. *Example illustration of a box frame wheelchair.*

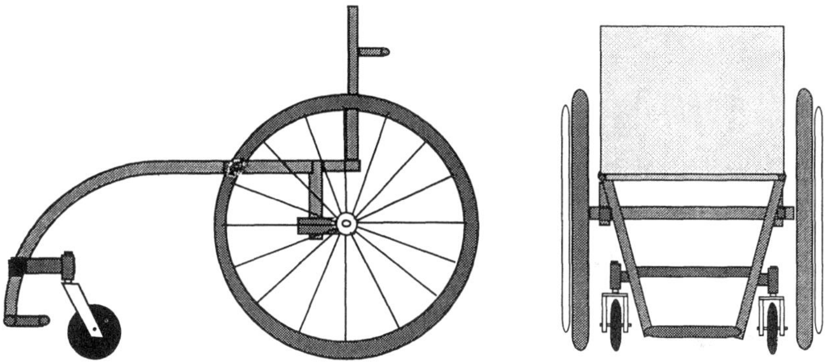

Figure 7.6. *Example illustration of a cantilever frame wheelchair.*

provided by the seat cushion, the wheels and the wheel mounting hardware. Many manufacturers do not triangulate, i.e., have tubes criss-crossing the frame; their box frame design allows some flexibility in the frame. The cantilever frame is based upon a few basic principles: (1) the frame can act as suspension and there is some flexibility purposely built into the frame; (2) there are fewer tubes and they are closer to the body which may make the chair less conspicuous; and (3) there are fewer parts and fewer welds which makes the frame easier to construct. Both basic frame types are very functional and have their merits. Some users prefer the lines and the feel of the box frame while others prefer the cantilever frame.

7.3.2. Folding mechanisms

There are three commonly used folding mechanisms used with wheelchair frames: (1) cross-brace; (2) parallel-brace; and (3) forward folding. Most common folding wheelchairs use variations on the cross-brace design. Each

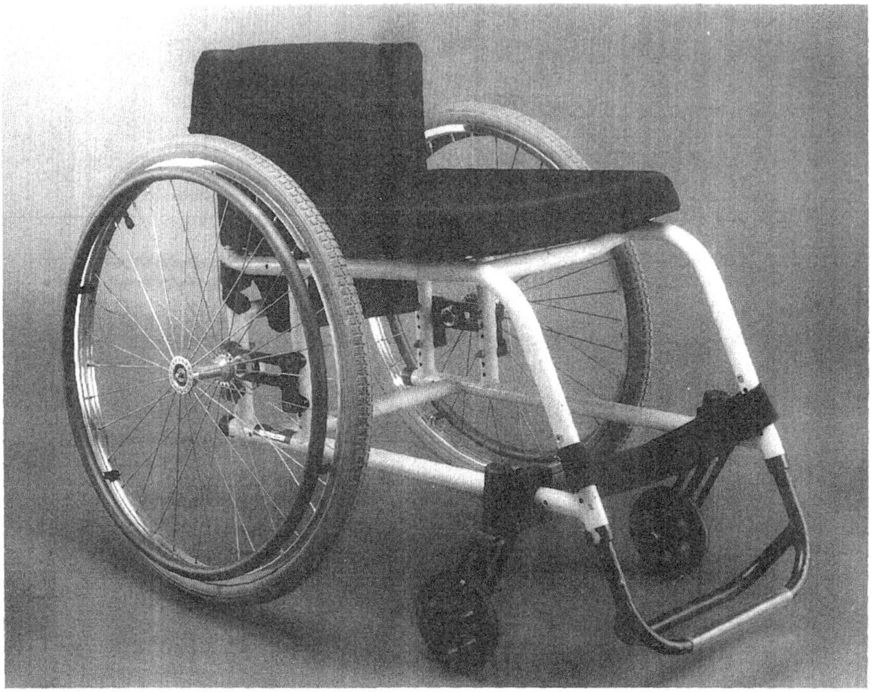

Figure 7.7. *Aluminum box frame rigid ultralight wheelchair used by many active wheelchair riders. (Photograph by S Parsons, CSUS Media center.)*

of these folding mechanism designs have their relative advantages and their application depends upon the user's preferences.

Wheelchairs are designed with either single or double cross-brace mechanisms. Double cross-brace designs add some stiffness to the frame. A cross-brace folding mechanism consists of two frame members connected in the middle attached to the bottom of a side frame member on one side of the chair and to the seat upholstery above the top side frame member on the opposite side (figure 7.9). The cross members are hinged at the bottom and are pinned together in the middle. When viewed from the back of the frames, the cross members form an X. The chair is folded by pulling upwards on the seat upholstery. When the seat is lifted the cross-members move upwards-pulling the frame together. The user's weight keeps the frame from folding when the wheelchair frame is extended. Cross-member folding mechanisms are simple and easy to use. However, the wheelchair may collapse when tilted sideways and the frame get taller when folded. Some chairs incorporate snaps or over-center locking mechanisms to reduce the problems of tilting.

The parallel brace folding mechanism lets the frame fold sideways by having the frame cross member hinge forward. Each cross member is hinged in the center and at each end (figure 7.10). When in the extended position

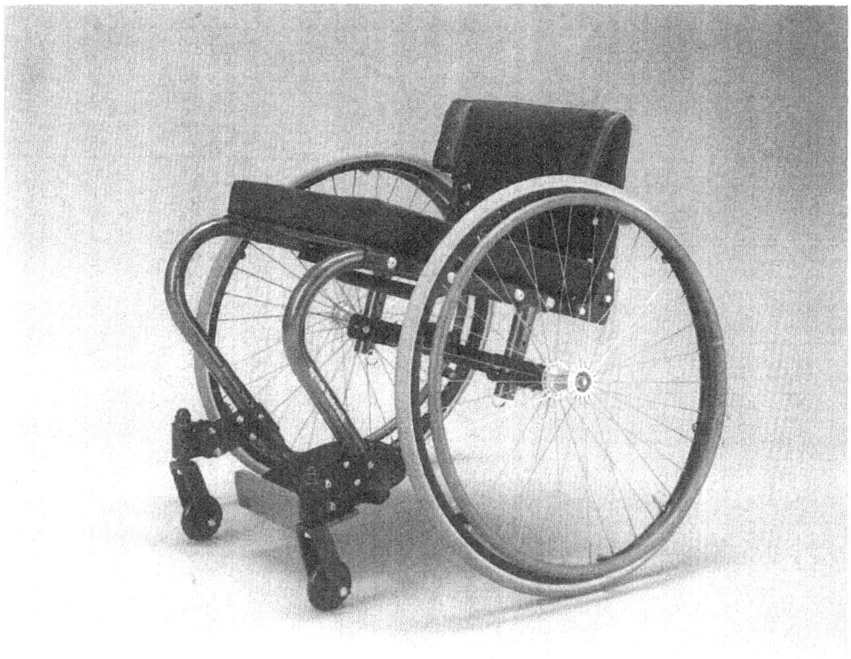

Figure 7.8. *Aluminum cantilever frame ultralight wheelchair with radical bends in side-frame tubes. (Photograph courtesy of Kuschall of America, Inc.)*

the center hinge is locked. The user releases the lock and pulls forward. The chair folds as the user pulls forward. The cross members can be locked with a pin or cam mechanism. The advantages of this design are that the frame behaves like a rigid frame in the extended position and the chair can be partially folded with the user in it. This could permit negotiation of some narrow entrances. Parallel brace mechanisms are more difficult to operate and maintain.

Forward-folding wheelchairs are not very popular, but have some nice features. Many ultralight wheelchairs incorporate forward folding backrests. The concept of a forward-folding wheelchair involves hinging the front end of the wheelchair and the backrest (figure 7.11). The backrest folds onto the seat and the front end folds under the seat. Forward-folding wheelchairs can be made very compact if the rear wheels are quick release. However, forward-folding wheelchairs require more operations to fold them.

7.4. MATERIALS

Most wheelchairs are made of either aluminum, steel, or titanium. Some chairs have been made of advanced composite materials, primarily carbon

(*a*)

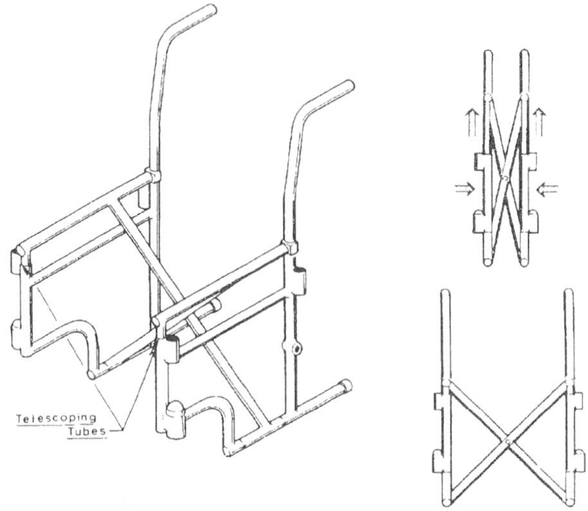

(*b*)

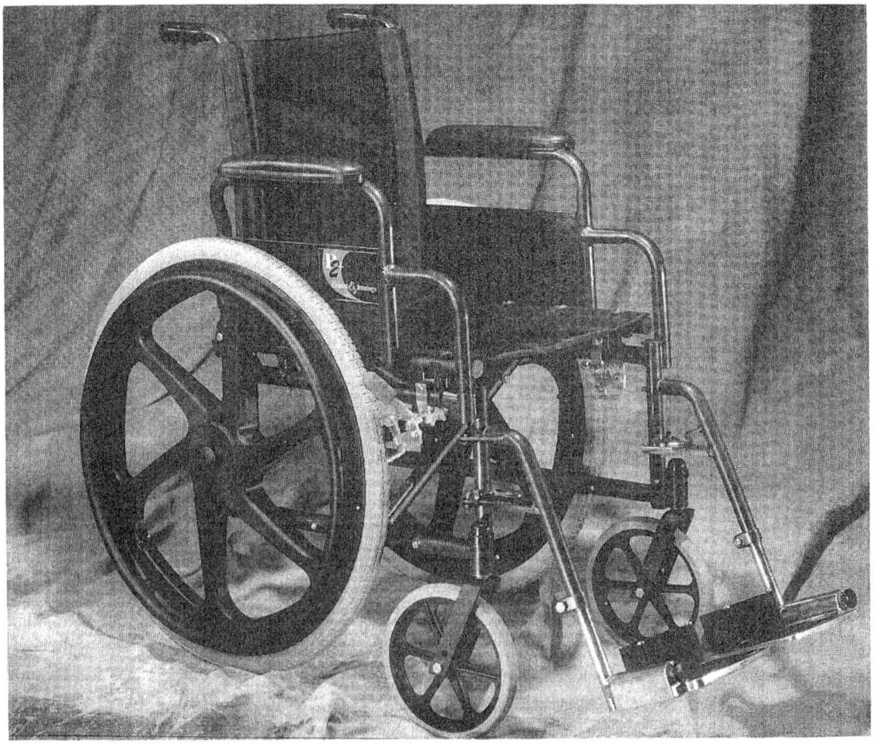

Figure 7.9. *(a) Illustration of folding cross-brace design wheelchair frame (figure reproduced with permission from Wheelchairs: A Prescription Guide 2nd edition, A Bennett Wilson, Jr, Demos, New York, NY, 1992). (b) Common lightweight wheelchair with folding cross-brace, and swing-away legrests. This wheelchair incorporates mag drive wheels and casters to reduce maintenance. (Photograph courtesy of Everest and Jennings International, Ltd.)*

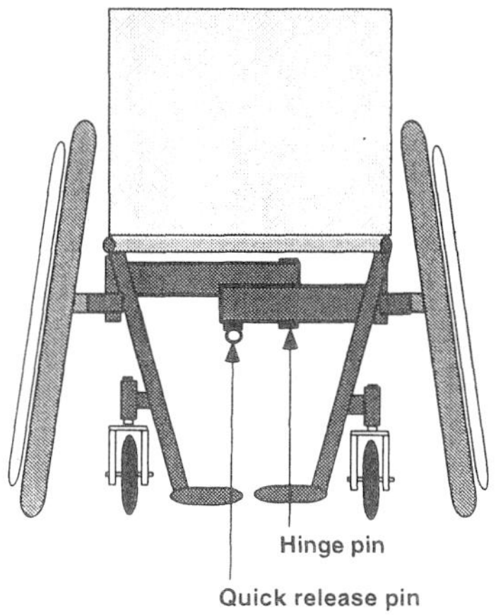

Figure 7.10. *Illustration of wheelchair with parallel-brace design (front view).*

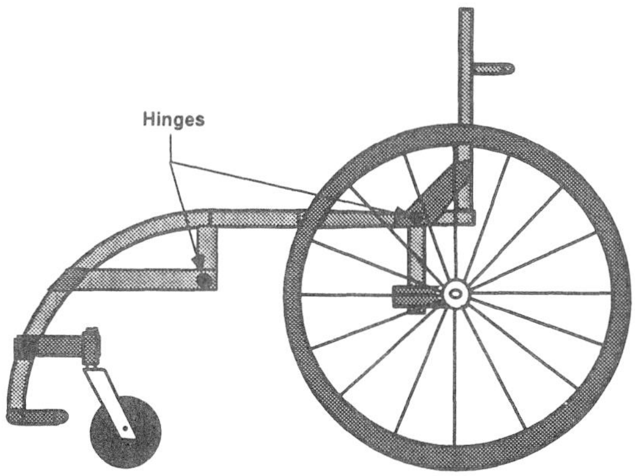

Figure 7.11. *Illustration of wheelchair with forward-folding design (side view).*

fiber and in the future carbon fiber frames will probably begin to become more available. All of these materials have their strengths and weaknesses.

7.4.1. Aluminum

Aluminum is a lightweight non-ferrous metal. Commonly aluminum wheelchairs are TIG (tungsten inert gas) welded. Aluminum is electrically welded together in a cloud of inert gas. Aluminum wheelchairs are sometimes bolted together using lugs (lugs are fittings made to match pieces of a joint). Most aluminum wheelchair frames are constructed of round drawn

6061 aluminum tubing. This is one of the least expensive and most versatile of the heat-treatable aluminum alloys. It has most of the desirable qualities of aluminum. It has good mechanical properties and high corrosion resistance. It can be fabricated using most standard techniques. If annealed it has good workability. In the T4 (solution heat treated, then naturally aged) condition fairly severe forming operations can be performed. Full T6 properties may be obtained by artificial aging. This type of aluminum (6061) is available, in clad form ('Alclad'), with a thin surface layer of high-purity aluminum to improve appearance and corrosion resistance. 6061 is used because of its availability, appearance, corrosion resistance, good strength and low weight.

7.4.2. Steel

Most steel wheelchairs are made of chromium–molybdenum alloy (4130 or 4140) seamless tubing commonly called chro-moly or mild steel (1010 or 1020). Chro-moly is one of the most widely used steels because of its weldability, ease of fabrication, mild hardenability and high fatigueability. Thin section may be heat treated to high strength levels. It may be treated, nitrided, for increased wear and abrasion. Chro-moly is usually either TIG welded, or brass welded. It has a high strength to weight ratio. Steel chairs are made of tubing 0.028–0.065 inches in wall thickness and diameters vary depending on the expected loads from between 0.25–1.25 inches.

7.4.3. Titanium

An increasing number of the more expensive wheelchairs are made of titanium. Titanium is a lightweight, strong, non-ferrous metal. Titanium wheelchair frames are TIG welded. Titanium is electrically welded together in a cloud of inert gas. It requires a highly skilled welder to make suitable welds with titanium. Titanium is the most exotic of the metals used in wheelchair production and the most expensive. Titanium requires special tooling and skill to be machined. It has very good mechanical properties and high corrosion resistance. It is resilient to wear and abrasion. Titanium is used because of its availability, appearance, corrosion resistance, very good strength and low weight. A drawback of titanium, besides cost, is that titanium, once worn or if flawed, may break rapidly (i.e., it has a tendency towards brittle fractures).

7.4.4. Advanced composites

Advanced composites have been in use in aerospace applications for a number of years. These materials include Kevlar and carbon fiber. These materials are now making the transition to wheelchair design (figure 7.12). Kevlar is an organic fiber which is yellow in color and soft to the touch.

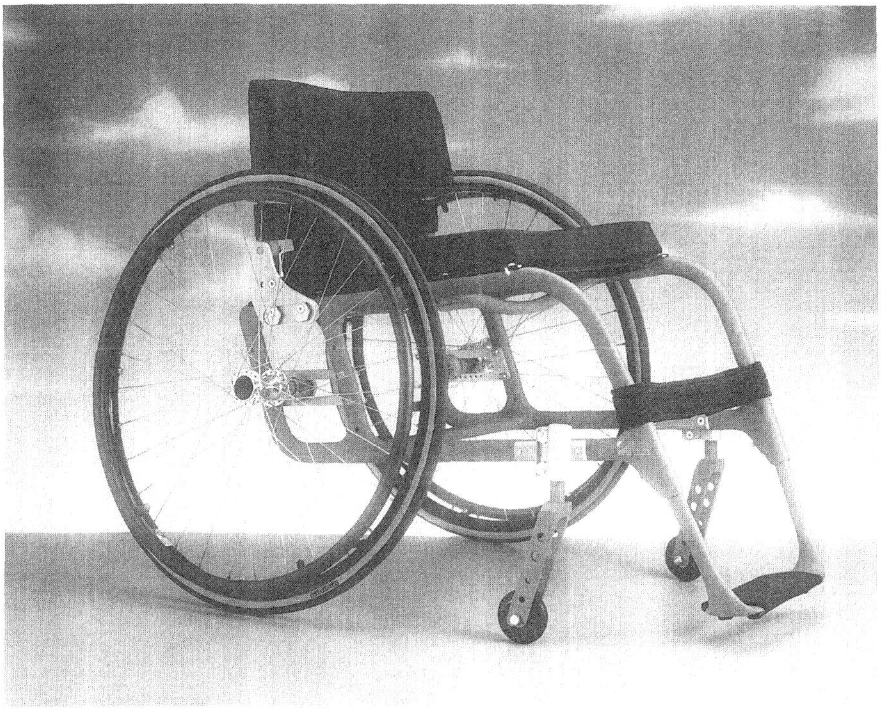

Figure 7.12. *Carbon-fiber composite rigid, box-frame, ultralight wheelchair with roller-blade casters. (Photograph courtesy of Everest and Jennings International, Ltd.)*

It is extremely strong and tough. It is one of the lightest structural fabrics on today's market. Kevlar is highly resistant to impact, but its compression strength is poor. Carbon fibers are made by changing the molecular structure of Rayon fibers by extreme stretching and heating. Carbon fiber is very stiff (high modulus of elasticity), very strong (high tensile strength) and has very low density (weight for a given volume). Composites come as cloth or yarn. Composite cloth is woven into bidirectional or unidirectional cloth. Unidirectional weaves can add strength along a particular direction. Composites must be bound together by resin or epoxy. Generally polyester resins or various specialty epoxies (e.g., safe-T-Poxy) are used. To achieve greatest strength a minimum amount of epoxy must be used while wetting all of the fibers. This is often achieved through a process called bagging. During bagging the thoroughly wetted piece is placed in a plastic bag and a vacuum is applied to remove the air from the bag and to pull the excess resin from the cloth. To increase the strength and stiffness of structural components, a foam (e.g., styrofoam, urethane, or PVC) core is used. The strengthening occurs because of the separation of the cloth layers (it now becomes more

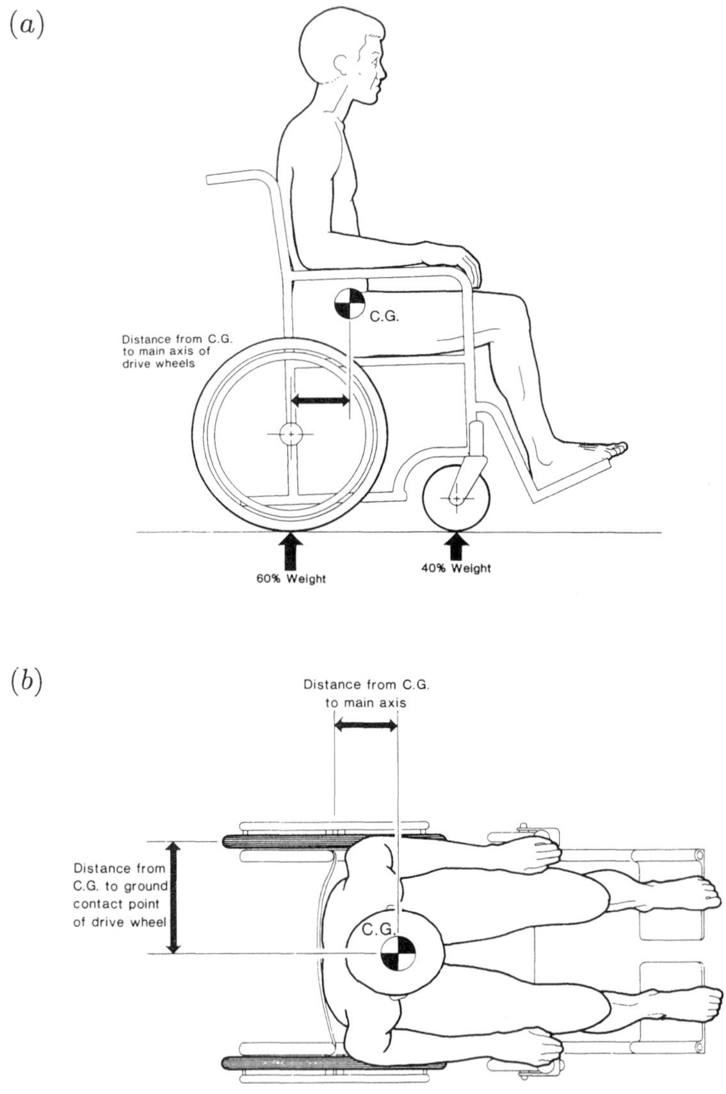

Figure 7.13. *The location of the COG for a hypothetical wheelchair with rider. (a) The location of the COG affects the distribution of mass on each of the wheels. Front casters typically exhibit higher rolling resistance than the larger rear drive wheels. A depot wheelchair may have 40% of the total weight placed on the front wheels whereas a sports chair may have only 15%. Shifting the COG may reduce stability in some directions. (b) The downhill turning tendency is related to the COG location with respect to the contact point of the rear drive wheels. The magnitude and distribution of the mass and the geometry of the chair are elements of the moment of inertia which affect the wheelchair dynamics. (Continued opposite)*

(c)

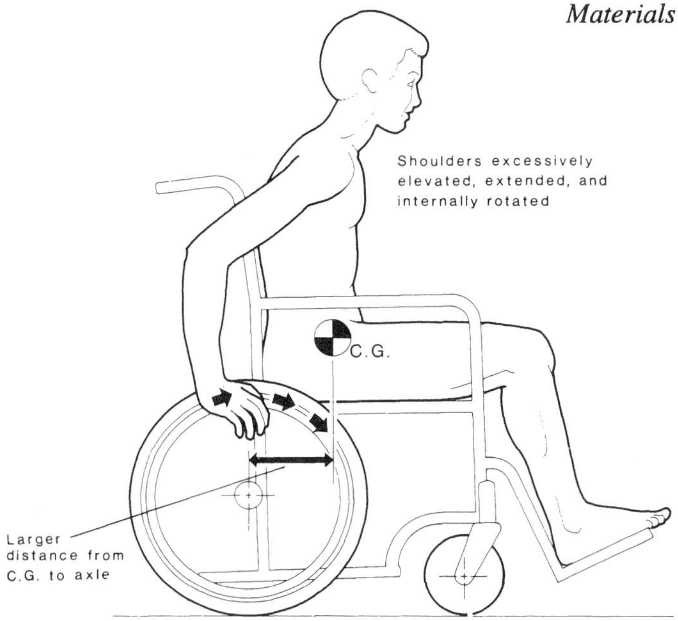

(d)

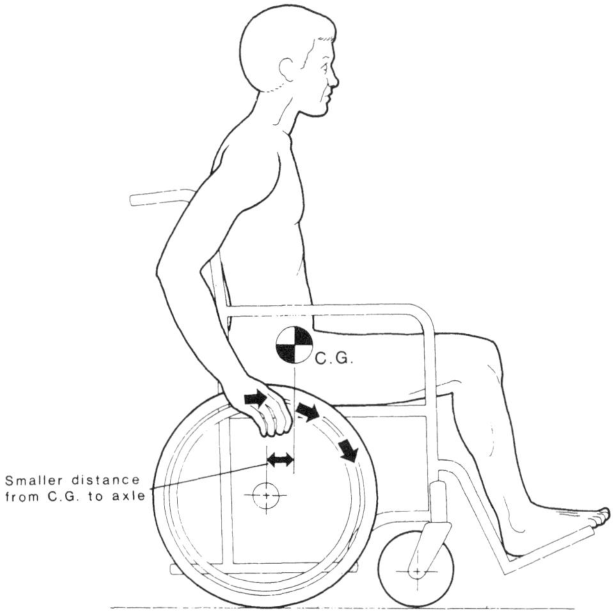

Figure 7.13. *Continued. (c) If the user sits too far forward or too low, the shoulders will be excessively elevated, extended, and internally rotated which may lead to rotator cuff injury. (d) When the user is seated properly, the shoulder position is more neutral and the propulsion phase is more balanced between horizontally and vertically directed motion. (Figure reproduced with permission from Choosing a Wheelchair System, Journal of Rehabilitation Research and Development—Clinical Supplement No 2, US Department of Veterans Affairs, Baltimore, MD.)*

like a tube than a flat sheet). Composite can be molded into elaborate shapes which opens a multitude of possibilities for wheelchair design.

7.5. THE WHEELCHAIR AND RIDER

Simplicity of form and function are the key guidelines to the design of a desirable wheelchair. The handling of a wheelchair is primarily dependent upon a few factors: the location of the center of gravity of the person and chair, the stiffness of the frame, wheels, and components and how well the pilot and chair mesh together. This last criterion is the most difficult to comprehend and the least understood.

7.5.1. Center of gravity (COG) location

Figure 7.13 illustrates the location of the COG for a hypothetical wheelchair with rider. The COG is typically located along the midline of the person and chair, and generally bilateral symmetry is assumed. The COG in the sagittal plane (fore and aft) should be located slightly forward of the rear axles. This depends on the user, as having the COG located near the rear axles makes the person and chair more likely to flip backwards (wheelie). However, having the COG near the rear axles has certain advantages: the downhill turning moment due to crown (slope for drainage on roadways and sidewalks) is minimized as the turning torque is related to the mass of the individual and the wheelchair and its location relative to the pivot point (the rear axles); the tendency for caster flutter is also reduced; the rolling resistance is reduced as most of the weight is born by the larger rear wheels; and the user-initiated turning ability is increased. The further the COG gets from the midpoint between the front and rear wheels and the closer it gets to the rear axles the greater the control effort and ability required to maintain balance. Thus in general the greater the functional ability of the user the closer the COG can be to the rear axles. There is going to be some variability due to personal preference and intended use. For most chair/user systems the mass of the system is dominated by the user. Thus the COG can be adjusted by adjusting the position of either the rear wheels with respect to the seat or vice versa. Generally this is an option left to the user. However, if the desired location of the COG is known, then the frame can be made lighter by designing fixed axle and seating structures.

7.5.2. Ride comfort and durability

Ride comfort is primarily a function of the following: frame, wheel and component stiffness, frame geometry, seat and cushion design and compliance. Ride comfort is an important issue because people must sit in the chair from 12 to 18 hours per day every day. Because some chairs are used by active users they may be used nearly ten miles per day over a variety

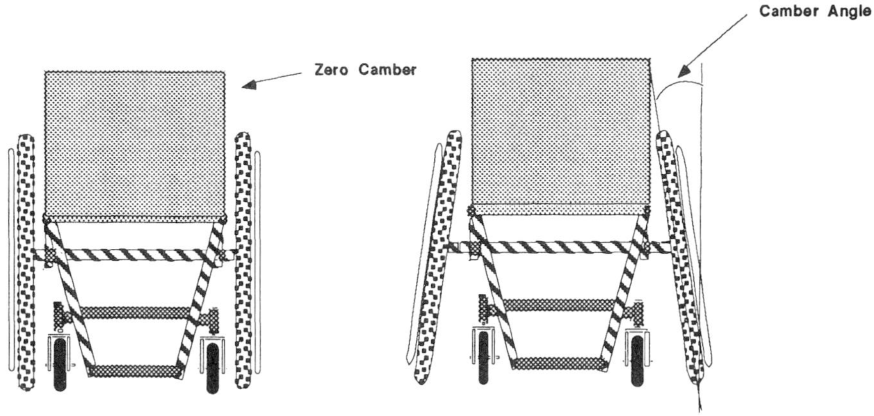

Figure 7.14. *Definition of rear wheel camber.*

of terrains (i.e., grass, carpet, gravel, concrete, asphalt, etc). The frame design must be durable but should have sufficient flexibility to withstand dropping off curbs (up to 35 cm high), and also not to break after several thousand miles of road vibration. Most wheelchair suspension comes from the cushion and the tires. Typically wheelchairs will use pneumatic tires and will have a foam, gel, or air cushion. The frame geometry can be used to reduce the effects of the impulse observed when going off a curb, or to minimize road vibration. Some suspension can be gained through the use of wheel camber (figure 7.14). By adding camber to the rear wheels the effective stiffness between the rolling surface and frame is reduced. Wheel camber has the additional advantages of increasing the footprint which increases lateral roll stability, forcing the bottom of the wheels to touch objects (i.e., doorways, walls, other chairs) first when skirting by them, thus protecting the user's hands, and enabling quicker turning and more natural stroke (arm) biomechanics. By optimizing the ride comfort of the wheelchair the time of use can be extended and the mobility of the user can be improved.

Durability is also a very important issue in wheelchair design. Wheelchairs may be used by demanding riders and/or receive little maintenance. The same wheelchair is often used for work or school and recreation. Therefore the chair must be designed to meet these demands. Typically a wheelchair will be used all day everyday for all major life activities (i.e., activities of daily living, work, school, recreation, sport) over a period of three to five years. The wheelchair must be designed to regularly withstand obstacles (i.e., rocks, curbs, bumps, dips, impacts) encountered during a variety of activities. Making the frame too flexible will cause it to absorb the energy of the rider and make it inefficient. Making the chair too stiff may cause it to fracture due to shock and will reduce ride comfort.

7.5.3. The user–wheelchair interface

The user–wheelchair interface is the most critical design factor, and is the least understood. It will require many more years of study before it can confidently be said that the user–wheelchair interaction is well understood. Essentially the user and the wheelchair must act as one. The chair is to be designed such that the user's potential is maximized. The chair should not limit the user. Experience tells us that there is a great deal of variability among individual preferences. However, for active wheelchair users there are some general concepts that are known to apply. The chair must become an extension of the user's body, much like an orthosis. This requires carefully matching critical chair dimensions to body dimensions, user ability, and intended use.

There are several critical seat dimensions to be considered during wheelchair design: seat height, seat angle, seat depth, seat width, back height and backrest angle. The seat base should be made of some stiff material to provide a rigid base of support for the cushion. Either a pliable material (e.g., canvas, nylon) stretched tightly across the side frames or a rigid base (e.g., plastic, aluminum, composite) works well. Seat height is defined as the distance between the lowest point on the seat base and the floor on a level surface. Seat height varies based upon desired use. In general while sitting upright with the hands resting on the top of the wheels, the elbow angle should be between 100 and 120 degrees for optimal mobility (figure 7.15).

Athletes who compete in sports where height is advantageous (e.g., basketball, field events) will choose higher seats while those requiring greater agility (e.g., tennis, slalom, racquetball) will choose lower seats. Seat heights vary from about 300 to 600 mm. The seat height is also dependent upon the total body length of the user. Users with longer lower-leg lengths will require higher seat heights to achieve sufficient clearance for the footrests. There is some flexibility when selecting seat height and even for taller individuals, because most active users prefer some seat angle (figure 7.16). By tilting the seat towards the back-rest the user fits more securely into the chair and the chair becomes more responsive to the user's body language. Seat angle also gives the user some pelvic tilt which provides greater trunk stability. Seat angle can be as little as zero degrees and as much as 20 degrees. Seat depth is determined from the length of the upper legs. Generally no more than a 75 mm gap should be between the front of the seat and the back of the knees when the person is in the wheelchair (figure 7.17). This will help to ensure broad distribution of the trunk weight over the buttocks and upper legs, without placing undue pressure behind the knee. Some gap is required to allow the user some freedom to adjust position. Seat width is determined from the width of the person's hips, the intended use, and whether the person prefers to use side-guards (pieces of plastic, aluminum or steel sheet placed between the seat and rear wheels to prevent clothing from rubbing on the rear wheels) or not (figure 7.18). Generally, the wheelchair should be as

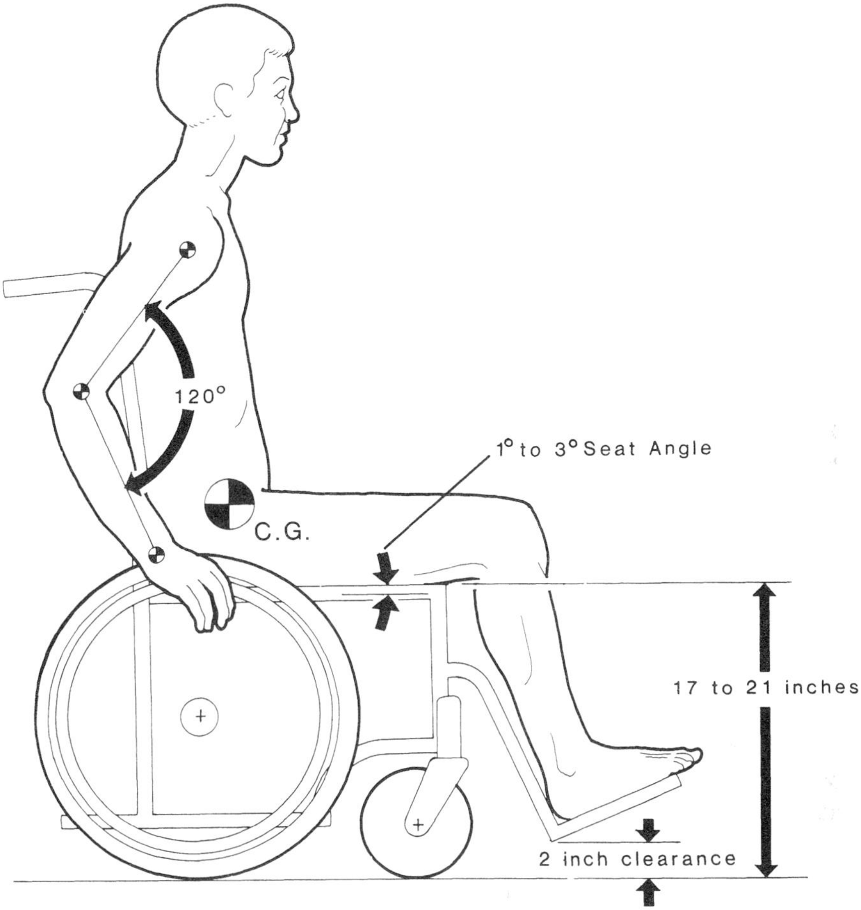

Figure 7.15. *Sagittal plane illustration of seat height and angle for proper posture and efficient propulsion. Safe and effective propulsion can be assisted by: placing the elbows at about 120 degrees of flexion when grasping the pushim at top dead center; placing the seat angle between one and three degrees; and setting the footrest ground clearance to 50 mm. The user's cushion must be in place when all measurements are taken. (Figure reproduced with permission from Choosing a Wheelchair System, Journal of Rehabilitation Research and Development, Clinical Supplement No 2, US Department of Veterans Affairs, Baltimore, MD, 1990.)*

narrow as possible. Thus a chair about one inch wider than the user's hips is desirable. Side-guards or armrests can help to keep clothing clean thus making the chair more acceptable to the user. Seat widths for adults typically range from 250 to 500 mm.

Backrest height, one of the most critical dimensions, is dependent upon the user's disability etiology, intended use and wheelchair skills. The comfort,

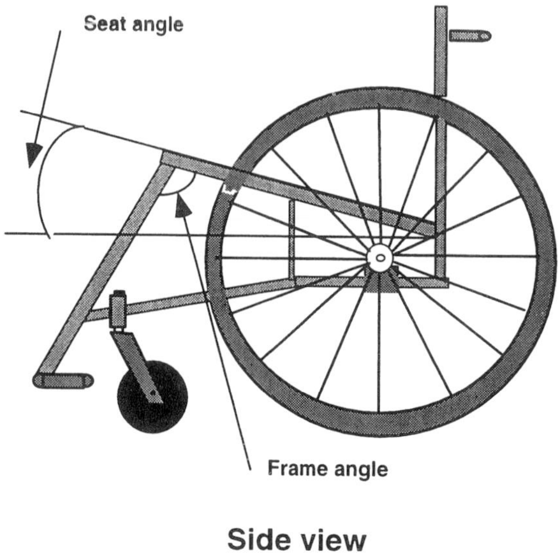

Figure 7.16. *Sagittal-plane schematic diagram of seat angle.*

stability and control of the wheelchair are influenced by backrest height. Generally, the lower the mobility impairment the lower the desired backrest height. Commonly, backrest height is made to be adjustable so that the user can tune it to his/her liking. A lower backrest height permits greater freedom of motion (leaning, turning) and is less restrictive (i.e., does not interfere with arms when pushing), but provides less support than a high backrest. Backrest heights may vary from 200 to 500 mm. The backrest should be made of a padded pliable material or when made of rigid material it not only should be padded but it must be mounted such as to conform to the user's changing body position. It is sometimes desirable to taper the backrest to provide for comfortable support of a user's well developed upper torso. The backrest angle is often set so that the backrest is vertical with respect to a level floor.

Most wheelchair users require support for their feet and lower legs. Support is provided by footrests. Footrests may be fixed, folding or swing-away (figure 7.19). The footrest must provide sufficient support for the lower legs and feet and hold the feet in the proper position to prevent footdrop. The feet must remain on the footrests at all time during propulsion and therefore some type of cradle is recommended. Some wheelchairs (primarily those with swing-away footrests) use foot stirrups behind the heels of each foot. However, for other wheelchairs it is probably best to use a continuous strap behind both feet as feet sometimes come over stirrups during active use, which is not possible with a strap. The feet should not be permitted to be pinched or trapped or scratched by the footrests during normal activities or

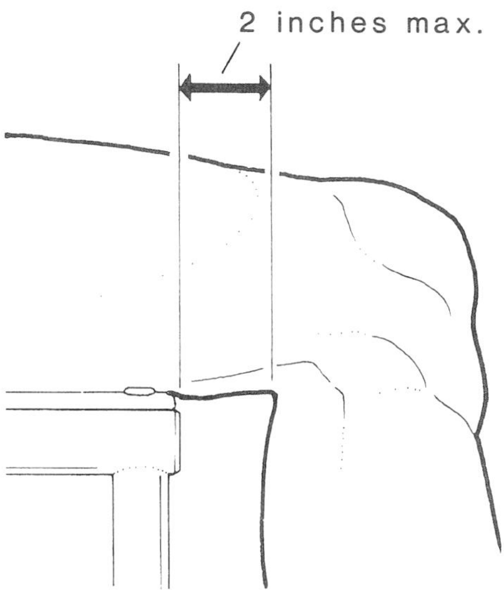

Figure 7.17. *The seat and thighs must be well supported to provide sufficient pressure relief. However, undue pressure should not be placed on tissue around the knees. (Figure reproduced with permission from Choosing a Wheelchair System, Journal of Rehabilitation Research and Development, Clinical Supplement No 2, US Department of Veterans Affairs, Baltimore, MD, 1990.)*

when transferring. The frame should be designed so that the feet sit firmly upon the footrests with shoes on without lifting the upper legs from the seat cushion. Care must be taken that sufficient ground clearance is maintained. The footrests are commonly placed between 25–500 mm from the ground. Often the footrests are the first part of the chair to come in contact with an obstacle (e.g., door, wall, another chair), therefore they must be durable.

The wheelbase and width affect handling and performance. The wheelbase affects caster flutter, rolling resistance, stability, controllability, and the obstacle performance. Specifically, a longer wheelbase chair becomes more stable, but less controllable. Therefore, a long chair may be less apt to tip over but the user may not be able to negotiate common terrain. Typically, active use wheelchairs are very controllable and maneuverable and hence less stable. This is because users prefer control over the chair to stability. A longer wheelbase makes negotiating obstacles like curbs easier as the angle of ascent for the center of gravity is reduced. This is an important consideration. A short wheelbase is more maneuverable and can get the user closer to counters, etc. However, a short wheelbase may move more weight over the front wheels which may increase rolling resistance and reduce the speed of onset of caster flutter. Obviously, trade-offs need to be made.

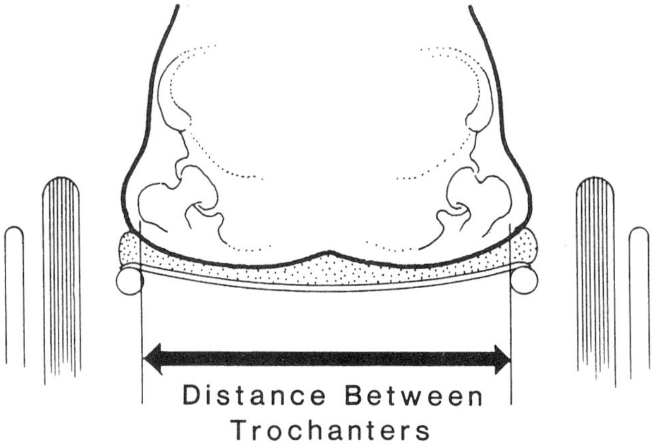

Figure 7.18. *The width of the wheelchair should be narrow so that the arms can be held close to the trunk while pushing. For many wheelchairs the trochanters should just fit inside the side frame of the wheelchair seat. (Figure reproduced with permission from Choosing a Wheelchair System, Journal of Rehabilitation Research and Development, Clinical Supplement No 2, US Department of Veterans Affairs, Baltimore, MD, 1990.)*

Typically, the wheelbase of wheelchairs is between 250 to 500 mm. The width of the footprint of the wheelchair is also an important design variable. Often the camber is variable (figure 7.14). Camber has several advantages: the footprint of the chair is widened creating greater side to side stability; camber allows quicker turning; camber helps to protect the hands by having the bottom of the wheels scuff edges, preventing them from hitting the area where the hands are in contact with the pushrims; and camber positions the pushrims more ergonomically for propulsion (it is more natural to push down and out). Camber angle can either be fixed or adjustable depending on the frame design. The width of the chair depends upon the width of the frame and the camber angle. Frame widths for adults range from 300 to 500 mm and camber angles are usually between 0 and 15 degrees, mostly between 7 and 12 degrees. Thus wheelchair widths vary from about 550 to 750 mm. Generally, for daily use the chair should be as narrow as possible without substantially diminishing the handling characteristics. The wheels should be offset enough from the seat to avoid rubbing against the clothing or body. Narrow chairs are easier to maneuver in an environment made for walking.

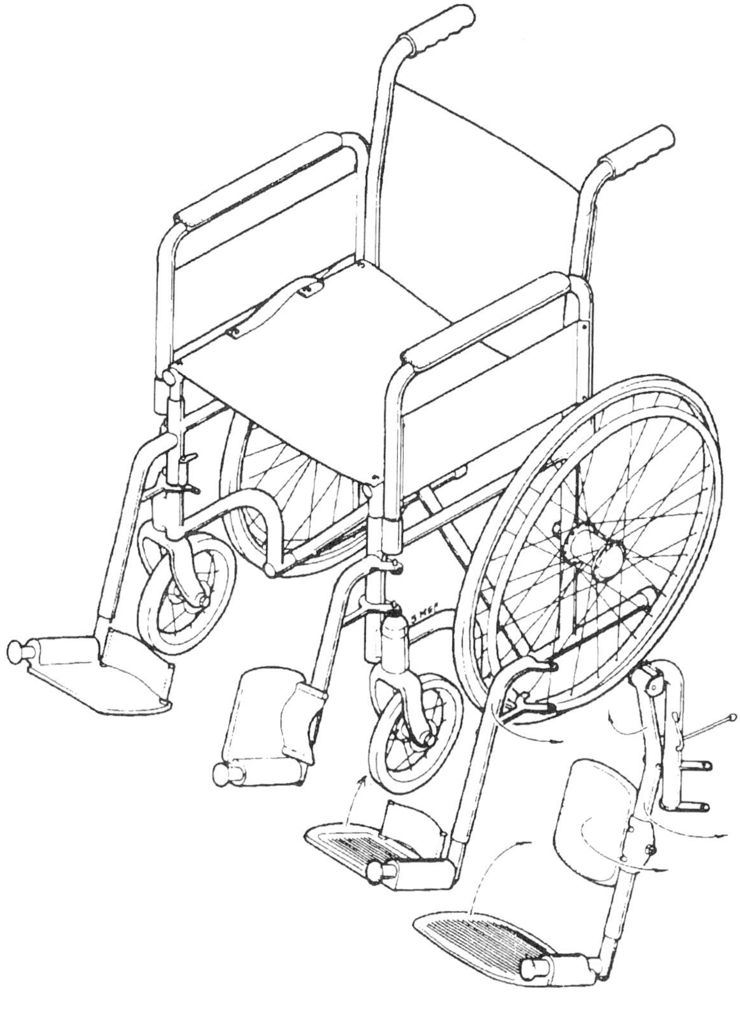

Figure 7.19. *Many different types of leg and footrests are used on manual wheelchairs. Rigid wheelchairs often use simple tubes across the front of the wheelchair. Folding wheelchairs often use foot and legrests which fold up and swing out of the way to ease transfers. Elevating legrests can be used for people who cannot maintain a 90 degree knee angle or who need their legs elevated for venous return. (Figure reproduced with permission from Wheelchairs: A Prescription Guide 2nd edition, A Bennett Wilson, Jr, Demos, New York, NY, 1992.)*

7.6. WHEELS AND CASTERS

Wheels and casters are available in a wide range of styles and sizes (figure 7.20). The front casters can range from 50 to 200 mm in diameter for manual wheelchairs designed for daily use. Casters are either pneumatic or polyurethane. Pneumatic casters offer a smoother ride at the cost of durability, whereas polyurethane casters are very durable. Most active users prefer 125 mm polyurethane casters for daily use. However, 200 mm polyurethane casters offer better ride comfort at the expense of foot clearance. Caster foot clearance is maximized with 50 mm 'Roller Blade' casters often used for court sports (e.g., basketball, tennis and racquetball). The styling of the casters often reflects a personal statement by the user. Rear wheels come in three common sizes 22, 24 and 26 inch. They come in two styles: spoked and MAG (figure 7.21). MAG wheels are typically made of fiber-reinforced ABS (acrilonitrile–butadiene–styrene) or PVC plastics and are die cast. MAG wheels require minimal maintenance and wear well. However, spoked wheels are substantially lighter, more active and are generally preferred by sports wheelchair users. Rear tires can be of two types; pneumatic or puncture proof (figure 7.21). Pneumatic tires can either use a separate tube and tire or a combined tube and tire (sew-up). Commonly, a belted rubber tire with a Butyl tube (65 psi) is used. However, those desiring higher performance prefer sew-up tires or Kevlar belted tires with high pressure tubes (180 psi). Puncture-proof tires are heavier, provide less suspension and are less lively than pneumatic tires. Pneumatic tires are preferred. Pushrims are metal rings attached to the drive wheels used to propel the wheelchair. Pushrims are commonly slick aluminum tubing bent into circles a few centimeters smaller in diameter than the drive wheels. Pushrims can also be coated or have protrusions attached (i.e., protruding pushrims) for use by people with limited grasp (figure 7.22).

The wheels have a profound effect on the performance of the wheelchair. After choosing the casters and rear wheels the chair must be designed to optimize the interaction of the wheels with the ground. Five critical performance factors need be considered: (1) caster flutter, (2) caster float, (3) tracking and (4) alignment.

7.6.1. Caster flutter

Caster flutter is the shimmy (rapid vibration of the front wheels) that may occur on some surfaces above certain speeds. Caster flutter results when there is not enough trail (figure 7.23). Trail determines the tendency of the front wheels to follow the path of the caster housing. It is desirable to have enough trail to prevent caster flutter for the maximal speed expected for the chair, yet not have the front wheels interfere with the placement of the rider's feet. This has become a rather complex trade-off as many active riders prefer to have their feet tucked in nearly directly below the knees. In order to achieve a long enough wheelbase to yield a reasonably stable base and have enough trail to minimize caster flutter, caster size, trail, and foot placement all have to be evaluated. Generally, trail of 63 to 75 mm will

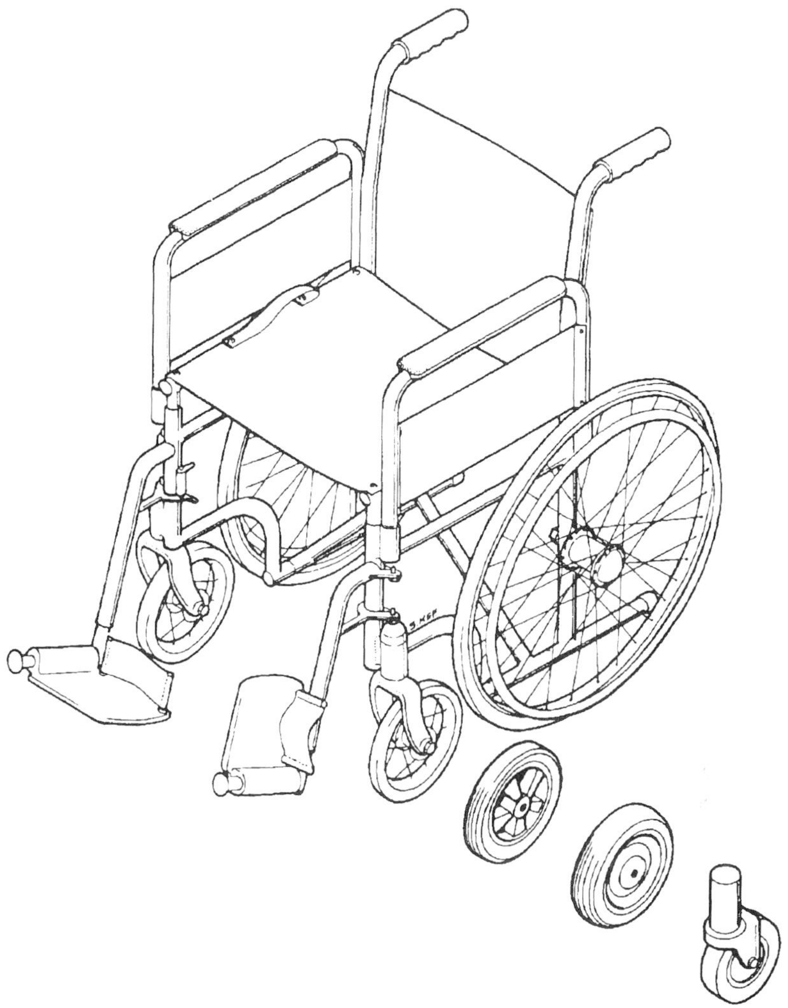

Figure 7.20. *Wheelchair casters come in a variety of styles and sizes. They can be pneumatic, solid or without tread. The widths of casters also vary. (Figure reproduced with permission from Wheelchairs: A Prescription Guide 2nd edition, A Bennett Wilson, Jr, Demos, New York, NY, 1992.)*

prevent noticeable flutter; however, this is dependent upon the size and type of wheel, weight distribution of the chair and rider and frame geometry. Simple alternatives to reduce caster flutter include using a viscous silicon-based grease in the caster housing bearings, or to place an oversized washer in between a bearing and the shaft of the caster fork (add a small amount of friction).

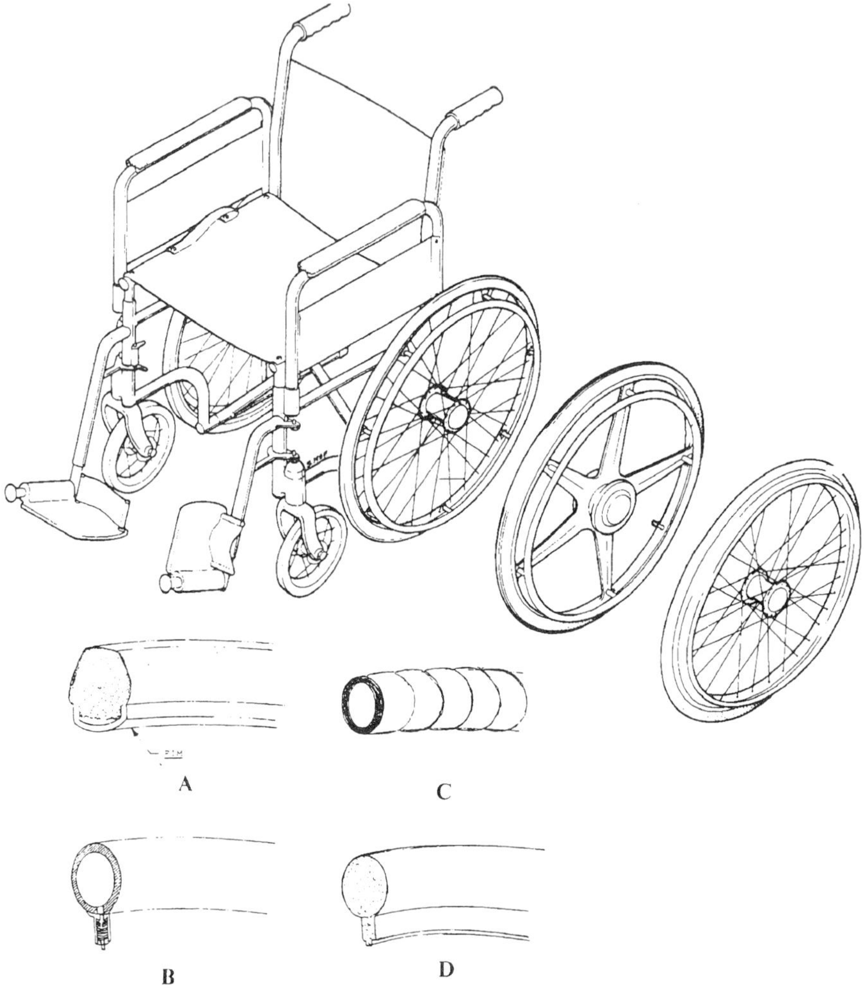

Figure 7.21. *Rear wheels and tires come in many types. Spoked wheels can be strong and lightweight. MAG wheels are low maintenance. Tires can be made of expanded rubber (A), tubular with pneumatic pressure (B), semi-pneumatic foam and rubber with valveless insert (C) and solid rubber tires (D). (Figure reproduced with permission from Wheelchairs: A Prescription Guide 2nd edition, A Bennett Wilson, Jr, Demos, New York, NY, 1992.)*

7.6.2. Caster float

When one of the casters does not touch the floor when on level ground, the wheelchair has caster float. Caster float decreases the stability and performance of the wheelchair. If the wheelchair has caster float a few items should be checked. Unequal rear-wheel camber will cause caster float (most of the weight of the rider/chair system is over the rear axles). Thus

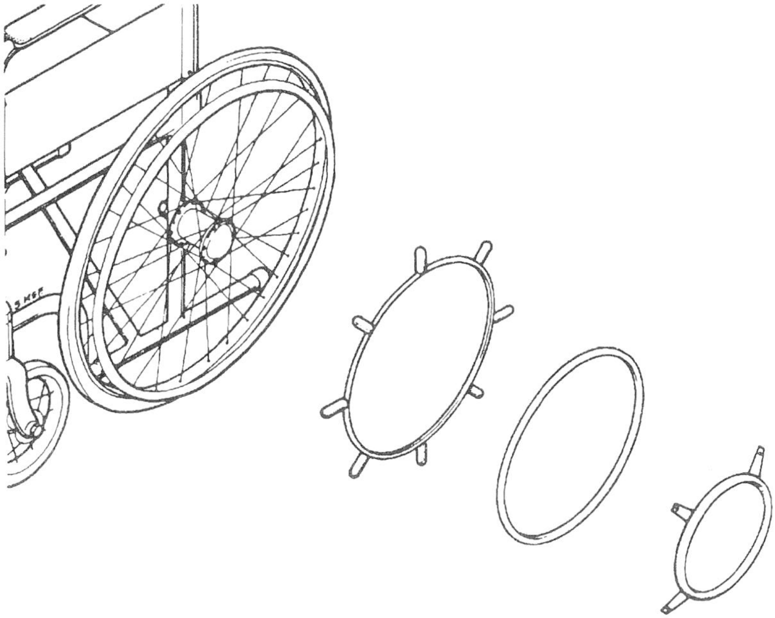

Figure 7.22. *Pushrims are use to protect the hands and propel the drive wheels. Normally aluminum or steel tubing is used to construct pushrims. Pushrims can be used with the natural metal finish or coatings can be added. Protruding pushrims can be used by people with limited hand function. (Figure reproduced with permission from Wheelchairs: A Prescription Guide 2nd edition, A Bennett Wilson, Jr, Demos, New York, NY, 1992.)*

correcting the camber may eliminate the problem. If both caster housings are not vertical, caster float will result. Setting the orientation of both housings to vertical with a level should eliminate caster float. One caster may be excessively worn compared with the other; replacing the casters will solve the problem. Otherwise, placing a spacer under the floating caster will solve the caster float.

7.6.3. Tracking

As manual wheelchairs use rear-wheel steering via differential propulsion torque, tracking is the tendency of the wheelchair/rider to maintain its course once control has been relinquished. Tracking is important, as the rider propels the handrims periodically (about every second) and if the chair does not track well it will drift from its course between pushes and force the rider to correct the steering. This will waste valuable energy and reduce control over the chair. In sports like basketball, tracking is extremely important. To ensure tracking does not present a substantial problem, all four wheels must be pointing in the forward direction when in their natural (resting) position.

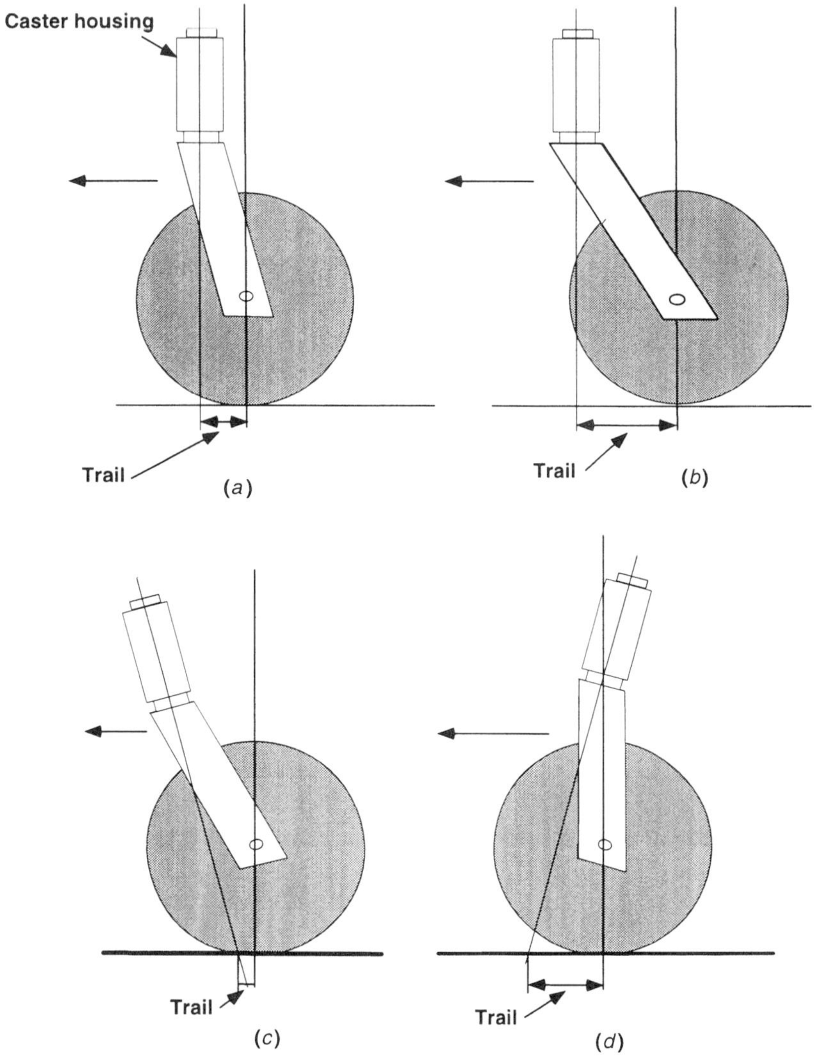

Figure 7.23. *Caster trail is required for a manual wheelchair to be capable of turning. Trail can be obtained through a variety of geometries. Commonly, the caster stem is adjusted to be vertical and the axle of the caster follows the center of the caster stem by about 50 mm.*

If the caster housings are tilted, such that they have a tendency to rest in other than a forward pointing direction, tracking will be affected. If the rear wheels are misaligned, tracking can present a problem. Tracking can be easily checked by rolling through a puddle of water and the allowing the chair to coast: the chair should maintain its direction.

7.6.4. *Wheel alignment*

Alignment generally refers to the orientation of the rear wheels with respect to one another. Typically, it is desirable to have the rear wheels parallel to one another without any difference between the distance across the two rear wheels at the front and back. Misalignment on the order of 2 mm can cause a noticeable increase in the effort required to propel the wheelchair. Alignment is adjusted in various ways depending on how the rear wheels are mounted to the frame. Most sports wheelchairs use either axle plates (figure 7.24(*a*)–(*c*)) or a tube with an axle insert pressed into it (figure 7.24(*d*)–(*f*)). If an axle plate is used then spacers are added (around the mounting) between the plate and frame to align the wheels and adjust camber. If an axle tube configuration is used, then alignment is made by rotating the tube until the desired alignment is achieved. In the latter case, camber is fixed.

7.7. COMPONENTS

Several components can be added to the basic chair design to provide riders with customized features. Some of the more common additional components are wheel locks, armrests, anti-tip casters, footrest rollers, side-guards and quick-release axles.

Wheel locks act as parking brakes and assist when transferring to other seats and when the rider wishes to remain in a particular spot (figure 7.25). They help to allow the rider to push things and to be more stable when desired. Armrests provide a form of support and are convenient handles when the rider leans to one side or the other. Anti-tip casters are small wheels mounted at the back of the wheelchair which help to prevent the wheelchair and rider from tipping backwards. They are used by many neophyte wheelchair users and by some basketball players. Footrest rollers are used in court games where the rider and wheelchair may tip forwards onto the footrests. The rollers help to prevent the wheelchair from flipping forwards. Side-guards are supports which come up the side of the wheelchair frame and help to prevent the hips and clothing from rubbing against the wheels. They also provide a more snug fit to the chair. Quick release axles allow the rear wheels (and sometimes the front casters) to be removed by simply pushing a button. Quick release rear axles are commonly just 12 mm diameter ball lock pins. Other accessories and components can be added to fit the rider's needs.

7.8. HUMAN FACTORS DESIGN CONSIDERATIONS

To properly design or specify it is important to understand the intentions and abilities of the user. The best designed wheelchair will not be successful if

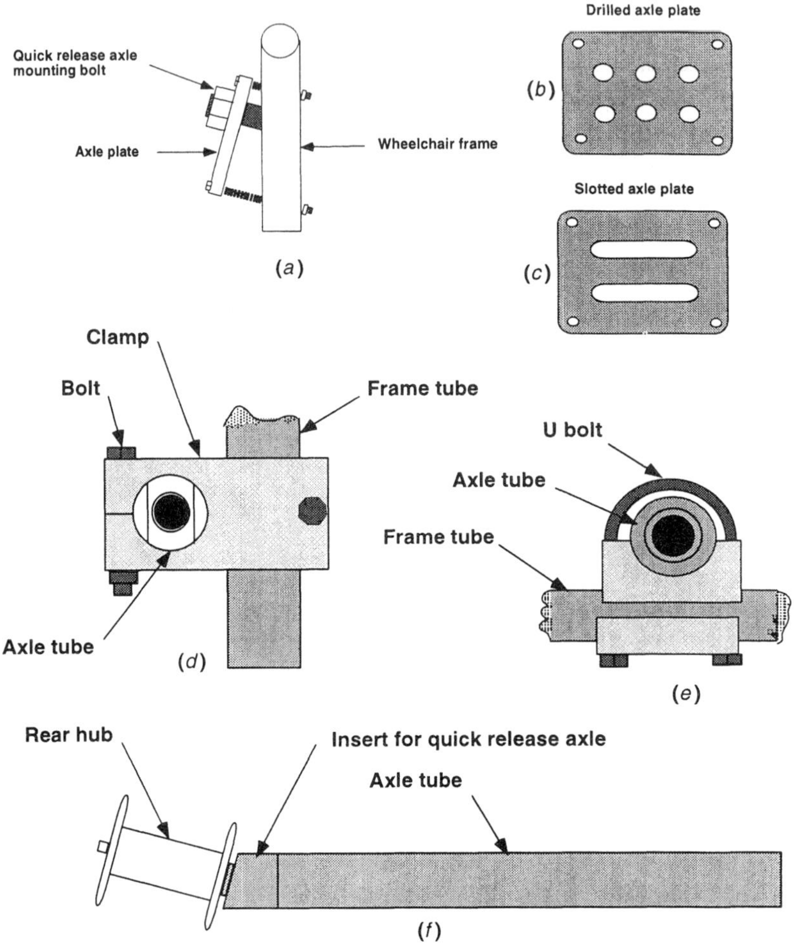

Figure 7.24. *Common rear-wheel mounting mechanisms.*

rejected by the user. Therefore the needs, desires, and abilities of the user must be ascertained and incorporated into the wheelchair design.

7.8.1. What are the intended uses?

The design of the wheelchair depends upon the intended uses of the wheelchair. It is no longer acceptable to stock a particular model of wheelchair and to provide it to all new wheelchair users. Manufacturers who do not follow the intended uses of the consumer will soon find themselves wanting for business. The intentions of the user must be ascertained *a priori* as must the activities and the environment of the wheelchair and the user.

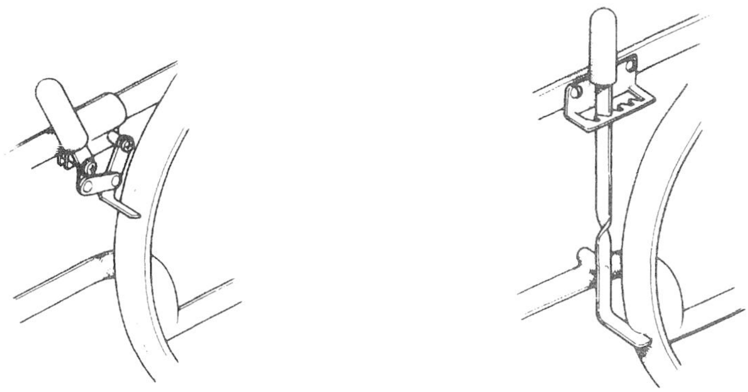

Figure 7.25. *There are a variety of wheel locks used to restrain wheelchairs when transferring or parked. High lock brakes are most common. Some wheel locks are over-center and are either engaged or disengaged (a). Other wheel locks allow selection of the braking force (b), these are sometimes called sweeper brakes as the wheelchair can still be pushed in some positions, which allows the user to push and then sweep a broom to clean a floor. (Figure reproduced with permission from Wheelchairs: A Prescription Guide 2nd edition, A Bennett Wilson, Jr, Demos, New York, NY, 1992.)*

This will assist with determining the geometry, components and durability of the wheelchair.

7.8.2. What are the abilities of the user?

The abilities of the user must be matched with the intended use of the wheelchair. For some users many of the desired tasks may be simply accomplished with existing technology; others will require custom products and yet others will not be able to achieve their goals with existing knowledge. The type of chair and how it is set-up depends upon the interaction between the user and the intended use (person, place, task). The real abilities of the user as related to the intended uses must be assessed.

7.8.3. What are the resources available?

Financial and physical resources are required to design and purchase wheelchairs. The limits of the available resources must be weighed against the abilities and desires of the user. Resources are always limited and new technology can be costly. The user should be provided the best possible product with the available resources. Assistive technology is considered to be medical equipment and thus extreme caution must be taken to ensure the safety of all potential users. Product liability can result in substantial penalties.

7.8.4. What are the existing products available?

The rehabilitation engineer must remain current with respect to new product developments. There is no sense in reinventing the wheel. Often an existing product can be adapted to meet the user's needs. Caution must be observed when modifying existing technology for applications not intended by the manufacturer. The rehabilitation engineer should evaluate possible problems which could result from any modifications. Often products in other areas may prove useful in the design and development of wheelchairs. The rehabilitation engineer should avoid becoming too narrowly focused.

7.9. FUTURE DIRECTIONS

Adaptability and adjustability will play key roles in the future of wheelchair design. Many users want to have one wheelchair to use to play racquetball, basketball, tennis, etc, and for use at work. The characteristics of the wheelchair need to be adaptable for this to be possible. Adjustability without compromising performance is also going to become more important. Adjustability might mean flexible manufacturing, or it might mean the development of simple means of personalizing the wheelchair. Each rider has his/her preferences that often require some adjustment after using the wheelchair for a while.

More needs to be learned about optimum stiffness for energy transfer, durability and ride comfort. There will be greater application of high-tech materials, analytical tools and manufacturing processes. The rider–wheelchair interface is only beginning to be understood and much more work is required in this area. Studies are being performed on durable and inexpensive high-performance tires. Front casters also require study. What are the desired rolling and suspension properties of front casters and how do they depend upon the position of the center of mass of the wheelchair and rider?

Some wheelchairs will become less conspicuous, blending into the body of the person. They will become more an extension of the person in form and function. Other wheelchairs will become more flashy. Some riders prefer to flaunt the wheelchair and therefore will demand outrageously appearing designs with space for finishes that will express the rider's personality. The designs of wheelchairs will focus on performance and cost. A low-cost sports wheelchair market will probably develop in response to the high-tech revolution. Many riders who cannot afford the latest in a molded composite wheelchair will still want a sports wheelchair and will opt for a lower-cost model. With composite technology coming to the commercial market, there is opportunity to allow the imagination greater latitude for design.

FURTHER READING AND REFERENCES

Asato K T, Cooper R A, Robertson R N and Ster J S 1992 SMART^{Wheels}: development and testing of a system for measuring manual wheelchair propulsion dynamics *IEEE Trans. Biomed. Eng.*

Baldini F D, Cooper R A, Robertson R N, Jones D and Widman L 1992 The economy of wheelchair propulsion in wheelchair athletes *Med. Sci. Sports Exercise* **24** 532

Brubaker C E 1986 Wheelchair prescription: an analysis of factors that affect mobility and performance *J. Rehab. Res. Dev.* **23** 19–26

Brubaker C E, McLaurin C A and McClay I S 1985 A preliminary analysis of limb geometry and EMG activity for five lever placements *Proc. RESNA 8th Ann. Conf. (Memphis, TN, 1985)* pp 350–2

Collins T J and Kauzlarich J J 1988 Directional instability of rear caster wheelchairs *J. Rehab. Res. Dev.* **25** 1–18

Cooper R A 1991 High tech wheelchairs gain the competitive edge *IEEE Eng. Med. Biol. Mag.* **10** 49–55

Cooper R A, Asato K T, Robertson R N and Ster J S 1992 2-Dimensional kinetic analysis of manual wheelchair propulsion with an improved SMART^{Wheels} *Proc. 14th Ann. IEEE/EMBS Int. Conf. (Paris, 1992)* vol 14 1544–5

Cooper R A, Baldini F D, Bennett P, Robertson R N, Langbein E and Monical S 1992 Prediction of pulmonary function in wheelchair users *Paraplegia* **31** 560–70

Cravotta D 1991 Mobility for the masses *Mainstream* July 11–7

Department of Veterans Affairs 1990 Choosing a wheelchair system *J. Rehab. Res. Dev.—Clin. Suppl.* No 2 March 1990

Gaines R F and La W H T 1986 Users' responses to contoured wheelchair handrims *J. Rehab. Res. Dev.* **23** 57–62

Gordon J, Kauzlarich J J and Thacker J G 1989 Tests of two new polyurethane foam wheelchair tires *J. Rehab. Res. Dev.* **26** 33–46

Heil T 1992 Design of a dynamic wheelchair brake *Proc. RESNA Int. '92 (Toronto, Ont., 1992)* pp 643–5

Hotchkiss R D 1993 Ground swell on wheels *Sciences* July/August 14–9

Hotchkiss R D and Pfaelzer P 1992 Measuring success in third world wheelchair building *Proc. RESNA Int. '92 (Toronto, Ont., 1992)* pp 618–20

Kauzlarich J J 1990 *A New Maintenance-Free Wheelchair Tire* (Washington, DC: RESNA) pp 67–8

Kauzlarich J J, Bruning T and Thacker J G 1984 Wheelchair caster shimmy and turning resistance *J. Rehab. Res. Dev.* **20** 15–29

Kauzlarich J J and Thacker J G 1987 A theory of wheelchair wheelie performance *J. Rehab. Res. Dev.* **24** 67–80

Kauzlarich J J and Thacker J G 1991 Antiskid Wheelchair Brake Design *Proc. RESNA 14th Ann. Conf. (Kansas City, MO, 1991)* pp 143–5

Lemaire E D, Lamontagne M, Barclay H, John T and Martel G 1991 A technique for the determination of center of gravity and rolling resistance for tilt-seat wheelchairs *J. Rehab. Res. Dev.* **28** 51–8

Masse L C, Lamontagne M and O'Riain M D 1992 Biomechanical analysis of wheelchair propulsion for various seating positions *J. Rehab. Res. Dev.* **29** pp 12–28

McLaurin C A and Brubaker C E 1985 A lever drive system for wheelchairs *Proc. RESNA 8th Ann. Conf. (Memphis, TN, 1985)* pp 48–50

Nashihara H and Shizukuishi K 1992 Development of indoor wheelchair mainly made of wood *Proc. RESNA Int. '92 (Toronto, Ont., 1992)* pp 533–5

Pettit D, Cooper R A and Bennet P 1992 Design and evaluation of a simple, inexpensive, ultralight wheelchair *Proc. 14th Ann. IEEE/EMBS Int. Conf. (Paris, 1992)* vol 14

Quickie Designs Inc. 1989 *The Perfect Fit. An Adjustments Guide to Your Manual Wheelchair (Fresno, CA)*

Szeto A Y J and White R N 1983 Evaluation of a curb-climbing aid for manual wheelchairs: considerations of stability, effort and safety *J. Rehab. Res. Dev.* **20** 45–56

Veeger D, Van der Woude L H V and Rozendal R H 1989 The effect of rear wheel camber in manual wheelchair propulsion *J. Rehab. Res. Dev.* **26** 37–46

Veenbaas R 1993 A method for choosing the main measures of a hand-propelled wheelchair system, based on anthropometric sources and multivariate statistics *Proc. 2nd Eur. Conf. Advancement Rehabilitation Technol. (Stockholm, 1993)* p 3.4

Watanabe K T, Cooper R A and Ster J S 1991 SMART^{Wheels}: a device for studying wheelchair propulsion dynamics *Proc. 13th Ann. IEEE/EMBS Int. Conf. (Orlando, FL, 1991)* vol 13 pp 1817–8

Wilson Jr A B 1992 *Wheelchairs: a Prescription Guide* 2nd edn (New York: Demos)

CHAPTER 8

POWER WHEELCHAIR DESIGN

8.1. INTRODUCTION

Power wheelchairs provide greater independence to thousands of people with severe mobility impairments. Just as among the ambulatory population, mobility among people with disabilities varies. Mobility is more a functional limitation than a disability-related condition. Powered mobility can have tremendous positive psycho-social effects on an individual.

Power wheelchairs began in the 1940s as standard cross-brace folding manual wheelchairs adapted with automobile starter motors and an automobile battery. The cross-braced wheelchair remained the standard for a number of years, until people realized that frames could be designed specifically for power wheelchairs. When the rigid power wheelchair frame was developed, space became available under the seat for electronic controls, respirators, communication systems, and reclining devices. By the mid-1970s, power wheelchairs had evolved to the point where people had acquired a significant level of mobility. However, power wheelchairs were often cobbled together and left much to be desired. During the 1980s research was begun to systematically improve power wheelchairs design. Such factors as noise, efficiency, braking systems, control systems, and interface systems were improved.

It is important to realistically assess the abilities and needs of an individual before a power wheelchair is prescribed. Some people are impaired to an extent that they would have no mobility without a power wheelchair. In such cases, the choice is obvious. However, some people may have limited upper-body mobility and may have the ability to propel a manual wheelchair for short distances. These people may liken using a power wheelchair to admitting defeat. However, a power wheelchair may provide greater mobility. In such cases it may be best to suggest a power wheelchair for longer excursions, and a manual wheelchair for in-home and recreational use.

Power wheelchairs are often used in conjunction with a number of other adaptive devices. For people with severe mobility impairments, power wheelchairs may be used with communication devices, computer access devices, respirators, and reclining seating systems. The integration of the user's multiple needs must also be considered when designing or prescribing a power wheelchair.

The power wheelchair itself consists of several subsystems. Power wheelchairs are also designed for different uses. These areas will be discussed in greater depth in this chapter.

8.2. CLASSES OF POWER WHEELCHAIRS

There are essentially three general categories of power wheelchairs: power bases, power wheelchairs and scooters. Power bases are self-contained mobile platforms upon which a seating system can be mounted (figure 8.1). Power bases are useful where contoured or customized seating systems are required. Power bases can also be used to construct motorized gurneys for individuals who are unable to sit. Power wheelchairs contain the seat and the power drive system. Power wheelchairs are used for individuals with moderate trunk balance who only require moderate seating support (figure 8.2). Some seating systems have been developed to provide custom seating in a standard power wheelchair seat. Scooters are powered mobile platforms with a seat, which may swivel, based upon them. Scooters often use a handlebar type steering system (figure 8.3). People with limited walking ability may benefit most from scooters.

Power bases and wheelchairs are further divided into three classes based upon their capabilities and intended use. Class 1 wheelchairs are designed to carry a 100 kg person at a maximum speed of 1.1 to $1.8\,\mathrm{m\,s^{-1}}$ on level ground. They must be capable of climbing a 6 degree slope, travel 10 km on a single charge, and climb a 25 mm obstacle. Class 1 wheelchairs are primarily designed for light-duty use. Specifically they are intended for indoor use. Class 2 wheelchairs must be capable of carrying a 100 kg person at a maximum speed of 1.8 to $2.7\,\mathrm{m\,s^{-1}}$ on level ground and they must be capable of negotiating a 9 degree slope. They must be able to carry a 100 kg person a minimum of 16 km on a single charge and negotiate a 50 mm obstacle. Class 2 wheelchairs are designed for indoor and moderate outdoor use (figure 8.4). Class 3 wheelchairs must meet the same criterion as Class 2 wheelchairs except they must be capable of performing these tasks carrying a 113 kg person. Class 3 wheelchairs are designed for active indoor and outdoor use.

8.3. MOTOR SELECTION

The electric motors are the heart of any power wheelchair system (figure 8.5). They convert the electrical energy of the batteries to mechanical work.

Figure 8.1. *Power wheelchair base with linear seating system attached. (Photograph courtesy Everest & Jennings International, Ltd.)*

The proper selection can lead to a more durable, functional and efficient power wheelchair system. There are several types of electric motors. Most power wheelchairs use armature controlled, permanent magnet direct current motors. However, other electric motors have been used in prototypes and may eventually become the norm. Proper motor selection depends on translating the demands of the application to the properties of the appropriate motor.

8.3.1. Figures of merit

Figures of merit (FOM) are electric motor parameters related to electrical and mechanical characteristics of the motor (figure 8.6). FOMs are used to select, specify and compare motors. The mechanical time constant and power

Figure 8.2. *Full size power wheelchair with 20 inch rear wheels and multiple v-belt drive system. (Photograph courtesy of Invacare Corporation.)*

rate are the most reliable FOMs as they are related to a motor's electrical, mechanical, and magnetic properties.

The motor constant, K_m, is an often-used property to describe a parameter which may be likened to the motor's low speed gain, it is a function of torque (T) and power (P)

$$K_m = \frac{T}{\sqrt{P}} \ \mathrm{N\,m\,W^{-1/2}}. \tag{8.1}$$

The motor constant can be very misleading. It was developed from a class of motors called 'torquers', which are designed for static or low-speed operations, less than 1500 rev min^{-1}. However, motor constants also appear for higher-speed motors.

If the equation for K_m is broken down, this dependence upon the torque constant (K_t)and winding resistance (r) becomes clearer

$$K_m = \frac{K_t}{\sqrt{r}} \ \mathrm{N\,m\,(rev\,min^{-1})^{-1}\,\Omega^{-1/2}}. \tag{8.2}$$

To get a high motor constant, one has to try and obtain maximum motor torque with minimum resistance, armature, or field. For a fixed motor size

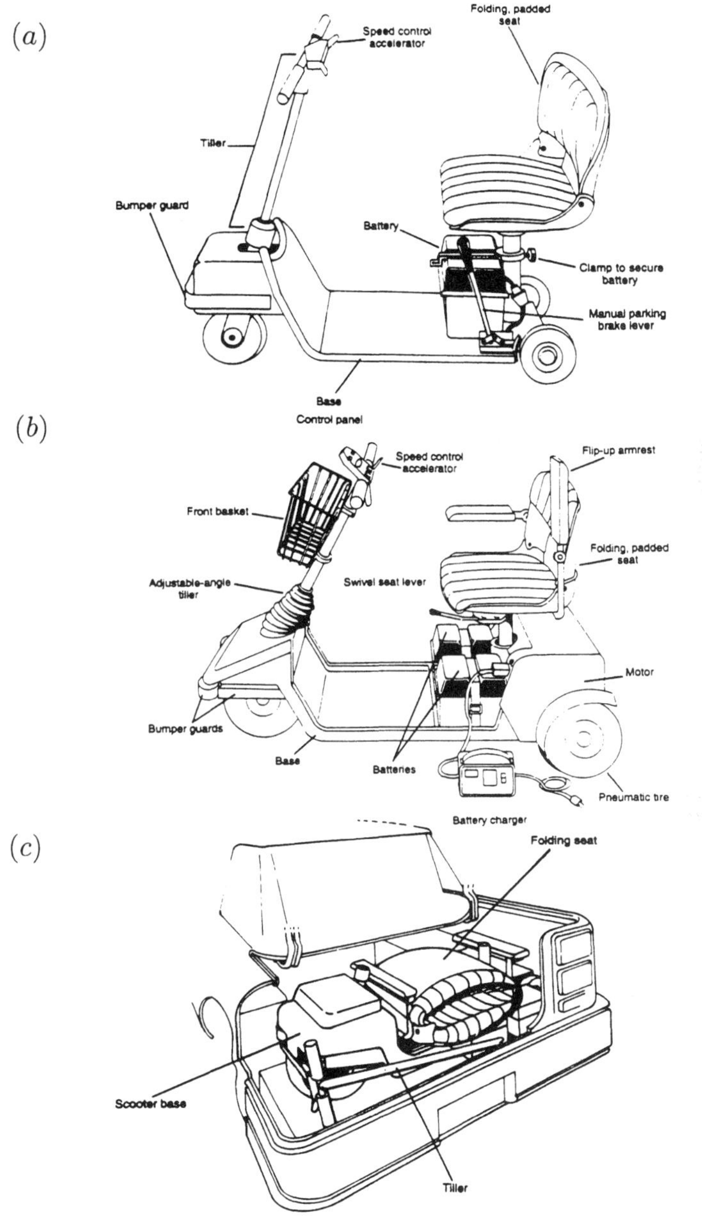

Figure 8.3. *Scooters may use three or four wheels and be either front- or rear-wheel drive. Front-wheel drive scooters are light duty and designed primarily for indoor use (a). Rear-wheel drive scooters are often designed for indoor and outdoor use (b). Scooters are flexible electric powered vehicles which provide mobility to people with rheumatic, respiratory, circulatory, or orthopaedic impairments which limit their ability to ambulate. (c) Scooters can be broken down to fit into the trunk of a full-size or hatchback automobile.*

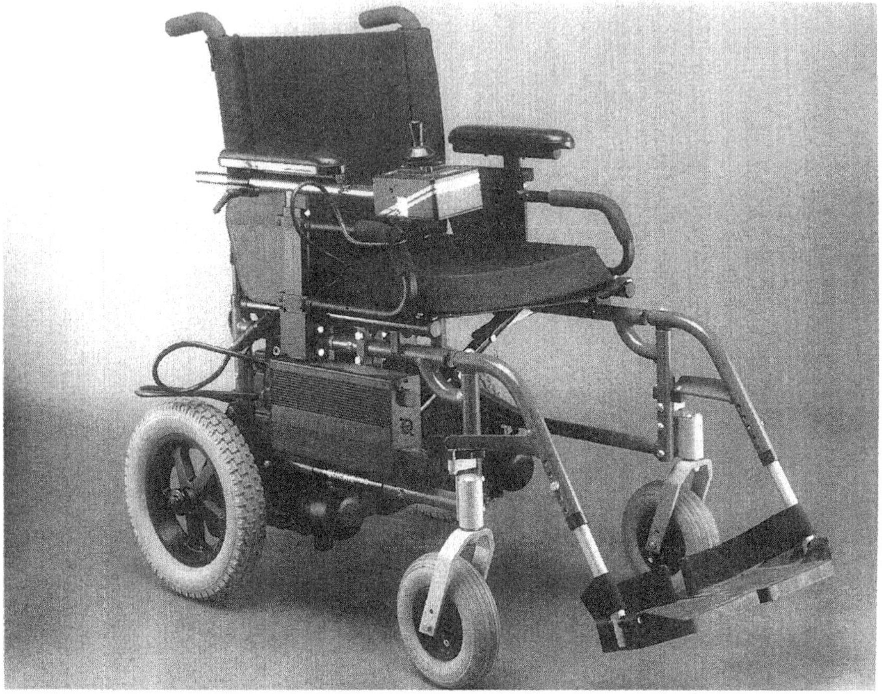

Figure 8.4. *Lightweight folding cross-brace power wheelchair with swing-away legrests,microprocessor control, and worm-gear drive. (Photograph by S Parsons, CSUS Media Services.)*

this is done by trading space in the windings for iron with space for copper. At low speeds this is fine. However, at high speeds this produces high iron losses. Therefore K_m has little meaning for motors which are rated above 1500 rev min^{-1}. Fair comparisons of motor constants can only be made if the motors are similarly constructed.

If the motor is considered to be a black box with wires going into one end and a shaft coming out of the other end, then we can evaluate the motor based upon its ability to transfer electrical power to mechanical power, power rate. The equation describing power rate is given

$$\dot{P}_m = \frac{T^2}{J} \quad \mathrm{W\,s^{-1}} \tag{8.3}$$

where T is the output torque, and J is the inertia of the motor. Examining the units of power rate leads to understanding of its significance. Ideally, all of the electrical power delivered to the motor would come out the other end as mechanical power. Power rate is a measure of the rate at which electrical power is converted to mechanical power. Some properties of power rate can

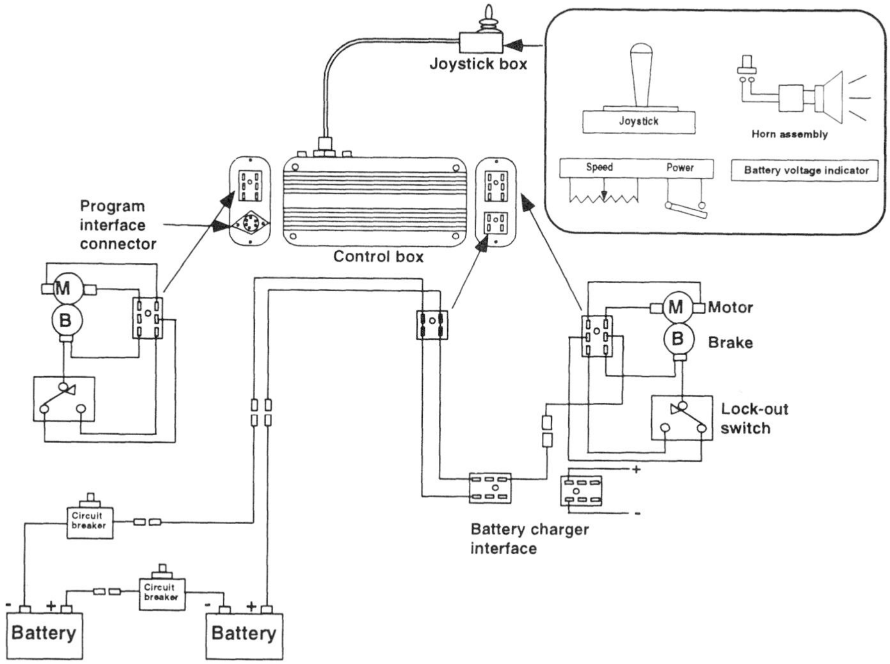

Figure 8.5. *Power wheelchair system schematic.*

FOM Rating Chart

Figure of merit	Electrical	Mechanical	Magnetic	Units	Symbol
Inertia	—	X	—	$kg \cdot m^2$	J
Electrical time constant	X	—	X	s	t_e
Mechanical time constant	X	X	X	s	t_m
Motor constant	X	—	X	$N\,m\,W^{-1/2}$	K_m
Acceleration	X	X	X	rad s^{-2}	α
Power rate	X	X	X	W s^{-1}	P_m

Figure 8.6. *Figures of merit for motor selection.*

be derived by rewriting equation (8.3), where I is motor current, r is the motor winding electrical resistance, and τ_m is the motor mechanical time constant

$$\dot{P}_m = \frac{I^2 r}{\tau_m} \quad \mathrm{W\,s^{-1}}. \tag{8.4}$$

If a motor has a small mechanical time constant (denominator), and a high thermal capacity (numerator), then the motor would have a high power rate, which is desirable for motors in most circumstances. Power rate is a generic property of electric motors. All motors have a power rate, and, in general, motors of any technology with a power range of less than 50 W are comparable. Power rate is absolute, and can thus be used in numerical calculations and need not be used for comparisons only. However, engineers must be aware of manufacture specmanship. Power rate for motors must be compared at the same speed and torque, hence care must be taken when comparing motors based upon manufacturer sales literature. In addition, power rate and rated power are a pair. Comparing power rates of 100 and 10 W motors is not valid.

8.3.2. System figures of merit

Electric motors are a component of the power wheelchair system. The selection of the proper motor is partially dependent upon certain characteristics of the system as a whole. Therefore, the system must be characterized in terms related to motor selection.

Static stiffness is a measure of how well the entire system holds its position when motionless. Often high static stiffness is desired. For power wheelchairs static stiffness helps to hold the wheelchair in place when power is removed. However, low static stiffness may be desirable if the power wheelchair is to be pushed without disengaging the drive system. It has units of a spring constant (force/distance) or (torque/radians). Static stiffness is the slope of the torque displacement curve as this curve crosses the zero axis.

Dynamic stiffness defines how closely a moving system follows a desired motion profile in the presence of disturbance torques. This is important for speed control of power wheelchairs. It has the same units as static stiffness. Speed control on power wheelchairs is used to have the chair drive at the same speed for the same input value (e.g. joystick position) regardless of the terrain or slope. Dynamic stiffness must be measured with known disturbance torques on the complete system. However, a good design rule of one tenth of the static stiffness is often acceptable.

Bandwidth defines the limiting frequency of motion commands to which a system can respond (joystick, controller, servo amp, motor, wheelchair). It is limited by the number of components in the command chain, and the bandwidth of those components. Bandwidth is not to be confused with the switching frequency of the servo amplifier. It is more important to have sufficient bandwidth to achieve sophisticated control (i.e., speed contours). For design purposes it is often assumed that motors in open loop systems have a bandwidth of about 150–250 Hz, whereas in closed loop systems a bandwidth of 50–100 Hz may be used.

Response time is the minimum time the system requires to respond to a command. It is approximately the reciprocal of the bandwidth. The faster the response time the better the control that can be achieved.

8.3.3. Motor control of powered wheelchairs

There are several motor types that may be suitable for use with power wheelchairs. Most current designs use permanent magnet direct current (d.c.) motors. These motors provide high torque, high starting torque and are the simplest to control.

Permanent magnet d.c. motors can either be controlled in what are commonly called current mode or voltage mode. These modes developed out of designs based upon controlling torque and speed, respectively.

Separately excited d.c. motors may have their place in power wheelchairs as well. Separately excited motors use electromagnets to provide the field flux instead of permanent magnets. The torque of these motors can be varied by controlling the field current. An advantage of separately excited motors is that they can achieve high power amplification. However, they typically do not have high torque capabilities.

Alternating current (a.c.) motors can be designed to be highly efficient and can be controlled with modern power circuitry. Because of the development and wide spread dissemination of switching d.c. converters, it is quite feasible to use a.c. motors with a battery supply. To date, a.c. motors have only been used in research on power wheelchairs. The output of the motor is controlled by varying the phase or the frequency.

A model of the motor is required if it is to be used in the design of a wheelchair controller (figure 8.7). This model can be used to determine the switching frequency of the servo amplifier, and the bandwidth of the power wheelchair.

The transfer function of the motor model can be derived from figure 8.7. The transfer function of the motor is based upon a linear approximation. In this case hysteresis, and the voltage drop across the brushes will be neglected. The input voltage can be applied to either the field or the armature terminals.

Motor air-gap flux (ϕ) is proportional to the field current (i_f) when unsaturated

$$\phi = K_f i_f. \tag{8.5}$$

Motor torque (T_m) is assumed linearly related to the air-gap flux (ϕ) and the armature current (i_a)

$$T_m = K_1 \phi i_a(t) = K_1 K_f i_f(t) i_a(t). \tag{8.6}$$

For the motor to behave linearly either the field current or the armature current must be held constant.

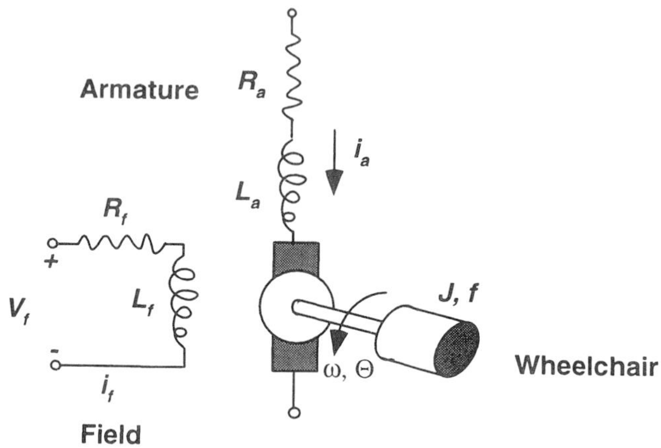

Figure 8.7. *Schematic of a d.c. motor model.*

The case of a field-controlled d.c. motor, which provides substantial power amplification, will be examined. In this case the armature current is assumed constant

$$T_m(s) = K_1 K_f I_f(s) I_a = K_m I_f(s) \qquad K_m = \text{motor constant.} \tag{8.7}$$

The motor/wheelchair model leads to a relationship between the field voltage and field current

$$V_f(s) = (R_f + L_f) I_f(s). \tag{8.8}$$

To be effective, the motor torque has to equal the load torque

$$T_m(s) = T_L(s) + T_d(s) \qquad T_L = \text{load torque} \qquad T_d = \text{disturbance torque.} \tag{8.9}$$

The load torque is a function of the inertia (J) and resistance (f) of the wheelchair system as seen by the motor

$$T_L(s) = Js^2\theta(s) + fs\theta(s) = Js\omega(s) + f\omega(s). \tag{8.10}$$

The torque equation can be rearranged to solve for the load torque

$$T_L(s) = T_m(s) - T_d(s). \tag{8.11}$$

A relationship between the motor torque and field current was developed earlier, in equation (8.7)

$$T_m(s) = K_m I_f(s) \tag{8.7}$$

where as before

$$I_f(s) = \frac{V_f(s)}{R_f + L_f(s)}. \tag{8.8}$$

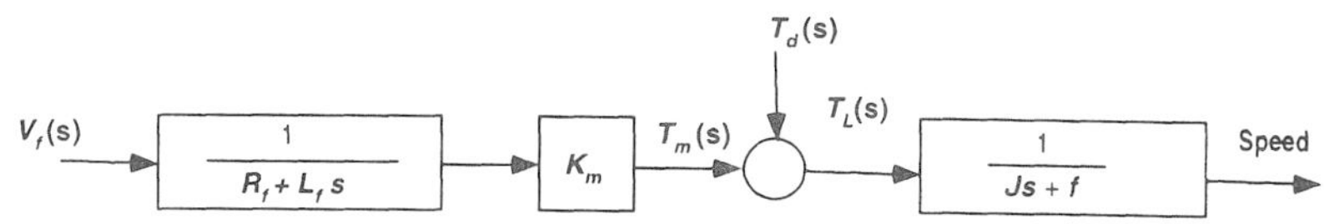

Figure 8.8. *Block diagram for a field-controlled d.c. motor.*

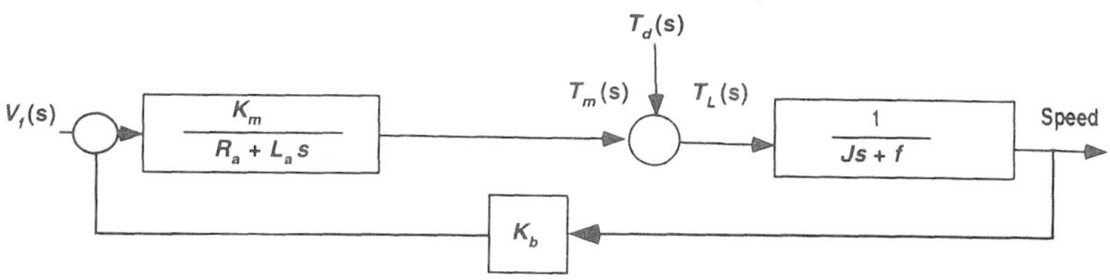

Figure 8.9. *Block diagram for an armature-controlled d.c. motor.*

Solving for the transfer function for field voltage to motor speed

$$T_L(s) = T_m(s) - T_d(s).$$

With the use of superposition, i.e., this gives ignoring disturbance torque

$$T_m(s) = K_m I_f(s)$$

$$T_L(s) = \frac{K_m V_f(s)}{R_f + L_f(s)}$$

$$Js\omega(s) + f\omega(s) = \frac{K_m V_f(s)}{R_f + L_f(s)}$$

$$(Js + f)\omega(s) = \frac{K_m V_f(s)}{R_f + L_f(s)} \tag{8.9}$$

$$\omega(s) = \frac{K_m V_f(s)}{(R_f + L_f s)(Js + f)}$$

$$\frac{\omega(s)}{V_f(s)} = \frac{K_m}{(R_f + L_f s)(Js + f)}$$

$$\frac{\omega(s)}{V_f(s)} = \frac{K_m/f R_f}{(1 + (L_f/R_f)s)(1 + (J/f)s)} \qquad \text{composite wheelchair.}$$

The motor speed for a field-controlled d.c. motor accounting for the disturbance torque is given

$$\omega(s) = \frac{(K_m/f R_f V_f(s)}{(1 + (L_f/R_f)s)(1 + (J/f)s)} - \frac{T_d(s)}{(Js + f)}. \tag{8.10}$$

The block diagram for a field-controlled d.c. motor is illustrated in figure 8.8.

The block diagram for an armature-controlled d.c. motor is illustrated in figure 8.9.

The motor speed for an armature-controlled d.c. motor accounting for the disturbance torque is

$$\omega(s) = \frac{K_m V_a(s) - (R_a + L_a s)T_d(s)}{s^2 J L_a + s(R_a J + L_a f) + (R_a f + K_m K_b)}. \tag{8.11}$$

Equations (8.10) and (8.11) can be used to model the motors to be used in the design of a power wheelchair. The motors are components of the larger system. To control the motors accurately and reliably a servo amplifier is required.

8.4. SERVO AMPLIFIERS

To implement a motor controller a servo amplifier is required to convert signal level power (volts at milliamps) to motor power (volts at amps). Typically a design requirement for series, shunt, and brushless motor drives is to control torque and speed, and hence power. Voltage control can often be used to control speed for both shunt and series motors. Series motors require feedback to achieve accurate control.

Either a linear servo amplifier or a chopper can be used. Linear servo amplifiers are not generally used with power wheelchairs primarily because of their lower efficiency than chopper circuits. A motor can be thought of as a filter to a chopper circuit, in this case the switching unit can be used as part of a speed and current control loop. The torque ripple and noise associated with phase control drives can be avoided by the use of high switching frequencies. The response of the speed control loop is likewise improved with increasing switching frequency.

Motor torque is proportional to the armature current in shunt motors and to the square of the current in series motors. The conduction loss of the motor and servo amplifier are both proportional to current squared. Optimal efficiency is achieved by minimizing the form factor ($I_{\text{rms}}/I_{\text{mean}}$). This can be done by increasing the switching frequency to reduce the amplitude of the ripple. Benefits of increased efficiency are increased brush life, gear life, and lower probability of field permanent magnet demagnetization.

Switching or chopper drives are classified as either unidirectional or bidirectional. They are further divided by whether they use dynamic braking. Typically power wheelchairs use bidirectional drives without dynamic braking. However, scooters may use unidirectional drives. The average voltage delivered to the motor from a switching drive is controlled by varying the duty cycle of the input waveform. There are two common methods of achieving this goal: (1) fixed pulse width, variable repetition rate, and (2) pulse-width modulation (PWM). Power wheelchair servo amplifiers typically employ PWM.

Fixed pulse width switching is used for some semiconductor switches where it is essential to maintain a minimum on period. However, this method generates a poor current waveform when starting the motor. When the switch is on the current rises to some high value due to the absence of a back electromotive force. Hence, the off time must be increased to restore a suitable average current. This produces undesirable torque variations and mechanical noise. In addition, the motor current is zero during the off cycle which can cause instability in closed loop systems, due to gain and phase characteristics of the amplifier–motor–load system.

PWM at a fixed frequency has no minimum on-time restriction. Therefore, current peaking and torque ripple can be minimized.

For analysis, a d.c. motor can be modeled as an RL circuit, i.e., resistor and inductor, in series with a voltage source. If the motor current is assumed continuous, then the minimum and maximum motor current can be represented by

$$I_{\mathrm{min}} = \frac{\mathrm{e}^{-(R/L)t_{\mathrm{off}}}\left(1 - \mathrm{e}^{-(R/L)t_{\mathrm{on}}}\right)}{1 - \mathrm{e}^{-(R/L)(t_{\mathrm{on}}-t_{\mathrm{off}})}}\frac{V_s}{R} - \frac{V_{\mathrm{gen}}}{R} \tag{8.12}$$

$$I_{\mathrm{max}} = \frac{\left(1 - \mathrm{e}^{-(R/L)t_{\mathrm{on}}}\right)}{1 - \mathrm{e}^{-(R/L)(t_{\mathrm{on}}-t_{\mathrm{off}})}}\frac{V_s}{R} - \frac{V_{\mathrm{gen}}}{R}. \tag{8.13}$$

Two basic design principles are used when designing switching servo amplifiers: (1) I_{max} should be limited to five times the rated current of the motor to insure that demagnetization does not occur; (2) the ripple, $(I_{\mathrm{max}} - I_{\mathrm{min}})/I_{\mathrm{avg}}$, should be minimized to improve the form factor, and reduce the conduction loss in the switching devices. To achieve low ripple, either the inductance has to be large or the switching frequency has to be high. Permalloy powder cores can be used to reduce core loss at frequencies above a few kilohertz. However, this comes at the cost of the electrical time constant of the motor, degrading the motor response time. Hence, raising the switching frequency is most desirable. A power MOSFET has the ability to switch rapidly without the use of load-shaping components.

Most power wheelchairs use switching servo amplifiers because of their efficiency and relative simplicity. Two types of components are used as switching elements in power wheelchair servo amplifiers: (1) silicon-controlled rectifiers (SCR); and (2) powder metal oxide semiconductor field effect transistors (MOSFETs).

8.4.1. Silicon-controlled rectifier (SCR) switching

An SCR acts like a voltage-controlled switch. An advantage to SCRs is that they are quite robust. They can withstand moderate temperatures and switch relatively high currents. However, they have a slow response time, i.e., they can switch at a maximum of about 2 kHz. Power wheelchair servo amplifiers with SCRs use typical switching frequencies of 500 to 1000 Hz. A SCR can be modeled using two transistors (figure 8.10).

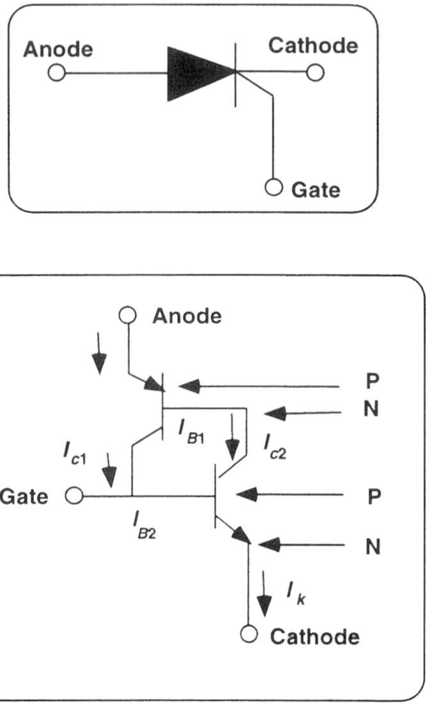

Figure 8.10. *Two-transistor model of an SCR.*

The SCR is either on or off depending upon the control voltage applied to the gate: in the on state the resistance is in the mΩ range and it is in the MΩ range when in the off state. The anode is connected to the voltage source, whereas the cathode is connected to the motor. A single SCR can be used to control a d.c. motor in one direction. The motor current is controlled by switching the SCR on and off. SCRs are used for either phase control or pulse width modulation. For powered wheelchairs, the entire battery voltage is applied to the motor through SCRs.

Four SCRs used in a full bridge are required for bidirectional control (figure 8.11).

The control circuitry must operate such that the SCRs are only turned on in pairs: 1 and 4 on with 2 and 3 off or vice versa. Otherwise vertical conduction can occur which will blow a circuit breaker and/or cause damage to the batteries. Using a bridge configuration, bidirectional control is achieved with a single source. Caution must be taken when changing directions. A small deadzone is added to the controller so that there is some time between one pair of SCRs turning off and the other pair turning on. A shunt resistor may be used to help the motor start smoothly.

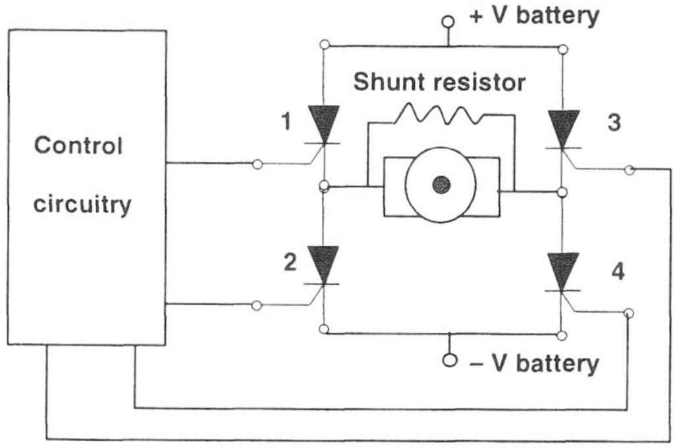

Figure 8.11. *Full H-bridge SCR d.c. motor control schematic diagram.*

8.4.2. Power MOSFET switching control

Power MOSFETs have the ability to switch rapidly, in the region of tens of kHz. Power wheelchair servo amplifiers use switching frequencies between 2 and 20 kHz. Power MOSFET servo amplifiers can achieve efficiencies in excess of 90%. This makes them very attractive for battery applications. The frequency limit of a MOSFET is dependent upon the reverse recovery charge of the free-wheeling diode.

The reverse recovery current of the free-wheeling diode adds to the motor current in the MOSFET, simultaneously approximately half of the supply voltage is maintained in the MOSFET during recovery. The diode recovery loss can be reduced by lowering $\mathrm{d}i/\mathrm{d}t$. However, this causes the MOSFET switching loss to increase. Optimum $\mathrm{d}i/\mathrm{d}t$, controlled by $\mathrm{d}V_g/\mathrm{d}t$, is achieved by minimizing the diode recovery and MOSFET switching losses.

Near the end of the diode reverse recovery period the diode also experiences a high voltage across it. As a MOSFET turns off, the drain–source voltage rises rapidly, which forward biases the free-wheeling diode. This is coupled with the high forward voltage seen by the diode for a brief time during turn-on. The high forward voltage is attributed to the turn-on delay in the diode, and the package inductance. The voltage drop can be around 50 V, depending upon $\mathrm{d}i/\mathrm{d}t$. Hence, the power MOSFET must be rated to handle this voltage in addition to the supply voltage. Typically, the supply voltage for a power wheelchair is 24 V.

The choice of the power MOSFET rating is also dependent upon the maximum die temperature under peak load conditions. The total power dissipation and the power–temperature derating curve are used to determine the maximum permissible case temperature. If the case temperature is high, multiple devices in parallel may be required. Power wheelchairs typically use MOSFETs rated at 100 V and 30 A continuous operation. Many use eight MOSFETs per bridge to further increase reliability.

For permanent magnet motors, those typically used in power wheelchairs,

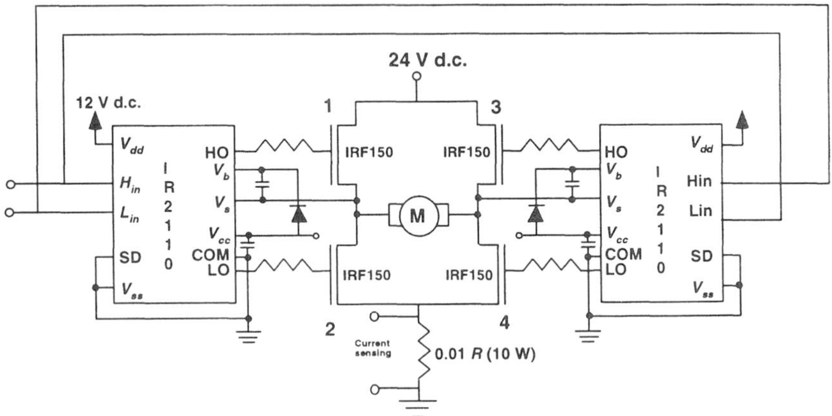

Figure 8.12. *Full-bridge chopper drive.*

controlled braking can be achieved by switching a MOSFET across the motor. This configuration can be used to provide current back to the supply, i.e., charge the battery during braking. The braking MOSFETs are pulse-width modulated to control armature current during braking, otherwise they act as free-wheeling diodes during motoring mode. Switching losses may be significant at ultrasonic frequencies, requiring the use of compensation circuitry.

A basic full-bridge circuit can be used for bidirectional control (figure 8.12). When driving the motor in the forward direction, MOSFETs 1 and 4 are off, MOSFET 2 is on, and MOSFET 3 is switched with a PWM signal. In this configuration MOSFET 1 acts as the free-wheeling diode. An alternative is to switch MOSFETs 2 and 3, in which case, the free-wheeling current flows via the supply through MOSFETs 1 and 4. However, the current decay during the off period is greater because the MOSFETs are subjected to the generated voltage plus the supply voltage. Hence, an external choke may be required. Another method of slowing turn-on is to add an inductor in series with the motor to limit di/dt.

8.5. MICROPROCESSOR CONTROL

Power wheelchairs can be controlled with simple analog or with digital microprocessor-based controls. A single microcontroller is used to control the wheelchair and both of its motors (figure 8.12). Microcontrollers from the INTEL 8000 or Motorola 6800 family are commonly used. Digital control offers flexible control and accommodation of various input devices (both analog and digital). An internal (resident) program maintains timing, coordination and contains the controls system algorithms. Motor voltage and current are sensed to limit power, to protect circuitry, and to achieve finer control.

Microcontrollers are used to tune the power wheelchair system to the user's individual needs. Many wheelchair controllers permit adjusting maximum speed, yaw velocity and acceleration, acceleration and deceleration rate, input filter parameters, and the input device deadzone. A number of methods have been used to vary controller parameters. An external control unit can be used which interfaces with the wheelchair controller. The wheelchair controller parameters are tuned through a series of switches or key strokes. Another approach is to use the user interface (e.g. joystick, sip-and-puff device, ultrasonic control device) to tune the wheelchair controller. This is accomplished by selecting the program mode, usually done by using a switch. Then a series of 'menus' are used to tune each parameter of interest.

Microcontrollers improve control not only by permitting repeatable tuning and control of certain system parameters, but also by allowing implementation of dynamic control algorithms. The most common control variable is speed. A typical speed control algorithm uses tracking, or contour, control to have the wheelchair automatically follow the desired speed profile set by the user, by varying the input device, regardless of terrain or slope. Hence the wheelchair/rider system will move at the same speed up an incline as it will down an incline for the same user-desired speed. The same is true when comparing a smooth surface to a rough surface. Speed feedback is done with an optical encoder in quadrature for direction, or by using the back electromotive force (EMF) of the motors. Heading is derived from using both tachometers or the back EMF from both motors.

For speed control to be functional, a model of the wheelchair and its critical components must be developed and then used to implement compensation. Typically power wheelchair motors are permanent magnet d.c. motors whose characteristic equations were derived in a previous section (R_a = armature resistance, L_a = armature inductance, K_v = motor voltage constant, ω_m = motor angular velocity, i_a = armature current, V_a = armature voltage, T_m = motor torque, J_m = motor inertia, K_t = motor torque constant:

$$V_a = i_a R_a + K_v \omega_m + L_a \frac{\mathrm{d}i_a}{\mathrm{d}t} \tag{8.14}$$

$$J_m \frac{\mathrm{d}\omega_m}{\mathrm{d}t} = i_a K_t - T_m. \tag{8.15}$$

The motor transients, associated with L_a, due to changes in armature current are assumed negligible, because the time constant of the chair is typically on the order of one hundred times larger.

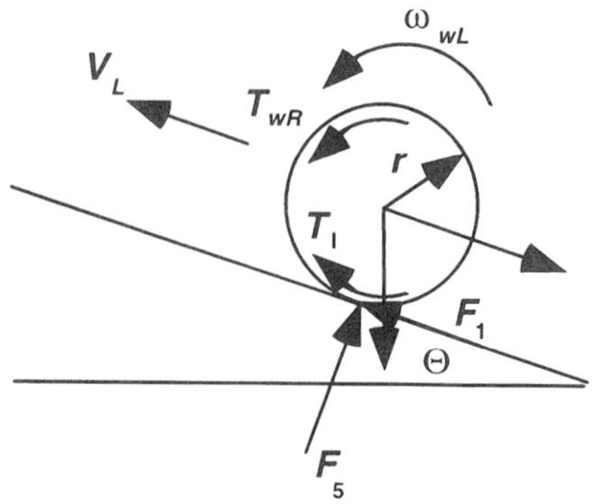

Figure 8.13. *Left-wheel free-body diagram on a slope.*

The wheels, bearing and rolling friction, have been shown to be a significant source of torque loss. The equations of motion for the left wheel can be obtained from its free-body diagram, J_W = wheel inertia, ω_W = wheel angular velocity, r = wheel radius, K_g = gear ratio, f = rolling friction coefficient, b_c = bearing friction constant (figure 8.13). The equations for the right wheel can be determined using symmetry

$$T_{wL} - T_f - F_1 r = J_w \frac{\mathrm{d}\omega_{wL}}{\mathrm{d}t} \tag{8.16}$$

where

$$T_{wL} = K_g T_{mL} \tag{8.17}$$

$$T_f = \operatorname{sign}(\omega_{wL}) \left[\beta_c (\omega_{wL})^{2/3} + F_5 f_r\right] \tag{8.18}$$

$$\omega_{wL} = \frac{\omega_{mL}}{K_g} = \frac{V_L}{r}.$$

One of the objectives of speed control is to make actual speed independent of terrain and slope. Therefore a model of a wheelchair system on a slope is required (figures 8.14 and 8.15).

Several simplifying assumptions are made with the model of a wheelchair on a side slope: only vertical forces act on the front castors; the front wheels are loaded equally; the center of mass is stationary with respect to the chair; the wheels do not slip; and changes in slope are gradual, i.e., $\mathrm{d}a/\mathrm{d}s$ and $\mathrm{d}f/\mathrm{d}s$ are near zero.

The three coordinate systems defined in figure 8.14 are related to each other by standard rotation transformations. Pitch (slope) angle, θ, and roll angle, ϕ, of the wheelchair/rider are related to the yaw angle (heading angle), β, of the wheelchair/rider, and to the incline angle, α, and the side slope angle, γ, of the ground by

$$V_{P\omega}(k) = C_{P\omega} V_{a\omega}(k). \tag{8.19}$$

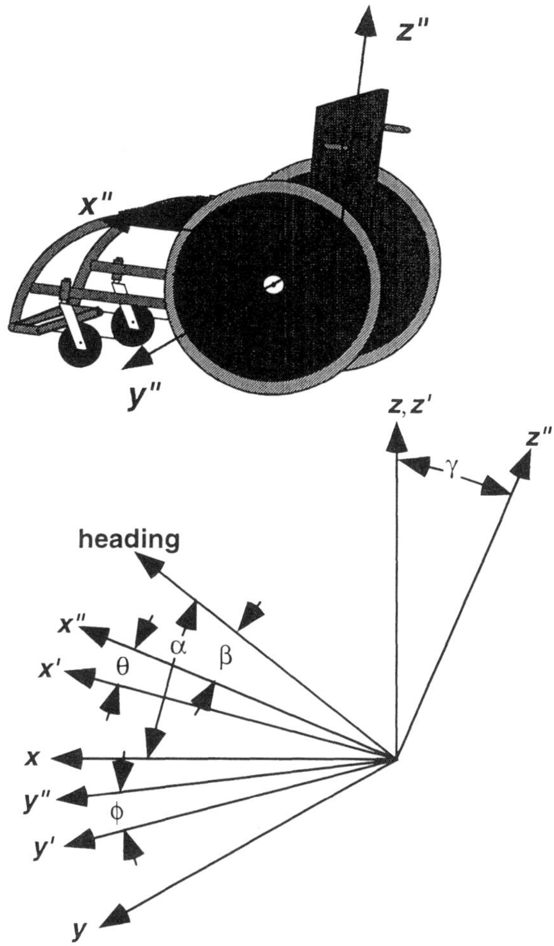

Figure 8.14. *Wheelchair on an inclined surface showing α and γ associated with the surface, and θ and ϕ associated with the wheelchair, and the chair heading angle β.*

Force balance equations for the wheelchair/rider system can be derived from figure 8.14, where M = mass of the wheelchair/rider system:

$$\begin{aligned} F_1 + F_3 - F_x &= M\frac{\mathrm{d}v_x}{\mathrm{d}t} \\ F_2 + F_4 + F_y &= M\frac{\mathrm{d}v_y}{\mathrm{d}t} \\ F_5 + F_6 + F_7 + F_8 - F_z &= M\frac{\mathrm{d}v_z}{\mathrm{d}t} = 0. \end{aligned} \tag{8.20}$$

The accelerations of the wheelchair/rider system along the x and y axes are

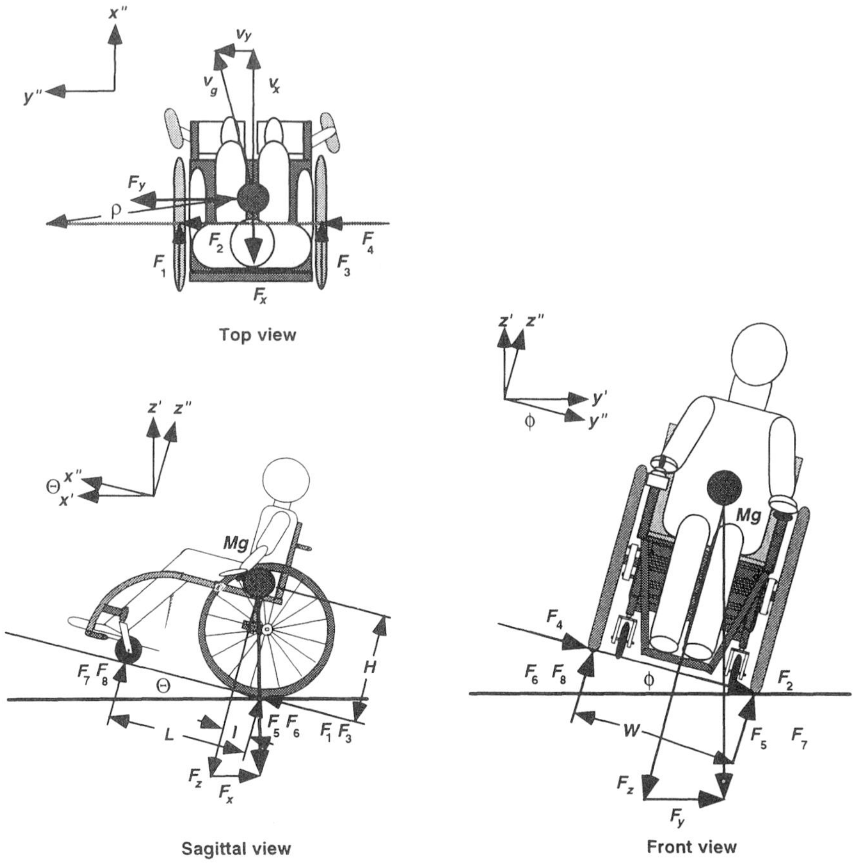

Figure 8.15. *Free-body diagram of a power wheelchair and rider on an inclined surface.*

related to the accelerations and velocities of the wheelchair's left, v_l, and right, v_r, wheels by

$$\begin{aligned}\frac{dv_x}{dt} &= \frac{(dv_r/dt + dv_l/dt)}{2} - \frac{l(v_r - v_l)^2}{W^2}\\ \frac{dv_y}{dt} &= \frac{l(dv_r/dt - dv_l/dt)}{W} - \frac{W(v_r - v_l)^2}{2}\end{aligned} \tag{8.21}$$

with l = distance the center of mass is forward of the rear axles, and W = width of wheelchair as measured between the rear wheels.

The forces along the x, y and z axes can be derived from the mass of the wheelchair/rider system and it orientation with respect to the earth

$$F_x = \frac{Mg\tan\theta}{(1+\tan^2\theta+\tan^2\phi)^{1/2}}$$
$$F_y = \frac{Mg\tan\phi}{(1+\tan^2\theta+\tan^2\phi)^{1/2}} \qquad (8.22)$$
$$F_z = \frac{Mg}{(1+\tan^2\theta+\tan^2\phi)^{1/2}}$$

where g = acceleration due to gravity.

Moment balance equations are also required to characterize the dynamics of the wheelchair/rider system,

$$(F_3 - F_1)\left(\frac{W}{2}\right) - (F_2 + F_4)l = I_z\left(\frac{\pi}{180}\right)\left(\frac{d\omega_z}{dt}\right)$$
$$(F_5 + F_7)\left(\frac{W}{2}\right) + (F_2 + F_4)H - (F_6 + F_8)\left(\frac{W}{2}\right) = 0 \qquad (8.23)$$
$$(F_7 + F_8)(L - l) + (F_1 + F_3)H - (F_5 + F_6)l = 0$$

where $I_{x,y,z}$ = moment of inertia about axes, $\omega_{x,y,z}$ = chair angular acceleration about the x'', y'', z'' axes, respectively, H = height of center of mass, L = wheel base of the wheelchair.

The linear accelerations of the left and right wheels as well as the angular velocity of the heading can be determined from the force and moment balance equations (8.20)–(8.23)

$$\frac{dv_L}{dt} = \frac{(A - B)}{2} \qquad \frac{dv_R}{dt} = \frac{(A + B)}{2}$$
$$\omega_z = \frac{d\beta}{dt} = 180(v_R - v_L)\left(\frac{W}{\pi}\right). \qquad (8.24)$$

The variables A and B are functions of the wheelchair/rider system parameters and are given for V_L and V_R both greater than zero by equation (8.25) given overleaf.

A simple quasi-static model, equation (8.26), of the wheelchair can be programmed in the microprocessor using Euler's method with equations (8.24) and (8.25)

$$v_L(k) = (A - B)\frac{\Delta t}{2} + v_L(k-1)$$
$$v_R(k) = (A + B)\frac{\Delta t}{2} + v_R(k-1) \qquad (8.26)$$

$$\beta(k) = \frac{180 W \Delta t}{\pi}(v_R(k) - v_L(k)) + \beta(k-1).$$

$$A = \frac{1}{\left[\frac{J_w + J_m K_g^2}{r} + \frac{Mr}{2}\left(\frac{Hf}{L} + 1\right)\right]\left[K_t K_g\left(i_{mR} + i_{mL}\right) - \beta_c\left(\left(\frac{V_R}{r}\right)^{2/3} + \left(\frac{V_L}{r}\right)^{2/3}\right) - \frac{rfF_z(L-l)}{L} - r\left(\frac{fH}{L} + 1\right)\left(F_x - \frac{Ml}{W^2}(v_R - v_L)^2\right)\right]}$$

$$B = \frac{1}{\left[\frac{J_w + J_m K_g^2}{r} + \frac{2r}{W^2}\left(HfMl + I_z + Ml^2\right)\right]\left[K_t K_g\left(i_{mR} - i_{mL}\right) - \beta_c\left(\left(\frac{v_R}{r}\right)^{2/3} - \left(\frac{v_L}{r}\right)^{2/3}\right) - \frac{2r}{W}(Hf + l)\left(\frac{MW}{2}\left(v_R^2 - v_L^2\right) - F_y\right)\right]} \tag{8.25}$$

This model is not intended to describe the complete dynamics of a wheelchair with rider, but to provide dynamic equations suitable for developing wheelchair controllers. Power wheelchair control has been dramatically improved by using microprocessor-based feedback controllers. Standard power wheelchairs use differential torque control with the rear wheels powered and the front wheels castering. Feedback control is commonly used to maintain the desired speed of the wheelchair/rider in the presence of disturbances, e.g. side slopes, changes in terrain, changes in the texture of the rolling surface. Power wheelchairs always use feedback control. However, without automatic speed control the user must assume the role of feedback sensor and controller. Automatic feedback control allows the user to select a desired speed and direction using the joystick, or other interface device, and the microprocessor senses the actual speed and heading and adjusts the servo-amplifier voltage appropriately. This makes controlling the wheelchair easier for the user, i.e., the input processing of the user is reduced. Automatic feedback control of the wheelchair requires some tuning so that the wheelchair provides the user with some of the feel of an open loop system, i.e., it may be desirable to sense low tire pressure or bearing failure. In addition, redundant sensors or sensing techniques should be used to ensure safe operation of the automatic control system. One approach may be to have the controller revert to open loop, direct proportional control, if catastrophic controller failures occur.

With joystick control, the joystick outputs a desired heading (V_{jw}) and speed (V_{js}) as voltages to either the servo amplifier or the automatic control system. With automatic control the difference between the desired heading and speed becomes the error signal for the controller. With open loop, manual control, the user continuously adjusts the joystick to attain the desired speed and heading. Proportional plus integral automatic feedback control is a common and practical approach for a power wheelchair.

The speed of the left and right wheels of the wheelchair must be sensed to provide effective feedback control. Two simple methods are commonly employed to measure the velocity of the rear wheels: for armature-controlled permanent magnet d.c. motors, the back EMF is proportional to the motor speed, or a simple optical or Hall effect sensor may be used. Direct measurement of the speed will likely lead to more accurate control. The actual linear velocity (V_{al}), and actual angular velocity (V_{aw}) of the wheelchair expressed in V are proportional to the velocity of the left (v_L) and right (v_R) wheels

$$V_{al} = C_v\,(v_R + v_l) \qquad V_{\omega l} = C_\omega\,(v_R - v_L)\,. \tag{8.27}$$

A simple decoupling structure may be applied between the outputs of the controller and the inputs to the motor servo amplifiers. The input to the left motor servo amplifier (V_{sL}) is the difference between the output of the linear velocity controller (V_{cv}) and the angular velocity controller ($V_{c\omega}$)

$$V_{sL} = V_{cv} - V_{c\omega}. \tag{8.28}$$

The input to the right motor servo amplifier (V_{sR}) is the sum of the output of the linear velocity controller and the angular velocity controller

$$V_{sR} = V_{cv} + V_{c\omega}. \tag{8.29}$$

The feedback parameters can be set using an external input to the microprocessor to limit the maximum linear and angular velocities of the wheelchair below the maximum open loop values. This is used on a number of power wheelchairs to tune the chair to more closely match the abilities of the user. Tuning can be accomplished in a number of ways, most notably by: using potentiometers which are accessible to a qualified technician using the joystick or user interface in a 'program' mode; and by using an external device that connects to the microprocessor controller. An added advantage of an automatic control system is that speed can be limited to within safe ranges for the user without compromising power required for steep hills and radical side slopes. Feedback control also permits independent limits on linear and angular velocity. This gives the user better control without dramatically compromising speed. Maximum linear and angular velocity are primarily adjusted by scaling the feedback gains C_V and C_ω.

Proportional integral (PI) control uses the sum of the scaled and integrated values of the error signal

$$V_{cv} = V_{Iv} - V_{Pv} \qquad V_{c\omega} = V_{I\omega} - V_{P\omega}. \tag{8.30}$$

The integral (V_{Iv}) and proportional (V_{Pv}) linear controller voltages result from

$$\begin{aligned} V_{Iv}(k) &= C_{Iv}\left[V_{js}(k) - V_{as}(k)\right] + V_{Iv}(k-1) \\ V_{Pv}(k) &= C_{Pv}V_{as}(k). \end{aligned} \tag{8.31}$$

The integral ($V_{I\omega}$) and proportional ($V_{P\omega}$) angular controller voltages result from

$$\begin{aligned} V_{I\omega}(k) &= C_{I\omega}\left[V_{j\omega}(k) - V_{a\omega}(k)\right] + V_{I\omega}(k-1) \\ V_{P\omega}(k) &= C_{P\omega}V_{a\omega}(k). \end{aligned} \tag{8.32}$$

The input voltages are left out of the proportional equations to avoid setpoint kick, which results in the desirable condition of low accelerations for small, perhaps unintentional, changes in joystick position. The wheelchair model is non-linear, therefore the gains of the PI system must be chosen using simulation or by iteration on the actual system. The gains must be selected to achieve the desired response time, precision, smoothness, and system stability.

The sampling time of the microprocessor-based control system affects performance of the closed loop system. The longer the sampling time, the lower the PI gains must be in order to ensure system closed loop stability. The faster the sampling rate the closer the system emulates its continuous time analog. Hence, the model response will more closely simulate the system response. However, sampling rate is limited by the speed of the microprocessor, the sensors, the actuators, and the control software. A sampling rate of 0.01 seconds is generally suitable for power wheelchair control, and is readily achievable.

8.6. SHARED CONTROL

The goal of shared control is to enhance the mobility of the users by off-loading some wheelchair driving functions to an automatic control system. This approach is sometimes called sensor-based control, because of the array's internal and external sensors used by the automatic control system (see figure 8.16). Some people may have sensory, cognitive and/or physical impairments that limit their ability to control a power wheelchair. A variety of strategies are employed to ameliorate the limitations imposed by physical, sensory, and cognitive impairments. Proper body positioning and support can enhance the driver's ability to maintain control and increase the range of driving tasks that can be performed. The selection of the interface device is also an area of much investigation and has led to expanded ranges of performance for power wheelchair drivers. Once the driver is properly positioned, supported, and has an input device that accommodates the driver's residual abilities, the driver may still be very limited in the activities which can be performed successfully. In this case, the parameters of the microprocessor controller are typically tuned to minimum or near minimum driving parameters (e.g. linear and angular speed, linear and angular acceleration). The controller input filter parameters may also be tuned to remove much of the signal provided by the driver. The result of detuning a classical PI microprocessor control system is a sluggish slow-moving wheelchair. Some people attain functional mobility using these methods, however, many do not. Shared control systems are quite successful in many automobile driving applications (e.g. automatic braking systems, adaptive suspension systems, power steering systems, cruise/speed control, automatic fuel injection). Some of this technology can be applied to power wheelchair driving.

8.6.1. Input conditioning

The user must provide some indication of the desired speed and direction for functional driving of a power wheelchair. Input signals can either be

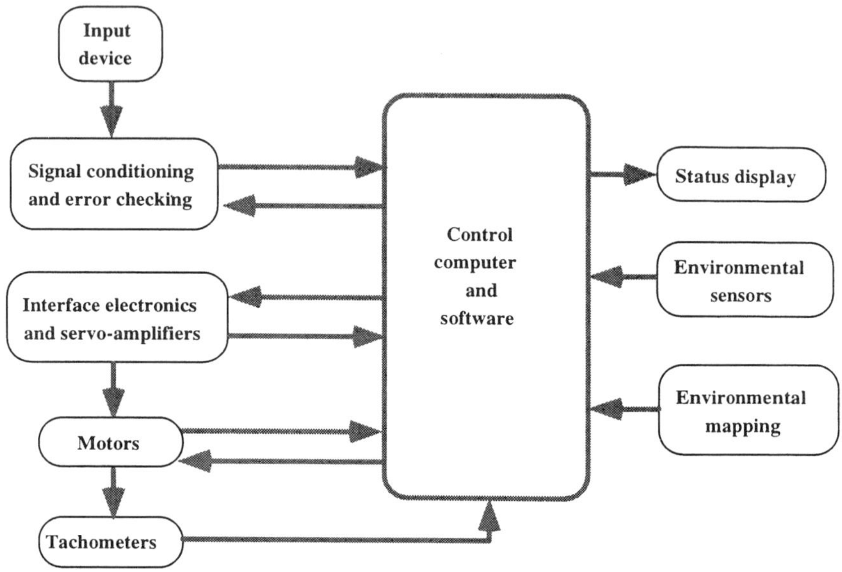

Figure 8.16. *Shared control system block diagram. There are three primary components: user and wheelchair internal drives and sensors; control computer; and external sensors. Information is shared among the various components and the wheelchair is directed accordingly.*

variable (i.e., continuous or discrete steps variable by the user) or latched (i.e., fixed selectable discrete steps). The degrees of freedom of the input device affect the sharing strategy (figure 8.17). Variable input signals are provided by devices such as joysticks, ultrasonic transducers, and optical sensors. Latched inputs are provided by switches, sip-and-puff devices, and voice recognition systems.

The input provided by the user can be input directly to the controller or it can be pre-conditioned. Pre-conditioning can be as simple as a low-pass filter or a sliding window signal averager or as complex as a neural network or fuzzy logic. More sophisticated routines are used to discern desired input signals from tremor or electrical noise (e.g. continuous or transient). Simple low-pass filtering or signal averaging techniques may eliminate signal power associated with user-determined desired direction or speed. The tremor of the driver and the driver's intended behavior can be learned and filtered to provide improved control. There are two basic approaches: driving behavior is recorded on an actual or virtual driving course; tremor data are measured while the input device is in the neutral position. Driving data can be used to train a neural network or program fuzzy logic (figure 8.18). Neutral position tremor can be measured and used as background noise to develop an adaptive noise cancellation filter (figure 8.19). The advantage that these schemes offer is that they accommodate the user's abilities.

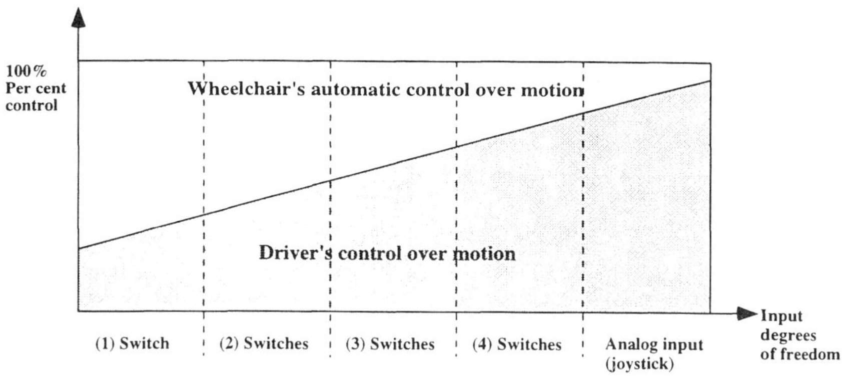

Figure 8.17. *Diagram illustrating the proportion of control effort assumed by the person and the automatic control system.*

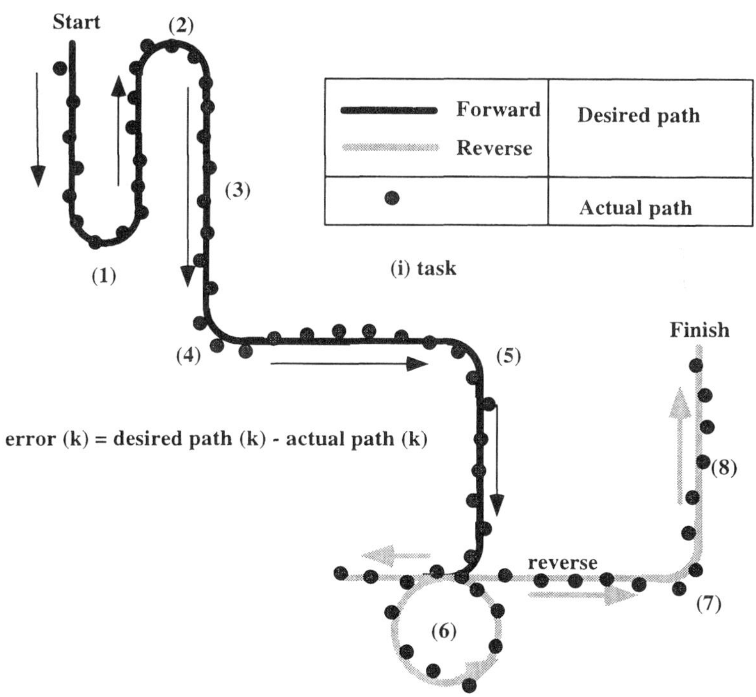

Figure 8.18. *Error signals developed while driving on a test course can be used to train a neural network or develop a fuzzy logic controller. Deviations from the actual path can be measured with both internal and external wheelchair sensors. Such courses can also be used to evaluate control systems and interface devices.*

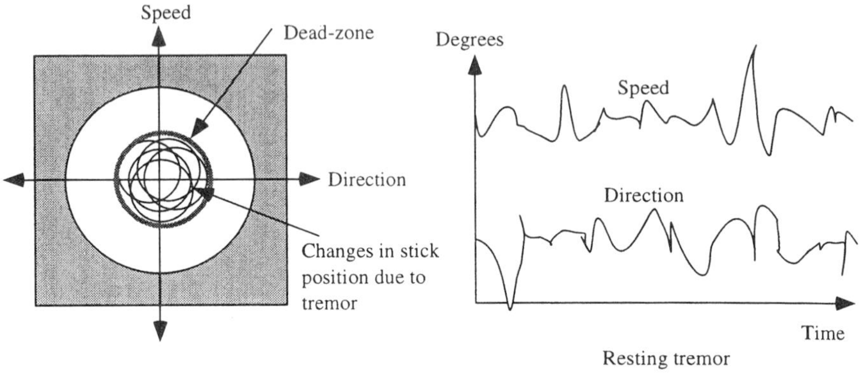

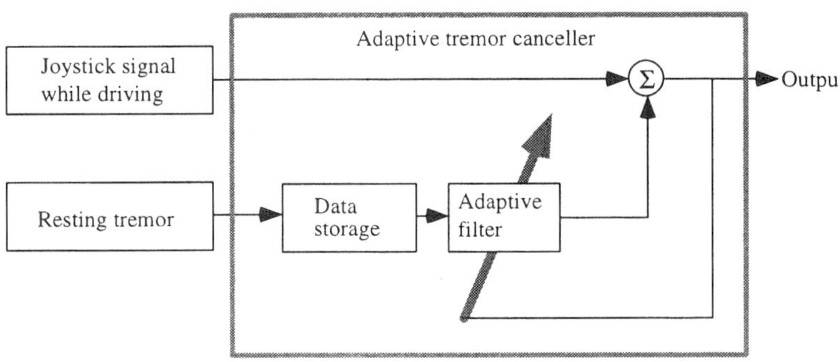

Figure 8.19. *Input devices often incorporate deadzones to prevent the chair from lunging forward if the input device is active on start-up. Tremor can be recorded within this deadzone to use with an adaptive tremor canceller during driving. This method provides the user with greater control than simple stationary filters.*

Pre-conditioning is also used to remove input device component drift, sensor imbalance, and to ensure gentle degradation of performance with component failure. A simple means of providing some protection in the event of component failure is to place a pull-up resistor on the output stage of the input device to place the control input in neutral when a failure occurs, and to incorporate a deadzone to accommodate component and sensor drift. Another approach is to use redundant components. By processing dual sets of direction and speed data, one IC may fail and the controller will still receive the correct joystick position (figure 8.20). This strategy implies that the microprocessor must discern between correct and erroneous signals. This can be accomplished by having the microprocessor maintain a history of direction and speed data, which can be used to determine whether the two direction and speed monitors begin to differ significantly. If the difference

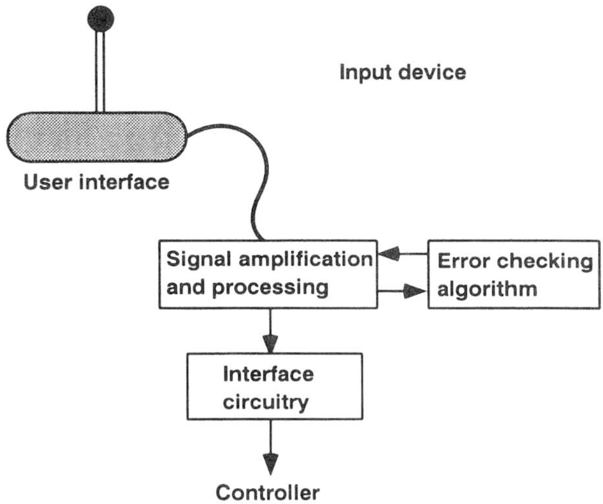

Figure 8.20. *Schematic diagram of an input device (joystick) with error checking (fault-tolerant) circuitry. Known responses can be fed through the signal amplification and processing hardware by the error checking circuit to determine if faults exist and then to determine how best to proceed.*

exceeds a threshold, then an error is signaled to the user. The controller can then begin to monitor chips to determine whether any have encountered stuck-at faults or randomly varying signals. These are the most common type of fault. This method can be used to determine many types of failure. Once the microprocessor determines that the input sensor has returned to the neutral position, it can carry out a dynamic diagnostic routine. This is accomplished by comparing the outputs of the redundant direction and speed circuits while the microprocessor generates input values across the input sensor's full range. If a circuit is determined to have failed, then the microprocessor can signal the user and switch to limp mode, allowing the user to drive to a repair station.

8.6.2. Control strategy

Sensor-based control involves three main components: internal sensors; external sensors; and a decision-making element. Internal sensors monitor the power wheelchair system status. Sensors can monitor state of wheel speed, battery charge, power stage current, bearing failure, motor temperature, and the primary microprocessor. Watch-dog processors are found in most power wheelchair controllers. External sensors are used to measure orientation, proximity, and relation to objects/hazards in the environment.

External sensors are used to sense potential hazards to power wheelchair users (e.g. stairs, curbs, steep hills) and to safely avoid them. External sensors include infrared detectors, ultrasound transducers, video, structured light,

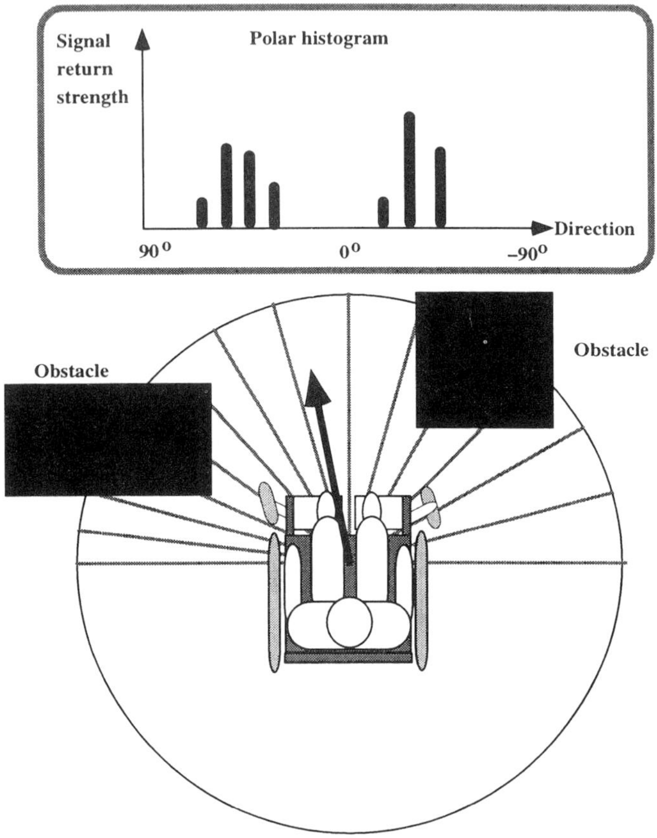

Figure 8.21. *The user can choose to pass through obstacles and the wheelchair's automatic control system can use environmental sensors to decide how to proceed with the user's commands. A simple histogram can be used to determine where obstacles are absent.*

magnetic sensors, and telemetry systems. Global positioning systems are also a possibility. External sensors may also help with navigation. For example, if the user has difficulty making fine directional adjustments, external sensors can be used to avoid obstacles (e.g. furniture, people, doorways) along the path chosen by the driver. An extension of this concept is to have the user select a destination, and have the power wheelchair determine and negotiate the path (figure 8.21). The control processing is shared between the driver and the automatic control system. The degree of sharing is dependent upon the driver's abilities, and is determined with actual and virtual driving tests.

Shared control can be based upon two models: human decision models and human behavior models. Human decision models are based upon inferencing (i.e., determining the intentions of the user). To use a human decision model for tracking, a model of how people extract control variables from sensory data is required. Deterministic models are not well developed for this task. Probabilistic models can be developed from training data (i.e., the behavior from previous trials can be used to identify characteristics of the driver that reveal his/her intentions). Probabilistic models use actual data to extract characteristics of the process of human driving. Based upon the values of the measures (e.g. direction changes, speed changes, time delays, body position, eye tracking) the probability that a person will institute a transition to each of the predefined modes can be determined. The mode selected will be the one that has the highest probability. This problem is well suited for fuzzy logic.

Human behavior models assume that driving behavior can be classified into a few basic tasks, such as high-speed driving, close-quarter maneuvering, and docking maneuvers. High-speed driving is used to characterize tasks associated with open areas where speed is high and obstacles are few and infrequent. Close-quarter maneuvering describes mobility with a large number of obstacles or transitions present (i.e., within rooms of furniture, or from one room to another) (figure 8.22). Docking maneuvers describe tasks required to approach and use a desk, counter-top, or other piece of furniture or equipment. The user selects a mode via a switch or other suitable input device.

8.7. FAULT-TOLERANT CONTROL

The power wheelchair and driver system is so complex that it is impractical to describe the dynamics by exact mathematical models. Adaptive controllers (e.g. fuzzy-logic based, least-squares based, neural-networks based) perform better in the presence of model uncertainty than classical feedback controllers, since they actively reduce uncertainty by on-line parameter estimation. Adaptive control strategies can become unstable in the presence of inevitable unmodelled dynamics and unpredictable external disturbances. Techniques have been developed which supervise the operation of advanced controllers and take appropriate action when unacceptable behavior occurs. Such strategies are called 'fault-tolerant supervisory control' and may be based upon heuristic rules designed to take appropriate action in the face of component failure or external disturbance (figure 8.23).

Power wheelchairs are crucial to the mobility of their drivers. If a component fails while driving, in some situations the driver can be placed at extreme risk. For example, when crossing a busy intersection, the wheelchair should allow the driver to proceed to safety in the presence of component failure or external disturbance. If a wheelchair user is driving along a sidewalk, an external disturbance should not guide the user into a busy street. In such cases, real-time decisions are made to continue driving with either short-term degradation or drastically reduced performance. The emphasis

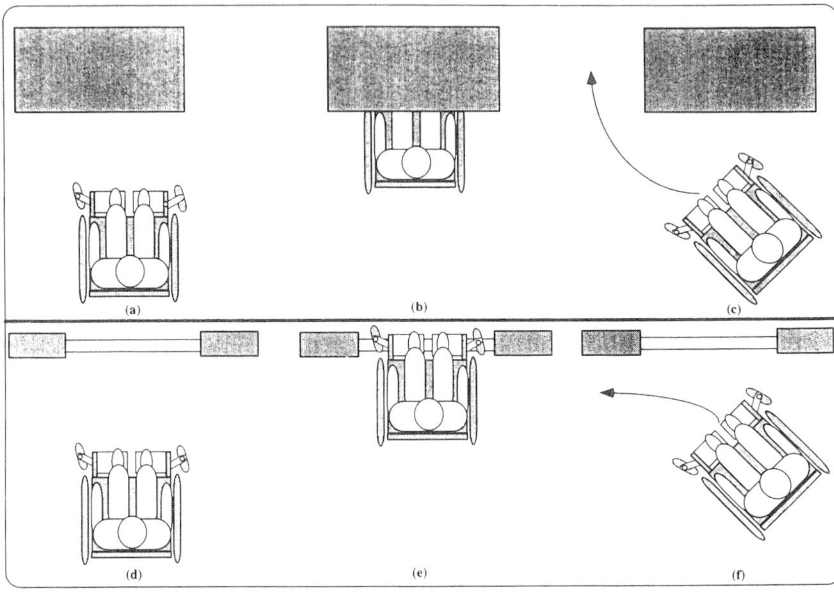

Figure 8.22. *Two classic problems face shared control systems: the docking problem and the passage problem. The user and wheelchair must communicate to determine whether docking is to occur or whether obstacle avoidance is preferred. The passage problem is similar, however, the wheelchair must recognize the doorway (absence of an obstacle) and safely negotiate through it.*

here is on providing the user control over the wheelchair and maintaining safety. This mode of control is called 'limp mode,' because it allows the user to drive the damaged wheelchair to safety (see figure 8.24). If the system is exposed to transient disturbances, it may attempt to reconfigure itself to achieve normal performance, after the transients have deceased. The microcontrollers available on power wheelchairs are under-utilized, and could be programmed to incorporate intelligent and fault-tolerant techniques.

8.8. INTEGRATED CONTROLLERS

Many power wheelchair users require multiple assistive devices (e.g. wheelchair, environmental control unit, communication device, computer access device). Integrated controls allow the user to control more than one assistive device through a single input device. Typically, this is the device chosen to control the power wheelchair. If the power wheelchair is optimized to the driver as described above, then the appropriate hardware and software protocol can be used to interconnect various devices. This

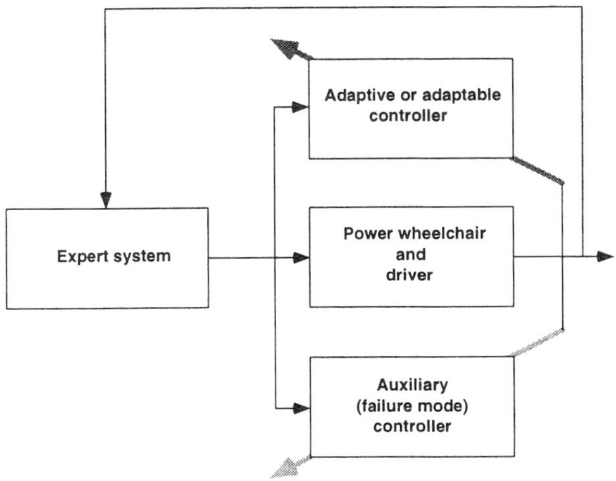

Figure 8.23. *Block diagram for a supervisory fault-tolerant control scheme. The wheelchair is modeled as four blocks: expert system; adaptive/adaptable controller; wheelchair and driver; and auxiliary controller. The power wheelchair and driver can act as a simple closed loop servo control system. The driver, wheelchair, and adaptive/adaptable automatic control system can provide more sophisticated and efficient control, which would be the primary control scheme. The auxiliary controller would activate upon detection of component failure or persistent system fault. The expert system determines the mode of operation and provides the driver with system status.*

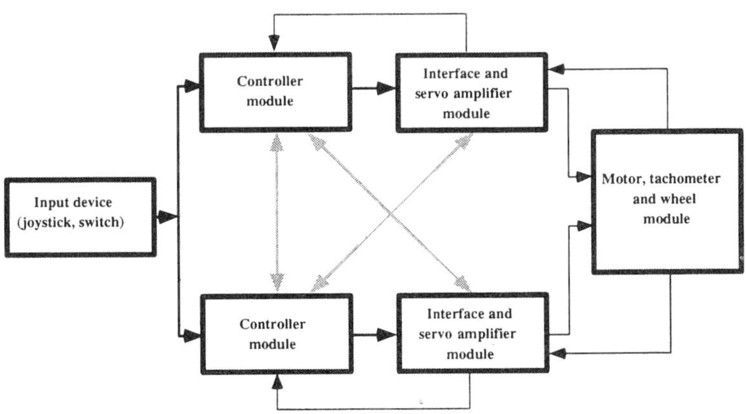

Figure 8.24. *Block diagram for a power wheelchair fault-tolerant control scheme. Black arrows illustrate paths during normal operation, gray arrows represent software-controlled interconnections used during 'limp mode' operation. Redundant controller modules improve reliability of the system.*

would provide the user with effective control of a variety of electronic assistive devices. Integrated controls is the term used to describe plug-and-play for assistive devices. Ideally, all electronic assistive devices would use compatible communication protocols and interconnect hardware. The need for integrated controls is illustrated by the walking–talking problem. Many power wheelchair users who use augmentative communication devices cannot operate both systems simultaneously (i.e., the walking and talking problem). Neither of these systems require constant vigilance, but they are limited by their input structure and inability to communicate over a common communications bus. If the power wheelchair and augmentative communication system both could be controlled with the same input device and communicate over a common data bus, then shared control would permit the user to toggle between driving and talking states. The two subsystems maintain speech and safe driving while the other subsystem is being controlled by the user. This case and others like it have prompted work on a worldwide assistive device electronic communication standard.

The RESNA Serial Interface Standards Committee and Trace Center have made significant advances towards developing a worldwide serial interface communications standard between power wheelchairs and other devices. This work and the work of European groups has formed the basis for a draft ISO standard (TC 173/SC-1/WG-7). The multiple master–multiple slave (M3S) initiative is an European Community (EC) effort to develop a standard interface between various electronic assistive technologies. M3S is based upon the CAN (controller area network) protocol used in many modern automobiles.

Many assistive devices include at least one input device (e.g. joystick, head-operated switch, extremity-operated switch, or voice recognition system) and at least one actuator (e.g. wheelchair, environmental control unit, robot, voice synthesizer). The M3S bus is suitable for general purpose applications with the proper systems and subsystems (see figure 8.25). The M3S standard is essentially the CAN bus with additional lines used to increase margins of safety and reliability. The advantage of M3S over current technology is the ability to simply interface multiple assistive devices and other technical products within the user's environment. The M3S bus permits control of multiple actuators with a single input device. This allows the input device to be optimized to the user's abilities, and subsequently used to control a number of actuators.

The M3S standard would specify a basic hardware architecture, a bus communication system, and a configuration method. A basic M3S bus consists of 7 wires: 2 power lines, 2 CAN lines, 2 safety lines, 1 shield, and 1 special support line (harness number). The M3S bus is intelligent in that it is easily configured to meet each user's individual needs. The intelligence in the system exists in the controllers of the individual input devices, actuators, and the control and configuration module (CCM). The

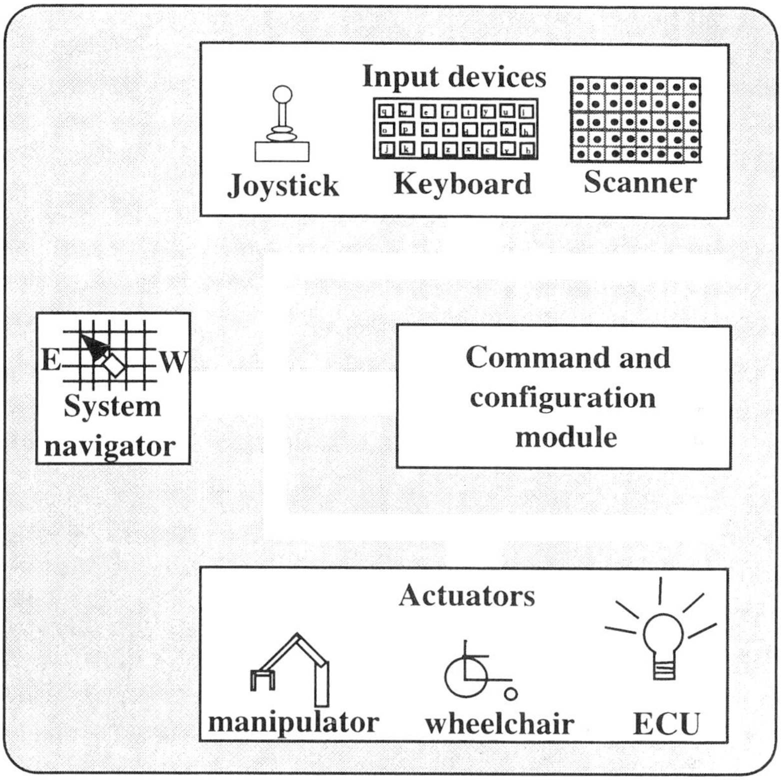

Figure 8.25. *Illustration of multiple master–multiple slave (M3S) architecture. Multiple input devices and multiple actuators can be on the M3S bus simultaneously. A system navigator works with the control and configuration module (CCM) to set priorities, perform hand-shaking, and for device identification. The CCM monitors and control communication between various components of the system.*

CCM deals with the processing necessary to configure the system and the ongoing need to monitor the safety of the system. The CCM is linked to a simple display that allows the user to select and operate each actuator. The CCM processor itself contains the master menu used to select individual actuators. The M3S system possesses the capability to access a wide range of devices via wireless link (infrared or radio) to a home bus.

8.9. ELECTROMAGNETIC COMPATIBILITY

Electric powered wheelchairs may be susceptible to electromagnetic interference (EMI) present in the ambient environment. Some level of EMI immunity is necessary to ensure the safety of power wheelchair users. While there are many advantages to digital and microcontroller-based designs, these components can be sensitive to electromagnetic radiation and electrostatic discharge (ESD). Many people attach accessories (e.g., car stereos, computers, communication systems) to their power wheelchairs, which share the batteries. Such interconnection may increase the susceptibility of other system components to EMI. A number of companies make electric powered devices designed to operate on power wheelchairs to provide postural support, pressure relief, environmental control, and motor vehicle operation. These devices may alter the EMI compatibility of power wheelchairs as provided by the original equipment manufacturer.

There have been some unsubstantiated cases of power wheelchairs behaving erratically for unexplained reasons. Recently, the US Food and Drug Administration (FDA) released a position paper on EMI compatibility testing for powered wheelchairs. Members of the American National Standards Institute (ANSI) and the RESNA sub-committee on Power Wheelchair Electro-Magnetic-Interference (EMC) and the International Standards Organization Technical Committee 173, Sub-Committee 1, Working Group 10 (ISO-TC-173-SC1-WG10) are currently developing an addendum to ISO 7176-part 14, power and control systems, of the wheelchair standards.

Field strengths have been measured at 20 $\mathrm{V\,m^{-1}}$ from a 15 W hand-held cellular telephone, and 8 $\mathrm{V\,m^{-1}}$ from a 1 W hand-held cellular telephone. Picofarad capacitors are currently being used on wheelchair motors to filter EMI. The US FDA has begun to require that a warning sticker be placed on the wheelchair. The FDA Center for Devices and Radiological Health (CDRH) has investigated several power wheelchairs and scooters and has found them to exhibit some degree of susceptibility to EMI. The CDRH found some wheelchairs to be susceptible to field strengths as low as 5 $\mathrm{V\,m^{-1}}$, which is equivalent to a 4 W hand-held radio transmitter at a distance of 2 m.

Power wheelchairs should be tested from 26 MHz to 4 GHz. CDRH and ANSI/RESNA recommend 1 MHz linear steps. Logarithmic steps may be a viable alternative, but require further experimentation and justification. With logarithmic stepping, more steps are used at lower frequencies than higher. This helps attain a high Q at the higher frequencies. Most wheelchairs currently respond at broad-band to EMI. Some respond at about controller or watch-dog processor clock frequencies. During a typical failure, RF usually goes through some circuit, hits a diode junction, gets rectified, and creates a d.c. offset that may be above logic level, creating an error. Modern motion control hardware and software may be applied along with good grounding/shielding practices to reduce susceptibility.

EMI testing is expensive and complicated. Standards are required to ensure that test results are valid and consistent among test laboratories. Calibration is done using a volume mapping within a closed cuboid. All sensors must be within −0 dB and +10 dB of the center sensor (reference), over 75% of the cuboid's surface, defining a uniform field. Testing is done in all environments (e.g. open-air, anechoic chamber, Gigahertz tester for electromagnetism (GTEM)). The EMI test methods must be sufficiently rigorous to ensure product safety, but must not be cost prohibitive. This balance requires a series of complicated trade-offs. Square wave modulation can be used to add some side-band. Generally, this represents worst case. The ANSI/RESNA sub-committee recommends amplitude modulation of 80% ± 20% with 1 kHz±10% rms frequency modulation. They also recommend wheelchairs be tested at a minimum of 30 $V\,m^{-1}$. Currently, a dwell time of two seconds is used by the FDA test laboratory. Testing should also be conducted at low field strengths (10 $V\,m^{-1} \pm 3$ dB) to determine whether there is an undetected response at higher field strengths (window-effect). Some wheelchairs have sheet metal components or enclosures that may affect the field strength around critical components (e.g. controller, joystick, significant portions of loom). The wheelchair must be exposed in the frontal and lateral planes. The wheelchair must be irradiated from the sides that provide maximum exposure of electronic components and from the wiring harness. No part of the wheelchair should be closer than 1 m from the floor and ceiling when testing in a shielded environment or anechoic chamber. The supports for the wheelchair must be non-conductive.

When the wheelchair is exposed to EM radiation, the brakes should not release nor should the wheels move with a wheel torque more than that experienced on a 1:6 slope with a 100 kg rider. Non-electrical contact methods (e.g. audio sensing, optical sensing) of measuring brake release or wheelchair movement are preferable. Contact methods are acceptable; however, a minimum of 10 kΩ impedance between the sensing device and wheelchair is required. A plastic tube, placed near the brakes, interfaced to a remote audio sensing circuit works well in the testing of many wheelchairs. The wheelchair must be tested at 40% to 60% of its maximum speed. Maximum speed is measured with the controller package set as provided by the manufacturer or contractor, with the speed input device (e.g. joystick) set to maximum for forward, and the input device speed trim pot or switch, if available, set to maximum. Nominal wheel speed may drift over the length of the test. This drift is primarily due to drop in battery voltage over the test interval. Wheel speed must be recorded between test intervals without EMI. The percentage change in wheel speed during exposure to EMI should be referenced to the nominal wheel speed for that test interval. The variation in absolute forward speed is to be within 30% of the nominal forward speed. The differential speed between the two wheels is to be within 30% of each other.

Table 8.1. *Standard power wheelchair battery group sizes (units are inches).*

Group number	Length	Width	Height
U1	$7\frac{3}{4}$	$5\frac{3}{16}$	$7\frac{5}{16}$
22NF	$9\frac{7}{16}$	$5\frac{1}{2}$	$8\frac{15}{16}$
24	$10\frac{1}{4}$	$6\frac{13}{16}$	$8\frac{7}{8}$
27	$12\frac{1}{16}$	$6\frac{13}{16}$	$8\frac{7}{8}$

8.10. BATTERIES

The battery energy storage system is recognized as one of the most significant limiting factors in powered wheelchair performance. Battery life and capacity are important. If battery life can be improved, the powered wheelchair user will have longer reliable performance from his/her battery. An increase in battery capacity will allow powered wheelchair users to travel greater distances with batteries that weigh and measure the same as existing wheelchair batteries. Most importantly, increases in battery capacity will enable the use of smaller and lighter batteries. Because batteries account for such a large proportion of both the weight and volume of current powered wheelchair systems, wheelchair manufacturers must base much of their design around the battery package.

Power wheelchairs typically incorporate 24 V d.c. energy systems. The energy for the wheelchair is provided by two deep cycle lead-acid batteries connected in series. Either wet cell or gel cell batteries are used. Wet cell batteries may require greater maintenance, but have the ability to store greater energy than gel cells at temperatures from freezing to room temperature. Wet cell batteries also cost about one half as much as gel cell batteries. Gel cells may be required for transport by commercial air carriers.

Battery technology for wheelchair users remains unchanged despite the call for improvements by power wheelchair users. This may be in part due to the relatively low number of units purchased, about 500 000 per annum, when compared to automotive applications, where about 6.6 million per annum can be sold by a single manufacturer. Wheelchair batteries are typically rated at 12 V and 30–90 A h capacity at room temperature. A power wheelchair draws about 10 A during use. The range of the power wheelchair is directly proportional to the ampere-hour rating for the operating temperature.

Batteries are grouped by size. Group size is indicated by a standard number. The group size defines the dimensions of the battery (table 8.1). The ampere-hour rating defines the battery's capacity.

It is important that the appropriate charger be used with each battery set. Many battery chargers automatically reduce the amount of current delivered to the battery as the battery reaches full charge. This helps to prevent damage to the battery from boiling. The rate at which wet and gel cell batteries charge is significantly different. Some chargers are capable of operating with both types of batteries. Many require setting the charger for the appropriate battery type. Most wheelchairs contain two 12 V batteries connected in series, and are charged simultaneously with a 24 V battery charger.

8.11. GEAR BOXES

The efficiency of a power wheelchair is affected by the drive system. An efficient drive system can extend the distance covered on the energy stored in the batteries. Pulse width modulated servo amplifiers can be designed to have high efficiencies, in excess of 80%. The motors of the wheelchair are often connected to the wheels through a gear box or pulley set. The pulley set or gear box decreases the speed of the motor and increases the torque seen at the wheels. This allows the motor to operate about its rated speed, where it is more efficient. However, mechanical losses in the gear boxes or pulley set are also causes of inefficiency.

Three common methods are used to reduce the speed of the motor and to increase the torque at the wheel: a worm gear right angle drive; a spur gear reducer; and a belt–pulley system. The worm gear drive although compact is not very efficient, because the sliding contact between the worm and the worm gear creates high friction forces. A spur gear uses a small pinion, driven by the motor, and a larger gear. Involute cutting of the gear teeth permits the pinion and gear combination to turn with rolling friction, rather than the constant sliding friction of the worm gear. However, the spur gear arrangement requires the motor to be orthogonal to the gears, which restricts folding the wheelchair frame. Efficiency of a gear train is affected by speed. The efficiency of a power wheelchair worm gear drive is about 70% compared with 80% for a spur gear drive. Belt drives vary from about 60 to 90%. Helical timing drive (HTD) belt drives though not commonly used on power wheelchairs, have been shown to be about 90% efficient on motorcycles.

Lubrication is also an important factor to consider when studying gear box efficiency. Many wheelchair gear boxes are designed for other applications, and are filled with high viscosity grease by the factory. This is because the gear drives are often designed for higher speed and load applications. At the relatively low speeds of wheelchair use, this high viscosity grease can cause excess drag. A polytetrafluoroethylene (PTFE) lubricant can be used to improve the efficiency of power wheelchair drives, because of its good lubrication properties at low speeds. PTFE is a solid lubricant; very small solid particles slide over one another with a very low coefficient of friction. As two surfaces slide over one another the PTFE particles act as microscopic ball bearings. Only small amounts of PTFE are required for lubrication, in contrast to total immersion used with standard grease. About a 5% improvement in performance can be gained using PTFE.

8.12. USER INTERFACES

Joysticks and switches can be used to effectively control a powered wheelchair or power base (see figure 8.26). The joystick is the most common control interface between the user and the wheelchair. Joysticks produce voltage signals proportional to displacement, force, or switch closures. Displacement joysticks are most popular. Displacement joysticks may use either potentiometers, variable inductors (coils) or optical sensors to convert displacement to voltage. Inductive joysticks are most common as they wear well, because the stick is not physically in contact with the windings, and they can be made to be quite sensitive. Joysticks can be modified to be used for chin, foot, elbow, tongue, or shoulder control. Typically, short throw joysticks are used for these applications. Force-sensing joysticks use three basic transducers: simple springs and dampers on a displacement joystick, cantilever beams with strain gages, and fluid with pressure sensors. Force-sensing joysticks which rely on passive dampers or fluid pressure generally require the user to have a range of motion within the normal values for displacement joystick users. Beam-based force-sensing joysticks require negligible motion, and hence may be used for people with limited motion abilities.

People who exhibit intention tremor require special control considerations. Signal processing techniques are often required to grant the user greater control over the wheelchair. Typically signal averaging or a low-pass filter with a cut-off frequency of below 5 Hz is used. The signal processing is typically incorporated into the controller.

Some people lack the fine motor control to effectively use a joystick. An alternative for these people is to use switch control or head position control. Switch control simply uses either a set of switches or a single switch and a coded input, i.e., Morse code or some other simple switch code. The input of the user is latched by the controller and the wheelchair performs the task commanded by the user. The user may latch the chair to repeatedly perform a task a specified number of times, e.g. continue straight until commanded to do otherwise. Switch control is quite functional, but it is generally slower than joystick control. Switch inputs can be generated in many ways. Typically low-pressure switches are used. The input can come from a sip-and-puff mechanism which works off a pressure transducer. A switch contact is detected when the pressure exceeds or drops below a threshold. The pressure sensor may be configured to react to pressure generated by the user blowing into or sipping from an input or by the user

(*a*)

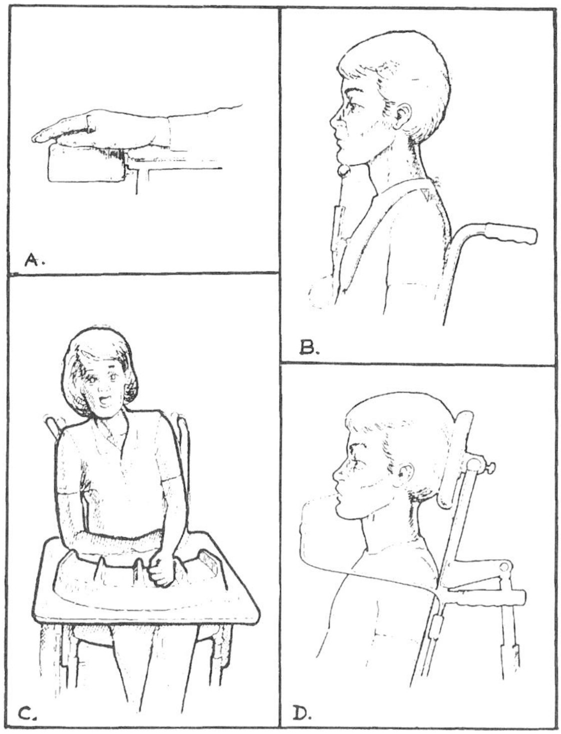

Figure 8.26. *(a) Control input devices for powered wheelchairs and power bases. The hand-operated joystick is most commonly employed (A). Chin-operated joysticks can be used to provide effective control for people with impaired arms, who are unable to reliably operate a hand joystick (B). When the user has use of his/her arms but lacks the fine motor control required to operate a joystick, then multiple switches can provide a useful input device (C). For the most severely motor impaired person, a sip-and-puff switch can provide a safe and effective input to the controller (D). (Figure reproduced with permission from Bennett Wilson A Jr 1992 Wheelchairs: A Prescription Guide 2nd edn (New York: Demos).) Continued opposite.*

simply interrupting the flow in or out of a tube. Sip and puff may also be used as a combination of proportional and switch control. For example, the user can put the control in the 'read speed' mode and then the proportional voltage output from the pressure transducer will be latched as the user-desired speed.

Simple switches of various sizes can be used to control the chair with many parts of the body. Switches may be mounted on the armrests or a lap tray for hand or arm activation, on the footrest(s) for foot activation, or on a headrest for head activation. The motion of the head can also be used for proportional control by using ultrasonic sensors. Ultrasonic sensors

(b)

Figure 8.26. *Continued. (b) Power wheelchair computer controller and user input devices. Inductive joysticks are used with both of these input devices. (Photograph courtesy of Invacare Corporation.)*

can be mounted in an array about the headrest. The signal produced by the ultrasonic sensors is related to the position of the head. Hence, motion of the head can be used to create a proportional control signal. Ultrasonic head control and switch control can be combined to give some users greater mastery over their power wheelchair. Switches can be used to select the controller mode, whereas the ultrasonic sensors give a proportional input signal.

A critical consideration when selecting or designing a user interface is that the ability of the user to accurately control the interface is heavily dependent upon the stability of the user within the wheelchair. Often custom seating and postural support systems are required for a user interface to be truly effective. The placement of the user interface is also critical to its efficacy as a functional control device.

FURTHER READING AND REFERENCES

Amori R D 1992 Vocomotion—An intelligent voice-control system for powered wheelchairs *Proc. RESNA Int. '92* pp 421–3

ANSI/RESNA 1994 Electromagnetic Compatibility Requirements for Powered Wheelchairs and Motorized Scooters (ANSI/RESNA Draft Standard) version 1.5 November 1994

Aylor J H, Byun H and Kauzlarich J J 1991 A user survey on backup power for powered wheelchairs *Proc. RESNA 14th Ann. Conf. (Kansas City, MO, 1991)* pp 234–6

Aylor J H and Johnson B 1981 *Proc. RESNA 4th Ann. Conf. (Washington, DC, 1981)* pp 87–90

Aylor J H, Thieme A and Johnson B W 1992 A battery state-of-charge indicator *IEEE Trans. Ind. Electron.* **39** 398–409

Bailey D M and DeFelice T 1991 Evaluating movement for switch use in an adult with severe physical and cognitive impairments *Am. J. Occup. Ther.* **45** 76–9

Baldwin J D and Thacker J G 1991 Structural reliability assessment techniques applied to tubular wheelchair frames *Proc. RESNA 14th Ann. Conf. (Kansas City, MO, 1991)* pp 237–9

Baumgartner E T and Yoder J D 1994 An automatically-guided powered wheelchair for the severely disabled *Proc. 17th Ann. RESNA Conf. (Nashville, TN, 1994)* pp 350–2

Bell D A, Levine S P, Koren Y, Jaros L A and Borenstein J 1993 Shared control of the NavChair obstacle avoiding wheelchair *Proc. RESNA 16th Ann. Conf. (Las Vegas, NV, 1993)* pp 370–2

Bell D A, Levine S P, Koren Y, Jaros L A and Borenstien J 1994 Design criteria for obstacle avoidance in a shared-control system *Proc. 17th Ann. RESNA Conf. (Nashville, TN, 1994)* pp 581–3

Bennett L 1987 Powered wheelchair bucking *J. Rehab. Res. Dev.* **24** 81–6

Bourhis G and Pino P 1993 Man/machine cooperation for the piloting of an intelligent electric wheelchair *Proc. 2nd Eur. Conf. Advancement Rehabilitation Technol.* p 13.1

Brown K E, Inigo R M and Johnson B W 1987 A adaptable optimal controller for electric wheelchairs *J. Rehab. Res. Dev.* **24** 87–98

Brubaker C E 1988 Survey of powered wheelchair problems and features *Proc. Wheelchair IV (Washington DC, RESNA, 1988)* pp 68–75

Carlson C W 1995 Understanding EMI *Team Rehab. Report* January pp 32–3

Chase J and D M Bailey D M 1990 Evaluating potential for powered mobility *Am. J. Occup. Ther.* **44** 1125–9

Choy T T C and Koo J T K 1992 An electric wheelchair with special features. *Proc. RESNA Int. '92 (Toronto, Ont., 1992)* pp 310–2

Cooper R A 1994 Observability and controllability of scooters *Proc. IEEE-EMBS 16th Int. Conf. (Baltimore, MD, 1994)* vol 16 488–9

Craig I and Nisbet P 1993 The smart wheelchair: an augmentative mobility 'toolkit' *Proc. 2nd Eur. Conf. Advancement Rehabilitation Technol.* p 13.1

Department of Veterans Affairs 1990 Choosing A Wheelchair System *J. Rehab. Res. Dev. Clin. Suppl.* No 2 March

Dunham J, Roberson G and Inigo R M 1993 A direct current to three-phase inverter for electric wheelchair propulsion *Proc. RESNA 16th Ann. Conf. (Las Vegas, NV, 1993)* pp 373–5

Fisher W E, Garrett R E and Seeger B 1988 Testing of gel-electrolyte batteries for wheelchairs *J. Rehab. Res. Dev.* **25** 27–32

Fiss J, Markmeller B, Luebben A J and Collins R A 1993 More patterns in power mobility repairs *Proc. RESNA 16th Ann. Conf. (Las Vegas, NV, 1993)* pp 376–8

Ford M R, Kauzlarich J J, Thacker J G 1992 Powered wheelchair gearbox lubrication *Proc. RESNA Int. '92 (Toronto, Ont., 1992)* pp 316–8

Ford M R, Thacker J G, Kauzlarich J J 1991 Improved wheelchair gearbox efficiency *Proc. RESNA 14th Ann. Conf. (Kansas City, MO, 1991)* pp 146–7

Grant D A and Gowar J 1989 *Power MOSFETS Theory and Applications* (New York, NY: Wiley)

Gregson P H and Kirby R H 1992 Development of a 'Smart' powered wheelchair: A progress report *Proc. RESNA Int. '92 (Toronto, Ont., 1992)* pp 424–5

Guerette P, Caves K, Nakai R, Sumi E, McNeal D and Hoffer M 1994 Determinating the appropriateness of integrated control of assistive devices *Rehab. R&D Progress Reports 1992–1993* pp 408–9

Hendriks J L,Rosen M J, Berube N L J and Aisen M L 1991 A second-generation joystick for people disabled by tremor *Proc. RESNA 14th Ann. Conf. (Kansas City, MO, 1991)* pp 248–50

Hoyer H and Hoelper R 1994 Intelligent omnidirectional wheelchair with a flexible configurable functionality *Proc. 17th Ann. RESNA Conf. (Nashville, TN, 1994)* pp 353–5

IEC 1984 Electromagnetic Compatibility for Industrial Process Measurement and Control Equipment (IEC) pp 801–3

Inigo R M, Shafik K T and Park C W 1990 An improved DC-DC converter for electric wheelchair propulsion *Proc. RESNA 14th Ann. Conf. (Kansas City, MO, 1990)* pp 148–50

Jaffe D L 1981 Smart wheelchair *Proc. RESNA 4th Ann. Conf. (Washington, DC, 1981)* pp 91–3

Jamieson J R and Nowack P F Model-based reasoning for industrial control and diagnosis *Control Eng.* **BME-39** 120–8

Jaros L A, Bell D A, Levine S P, Borenstein J and Koren Y 1993 NAVCHAIR: Design of an assistive navigation system for wheelchairs *Proc. 16th Ann. RESNA Conf. (Las Vegas, NV, 1993)* pp 379–81

Johnson B W and Aylor J H 1981 Modelling of wheelchair dynamics for the design of a microcomputer-based controller *Proc. IEEE-IECI Ann. Conf.* pp 70–5

Junkman B C, Aylor J H and Kauzlarich J J 1988 Estimation of battery state-of-charge during charging using the charge recovery process *Proc. RESNA 11th Ann. Conf.* pp 280–1

Kauzlarich J J and Dwyer M A 1982 Test of nickel-zinc battery for wheelchairs *Proc. RESNA 5th Ann. Conf.* p 110

Kauzlarich J J, Thacker J G and Ford M R 1993 Electric wheelchair drive train efficiency *Proc. RESNA 16th Ann. Conf. (Las Vegas, NV, 1993)* pp 310–2

Kauzlarich J J, Ulrich V, Bresler M, Bruning T 1983 Wheelchair Batteries: driving cycles and testing *J. Rehab. Res. Dev.* **20** 31–43

Korba L, Park G, Farley R, Durie N, Roy O Z 1990 Development of a wheelchair controller: conversion to a microcontroller *Proc. RESNA Press (Washington, DC, 1990)* pp 199–200

Lavanchy C 1992 Comparative evaluation of major brands of lead-acid batteries *Proc. RESNA Int. Conf. (Toronto, Ont., 1992)* pp 541–3

Lefkowicz A T and Wierwille W W 1992 Validation of a PC-based perspective-view wheelchair simulator *Proc. RESNA Int. '92* pp 415–7

Lipskin R 1970 An evaluation program for powered wheelchair control systems *Bull. Prosthet. Res.* pp 121–9

Martin D L and Aylor J H 1993 Improving wheelchair efficiency through intelligent diagnostics *Proc. RESNA 16th Ann. Conf. (Las Vegas, NV, 1993)* pp 367–9

Mohamadi M C, Aylor J H and Schwab A J 1992 A fault-tolerant optical joystick control integrated circuit for a powered wheelchair *Proc. RESNA Int. '92 (Toronto, Ont., 1992)* pp 307–9

Nisbet P D and Craig I 1994 Mobility and mobility training for severely disabled children: results of the 'Smart' wheelchair project *Proc. 17th Ann. RESNA Conf. (Nashville, TN, 1994)* pp 341–3

Orpwood R 1993 A powered mobility aid for pre-school children with cerebral palsy *Proc. 2nd Eur. Conf. on the Advancement of Rehabilitation Technol.* p 9.1

Pence R, Inigo C and Inigo R M 1991 A new method for electric wheelchair propulsion *Proc. RESNA 14th Ann. Conf. (Kansas City, MO, 1991)* pp 151–3

Philipson L and Heckathorne C W 1981 The use of a microprocessor as a versatile controller for a powered wheelchair *Proc. RESNA 4th Ann. Conf. (Washington DC, 1981)* pp 84–6

Powell F and Inigo R M 1992 Pressure sensitive joystick and controller for front wheel steering wheelchairs *Proc. RESNA Int. '92 (Toronto, Ont., 1992)* pp 304–6

Powell F and Inigo R M 1992 Microprocessor-based DC brushless motor controller for wheelchair propulsion *Proc. RESNA Int. '92 (Toronto, Ont., 1992)* pp 313–315

Ramamurthi K and Agogino A M 1993 Real-time expert system for fault-tolerant supervisory control *ASME J. Dyn. Systems Meas. Control* **115** 219–25

Riley P O and Rosen M J 1987 Evaluating manual control devices for those with tremor disability *J. Rehab. Res. Dev.* **24** 99–110

Schauer J, Kelso D P and Vanderheiden G C 1990 Development of a serial auxiliary control interface for powered wheelchairs *Proc. RESNA 13th Ann. Conf. (Washington, DC, 1990)* pp 191–2

Schlemer M R, Johnson D and Granic J 1993 Powered wheelchair mobility simulator *Proc. RESNA 16th Ann. Conf. (Las Vegas, NV, 1993)* pp 357–9

Scooters 1991 *Product Comparison and Evaluation* REquest, Rehabilitation Engineering Center, National Rehabilitation Hospital

Scott C M and Prior R E 1978 Mobility engineering for the severely handicapped *Bull. Prosthet. Res.* BPR **10** 248–55

Shung J B, Stout G, Tomizuka M and Auslander D M 1983 Dynamic modeling of a wheelchair on a slope *ASME J. Dyn. Systems Meas. Cont.* **105** 101–6

Shung J B, Tomizuka M, Auslander D M and Stout G 1983 Feedback control and simulation of a wheelchair *ASME J. Dyn. Systems Meas. Cont.* **105** 96–100

Silberberg J L 1993 Performance degradation of electronic medical devices due to electromagnetic interference *Compliance Eng.* pp 25–39

Stefanov D H 1994 Powered wheelchair controller with dynamic stability checking *Rehabilitation R&D Progress Reports 1992–1993* pp 410–1

Stout G 1979 Some aspects of high performance indoor/outdoor wheelchairs *Bull. Prosthet. Res.* BPR **10** 330–2

Teft D, Furumasu J and Guerette P 1992 Cognitive predictors of successful powered wheelchair control in very young children *Proc. RESNA Int. '92* pp 412–4

Thacker J 1978 Wheelchair Dynamics *University of Virginia REC Ann. Report* pp 41–4

Ulrich V, Bresier M and Kauzlarich J J 1980 Wheelchair battery testing *Proc. RESNA 3rd Ann. Conf.* pp 132–3

van Woerden J A 1993 M3S: A general purpose interface for the rehabilitation environment *Proc. 2nd Eur. Conf. Advancement Rehabilitation Technol. (Stockholm, 1993)* p 22.1

Wakaumi H, Nakamura K and Matsumura T 1992 Development of an automated wheelchair guided by a magnetic ferrite marker lane *J. Rehab. Res. Dev.* **29** 27–34

Whitmeyer J J 1993 A modification of joystick assemblies to limit the excursion and force applied by the user *Proc. RESNA 16th Ann. Conf. (Las Vegas, NV, 1993)* pp 360–1

Widman L M, Cooper R A, Robertson R N, Ster J F and Grey T 1993 Evaluation of an isometric joystick for power wheelchair control *Proc. 16th Ann. RESNA Conf. (Las Vegas, NV, 1993)* pp 364–6

Wilson A B Jr 1992 *Wheelchairs: A Prescription Guide* 2nd edn (New York, NY: Demos)

Winkler F W and Zuber D 1991 Development of an analog controller interface to relate steered front wheels to powered rear wheels *Proc. RESNA 14th Ann. Conf. (Kansas City, MO, 1991)* pp 140–2

Witters D and Weigle R 1993 *FDA/CDRH proposed EMC addition to ANSI/RESNA WC/14* US Food and Drug Administration, 13 June 1993

Witters D M and Ruggera P S 1994 Electromagnetic compatibility (EMC) of powered wheelchairs and scooters *Proc. 16th IEEE-EMBS Int. Conf. (Baltimore, MD)* vol 16 pp 894–5

CHAPTER 9

POSTURAL SUPPORT AND SEATING

9.1. SEATING AND POSTURAL SUPPORT SYSTEMS

Wheelchair users often neglect specialized seating and postural support systems. Proper seating can improve a person's stamina, and provide greater comfort. Developing or selecting the right seat requires some knowledge of biomechanics as well as familiarity with available products. A person's anthropometry, body posture, and the necessary supportive forces must be evaluated to properly design or select a seating system. Proper seating position should place all joints at or near their neutral position while seated (i.e., hip, knee, and elbow angles at 90 degrees). Many wheelchair users have abnormal body postures due to their physical impairment, their years of sitting improperly, or both. Problems with abnormal pelvic tilt, pelvic obliquity, and pelvic rotation must often be addressed. Abnormal or exaggerated hip and trunk postures (e.g., exaggerated kyphosis, exaggerated lordosis, rotation, and scoliosis) must also be ameliorated. Some individuals may have compound curvature of the spine as well as abnormalities of the pelvic and rib anatomy. Designing a seating system requires knowledge of the deforming forces and moments present, the realistic limits of correction which can be achieved, and knowledge of materials and their properties.

A seating system should enhance the position of the user in one way or another and provide stability for proper posture. A seating system is a static device used to align and hold a person in a manner that improves seated posture and alignment while maintaining or improving function. The shape and firmness of the pads or straps used to control body position influences the efficacy of the seating system. A functional seating system has to provide some degree of movement by the user with the incident reduction of maximal postural control.

Firm supports generally provide the greatest postural support. A firm lateral thoracic support attached by a metal bracket to the backrest canes will often provide greater support than will a contoured foam backrest. Some people require moderate pelvic restraint or support. Rigid pelvic restraints provide maximal support often with a significant reduction in desirable body movement. Flexible restraints provide some pelvic support while permitting desirable body movement. The selection of the proper restraint system depends upon the degree of support required by the user, and the user's functional movements. Maximal control of body position can often be achieved by using flat stabilizers, in place of contoured surfaces. For example, a flat lateral pelvic stabilizer can block lateral movement more effectively than a contoured seat-base lateral support surface.

9.1.1. Medical implications of wheelchair seating

Proper seating is a medical necessity for everyone who uses a wheelchair. There are several primary medical implications which are directly related to seating. Pressure sores are one of the most dramatic results of improper seating, and they exact a costly toll on both the wheelchair user and the healthcare system. A pressure relieving cushion is the starting point in reducing the risk of developing decubitis ulcers via chronic wheelchair use. A good cushion must provide sufficient pelvic support to reduce the risk of pressure sores over the ischial tuberosities due to pelvic obliquity. If the pelvis exhibits abnormal posterior pelvic tilt, then the coccyx may bore into the seating surface and the ischia are pushed forward creating shear forces. This situation often results in ulcers and tissue necrosis over the coccyx and ischial tuberosities. A properly supporting cushion and backrest can help to ameliorate these posturally induced injuries.

Lower back and pelvic pain are the most commonly experienced forms of chronic pain experienced by wheelchair users. Poor posture can exacerbate both of these conditions which are often treatable with a custom seating system. Scoliosis, exaggerated lordosis, and exaggerated kyphosis are the most common spinal deformities among wheelchair users. These deformities are often caused by poor pelvic alignment and improper support of seat and back structures found on many wheelchairs. Proper seating can often prevent or retard the progression of these deformities.

Improper seating posture can also lead to the loss or reduction of hip range of motion. In severe instances this can lead to an inability to sit upright in a wheelchair. Many wheelchair users with severe impairments have reduced respiratory function. Seating which causes or does not inhibit forward flexion of the trunk causes decreased chest expansion and diaphragm excursion. This adversely affects respiratory capacity and speech function. Respiration and speech are also affected by head and neck position. Poor skeletal position can affect a person's ability to eat and swallow properly. In addition, decreased blood flow to the abdominal cavity due to trunk flexion can adversely affect digestion.

Proper seating posture can provide maximal ability to see and interact with the environment. This is of particular concern for people with exaggerated kyphosis, who are forced to sit with a 90 degree seat to backrest angle.

When the trunk and limbs are not properly supported fatigue will develop more rapidly, reducing the person's active time. Functional independence requires adequate stability to promote maximal upper-extremity function. Appropriate seating can promote improved reach, wheelchair propulsion, and self-care.

9.1.2. Seated anthropometry

Several specialized measurements are used to define sitting. Functional relations are based upon anthropometry of seating. These measurements can be used to define wheelchairs, seating systems, and workspace requirements.

Sitting height is the vertical distance from the floor to the horizontal midsection of the back portion of the thigh while in contact with the seat, with a knee flexion angle of 90 degrees, and the feet resting flatly on the footrests. Elbow height is the vertical distance from the floor to the posterior tip of the olecranon when the arm is flexed at 90 degrees at the elbow and the shoulder is in its neutral position. Thigh height is the vertical distance from the floor to the highest part of the thigh while in the seated position. Patellar height is the vertical distance from the floor to the superior tip of the patella while seated. Orbital height is the vertical distance from the floor to the orbit when sitting with the spine erect. Shoulder height is the vertical distance from the floor to the superior aspect of the acromion while seated. Hand-grip height is the vertical distance from the floor to the midpoint of the hanging fist while seated.

Internal seating depth is the sagittal distance from the posterior aspect of the popliteal fold to the posterior aspect of the buttock. External seating depth is the sagittal distance from the anterior aspect of the knee to the posterior part of the buttock. Abdominal depth is the sagittal distance from the anterior abdominal wall to the posterior part of the buttock.

Buttocks width is the maximum transverse distance across the buttocks. Shoulder width is the maximum transverse distance across the shoulders. External elbow width is the maximum transverse distance between the tips of the olecrani when the arms are abducted to 90 degrees.

Wheelchair seating measurements vary somewhat from those used for seated anthropometry. Some simplification is employed because of tasks specific to wheelchair prescription and design (figure 9.1). These measurements are used to determine seating and wheelchair dimensions.

9.1.3. Standard wheelchair and seating dimensions

When selecting a wheelchair, seating needs should be given a high priority, especially for power wheelchairs. Power-base wheelchairs are typically narrower and have a higher seat height than standard power wheelchairs. Users of wide wheelchairs may have difficulty in negotiating doorways, and

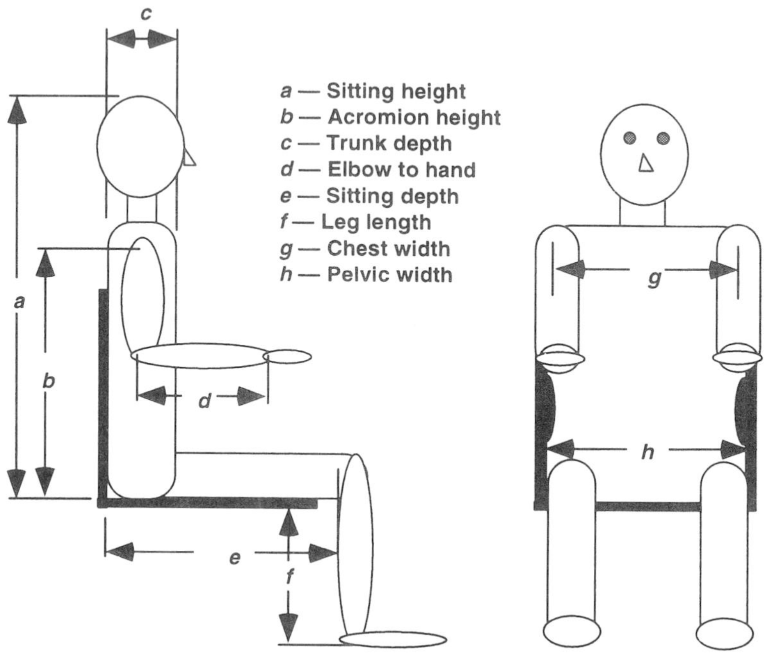

Figure 9.1. *Anthropometric measurements used in seating.*

turning in small spaces. Tall wheelchairs may create difficulty in getting into vehicles and when using tables. Some power wheelchair users may require architectural modifications to their home and workplace. Standards have been developed to make measurements consistent throughout the world. Overall wheelchair dimensions are illustrated in figure 9.2.

Some wheelchairs are foldable. Folding sometimes makes it easier to stow the wheelchair in the trunk of a car or luggage compartment of an airplane. Folding power wheelchairs often require removal of the battery assembly which can be heavy. Once the chair has been folded, the entire chair or its major components may be difficult to lift because of their size and weight.

The ANSI/RESNA (American National Standards Institute) and ISO (International Standards Organization) specify test procedures for measuring wheelchairs and definitions for describing wheelchairs. ISO 7167-07 and RESNA WC-07 describe measurement of wheelchairs and seating systems. According to these standards the weight of the wheelchair includes typical armrests, legrests, wheels, and casters. The weights of detachable parts (e.g., armrests, footrests) are also reported to give the consumer and prescriber information about the strength required for lifting the wheelchair.

The type of upholstery fabric may vary substantially from manufacturer to manufacturer or model to model. Many materials are listed by manufacturers for seat upholstery: boat cloth, cordura, leather, nylon, polypropylene, velour,

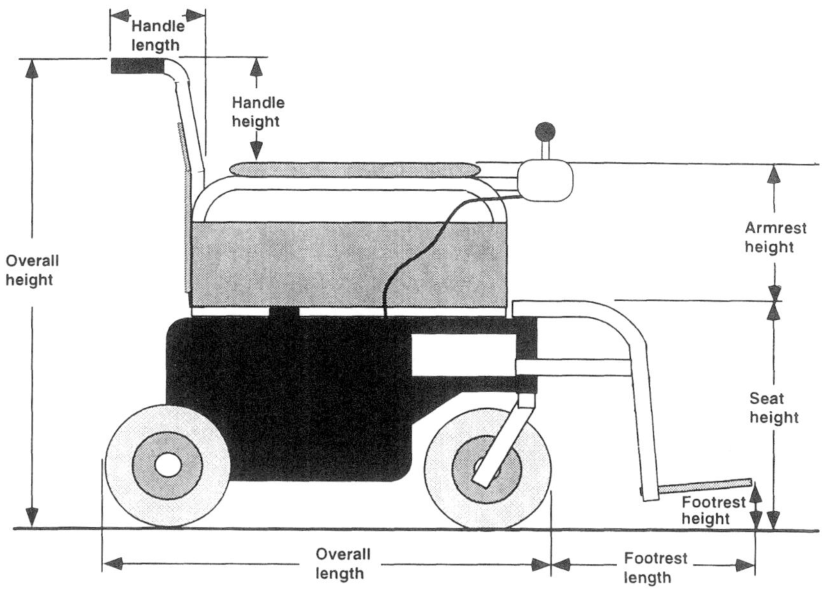

Figure 9.2. *Standard wheelchair overall dimensions.*

and vinyl. ANSI/RESNA and ISO standards require upholstery to meet flammability standards for use on wheelchairs.

Choosing the correct size wheelchair to meet the user's particular needs is a very important decision. Many users require some degree of customization of their wheelchair once they have tried it. Many manufacturers and designers accommodate a variety of options and accessories in their designs. It is often difficult to select the proper wheelchair because of the differences in the way manufacturers measure and report wheelchair and seating dimensions. The dimensions of the upholstery and frame structure may be quite different from the space usable by the wheelchair rider. ANSI/RESNA and ISO have developed standards for measuring dimensions meaningful to the user, clinician, and engineer.

Wheelchair measurements should be made with the chair loaded. This helps to simulate the dimensions of the chair when occupied. ANSI/RESNA and ISO standards specify using a seating dummy referred to as the reference loader gage (RLG) or Bratgaard dummy (figure 9.3). The RLG consists of seat portion and back portion which are hinged together and weighted to simulate the approximate weight distribution of a wheelchair user. The seat portion includes two curved sections which simulate legs. By shaping the upholstery like a person, the RLG creates reference planes which form a basis for many seating measurements (figure 9.4).

The seat angle represents the angle made by the base of the RLG when placed in the wheelchair to a flat horizontal surface. A positive seat angle indicates that the seat tilts downwards towards the backrest. Seat depth is

Figure 9.3. *Reference loader gage used to measure wheelchair seating dimensions. This device helps to provide consistent results across manufacturers and models of wheelchairs. It is based on using a human analog for seating measurements. (Photograph courtesy of Center for Assistive Technology, University of Pittsburgh Medical Center.)*

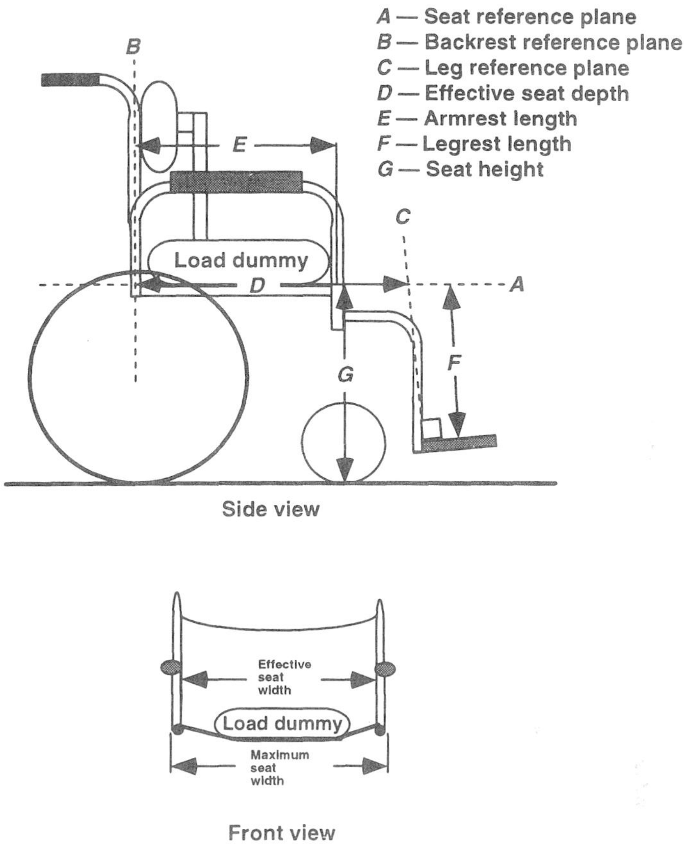

Figure 9.4. *Wheelchair seating dimensions.*

defined as the full seating depth available and not the length of the seat upholstery. Seat depth is measured from the loaded backrest reference plane (measured from the RLG) to the intersection of the leg reference plane. Seat height is the distance from the front edge of the loaded seat to the floor. This measurement is especially important for people who have difficulty transferring or when the wheelchair is intended to be foot propelled. The seat affects the user's positioning, weight distribution, and ability to maneuver. Forward sloping seats make it easier to sit straight and to transition from seating to standing, but they may make it more difficult to maintain seated balance.

Backrest height is the distance from the center of the top edge of the loaded backrest upholstery or pad to the level floor. Backrest angle is measured relative to vertical, and a positive angle indicates that the backrest is tilted towards the rear of the chair. The backrest angle and seat angle can be used to determine the seat-to-back-angle which is influenced by the user's flexibility

$$(\text{backrest angle}) + 90^\circ - (\text{seat angle}) = \text{seat-to-back angle}. \qquad (9.1)$$

Armrest come in a variety of styles and sizes. There are removable, swing-away, or pivoting armrests. There are also desk style or full-length. Armrests must be tested for minimal pull-out strength in case someone attempts to lift the occupant and wheelchair by the armrests. Armrests must also undergo impact testing. Armrest length defines how far the front edge of the armrest structure extends from the backrest cane (seat post). The measurement is important as it helps to determine how close the user will be able to get to a table or desk. Armrest height is the height of the top edge of the armrest to the top of the loaded seat surface. A range of values are reported for adjustable armrests.

Footrests and legrests also come in a variety of styles and sizes. The footrest-to-seat dimension, i.e., the legrest length, determines the range of leg lengths that can be accommodated for a given wheelchair. Leg-to-seat angle is used to indicate the angle between the lower leg and upper leg while seated in the wheelchair. This angle can be critical since many wheelchair users may not possess the flexibility to sit in some wheelchair configurations. Typically, the lower-leg angle is measured relative to the seat reference plane. The leg-to-seat angle is simply 180 degrees minus the lower-leg angle. For example, if the wheelchair has an 85 degree front end than it will have a 95 degree leg-to-seat angle ($180 - 85 = 95$).

Combining the design of wheelchair test dummies (ISO 7176/11) and of the reference loader gage device (ISO 7176/7) can make testing simpler and more meaningful. There are four dummies used for wheelchair testing (25, 50, 75, 100 kg). Dummies can be used for all tests where the wheelchair is to be loaded. The RLG is used during measurement of seating dimensions.

The test dummies described in the ISO/RESNA standards are linear and may not load wheelchairs realistically. On some wheelchairs the seat and back of the dummy are supported by the wheelchair's seat and backrest frame members. On other wheelchairs the dummy is supported by the upholstery. The inconsistency in the manner in which the dummy interfaces to the wheelchair can alter fatigue test results.

The ISO/RESNA test dummy should be similar anthropometrically and anthropomorphically to a human wheelchair rider and should be much more durable than any wheelchair being tested. The dummy should load the wheelchair similarly to a human wheelchair user. The seat and back of the dummy should be contoured similarly to a human. However, the seat and back of the contoured dummy must be durable enough to withstand numerous fatigue tests. The feet of the dummy should be able to rest properly on the footrest(s) of a wide variety of wheelchairs. The back should lean properly against the backrest and the base should sit properly on the

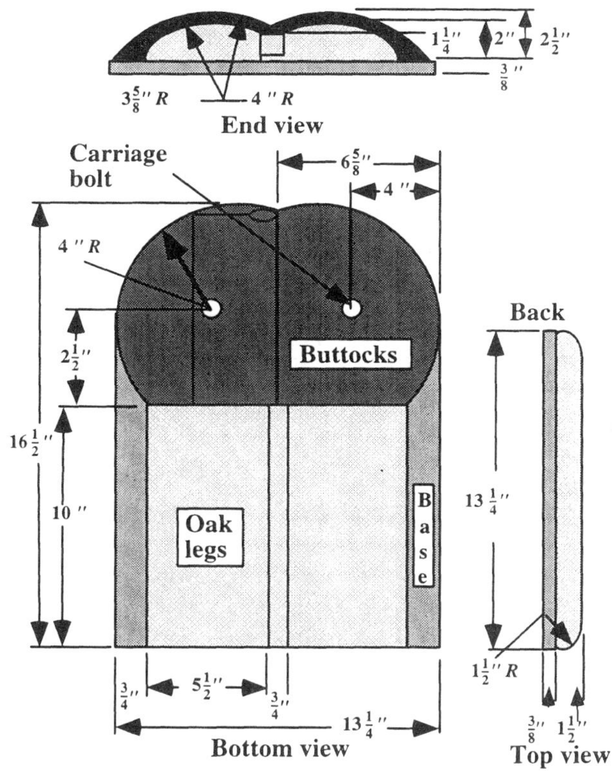

Figure 9.5. *Contoured test dummy schematic diagram.*

wheelchair seat. The design suggested in the standard has been observed to shear the seat upholstery on some wheelchairs, or to sit directly on the frame for others.

A test suitable for making seating measurements as well as impact and fatigue testing can be constructed from 9 mm aluminum (6061T6) plate. This material should be used because it is a relatively inexpensive structural aluminum, which is easy to machine, and is widely available. Steel plates can be used for additional mass. Steel is very durable, is readily available and is easy to work with. The buttocks of the dummy can be constructed from sections of ten-pin bowling balls. Bowling balls are readily available, low cost, and precise. The thighs and back contour of the dummy should be constructed of red oak. The thighs and back are bolted to the dummy base and back, respectively. Red oak is a commonly available, inexpensive hardwood. The basis for the dummy is constructed as per ISO/RESNA standards. The seat is modified by adding slices of a bowling ball to the posterior end of the dummy. Slices from opposite sides of the bowling ball are mated and bolted to the rear portion of the dummy base (figure 9.5).

These design specifications developed out of experience gained by testing wheelchairs to current ISO and RESNA/ANSI standards. Upon testing a variety of styles of wheelchairs from several manufacturers, it was discovered that the dummy did not sit consistently on all types and sizes of wheelchairs. It was hypothesized that the results of some wheelchairs may be biased because of dummy loading. Sometimes the dummy would sit on the seat support tubes, at other times it would sit just between the seat support tubes (often shearing the upholstery), or it would sit between the seat support tubes. This created a variety of different load cases, making comparison of fatigue results difficult. In addition, the loader gage used for seating measurements can be expensive, and adds to the number of test dummies required to evaluate wheelchairs. Use of the same dummy for fatigue testing and seating dimensions would help to make results more meaningful and reduce the amount of specialized test equipment required by testing centers.

Choosing the wrong size wheelchair can lead to user discomfort, decreased independence, injury, and pressure sores. The wheelchair and seating system must be chosen or designed together to avoid incompatibility. The improper wheelchair and seating system can make it difficult or impossible for the user to control the wheelchair. A seat that is too short provides insufficient leg support which may result in an improper weight distribution, leading to discomfort, tissue irritation, and possibly decubitis ulcers. A seat that is too long may place excessive pressure on the popliteal fossa (tissue behind the knees) disrupting circulation and limiting range of motion at the knees. Clothing affects effective seat width. Usually a narrow seat is best for propulsion. However, if the seat is too narrow the user may press against the side frames, wheels, or armrests. A narrow seat may create difficulty in transferring. A seat that is too wide can make it difficult if not impossible to push the chair properly, or may make it difficult to maintain balance and position to operate a joystick. A high backrest, in the middle of the shoulder or higher, is best if the user has very poor trunk stability and does not need to reach behind the wheelchair. The user's feet should be supported so that there are 90 degrees of flexion at the knees and ankles. If the foot support is too high, the user's knees will be raised decreasing the knee angle. This may also cause the person's pelvis to rotate so that the user assumes a sacral sitting position, increasing the risk of pressure sores. The thighs must be fully supported to ensure proper positioning and for postural stability. When the foot supports are too low excessive pressure may be placed upon the thighs.

9.1.4. Standard seating and postural support hardware

A number of devices have been designed to provide seating and postural support for wheelchair users. Some people require custom supports, whereas others can use simple off-the-shelf technology. Knowledge of the principles

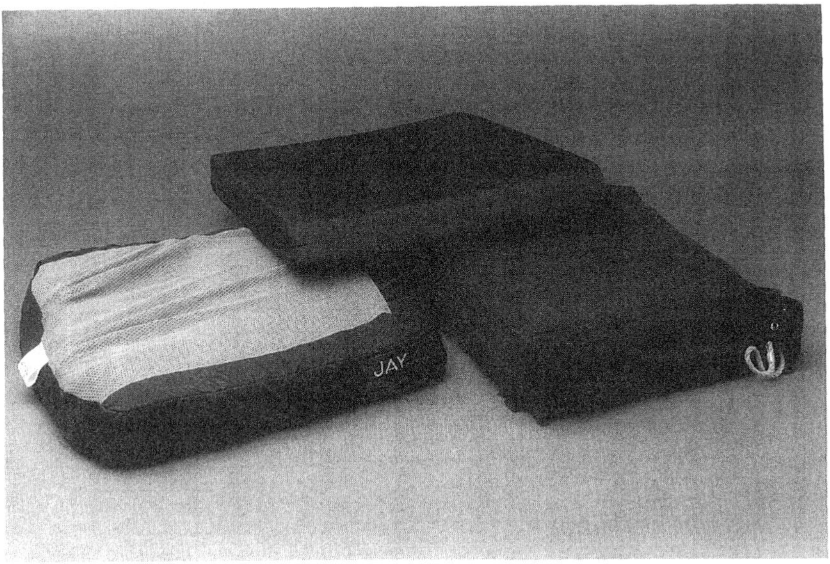

Figure 9.6. *Examples of some common linear cushion systems. (Photographs by S Parsons, CSUS Media Services.)*

behind the designs of various products can increase the likelihood of successfully matching the consumers needs to the appropriate seating technology. Children and adults sometimes have different seating and postural support needs. It is often desirable to design or select seating hardware for children which has the ability to accommodate growth. Products are available from several major manufacturers which allow enlargement of the seat width and seat depth. Legrests are commonly adjustable on most wheelchairs.

There are two basic types of seating systems used with wheelchairs: linear systems, and contoured systems. Linear seating systems are planar in that the seat and back surfaces are flat and only conform to the weight of the user (figure 9.6). For many wheelchair users a simple linear seating system or a standard contour is very effective. Foam, gel, or dry flotation (air) cushions are the most common (figure 9.7). Various densities and types (polyurethane, urethane, T-Foam, Sun-Mate) of foam are commonly used in linear seating systems. Foam has been shown to offer the lowest maximum pressure over the seating surface when the appropriate densities and contours are used. However, foam has a tendency to deteriorate at an undesirable rate. Gel cushions use an electrolyte gel in a closed plastic or latex pocket. The gel conforms to the body and provides a nearly even pressure distribution. The pressure distribution can be altered by using a stiff base (e.g., plastic or foam) to provide some contouring. Gel cushions suffer from excessive weight for many users, and their properties are subject to changes in temperature.

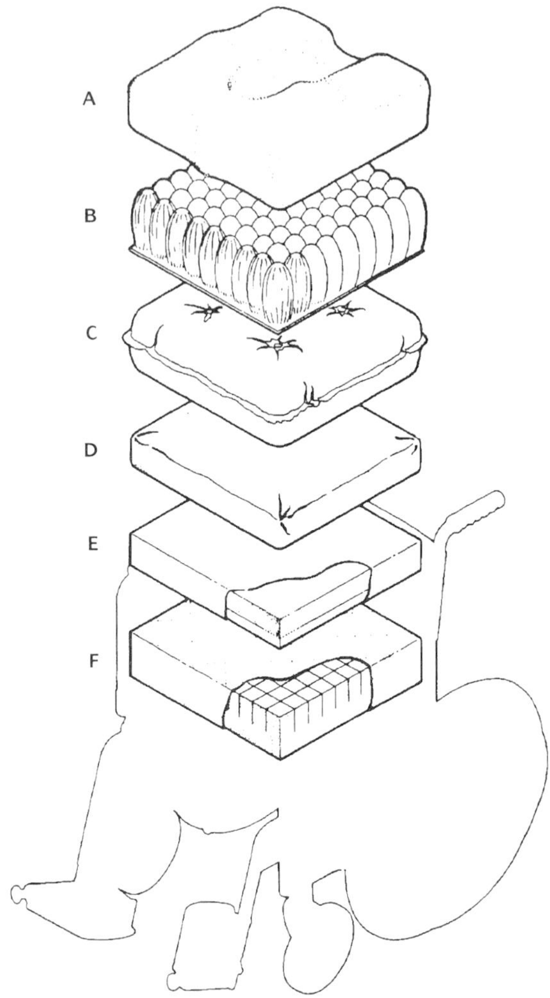

Figure 9.7. *Various common seat cushions used by individuals who do not require custom contoured cushions or specialized seating systems: (A) contoured foam, (B) air flotation, (C) liquid flotation, (D) gel flotation, (E) viscoelastic foam, (F) sectioned linear foam. (Figure reproduced with permission from Bennett Wilson A Jr 1992 Wheelchairs: A Prescription Guide 2nd edn (New York: Demos).)*

Air flotation cushions also create an even pressure distribution and are lightweight. The air is baffled (foam or molded baffles are used) to provide greater postural support for the user (figure 9.8). The pressure distribution of air flotation cushions can using multiple cells and/or a contoured base. Some air flotation cushions suffer from postural stability problems and may develop leaks.

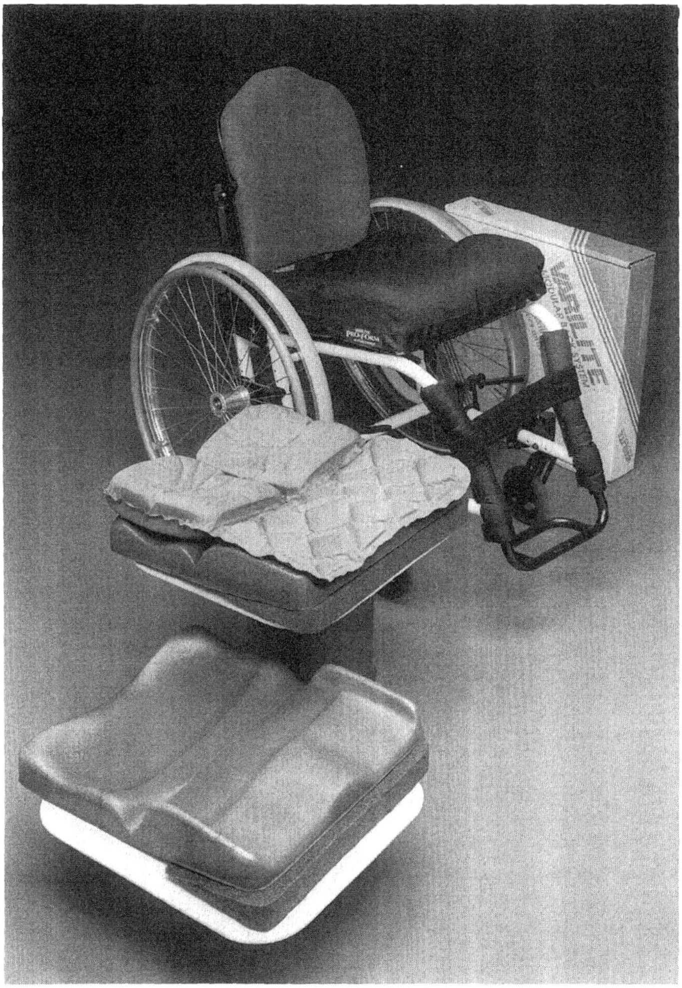

Figure 9.8. *Multiple component air flotation seating system with contoured base, and anterior wedge. (Photograph courtesy Varilite Corporation.)*

Seat bases may either be rigid or a cloth sling. In either case the seat should be stiff and provide a stable base for the cushion and user. Sling seats (seat and backrests) are constructed of synthetic materials with high tensile strengths. Nylon, cotton canvas, Kevlar, or similar materials are used to provide the structural component of the seat or backrest sling. In many instances the structural material may be enclosed in a vinyl, naugahyde, or other water resistant case, which acts as the interface to the user. Vinyl and naugahyde deform substantially under load and are not appropriate as seating structural materials. The exterior shell is often vacuum formed under

Figure 9.9. *Multicomponent air flotation backrest. (Photograph courtesy Varilite Corporation.)*

heat to form a bond between the shell material and the structural material. Seats and backrests may be attached to the wheelchair frame in a variety of ways. One of the more common means is to sew a pocket into each lateral edge of the seat base and backrest upholstery. When attaching the base and backrest upholstery to the frame, a metal (steel or aluminum) slat is placed into the upholstery pocket. Screws are then inserted through the slat and into the frame. For backrests it is not uncommon to insert the backrest canes through the upholstery and to use grommets and oversized contoured washers to secure the backrest upholstery to the canes. A sufficient number of attachments points should be used to prevent the backrest upholstery from sliding down the canes in the event of a screw or grommet failure, or an upholstery tear. Other methods can be used to resist having the backrest

Figure 9.10. *Cushion edge finish definitions.*

Figure 9.11. *Common seat cushion profiles.*

upholstery slide down the canes in the event of component failure. The backrest upholstery can be sewn in a catenary so that equal tension is applied to the canes across the entire surface of the backrest, or a stiffening material (high density foam) can be sewn into the upholstery. Seat base and backrest upholstery can also be attached using webbing or lacing.

Rigid seat bases and backrests have the advantage that their properties change little over time, therefore the cushion is always properly supported. However, a rigid seat base may weigh more, but as the application of composite materials becomes more widespread this should no longer be an issue. Aluminum, steel, fiberglass, Kevlar, carbon fiber, and several plastics are used in making seat bases and backrests. Most wheelchairs are available in common sizes, hence seat bases and backrests only need to be made for the sizes available. Rigid seat bases may incorporate some contouring to assist with proper seating positioning, with hip supports, or may be flat. Rigid or semi-rigid backrests may provide greater lumbar and lateral support than sling seats (figure 9.9). Rigid seat bases, and backrest supports form the foundation for many contoured seating systems, and they are becoming more popular.

Proper seating can be achieved for many people using standard seating and positioning equipment. The pelvis is often the foundation of the properly aligned seat. The structure, construction, and dimensions are critical in prescribing or designing a positioning pad or seating cushion. Seat cushions may have a squared edge, rolled edge, or beveled edge (figure 9.10).

Rolled or beveled edges are used to allow the knees to flex for people with contractures. The sagittal shapes of common seat cushions are shown in figure 9.11.

Seat cushions commonly include a cut-out for the ischial tuberosities, an abductor pommel, and hip adductor contours (figure 9.12). For people who require additional positioning or postural support, hip pads and/or hip belts can be used. For people with severe leg abduction or adduction, abductor or adductor pads can be attached to the seat pan. A cylinder of low to medium

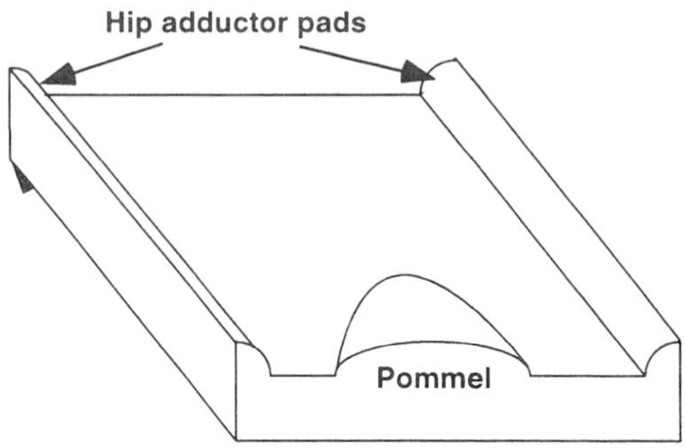

Figure 9.12. *Illustration of a cushion with hip adductor pads and pommel.*

density polyurethane foam on a stainless steel or chromed mild steel pillar will provide adequate support for even the most severe cases. Abductor or adductor pads can be prescribed or designed to be detachable or flip-down for easier transfers. Often a single cylinder is adequate; however, some people can benefit from two cylinders which can be independently laterally adjusted. A scissor board (i.e., a wedge of foam adhered to a semi-rigid material) with polyurethane foam can be used to reduce the risk of pressure sores at the knees when a narrow lateral separation is desired.

Backrests may also be contoured to provide additional support for those people who need it (figure 9.13). Two common approaches used are a flat back base with additional supports or a back base with curved sides. The width and height of the back base should be determined based upon the user's mode of mobility (e.g., manual, attendant propelled, power) and postural support needs. For people who use manual wheelchairs the backrest should follow the individual's frontal contour and the height should not extend above 4 cm below the scapula. The height should be sufficient to provide comfortable support and to provide a base to push against. Generally, the higher the degree of impairment the taller the backrest must be to provide adequate support. Contour can be added to backrests using carved foam, foam-in-place, standard contours, or custom molded/carved contours. People with significant loss of lateral stability may benefit from fixed or adjustable lateral trunk supports. A depth of between 125 to 200 mm is sufficient for most people. The height and width depend on the individual's anatomy. When using lateral supports many backrest plates require reinforcement. Scoliosis support systems can provide additional support for proper positioning, greater comfort and better control over the wheelchair. Scoliosis supports should be prescribed or designed to be independently adjustable (i.e., inferior–superior, fore–aft, medial–lateral). Supports can be designed or prescribed to wrap-around the trunk anteriorly, and may include a strap which wraps around the trunk for greatest support. Often the wrap-around scoliosis supports incorporate

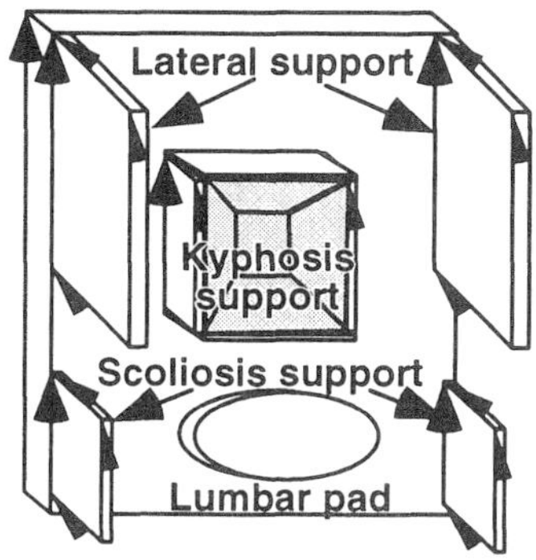

Figure 9.13. *Illustration of backrest with postural support devices.*

a quick adjust mechanism (e.g., quick release pin, knob head bolt) and a simply operated strap locking mechanism (e.g., Velcro, ring lock, or cam lock) to aid in transferring in and out of the wheelchair. Lateral and scoliosis supports require padding (i.e., typically 8 to 25 mm is sufficient depending upon the applied force) and the person should be isolated from all fasteners.

Some people who use wheelchairs require kyphosis support devices. A backrest box is a means of providing appropriate seating for many people who require kyphosis support. Kyphosis backrest boxes are commonly made of a composite material (e.g., fiberglass, carbon fiber, Kevlar). The kyphosis backrest box is filled with foam (i.e., typically a laminate of various types and densities of foams) which is cut out to match the placement of the kyphotic hump. This method relieves the hump so the person's back is positioned neutrally. Foam-in-place methods can be used instead of contouring the foam so that not only the kyphotic hump is accommodated, but the entire back as well.

Systems can be designed to provide chest and shoulder support to help people remain erect. An H-strap system often can provide adequate postural support. With an H-strap system one strap wraps anteriorly around the chest under the arms while two straps originate in the center of the chest and go over the shoulders to keep the trunk erect. Portions of the straps in contact with the person must be padded. A D-ring, Velcro, or snap-buckle can be used to secure the H-strap system in place.

People with extreme spasticity often have difficulty maintaining proper posture in their wheelchairs despite having adequate postural musculature. Side-wing controls or wrap-around backrests can be used to assist people with spasticity to maintain proper posture. Side-wings or wrap-around backrests are barriers added to the side of the wheelchair to confine the user within the backrest. Typically, a padded wall along the entire length of backrest of 100 to 150 mm high is effective.

A lumbar support can be added to the backrest of many wheelchairs for people who have difficulty maintaining the natural curvature of the spine. A

75 to 100 mm thick piece of contoured foam the width of the back is placed in the lumbar area. Custom devices and systems can be designed for people with unusual postural support needs.

In extreme circumstances people require head and neck supports (figure 9.14). Head and neck supports often provide the necessary stability required to control a power wheelchair, environmental control unit, or communication device. Extensions are designed and added to the backrest structure which supports the head and neck. The degree of support required determines the type of head and neck support employed. For people who require minimal head and neck support a simple flat headrest is sufficient. This type of headrest prevents them from extending beyond the plane of the backrest, which is especially important if the seating system is tilted or reclined. To assist in transferring, a flip-down headrest can be used which can be moved out of the way without being detached from the backrest. Removable headrests can also be prescribed or designed. Curvature can be added to the headrest to keep the user's head within the confines of the headrest. A winged headrest includes foam wedges on either side of center to provide slight lateral head support. A vertical foam wedge on the headrest can help to keep the head upright rather than extended, and gives slight lateral support. A neck yoke, a U-shaped support, with foam padding, can be used for lateral head support and to create neck flexion. Another means of providing moderate lateral support and neck flexion is to use an occipital headrest (i.e., a stiff U-shaped support padded with foam to fit the contour of the occipital region). Lateral head supports, padded arms extending on either side of the head, can be used to provide a high degree of support. For people with hydrocephalus a head sling can be used to relieve the head. A head sling is a padded shelf of molded material (e.g., plastic or composite) at the top of the backrest. Often the shelf is reinforced with a steel frame.

9.1.5. Contoured seating and postural support hardware

Contoured seating systems are molded to fit the anatomy and anthropometry of the user. Contoured seating systems may be made by hand shaping foam and plywood, by using a bead bag system to create a plaster mold or panograph of the seating contour, or by using foam-in-place techniques. The objective of contoured seating systems is to provide optimal postural support and control while minimizing pressure in sensitive areas. Custom contoured cushions accommodate the natural shape of the person and minimize tissue distortion that may cause discomfort or pressure sores. Contoured cushions provide postural support and positioning. The user should settle comfortably into the cushion and be supported statically and dynamically. The cushion should also protect the user against sudden impacts, such as going off a curb.

Plywood and foam is perhaps the simplest and most time consuming of the methods for developing contoured seating. However, plywood and foam

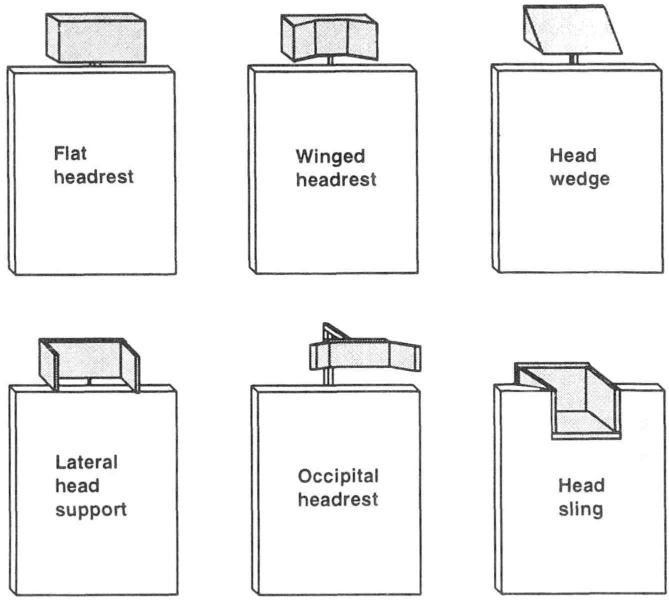

Figure 9.14. *Wheelchair head and neck supports.*

requires a great deal of skill on the parts of all involved (i.e., engineer, technician, clinician). Typically a piece of 6 to 12 mm interior plywood is cut to the shape of the seat base; in some instances a section is removed around the ischial tuberosities. A base of polyurethane, urethane, T-foam or similar material is glued to the plywood base. The thickness of the foam base depends upon the weight of the individual, the amount of fleshy tissue over the seating surface, and the health of the individual. Typically 50 to 150 mm of foam are used. Various densities of foam are used to provide support in some areas and pressure relief in others. Low density foam is used around bony prominences which carry significant weight (e.g., ischial tuberosities, coccyx). The foam is contoured using carving tools, sand paper, and finally high-density foam for smoothing. The foam is shaped until the client and clinician are satisfied that adequate pressure relief and postural support have been achieved. This process is sometimes aided by having the client lie prone on the edge of a table with his/her legs hanging over the edge. A clinician can then lay-up plaster-of-Paris over the person's thighs, buttocks, and lower back to create a mold of the person's seated shape. The mold can then be used to create a positive or male mold of the person's seated shape, which can be used when contouring a seating system.

Foam-in-place is another method of directly developing a custom contoured seating system (figure 9.15). Foam-in-place requires a fair degree of technical and clinical ability to be successfully applied. A rubber, latex,

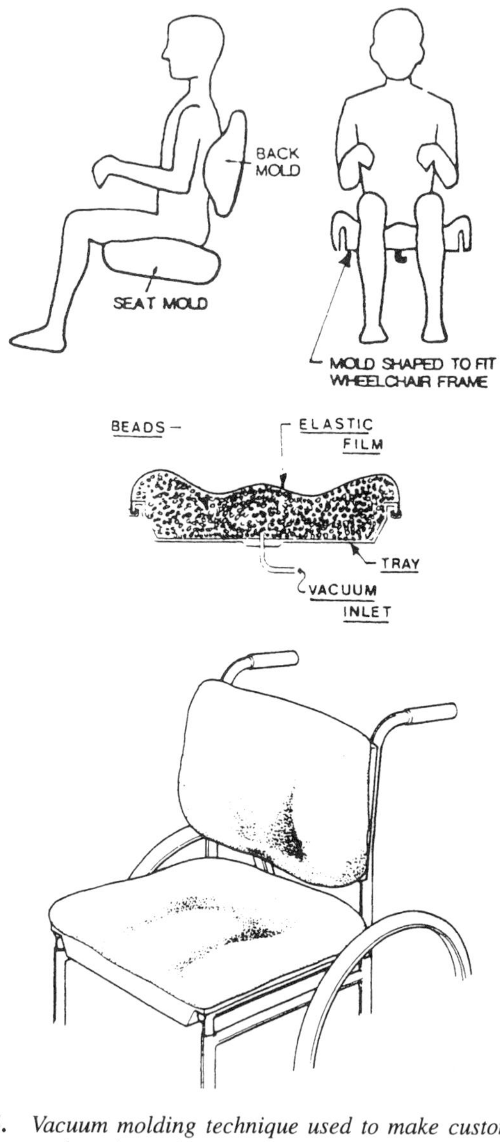

Figure 9.15. *Vacuum molding technique used to make custom contoured seats. A latex or other elastic film is used to form a sheath around styrofoam or other flexible beads. A vacuum is drawn to make the seat and backrest bead bags moldable. The clinician shapes the bags to provide appropriate cushioning and postural support. When the person is properly aligned, either foam is injected to solidify the cushions, or the person is removed and a plaster cast is made for shaping a cushion and back support. (Figure reproduced with permission from Bennett Wilson A Jr 1992 Wheelchairs: A Prescription Guide 2nd edn (New York: Demos).)*

plastic, or similar bag is filled with a liquid foam which when molded will form the contoured cushion. In some cases a 25 to 50 mm piece of foam is placed under or adhered to the bottom of the bag to provide a cushion base. A thin layer of foam (i.e., 8 mm thick medium-density foam) may be placed on top of the bag to provide a smooth finish. Usually the engineer or technician works from the side of the wheelchair, this helps to minimize the mess and to shape the cushion with the clinician. The foam base is mixed with a catalyst and poured into the bag. Time is of the essence so it is best to make several dry-runs with all parties involved. The client should wear thick clothing for protection from the 110–120 °F heat produced during the foam curing process. The client is placed upon the bag filled with liquid foam. The clinician and engineer/technician shape the foam to provide desired pressure relief and postural support. The client must be held in the desired position until the foam cures enough to maintain shape, at which time he/she can be carefully removed from the seat. When the foam is completely cured any excess material can be trimmed away. The bag is either separated from the foam in which case a vinyl cover is vacuum formed over the cushion or, if suitable, the bag is cleaned and then glued to the foam with a permanent adhesive.

Contour measurements can be made using vacuum forming or seating contour measurement systems. Typically such systems are used to create a contour representative of the client's seating profile. Vacuum forming methods are typically based on a flexible bag (e.g., latex, rubber) which is filled with Styrofoam or foam rubber pellets (figure 9.15). The bag is connected to a vacuum pump. The person to be molded is seated on the bead-bag in approximately the desired position. Air is drawn from the bag which becomes ever more rigid. While air is being drawn from the bag the clinician and rehabilitation engineer/technician form the bead-bag around the client pushing and pulling on the bag until the client is in the desired position. Air continues to be withdrawn from the bag until it becomes stiff, at which point the client can be removed from the chair. Vacuum forming systems are cost effective, and relatively simple. Contour measurement systems are more effective when they mimic wheelchair seating (i.e., legrests are attached, armrests are used, backrest is at the proper angle and of appropriate height). Once a male mold of the client's contour is completed, it must be transferred from the bead-bag to something that can be used to create a cushion. Many commercial contoured cushions are created using computer numerically controlled (CNC) foam milling machines (figure 9.16). These machines use arrays to represent x, y, and z coordinates of the seating contour. Hence, some companies have developed mechanical or electromechanical panographs to convert the male mold to a graph or directly to a computer file. A panograph may simply consist of a set of bars with rounded tips which glide over the mold surface. The other ends of the shafts are connected to a

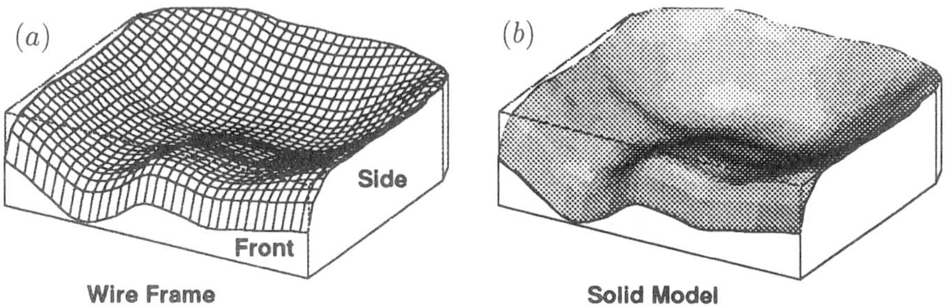

Figure 9.16. *Computer rendering of contoured foam seating. (a) A wire-frame representation of the contoured cushion. (b) A solid model of the same cushion. These representations can be used to visualize and modify the cushion before computer-controlled carving of the foam.*

pen or transducer. The sensors (fingers) of the panograph are indexed over the surface of the mold until a suitable representation has been created.

Systems have also been developed which directly measure the client's seating contour (figure 9.17). This has successfully been accomplished in a number of ways. A simple method used to directly measure a client's seating contour is to punch an array of holes into a piece of foam. Through the holes a plunger with a conforming head is placed through each hole in the foam. Each plunger is attached to its own stiff cable which runs through an incompressible flexible housing, on the other end of the cable there is either a pen or transducer to measure displacement. As the client sits on the sensor, the foam compresses and the plungers recess to create a portrait of the client's seating contour. If the foam is well characterized then the contour can be translated to an equivalent contour using a different foam with known density and compression properties. Each client must sit properly and comfortably on the shape chair and shape sensor. Seat tilt angle, backrest angle, back height, footrest position, and armrest position must all be set properly to yield satisfactory results (figure 9.18). Several sensors must be used to get an accurate contour profile; in some cases over 100 sensors are used to measure deflection.

More advanced systems have been developed which use linear variable displacement transformers (LVDT), linear potentiometers, or linear motors. System using LVDTs or potentiometers often use foam or coil springs to provide the necessary seating compliance. The pre-load force on coil springs can be set to simulate foams of different densities and compression properties. Coil springs behave linearly over a fairly large range, whereas foam does not. Linear motors offer greater flexibility than other seating pressure measurement systems as linear motors can be operated to behave like springs or foam. Linear motors can also be operated in closed loop to apply or withdraw pressure at desired points or to alter the shape of the

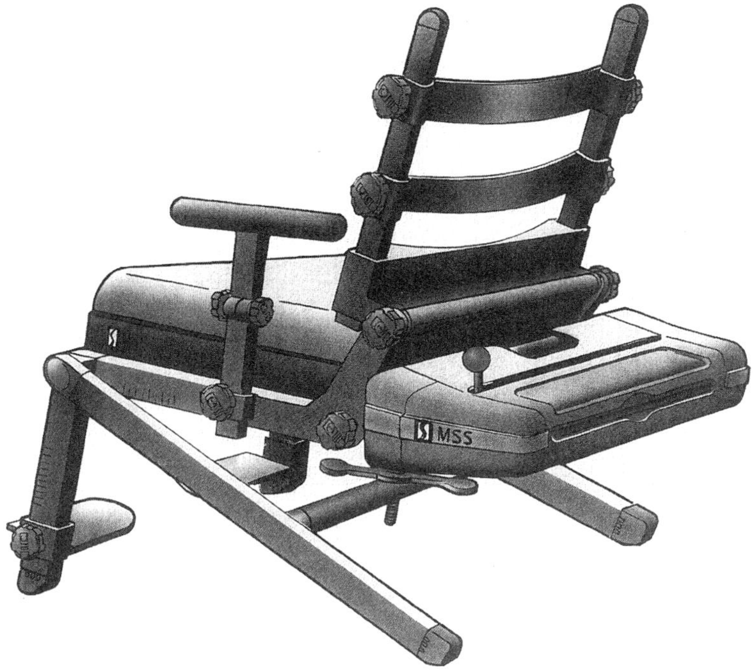

Figure 9.17. *Mechanical shape sensor seating contour measurement system. (Figure courtesy of Pin & Dot products.)*

contour. Position and/or force transducers are required to create a closed loop contour measurement systems. Direct seating contour measurement systems are most appropriate for people who only require moderate seating and postural support. Such systems do not offer the same degree of postural support as bead-bag systems.

9.2. DISTRIBUTION OF STRESSES IN SOFT TISSUES

When an individual sits on a cushion a number of activities take place. The interaction between the cushion and body tissue determines the user's comfort, function, and clinical safety. Distribution of stresses within the seating tissue affects the safety and effectiveness of the cushion. Wheelchair cushions are designed to provide pressure distribution for safe long-term seated posture, postural support, protection from vibration, and protection from shock. Poor distribution of stresses can lead to skin breakdown through a number of means. Normal stress is defined as force divided by the area

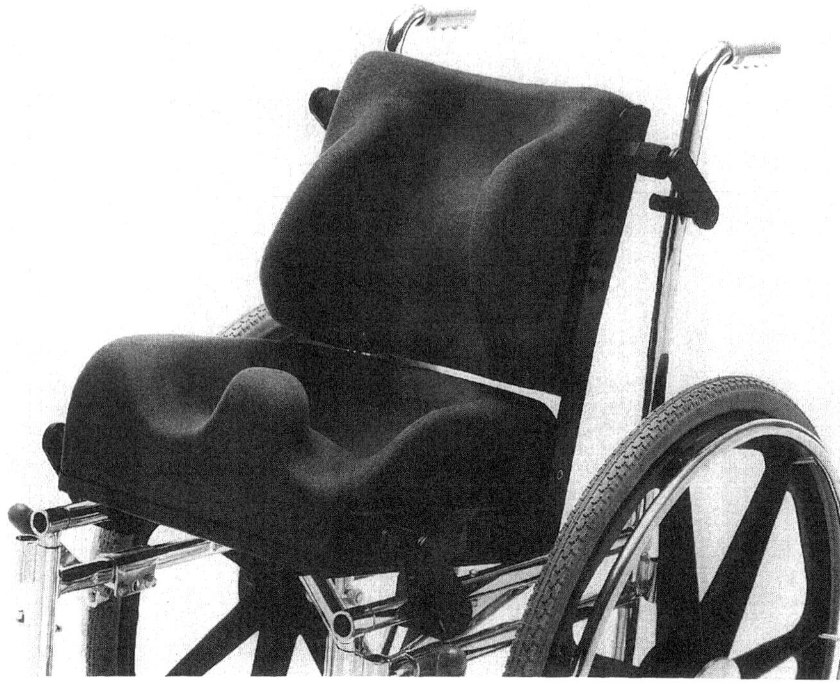

Figure 9.18. *Custom contoured seating system made from bead-bag mold. (Photograph courtesy Pin& Dot products.)*

over which it is applied:

$$\sigma_{\text{normal}} = \frac{F}{A}. \tag{9.2}$$

High stresses can occur with large forces or with small areas. The tissue either compresses or stretches in response to normal stress, see (figure 9.19). Localized stresses are a consequence of sitting.

Only when the body is suspended are the internal stresses minimized. Therefore the stress must be distributed over the tissue based upon the tissue's ability to withstand stress. Normal stresses act perpendicular to the skin, whereas shear stresses act parallel to the skin (figure 9.20). Sitting causes both normal and shear forces to exist within the seating tissue. Shear stress is applied force divided by the cross-sectional area

$$\sigma_{\text{shear}} = \frac{F}{A_c}. \tag{9.3}$$

Normal stress over bony prominences can cause a decrease in blood flow (ischemia) which causes a lack of oxygen, anoxia, and nutrients. Anoxia

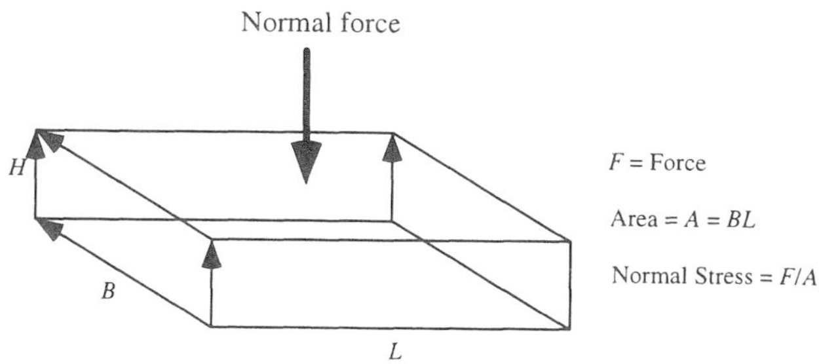

Figure 9.19. *Normal forces, commonly known as pressure, cause materials to compress in response to compression forces and expand in response to tension forces.*

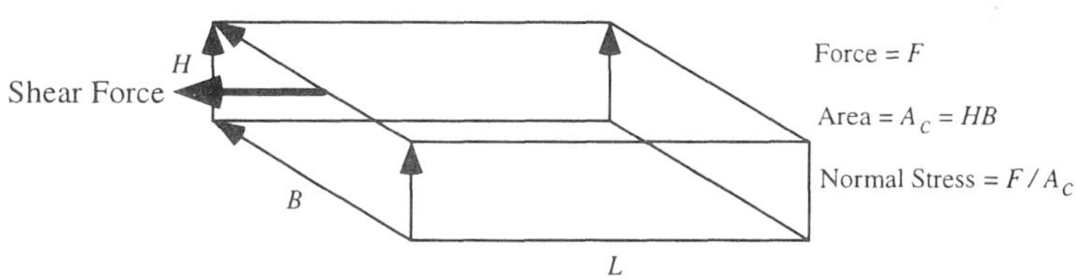

Figure 9.20. *Shear stresses act orthogonally to normal stresses and cause little change in tissue thickness.*

and lack of nutrients promote tissue death (necrosis). When normal stresses are applied capillaries are pinched off (occluded) when external pressure exceeds the internal tissue pressure. Capillary blood pressure is in the range of 32 mm Hg as measured in the finger nail beds of healthy subjects, but can be as low as 12 mm Hg. Friction and shear can cause skins abrasions. Shear also causes strain within the body tissue and can cause capillary occlusion. When shear is present, then tolerance for normal stresses is reduced.

9.3. SEATING PRESSURE MEASUREMENT

Seating pressure is often associated with compression of body tissue, muscle, and fat. However, a large percentage of the body is fluid (about 98%) which is incompressible. Body fluid under pressure within tissue, fat, and muscle is in shear. For example, compressing a balloon filled with water against a hard table causes the balloon to flatten. This is because the water moves parallel (shear) to the compression force. When a person sits on a flat surface, the fluid in the buttocks does not compress, it moves parallel to the exerted force. The buttocks flatten at the center and extends outward at the sides. The shear pressure exerted on the tissue and the distortion of the buttocks are related. To minimize distortion of the buttocks and thus damage to the tissue, the resting shape of the buttocks has to be maintained. Several approaches have been investigated into the optimal pressure distribution for wheelchair seating.

Static seating pressure measurements have long been applied in the field of rehabilitation. There are two major types of pressure monitors. Simple single-point pressure static pressure monitors, and pressure mapping systems. Systems are currently available that can measure from 64 to 2000 data points in a single second. In order to have a repeatable sensor orientation, many systems attach their sensors to a flexible membrane. The stiffness of the membrane may affect the ability of the sensors to conform to the surface of the buttocks and thus to measure normal force on contoured surfaces. To effectively analyze pressure distribution on a seated surface, time variations in pressure changes must be measured.

Dynamic pressure measurement is required as people move over time, and pressure varies as they move. Pressure must be measured over time for various activities before the time-pressure profile should be used to develop appropriate cushion and seating systems. Seat interface pressure changes as a person pushes a wheelchair. Pressure may vary from 50 mm Hg to 200 mm Hg for a single point on the seat during the propulsion recovery phase. Pressure measurement systems should enable clinicians and designers to make informed decisions about the effectiveness of products.

9.3.1. Bladder pressure sensors

Bladder sensors readily conform to the curvature of human soft tissue, have good repeatability, and are inherently insensitive to shear forces and temperature changes. Ideally, the sensor would not induce any changes in the tissue or tissue–cushion interface. Seating pressure measurements typically range from 0 to 1500 mm Hg. Some researchers have placed sensors over the ischial tuberosities or have taped sensors directly over the skin.

Switch-activated bladder pressure sensors rely on an interior switch to indicate pressure equilibrium. An air bladder is placed at the skin–cushion interface. The client is appropriately positioned, then the air bladder is inflated and the pressure in the bladder increases. The switch opens when the pressure in the air bladder equals that of the tissue. A pressure gage can be used to indicate pressure upon switch open. This process can also be reversed (i.e. deflate the bladder until the switch closes).

An array of bladder sensors can be made by using two sheets of air-tight flexible material molded to form a set of pockets forming a grid of switch-activated pressure sensors. The pressure in each pocket can be measured to yield an estimate of the seating pressure distribution. However, the pad may act as an air flotation cushion reducing the pressure gradient, especially around high-pressure areas.

Bladder sensors can also be integrated with other pressure sensors to produce a pressure contour. The Oxford air bladder system consists of a 3 × 4 matrix of sensors. A microprocessor controlled system inflates each of the bladder sensors individually. Each sensor is multiplexed to the microprocessor and a semiconductor pressure transducer. A vacuum pump holds each sensor deflated. The sensor of interest is activated and begins to inflate with a constant mass flow. When the pressure in the bladder exceeds that of the external tissue it begins to inflate, producing a change in volume and reducing the rate at which internal pressure increases. The pressure is recorded at this inflection point, the bladder is deflated and the process is repeated by the next sensor. A larger number of bladders is required to achieve a useful pressure contour.

Water can be used as an incompressible fluid to transmit seat interface pressure to a remote pressure transducer. Flexible, typically non-elastic, bladders filled with water are connected to a diaphragm-type pressure transducer. When force is exerted on the active bladder, the pressure developed is transmitted via a capillary to the transducer which produces a proportional voltage. Caution must be used to avoid overfilling the bladders which would create stresses on the walls of the bladder and produce a non-linear response. The height of bladder sensors may cause the person's weight to shift and cause measurement error.

9.3.2. Conductive polymer force sensing resistors

Devices for measuring pressure can be useful in identifying localized areas of high pressure. However, the skin pressure is lower than the interior pressure around bony prominences, which is where most ulcers begin.

There have been two recent innovations that may dramatically improve cushion design and prescription. The first is the development of low-cost, low-profile dynamic force sensors which can be arranged in an array to yield a model of the pressure distribution on the surface of the buttocks. The second is the development of non-invasive techniques for measuring the distribution and geometry of tissue and to model tissue behavior under various loads (e.g., finite-element analysis). The combination of a sensor array and a model of tissue behavior could lead to an understanding of how cushions need to be designed to prevent pressure sores.

Recently, conductive polymer sensors have become commercially available which, with some modification, may be suitable for measuring cushion interface pressure. Conductive polymer resistive force sensors (Interlink) with some adaptations may make them suitable for measuring cushion interface pressure.

Conductive polymer resistive force sensors are inexpensive, lightweight, and small. However, they are sensitive to how pressure is applied, and their response is non-linear. In addition, they have been used in some applications without knowledge of how they operate. This may produce erroneous results. The development of a reasonably priced normal force sensing mat or pressure

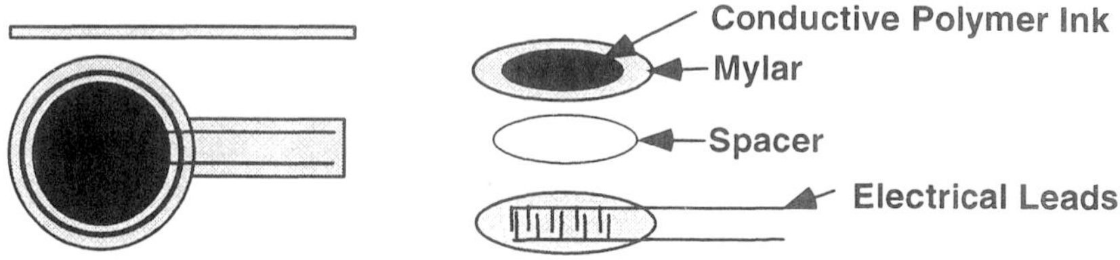

Figure 9.21. *Basic conductive polymer force-sensing resistor.*

sensing pants may yield the external loading data required for finite element or other models of the internal distribution of pressure within the buttocks. This could lead to improvements in cushions and reductions in pressure sores. However, the properties and limitations of these sensors need to be understood before they are applied. The efficacy of the data is dependent upon the quality of the sensor.

Conductive polymer resistive force sensors consist of three main parts: (1) a round piece of Mylar coated with a circle of a conductive polymer ink; (2) a thin metallic ring with an inside diameter slightly larger than the diameter of the conductive polymer; and (3) another round piece of Mylar with traces printed onto it in a pattern of interlocking fingers (figure 9.21). The leads come from the piece of Mylar with the fingers printed on it. Without any force upon the sensor, the resistance is nearly infinite (the fingers do not touch and the air gap prevents conduction through the polymer ink). As pressure is applied the conductive polymer ink begins to touch the traces of the opposite Mylar film and the resistance between the leads drops (the resistance of the polymer ink remains constant). As more pressure is applied the contact area increases and the resistance drops further. However, the contact area depends upon where the pressure is applied; there is greater contact with fingers in some areas than there is in others. The range is typically between 12 MΩ and 100 Ω. This range is large enough to be measured by very simple circuitry.

Resistances of five pressure transducers were measured over a range of 0 to 25 kg. Force was applied uniformly across the entire sensing surface using an electronically controlled press and it was measured using a force gage. The force on each transducer produced a resistance which was measured using a digital multimeter. At least three trials were run for each transducer. Each trial started at zero and force was slowly increased to 25 kg. A resistance reading was recorded for every 450 g. The ensemble mean for each transducer is plotted in figure 9.22.

9.3.3. Capacitive pressure mats

Capacitive sensors can be made to be very sensitive, linear, and inexpensive. Typically, larger capacitance requires simpler circuitry, but implies larger

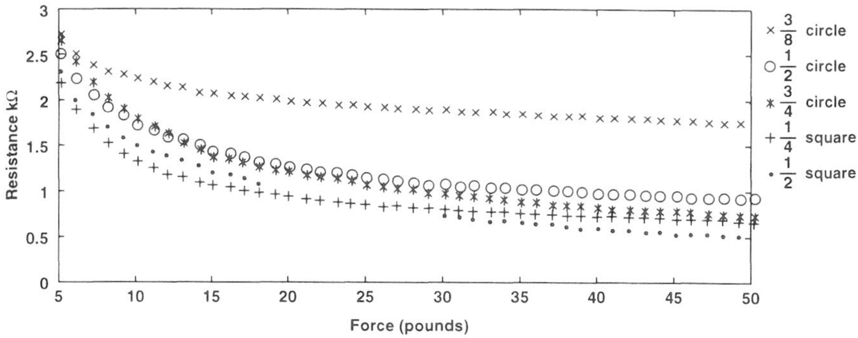

Figure 9.22. *Calibration curves for various conductive polymer resistive force sensors.*

physical size. Individual capacitive transducers do not show significant promise for seat interface pressure measurement. However, a mat of woven conductive strips may be a solution. A capacitor is a charge storage device which can be described by three physical parameters

$$C = \frac{\varepsilon A}{d} \tag{9.4}$$

where ε describes the dielectric material, A is the effective plate area, and d is the separation distance between the plates. Typically, a capacitive transducer works on the principle of varying A or d. A method of developing a useful capacitive transducer pressure sensing mat was proposed by Babbs *et al* (1990) who used two orthogonal arrays of a ribbon-like conductor. The arrays are separated by low-hysteresis foam rubber. Current is sensed between a pair of conductors on either side of the foam with a 5 kHz voltage source. All unused conductors are grounded to avoid cross-talk while measurements are made. The foam compresses as force is applied which brings the ribbon-like conductors closer together, increasing the capacitance.

9.4. CONTROL INTERFACE INTEGRATION

The joystick is the primary control interface between the user and wheelchair. Most joysticks are termed proportional in that their output is proportional to the shaft position. Most joysticks can be classified as inductive or resistive. Change in stick position either creates a change in inductance or resistance, respectively. Other joysticks use switches or optical sensors to convert shaft position to an electronic signal. Optical joysticks usually require the user to interrupt a light source which acts like a switch closure.

When a person does not have sufficient hand control to use a joystick with their hand or arm, other anatomical control sites must be considered

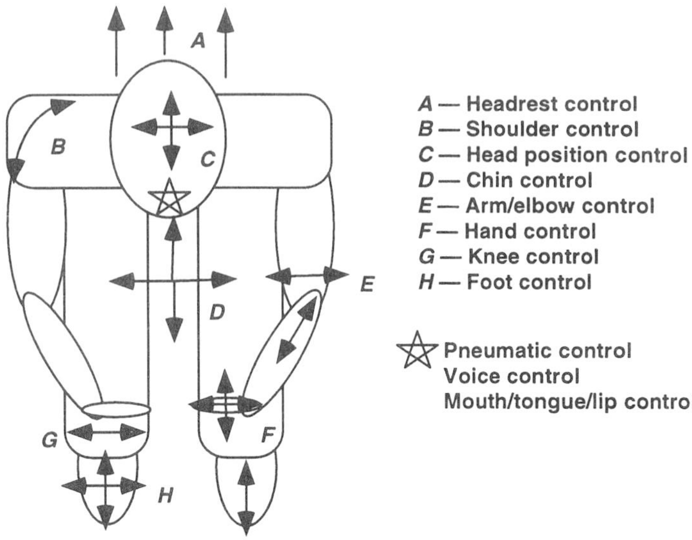

Figure 9.23. *Anatomical control sites.*

along with the possible reduction in wheelchair control. A joystick can be modified to be used with a chin, foot, elbow, tongue, or shoulder. Joysticks used in these instances are typically short-throw joysticks which require less motion than conventional joysticks to reach full-scale, typically less than 6 mm. Other types of controls and control sites are illustrated in figure 9.23.

The disability etiologies of power wheelchair users are diverse. Many marginal walkers or manual wheelchair users use a scooter or power wheelchair for longer distance, and can independently transfer in and out of their wheelchair. These users may have neuromuscular or muscular deficits that cause weakness rather than loss of control or function. Some of these users may have little or no difficulty using a joystick.

However, a large number of power wheelchair users are severely impaired and rely solely on their power wheelchair for their mobility. Choosing a control interface and site requires substantial evaluation and consideration. Third-party providers often require evidence of a person's need and ability to use a power wheelchair before agreeing to payment. Some users cannot meet performance standards set by third-party providers, despite adaptations being available. Excessive intention tremor, limited range of motion, and spastic rigidity are some of the disabilities which can render joystick control unacceptable. These conditions can be manifestations of a wide variety of physiological or neurological conditions such as cerebral palsy, multiple sclerosis, Parkinson's disease, Friedrich's ataxia, traumatic brain injury, and cardiovascular incident. Intention tremor is described as a 2 to 4 Hz oscillation when postural maintenance or purposeful movement is attempted. When people with tremor drive a power wheelchair, the oscillations of the arm or hand are transferred to the joystick and in turn to the wheelchair. This can cause difficulty in controlling the wheelchair especially in cramped or crowded spaces.

Limited range of motion is characterized by an intact sensory system with impaired motor control, weakness, or pain. This can be a result of neurological or neuromuscular impairment, arthritis, or arthrogryposis. In this situation, a combination of a short-throw joystick and properly adjusted motor controller helps to compensate for the limited movement.

Spastic hypertonia is a motor disorder characterized by a velocity dependent increase in tonic reflexes with exaggerated tendon jerks. Durability is a primary concern when selecting a control interface for a person with this type of disability. Many joysticks are not robust enough to withstand the constant stress inflicted by strong spasms. In many instances switch control devices are used.

When a person is unable to adequately operate a standard joystick, an approach may be to modify the handle and position of the joystick. If this method is not successful in providing a reliable means of controlling the wheelchair either the site (e.g., arm, hand, foot) must be abandoned or the interface (e.g., joystick). The process of selecting the proper interface and control site relies heavily on experience. Several interface devices may be used as alternatives to the joystick. Chin joysticks are used for people with very severe impairments. Typically, a short-throw joystick with a cup on the handle is mounted near the user's chin. In many cases a collar harness is used to hold the joystick close to the user's chin. This may be unattractive, interfere with speech, and small motions may lead to large wheelchair motions. Sip-and-puff (pneumatic) control uses hard puffs and sips to control backward and forward motions, and soft puffs and sips for directional control. Personal hygiene, and cleanliness are of concern with such systems, and such systems do not offer proportional control.

Headrest switches may be embedded into or mounted onto a headrest to provide directional control. During the control phase, the user's ability to use the headrest for postural support is restricted, and the switches do not provide proportional control. Ultrasonic head control is achieved by using the location of the head in the plane of the sensor as a proportional control signal. Many people have difficulty with head movement due to inertia, and with sudden movements when the user's attention is averted. Mouth, tongue, and lip control use microswitches mounted in or near the mouth. This may result in personal hygiene and speech problems.

Arm/elbow control, using what is called a balanced forearm orthosis, can be used to activate switches or for proportional control. The user must have sufficient arm control to manage inertial movement. Elevation/depression and protraction/retraction of the shoulder can be used to activate switches. This type of system is limited, requires extensive training, and can be cumbersome.

Voice-activated controls may be used on power wheelchairs. Voice recognition systems tuned to the user's voice and a set of wheelchair commands can provide some people independent control over their power wheelchair. Voice-activated control systems sometimes experience difficulty in noisy environments, can be expensive, and do not provide proportional control.

Isometric joysticks provide a control signal proportional to the force applied by the user. The stick or shaft remains stationary and force is determined by a force transducer. Isometric joysticks may reduce the effects of tremor, be more durable, and function well for people with restricted range of motion.

9.5. MULTI-CONFIGURATION SEATING AND POSTURAL SUPPORT SYSTEMS

Some people with severe mobility impairments and sensitive soft tissue may require adjustable or multi-configuration seating systems. There are three basic styles of multi-configuration seating and postural control systems: reclining seats; tilt-in-space (figure 9.24); and stand-up systems. Reclining wheelchairs allow the user to change sitting posture through the use of a simple interface (e.g., switch) (figure 9.25). Changing seating posture can extend the amount of time a person can safely remain seated without damaging tissue or becoming fatigued. Determining the optimal range of seating postures is difficult and is best approached by clinical teams. Multi-configuration wheelchairs assist in performing pressure relief. Changing seating position redistributes pressure on weight-bearing surfaces, alters the load on postural musculature, and changes circulation. Changing position can also facilitate respiration. Elevating the legs while lowering the torso can improve venous return, and decrease fluid pooling in the lower extremities.

Tilt-in-space systems are often characterized by the location of their pivot mechanisms. When the location is near the backrest the chair is said to have a rear-pivot system. When tilting, the rear-pivot system raises the user's knees above his/her whole body center of gravity. The pivot point can also be placed under the front edge of the seat base, in which case the chair is said to have a forward-pivot system. With forward-pivot systems the user's buttocks are lowered to effect tilting. A counter-balance effect, which reduces the forces required to adjust position, can be obtained by placing the pivot near the center of the seat (i.e., center-pivot system). Springs and dampers can be used to provide a suspended seating systems. The springs and dampers can be adjusted to create a floating pivot system. Floating-pivot systems can provide a wide variety of tilt angles. Tilt-in-space systems must be stabilized in their resting positions. Three common methods are used to maintain seating position with tilt-in-space systems: mechanical locking

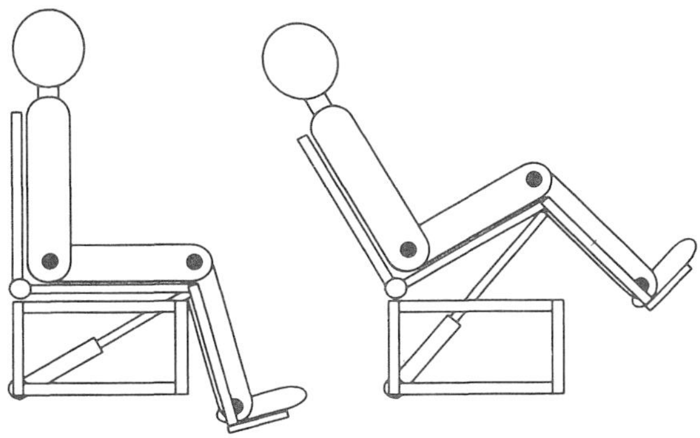

Figure 9.24. *Schematic diagram for tilt-in-space mechanism. The user remains in the same position; however his/her orientation changes. This allows pressure to be redistributed.*

pin; cable lock; and adjustable gas springs. All of these locks have been proven effective. There are some difficulties with tilt-in-space wheelchairs. The wheelchairs are heavier than standard wheelchairs, they are less stable, and can require greater turning diameter when reclined.

A potential problem with multi-configuration seating systems is that the body may not remain in a stable position after transitioning through several seating orientations. Sliding or stretching during reclining or tilting can cause undesirable shear forces, excite spasticity, and bunch clothing. If the joints of the wheelchair do not follow the anatomical paths of the user shear forces will result. Reclining systems which attempt to follow anatomical joint centers are called low-shear. The forces can be minimized by allowing the seat and backrest to freely translate while rotating about a hinge joint located near the user's hip; these systems are called zero-shear. Without an anti-shear mechanism up to 11 cm of displacement can take place between the person's back and the wheelchair's backrest. Displacement may also take place at the knee if the axes of rotation of the anatomical knee and wheelchair do not coincide. This does not usually present a problem because of the low mass of the lower limbs which move in response to the wheelchair. The length of the legrests present a greater potential problem. If the legrests are too short then high forces might be exerted on the bottom of the feet in the reclined position. Therefore, legrest length should be carefully adjusted and verified throughout the range of motion.

9.5.1. Stand-up wheelchairs

There are four basic types of stand-up wheelchair: manual drive with manual lifting mechanisms, manual drive with power lifting mechanisms, electrically powered drive with powered lifting mechanisms, and fixed standing position.

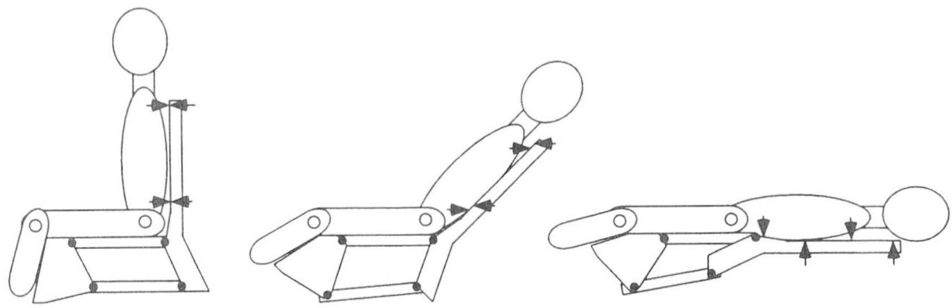

Figure 9.25. *Schematic diagram for reclining wheelchair seat. Arrows indicate the displacement between the user's back and the wheelchair backrest. Unless the back slides or the wheelchair's pivots follow the body's anatomical joint centers, shear forces will result.*

The first three types of stand-up wheelchair change configuration from sitting to standing and vice-versa. During this transition process a significant portion of the rider's body weight must be supported by the lifting mechanism. The seat of the wheelchair which supports the rider is often rotated about an axis parallel to the user's knees, causing the user to assume a standing position (the user's knees are secured via a padded support) (see figure 9.26). Standing systems may be categorized as prone standers, supine standers, vertical standers, and units that promote sit-to-stand. Stand-up wheelchairs incorporate the sit-to-stand category. The others may all be classed as fixed or mobile standing frames. Prone standers position the user so that some of his/her weight is supported by the anterior surface of the body. This type of system is often used with children. Supine standers position the user so that his/her weight is partially supported by the posterior surface of the body. Vertical standers allow the user's anatomical limbs to supports his/her body weight through the use of restraints at the hip, knees, and ankles. Sit-to-stand wheelchairs provide greatest mobility because the user can propel the wheelchair in the seated position and in some cases in the standing position. These devices combine mobility and the ability to adjust posture.

Stand-up wheelchairs can be modeled as quasi-static systems from the perspective of restraint design and positioning (see figure 9.27). Simple force and moment balance methods can be used to analyze stand-up wheelchair restraining forces, as given in the following.

Body and chair forces with lower lag and trunk held vertical, which is commonly the case with stand-up wheelchairs:

$$F_{\text{chest}} = F_{\text{chest}_x}\hat{i}$$

$$F_{\text{back}} = F_{\text{back}_x}\hat{i}$$

$$F_{\text{trunk}} = F_{\text{trunk}_x}\hat{i} + F_{\text{trunk}_y}\hat{j}$$

$$F_{\text{belt}} = F_{\text{belt}_x}\hat{i} + F_{\text{belt}_y}\hat{j}$$

$$F_{\text{gluteal}} = F_{\text{gluteal}_x}\hat{i} + F_{\text{gluteal}_y}\hat{j}$$

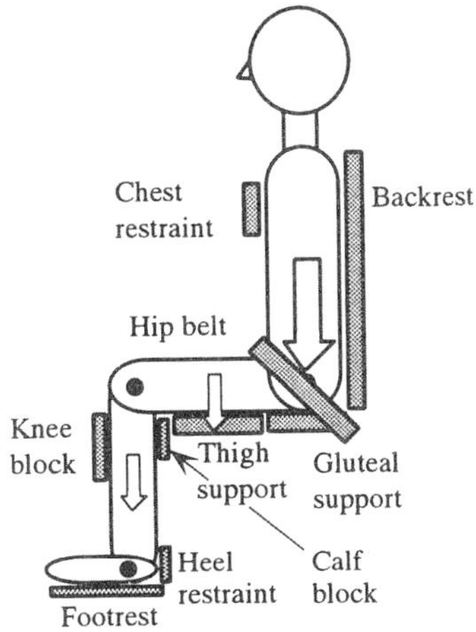

(*a*)

(*b*)

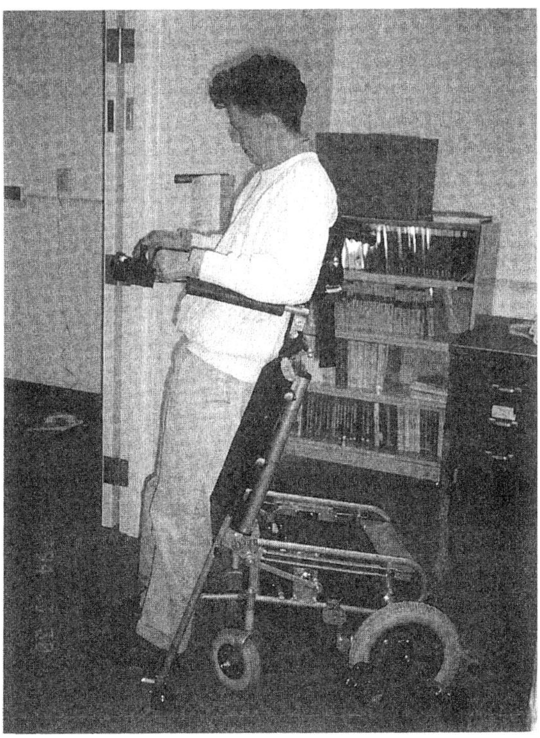

Figure 9.26. *(a) A sagittal-plane schematic diagram for a stand-up wheelchair seating and postural support system. (b) Stand-up wheelchair used to provide upright posture. Stand-up wheelchairs may provide physiological and psychological benefits. This chair uses a powered elevating system.*

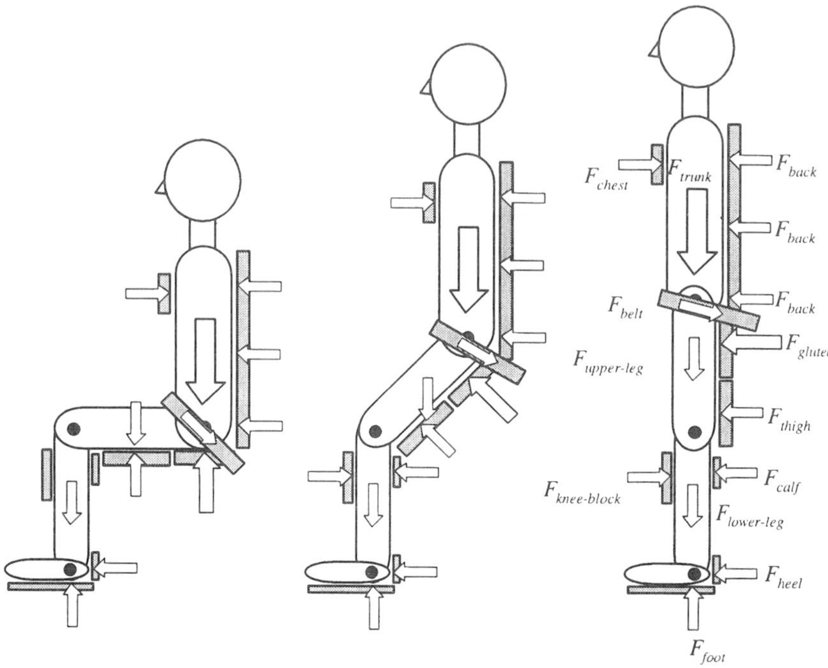

Figure 9.27. *Sagittal-plane schematic diagram showing force acting on body, seating system, and postural supports. (Photograph by R A Cooper.)*

$$
\begin{aligned}
F_{\text{thigh}} &= F_{\text{thigh}_x}\hat{i} + F_{\text{thigh}_y}\hat{j} \\
F_{\text{upper-leg}} &= F_{\text{upper-leg}_x}\hat{i} + F_{\text{upper-leg}_y}\hat{j} \\
F_{\text{knee-block}} &= F_{\text{knee-block}_x}\hat{i} \\
F_{\text{calf}} &= F_{\text{calf}_x}\hat{i} \\
F_{\text{lower-leg}} &= F_{\text{lower-leg}_y}\hat{j} \\
F_{\text{heel}} &= F_{\text{heel}_x}\hat{i} \\
F_{\text{foot}} &= F_{\text{foot}_y}\hat{j}.
\end{aligned}
\tag{9.5}
$$

Sum of the forces in the vertical direction (y-axis):

$$
0 = \sum F_y\hat{j} = F_{\text{trunk}_y}\hat{j} + F_{\text{belt}_y}\hat{j} + F_{\text{gluteal}_y}\hat{j} + F_{\text{thigh}_y}\hat{j} + F_{\text{upper-leg}_y}\hat{j} + F_{\text{lower-leg}_y}\hat{j} + F_{\text{foot}_y}\hat{j}. \tag{9.6}
$$

Sum of the forces in the horizontal direction (x-axis):

$$
0 = \sum F_x\hat{i} = F_{\text{chest}_x}\hat{i} + 3F_{\text{back}_x}\hat{i} + F_{\text{trunk}_x}\hat{i} + F_{\text{belt}_x}\hat{i} + F_{\text{gluteal}_x}\hat{i} + F_{\text{thigh}_x}\hat{i} + F_{\text{upper-leg}_x}\hat{i} + F_{\text{knee}_b\text{lock}_x}\hat{i} + F_{\text{calf}_x}\hat{i} + F_{\text{heel}_x}\hat{i}. \tag{9.7}
$$

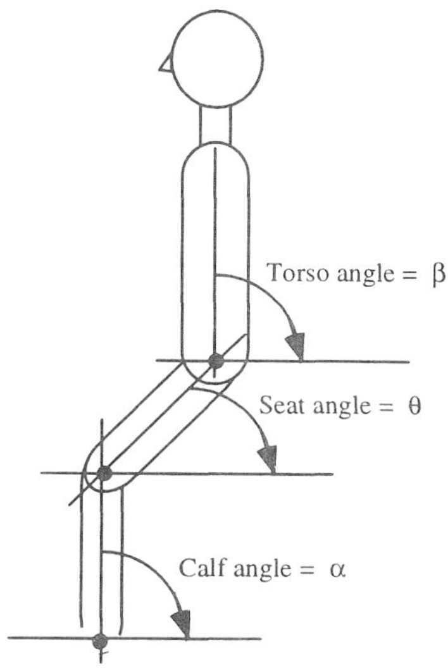

Figure 9.28. *Body reference angles for a static model of a stand-up wheelchair.*

The components of some of these forces can be resolved if simplifying assumptions are made. If it is assumed that there are no shear forces between the body and any support or restraints surface and that the user can generate no joint moments then simple trigonometry can be used to estimate body static forces (figure 9.28).

The knee forces are often thought to be most critical when transitioning to the standing position or when standing. Knee and seat forces can be estimated by assuming the hip restraint provides no weight bearing support and that the chest restraint and backrest forces cancel each other. The letter M is used to denote the mass of a segment with the subscript denoting the particular body segment, g represents acceleration due to gravity.

Ankle

$$\begin{aligned} &F_{\text{heel}} - F_{\text{ankle}} \cos\alpha = 0 && (x) \\ &0.43 M_{\text{lower-leg}} g + F_{\text{ankle}} \sin\alpha = F_{\text{foot}} && (y) \end{aligned} \tag{9.8}$$

Knee

$$\begin{aligned} &F_{\text{ankle}} \cos\alpha - F_{\text{knee}} \cos\theta + F_{\text{knee-block}} = 0 && (x) \\ &F_{\text{ankle}} \sin\alpha + F_{\text{knee}} \sin\theta = 0.57 M_{\text{lower-leg}} g && (y) \end{aligned} \tag{9.9}$$

Hip

$$\begin{aligned} &F_{\text{knee}} \cos\theta - F_{\text{hip}} \cos\beta = 0 && (x) \\ &F_{\text{knee}} \sin\theta + F_{\text{hip}} \sin\beta = 0.56 M_{\text{upper-leg}} g + M_{\text{trunk}} g. && (y) \end{aligned} \tag{9.10}$$

The gravitational forces of the segment mass contributions were derived as percentages of total body mass by Dempster in 1955. The constants in equations (9.8)–(9.10) are the percentages of total segement length to the center of mass of each segment. For example, the distance from the knee to the point of application of the gravitational force of the lower leg is 43% of the total length. When the total gravitational force is transferred to the joints, 43% of this force can be found in the ankle and 57% in the knee.

When fully seated the knee block applies only the pre-load force to the knees. When fully erect the knee block again applies little or no force, and the body weight is supported through the knees. If $\alpha = 90°$, $\theta = 90°$, and $\beta = 90°$, then $F_{\text{foot}} = Mg$, where M is the mass of the entire body.

Standing wheelchairs are commonly believed to improve bladder and bowel function, reduce incidence of decubitis ulcers, reduce muscle contractures and osteoporosis, and improve venous return. Standing wheelchairs may provide psychological benefits as well. However, there are some contraindications for stand-up wheelchairs. They should not be used by people with severe osteoporosis or unusually poor circulation. Some people with contractures may be incapable of standing. Stand-up wheelchairs cost substantially more than most common manual or power wheelchairs. However, they provide additional benefits to most users. The design of stand-up wheelchairs continues to improve and they should become more widely used.

FURTHER READING AND REFERENCES

Agrawal V P and Chandra R 1979 Optimization of a chair mechanism for partially disabled people for sitting-standing and sitting-lying motions *Med. Biol. Eng. Comput.* **17** 671–82

Andersson G B J and Oertengren R 1974 Lumbar disc pressure and myoelectric back muscle activity during sitting. III Studies on a wheelchair *Scand. J. Renab. Med.* **3** 122–7

Andersson G B J, Murphy R W, Oertengren R and Nachemson A L 1979 The influence of backrest inclination and lumbar support on the lumbar lordosis in sitting *Spine* **4** 52–8

Axelson P 1993 Wheelchair comparison *Sports 'N Spokes* **18** 34–40

Babbs C F, Bourland J D, Graber G P, Jones J T and Schoenlein W E 1990 A pressure sensitive mat for measuring contact pressure distributions for patients lying on hospital beds *Biomed. Instrum. Technol.* **24** 363–70

Brienza D M, Chung K-C, Brubaker C E and Kwiatkowski R J 1993 Design of a computer-controlled seating surface for research applications *IEEE Trans. Rehab. Eng.* **BME-1** 63–6

Brienza D M, Chung K C and Inigo R M 1988 Design of a computer aided manufacturing system for custom contoured wheelchair cushions *Proc. Int. Conf. Am. Assoc. Rehabilitation Therapy* pp 312–3

Brienza D M, Brubaker C E and Inigo R M 1989 A fiber optic force sensor for automated seating design *Proc. 12th Ann. RESNA Conf. (New Orleans, LA, 1989)* pp 232–3

Brienza D M, Gordon J and Thacker J 1989 A comparison of force transducers suitable for an automatic body support contour system *Proc. 12th Ann. RESNA Conf. (New Orleans, LA, 1989)* pp 238–9

Bush C A 1969 Study of pressures on skin under ischial tuberosities and thighs during sitting *Arch. Phys. Med.* **50** 207

Chung K C, DiNello A M and McLaurin C A 1987 Comparative evaluation of pressure distribution on foams and contoured cushions *Proc. 10th Ann. RESNA Conf. (San Jose, CA)* pp 323–5

Cooper R A 1991 High tech wheelchairs gain the competitive edge *IEEE Eng. Med. Biol. Mag.* **10** 49–55

Cooper R A 1991 A Comparison of pulmonary functions of wheelchair racers in their racing and standard chairs *Proc. 14th Ann. RESNA Conf. (Kansas City, MO, 1991)* pp 245–7

Cooper R A, Ward C and Ster J F 1992 Characterization and development of a resistive sensor for measuring cushion interface force *Proc. 15th Ann. RESNA Conf. (Toronto, Ont., 1992)* pp 627–9

Dempster W 1955 Space requirements of the seated operator *WADC Technical Repoı 55–159, Wright-Patterson Air Force Base, OH*

Ferguson-Pell M W 1980 Design criteria for the measurement of pressure at body/suppoı interfaces *Eng. Med.* **9** 209–14

Ferguson-Pell M W 1990 Seat cushion selection *J. Rehab. Res. Dev.—Clin. Suppl.* No : 49–73

Ferguson-Pell M W and Cardi M 1992 Pressure mapping systems for seating an(positioning applications: technical and clinical performance *Proc. RESNA Int. '9: (Toronto, Ont., 1992)* pp 219–21

Ferguson-Pell M W and Cardi M D 1993 Prototype development and comparativ(evaluation of wheelchair pressure mapping system *J. Assistive Technol.* **5** 78–91

Ferguson-Pell M, Cochran Gvan B, Palmieri V R and Brunski J B 1986 Developmen of a modular wheelchair cushion for spinal cord injured persons *J. Rehab. Res. Dev* **23** 63–76

Gilsdorf P, Patterson R and Fisher S 1991 Thirty-minute continuous sitting forc(measurements with different support surfaces in the spinal cord injured and able-bodied *J. Rehab. Res. Dev.* **28** 33–38

Gilsdorf P, Patterson R, Fisher S and Appel N 1990 Sitting forces and wheelchaiı mechanics *J. Rehab. Res. Dev.* **27** 239–46

Goosens R H M and C J Snijders 1995 Design criteria for the reduction of shear forces in beds and seats *J. Biomech.* **28** 225–30

Grady J H and Kuipers I 1993 Modular systems in wheelchairs: purpose in relation to function *Proc. 2nd Eur. Conf. on Advancement of Rehabilitation Technology (Stockholm, 1993)* p 33.1

Grieco A 1986 Sitting posture: an old problem and a new one *Ergonomics* **29** 345–62

J T Hunt 1993 Standing tall *Team Rehab. Report* 17–20

ISO Working group 9 1994 *Stand-up Wheelchair Draft Standard* WG 09–01 April

Katz R T and Rymer W Z 1989 Spastic hypertonia: mechanisms and measurement *Arch. Phys. Phys. Med. Rehab.* **70** 144–55

Keegan J J 1953 Alterations of the lumbar spine related to posture and seating *J. Bone Joint Surg.* **35-A** 589

Kwiatkowski R and Inigo R 1992 The design of a computer aided seating system *Proc. RESNA Int. '92 (Toronto, Ont., 1992)* pp 216–8

Kwiatkowski R J and Inigo R M 1993 A strain gage force sensor for a computer aided seating system *Proc. RESNA 16th Ann. Conf. (Las Vegas, NV, 1993)* pp 292–4

Laenger C J, Lee J H 1991 Small improvements in seating hardware *Proc. RESNA 14th Ann. Conf. (Kansas City, MO, 1991)* pp 353–5

Lim R, Sirett R, Conine T A and Daechsel D 1986 Clinical trial of foam cushions in the prevention of decubitis ulcers in elderly patients *J. Rehab. Res. Dev.* **25** 19–26

Noble P C, Goode B, Krouskop T A and Crisp B 1984 The influence of environmental aging upon the loadbearing properties of polyurethane foams *J. Rehab. Res. Dev.* **21** 31–8

Perr A, Lincoln A and McGovern T 1992 Pressure distribution on contoured wheelchair cushions utilizing the electronic shape sensor *Proc. RESNA Int. '92 (Toronto, Ont., 1992)* pp 213–5

Riley P O and Rosen M J 1987 Evaluating manual control devices for those with tremor disability *J. Rehab. Res. Dev.* **24** 99–110

Sacks A H 1989 Theoretical prediction of a time-at-pressure curve for avoiding pressure sores *J. Rehab. Res. Dev.* **26** 27–34

Seelen H A M 1993 Dynamic sitting posture and impaired postural control in seated spinal cord injured people *Proc. 2nd Eur. Conf. Advancement Rehabilitation Technol. (Stockholm, 1993)* p 33.2

Sengupta D, Sherwood A M and McDermott M 1974 Comparative evaluation of control surfaces for disabled patients *27th ACEMB (Philedelphia, PA, 1974)* 356

Shapcott N and Bar C 1990 Seating simulation as an aid to assessment *Proc. RESNA 13th Ann. Conf. (Washington, DC, 1990)* pp 111–4

Shapcott N, VanNote D, Kelly C, Bouge T and Lenker J 1991 A vinyl vacuum forming for covering FIP systems *Proc. RESNA 14th Ann. Conf. (Kansas City, Mo, 1991)* pp 367–9

Shaw C G 1993 Seat cushion comparison for nursing home wheelchair users *J. Assistive Technol.* **5** 92–105

Shaw G 1985 Rigid pelvic restraint *Proc. RESNA 8th Ann. Conf. (Memphis, TN, 1985)* pp 409–11

Shields R K and Cook T M 1988 Effect of seat angle and lumbar support on seated buttock pressure *J. Am. Phys. Ther. Assoc.* **68** 1682–6

Sopsato B A, Chung K-C and McLaurin C A 1990 Prescribing customized contoured seat cushions by computer-aided shape sensing *Proc. RESNA 13th Ann. Conf. (Washington, DC, 1990)* pp 103–4

Sprigle S and Cron L 1992 Development and evaluation of wheelchair cushion prototypes *Proc. RESNA Int. '92 (Toronto, Ont., 1992)* pp 519–521

Sprigle S and Schuch J Z 1993 Using seat contour measurements during seating evaluations of individuals with SCI *J. Assistive Technol.* **5** 24–35

St-Georges M, Valiquette C and Drouin G 1989 Computer-aided design in wheelchair seating *J. Rehab. Res. Dev.* **26** 23–30

Sunita M M and Sprigle S H 1993 Effects of undeformed contour and cushion stiffness at the buttocks–seat interface *Proc. RESNA 16th Ann. Conf. (Las Vegas, NV, 1993)* pp 286–8

Tchang F 1993 Guide to research results: seating materials and fabrication methods of 12 North American centers *Proc. RESNA 16th Ann. Conf. (Las Vegas, NV, 1993)* pp 268–70

van Eijk D and Bulsink D 1993 Design of a series of positioning systems for handicapped children *Proc. 2nd Eur. Conf. Advancement Rehabilitation Technol. (Stockholm, 1993)* p 9.2

Warren C G 1982 Reducing back displacement in the powered reclining wheelchair *Arch. Phys. Med. Rehab.* **63** 447–9

Warren C G 1990 Powered mobility and its implications *J. Rehab. Res. Dev.–Clin. Suppl.* No 2 74–85

Winter D A 1990 *Biomechanics and Motor Control of Human Movement* (New York, NY: Wiley)

Woo A 1992 Standing options *Team Rehab. Report* **3** 39–44

Zollars J A and Axelson P 1993 The back support shaping system: an alternative for persons using wheelchairs with sling back upholstery *Proc. RESNA 16th Ann. Conf. (Las Vegas, NV, 1993)* pp 274–6

CHAPTER 10

PROSTHETICS AND ORTHOTICS

10.1. INTRODUCTION

The majority of amputations performed in the western world are the result of medical conditions: primarily, occlusive diseases of the arteries to the limbs and diabetes. For the most part these amputations occur in older people. Temporary prosthetics used immediately post-operation have been one of the most significant advances in the rehabilitation of people with amputations. Automated fabrication using computer aided design/computer aided manufacturing (CAD/CAM) promises to provide functional, cost-effective, early-fit definitive prostheses. Prostheses are designed to be lightweight, durable, energy-storing, and responsive to residual limb changes. Most success is related to early and routine prosthesis use.

A prosthesis is a device which replaces a part of the function of a missing limb. A prosthesis should fit comfortably, be lightweight, and look realistic. A prosthesis cannot replace the function of a missing arm or leg. They do not automatically adjust to changes in body or limb weight, or other physiological, anatomical, or kinesiological changes to the body.

An orthosis is a device applied to the exterior of the body to stabilize or enhance motion of a limb or joint. Orthoses are also designed to reduce the stress on body parts and are used to improve or restore function to an impaired limb or body part. Splints and braces used to stabilize limbs are examples of common orthoses. The proper application of engineering materials, and components can improve limb function for a person with a mobility or manipulation disability. Orthoses are designed to perform several primary functions. Protection of diseased or injured limbs is often provided with an orthosis. Among many people with disabilities, orthoses are used to prevent or correct skeletal deformities. Orthoses are used to stabilize structurally unsound joints. People with mobility or manipulation impairments can use orthoses to improve function of their impaired limbs.

Orthoses are identified by the joints they include. Hence an orthosis which incorporates the ankle is referred to as an ankle–foot orthosis (AFO), and an orthosis which incorporates the knee and ankle is referred to as a knee–ankle–foot orthosis (KAFO). Familiarity with orthoses terminology facilitates effective communication between members of the orthotics team.

10.1.1. Causes of amputation

Most amputations result from peripheral vascular disease, diabetes, and related conditions. Many amputations involve the lower limbs and are related to the general health of the individual. Many people who require amputations are 60 years of age or older. Muscles lose their elasticity with passing years, and changes to the cardiovascular system may cause a decrease in blood supply to the muscles and periphery. Moderate physical activity can slow the aging process, and improve or maintain health. This population poses a challenge for healthcare and rehabilitation professionals.

Accidents, primarily automobile and motorcycle, account for a modest percentage of amputations, as does cancer (bone cancer and tumors). Disarticulation at the next major proximal limb is sometimes required during the treatment of bone tumors. Direct trauma (e.g., train wheel, power saw) may lead to direct amputation of a limb. Crushing of limbs which causes damage to bones and blood vessels may require amputation. Osteomyelitis or other infection may progress beyond control necessitating amputation long after the initial trauma. A small segment of the population of people with amputations is the result of congenital limb loss and congenital limb deficiencies. In some areas of the world, amputations related to war and natural disaster remain high.

Residual lower limbs of children with congenital limb deficiences will often allow partial or full end weightbearing. The likelihood of full end weightbearing increases if surgery was not required, and there is appropriate soft tissue at the distal end. Prosthetic components for children and young adolescents are limited in comparison with those for adults. Prosthetic fittings for children must be maintained and checked on a regular basis to accommodate growth.

Proper rehabilitation of a person who has experienced an amputation requires a multidisciplinary team of professionals. The patient must be an active participant in all steps of the planning and recovery process. Primary members of the rehabilitation team are the patient, surgeon, nurse, physical therapist, prosthetist, and biomechanist. When the amputation is elective, physical conditioning should begin prior to surgery when possible. Improved flexibility, strength and endurance can increase the likelihood of a successful recovery from an amputation. It is also good to begin to discuss and to set realistic rehabilitation goals prior to surgery. Often discussions with people who have recovered from similar amputations can improve a patient's perspective.

10.1.2. Post-operative treatment for amputations

The surgeon who performs the amputation is responsible for selecting the level of limb removal, and for reconstructing the residual soft tissues and bone. Technical details of amputation surgery are critical to residual limb function. Typically, surgeons try to maximize residual limb length and stabilize residual muscle. Surgeons must also determine treatment of bone, blood vessels, nerves, and subcutaneous tissues. The muscle and remaining joints of the residual limb act to assist in control of the prosthesis. Muscle action through the tendon and aponeurosis attachments to bone also provide a sense of proprioception (i.e., a sense of position and movement) and are very important to prosthetic control.

Emphasis upon immediate and progressive physical movement and prosthetic applications has accelerated the rehabilitation of people of all ages experiencing amputations. The management of an amputation involves working towards uneventful wound healing and the immediate commencement of the rehabilitation process. Rigid dressings allow the patient to move comfortably early in the post-operative period. Rigid dressings also provide protection and recovery for the tissues, a clean wound healing environment, and pain control. The rigid dressing forms a receptacle which can serve as a socket for an immediate or early post-operative prosthesis.

For people who have undergone lower-limb amputation, limited weightbearing using crutches or a walker may begin soon after surgery, with moderate increases in the amount of weight as the wound heals. Full weightbearing is not permitted until a temporary socket, immediate post-operative fitting (IPPF), can be fitted. Temporary prostheses are fitted immediately post-surgery to minimize loss of sensory motor coordination, muscular strength, and endurance. Elastic bandages or an elastic stockinette are used to wrap the residual limb as a post-operative method to prepare it for prosthetic fitting. The bandage is used to apply higher compression distally and decreasing pressure proximally. The pylon and foot can be attached to the plaster cast for early limited weightbearing. Soft dressings are easier to apply, but are not as effective as rigid dressings.

10.1.3. Medical applications of orthoses

Orthoses may be used in the immediate treatment of some injuries and diseases. Plaster casts and orthotic devices may be used to immobilize a limb or joint to promote fracture healing. Soft tissue injuries (e.g., sprains, strains) may be treated with orthoses to immobilize or stabilize joints and limbs. Many temporary orthoses are used to stabilize knee and ankle injuries. Temporary orthoses promote healing while retaining a greater degree of residual function than other treatment methods. Orthoses can also be used to reduce pain associated with some diseases and injuries. In the latter stages of rheumatoid arthritis, an acute chronic inflammatory joint disorder, or carpal tunnel syndrome, an acute repetitive strain injury to the innervation of the hand, an orthosis can be used to immobilize the joint and hence reduce pain. Orthoses are preferred to plaster casts in some cases because they are removable, which helps to maintain range of motion, reduce contractures, and promote circulation.

Orthoses may be used to prevent deformations of joints associated with many medical conditions. Some medical conditions when left untreated may lead to substantial skeletal deformities and significant problems with body alignment and function. An orthosis may be used to prevent such deformities from forming, often eliminating the need for extensive surgical intervention. Deformations caused by osteoporosis, scoliosis, and diseases of the peripheral nerves can be reduced or prevented by the proper application of orthoses.

Soft tissue injury or supporting muscle paralysis may result in joint instability. After a severe joint injury, surgical restructuring of the knee may require an orthosis while the person participates in vigorous activity. Many people must use a knee orthoses while participating in sports. Specially designed orthoses are used which can provide adequate protection of the joint, allow the person full participation in the desired activity, and that can withstand the applied stresses.

Some injuries (e.g., spinal cord injury, brain injury), diseases (e.g., poliomyelitis, muscular dystrophy), or congenital disorders (e.g., arthrogryposis) may cause weakness or severely limited motion of limbs and joints. Orthoses can be used to enhance function of impaired limbs and joints. Stabilization or immobilization of impaired joints can significantly improve an individual's ability to use the limb. Many orthotic devices have been developed to ameliorate disability and enhance function of impaired limbs and joints.

10.1.4. Prosthesis fitting

It is essential to have a comfortable artificial limb that is suited for the activities a person wishes to participate in. Special adaptions for and modifications to prostheses are required for particular activities. The socket is the most critical component of any prosthesis because it comes in contact with the residual limb. The socket must be comfortable and functional. The residual limb must fit so that is is properly supported in the socket. The prosthesis must also be suspended from the body properly. If suspension is not adequate, pistoning can occur between the socket and limb interface, resulting in skin irritation that can lead to debilitating conditions.

A total surface-bearing socket provides for full contact and distributes load over the greatest surface area. Silicone suction sockets can provide good suspension and reduced socket interface shear forces. Blisters and

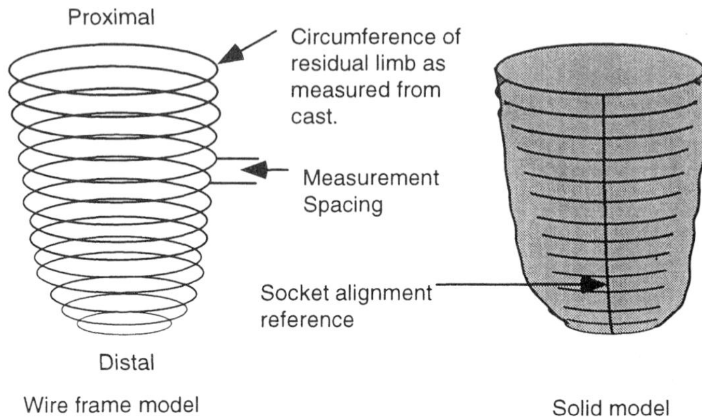

Figure 10.1. *Illustration of CAD-CAM above knee (AK) socket design.*

other irritations to the residual limb can be ameliorated by regular use of products that reduce shear forces and cushion vulnerable areas. Practicing good hygiene on the residual limb and socket will help to prevent skin problems. The proper number of socks can help to prevent skin problems, as can changing wet socks. Silicone can be used to add padding and damping to the socket to reduce shear forces associated with skin breakdown. Silicone end pads also help to absorb impact and reduce risk of injury to the distal end of the residual stump.

The socket must fit the shape of the residual limb as muscle and fat slowly atrophy. Fluid changes in the residual limb may affect prosthesis fitting. Fitting is often done by casting the residual limb. The cast of the residual limb is used to make a socket for the prosthesis. A plaster model is made from the cast, and modified to accommodate the needs of the user. The model can be used directly by a skilled prosthetist or the model can be digitized for computer-aided design. A skilled prosthetist sculpts the model by adding or removing plaster from the model to achieve the desired socket. Computer-aided design uses software to smooth, and sculpt the computer model of the residual limb (figure 10.1). A computer model is generated by placing the cast of the residual limb in a fixture. Once held in place, a probe (e.g., linear variable displacement transformer, linear optical encoder) is inserted into the approximate center of the cast (i.e., as viewed from the top) nearest the distal end. The probe is rotated one complete revolution (i.e., 360 degrees) to measure the radius at each sampling level. This process is repeated at fixed measurement spacings until the bottom of the cast is reached. The information obtained through the digitizing process is used to generate a wire-frame model of the residual limb. Computer modeling permits data from each patient to be stored for use on a future socket. The final version of the residual limb model can be output from a computer to a computer-controlled milling machine. The milling machine carves a plaster

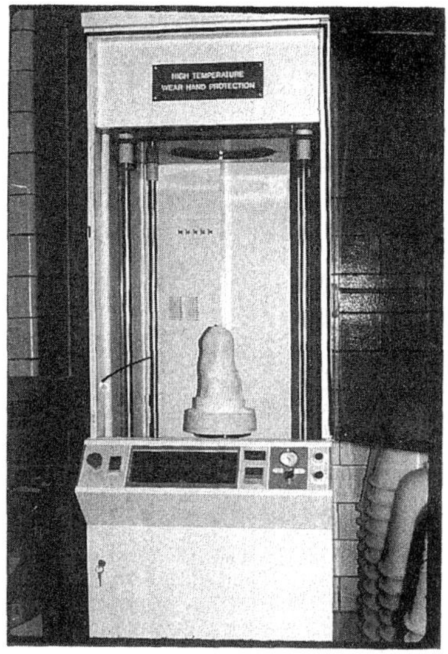

Figure 10.2. *Vacuum molding machine for the construction of lower-limb prosthetic sockets. Plaster male molds are placed in the machine, then the machine is heated and a vacuum is pulled. A sheet of translucent polypropylene is gradually pulled over the mold to produce a socket. (Photograph by R A Cooper.)*

or stiff polyurethane foam blank into a male mold of the desired socket. The male mold is placed in a thermal vacuum molding machine. Translucent polypropylene is formed around the mold to form a socket (figure 10.2).

Marks are often placed on the residual limb, and on the socket. Translucent polypropylene or transparent acrylic are useful in checking the alignment of the residual limb and for developing a socket (figure 10.3). Proper alignment increases the stability of the prosthesis, and reduces the pressure on the residual limb.

Useful CAD-CAM tools must give prosthetists flexibility in shaping the socket. It is also important for the CAD-CAM system to yield accurate and consistent measurements of the residual limb. A variety of graphical representations and views should be available with CAD-CAM systems. Many physicians prefer detailed measurements of the residual limb as the starting point for their socket designs, and will disregard the basic shapes suggested by some CAD-CAM systems. Instead of casting, where each prosthetist introduces some variability, a non-invasive imaging process which records the external shape and the internal tissue structure of the residual limb may make significant improvements in prosthetic design. With greater knowledge of residual limb tissue properties, and how these

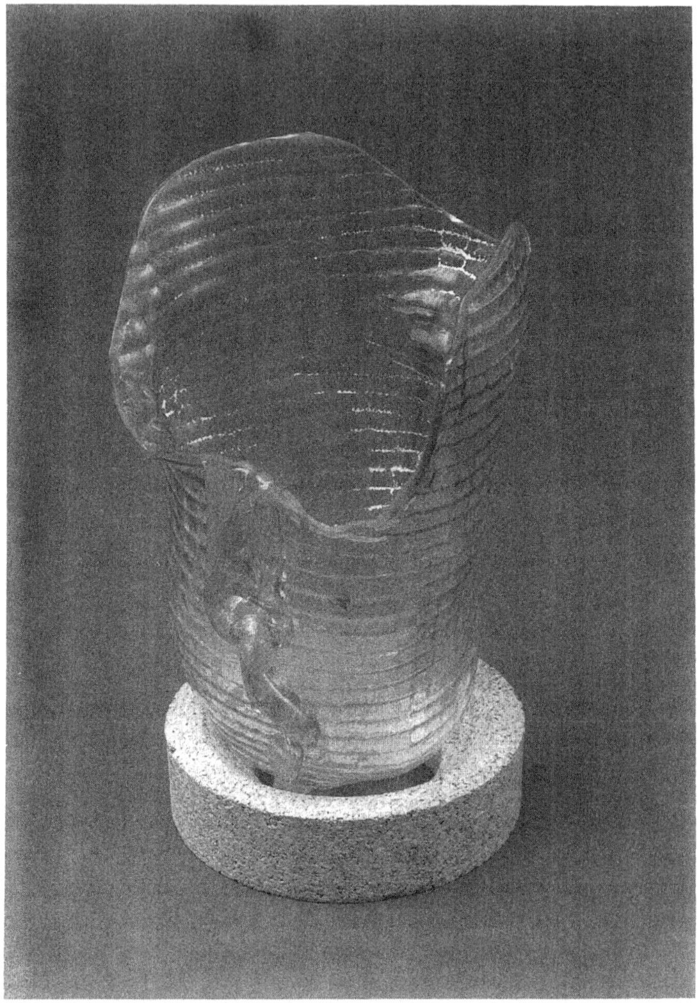

Figure 10.3. *Above the knee prosthetic socket made with computer aided design and manufacturing. (Photograph by S Parsons, CSUS Media Services.)*

properties relate to socket design, computers will be programmed to use this information to assist with the development of prosthetic sockets. Ultrasound, and computed tomography show some promise of furthering this aspect of CAD-CAM socket design. However, less expensive alternatives are desirable.

10.1.5. Prosthesis and orthosis design considerations

The weight of an entire below knee (BK) running prosthesis can be under three pounds. Components fabricated from composite materials (e.g., Kevlar, fiberglass, Spectrlon epoxy, acrylic resins, graphite), plastics (e.g., nylon, translucent polypropylene), and metals (e.g., titanium, aluminum) can provide strong, lightweight limbs. Suspension systems lighten the sensation of prosthesis weight, and make the prosthesis more efficient. A latex rubber or neoprene sleeve can be added to a supercondylar or patellar tendon bearing (PTB) style prosthesis to add suspension.

Prostheses are constructed using a crustacean (i.e., exoskeleton) load-bearing shell, and/or a skeletal structure. Crustacean prostheses are often made of thermoplastic polypropylene or composite materials (e.g., fiberglass, carbon fiber, Kevlar). The exoskeleton is made by vacuum forming thermal plastic over a plaster mold or by using hand lay-up of composite materials. Composite materials result in strong lightweight prostheses.

Skeletal prostheses use tubular and/or plate structures. This type of prosthesis can be covered with foam and latex to give the prosthesis a realistic appearance. Aluminum and carbon fiber are often used to form the skeletal structures. Lower-limb prostheses sometimes include shock absorbing systems and/or energy storing systems. This makes running or other strenuous activities more efficient and less painful. Skeletal prostheses are often lighter in weight, and more efficient than crustacean prostheses.

A prosthesis can often be made substantially lighter than the limb it replaces. A typical below the knee prosthesis can weigh less than three pounds. The size of a prosthesis should be near the size of the human limb. This makes dressing easier, and attracts less attention. Socks and stockings come in a multitude of colors, and some companies specialize in creating realistic appearing prostheses.

Bone density and strength may be reduced among people with impaired limbs. In addition, tendons and joint connective tissue may be weaker. It is important when designing orthoses to understand the properties of body tissue. An improperly designed orthosis can create joint or bone deformities. Body joints are often more complex in their movements than they appear upon casual observation. A single joint may incorporate several degrees of freedom (i.e., motion in several planes). Joints often combine rotation and translation (migration) and the joint center of rotation is not necessarily at the anatomical center of the joint. The interaction of muscle and connective tissue with the joint is also very important. Often orthoses are designed to oppose spastic muscle, or to augment function of limited muscle activity. Kinematic and kinetic analysis of the joints, limbs, and their interaction with orthoses is an important component of appropriate orthosis design.

Effective orthoses are lightweight, strong, durable, and simple. Fitting and cosmesis are also very important factors for orthosis design. People have a tendency to reject an orthosis which causes irritation to the body where it is attached, or to other areas with which it may come into contact. Material selection influences the orthosis design. Cuffs can be effectively made with polyvinylchloride (PVC), composite materials (e.g., fiberglass, carbon fiber), or shaped metal. Plastic or composite cuffs are easiest to form intricately for a close and comfortable fit. Excessive tightness of the orthosis can cause tissue breakdown and edema. Titanium, aluminum, and steel are used as structural components. Leather and nylon are effective strap materials. The orthosis must be easily donned and doffed. Potential for damage to apparel must be minimized. When an orthosis is used to stabilize or create a deforming force, a three-point pressure system is used. The orthosis is designed so that the middle point of the three-point system applies a restoring force at the vertex of the deformity.

10.2. UPPER-EXTREMITY PROSTHESES

Upper-extremity prosthetics consist of two primary components: the terminal device and the crane. The terminal device, or end-effector, provides some of the functions performed by a human hand. Many specialized terminal devices are used to optimize effectiveness of the prosthesis for a given task. The crane is used to position the terminal device in the desired location. The crane and terminal device must be capable of generating different forces for various positions related to common activities. Personal care and hygiene activities (e.g., teeth brushing, shaving, dressing, toileting) may have different requirements than work- or school-related activities (e.g., writing, lifting, sorting). People commonly lift and manipulate objects at floor, desk, shoulder, and overhead height. Designers must consider task requirements of prosthetic users in their designs.

10.2.1. Amputation level classification

There are four basic levels of upper-extremity amputations each related to the anatomy of the residual limb. For people with wrist amputations or with long residual limbs, voluntary rotation of the terminal device is possible through the use of flexible straps between the socket and the humeral cuff harness. Below the elbow amputations require different prosthetic devices depending upon the length of the residual limb. For a longer residual forearm, a socket which extends along the ulnar surface with a cutaway for the radial aspect near the proximal brim is used. This provides support for loads at the terminal device and for 90 degree elbow flexion. Short residual forearms are sometimes fitted with a socket and step-up elbow hinges. The step-up hinge amplifies the range of motion of the prosthesis at the cost of reduced torque. Many people prefer self-suspended prostheses because they are easier to don/doff. Above elbow prostheses incorporate a body-controlled elbow lock,

Table 10.1. *Relationship between residual limb and prosthesis.*

Amputation level	Prosthesis	Control source
Forequarter	Shoulder plate	Chest expansion
Shoulder disarticulation	Shoulder cap	Scapular protraction
Short above elbow	Locking elbow	Scapular protraction and humeral flexion turntable
Standard above elbow	Locking elbow	Humeral flexion turntable
Elbow disarticulation	Outside elbow	Elbow flexion
Very short below elbow	Step-up hinge	Elbow flexion
Short below elbow	Manual wrist	Forearm pronation, rotation, and supination
Medium below elbow	Step-up pronation	Forearm pronation and supination
Long below elbow	Screwdriver	Forearm pronation and supination with fit socket
Wrist disarticulation	Terminal device	Forearm pronation, rotation and supination with fit socket

a cable-controlled elbow or electrically operated elbow. A turntable between the socket and humeral section of the arm provides voluntary rotation. When the residual arm is not sufficiently long to provide direct control, a shoulder disarticulation or forequarter prosthesis is used. Excursion multipliers or external power sources are required to provide useful motions.

The design of the prosthesis is directly related to the residual function, and the functions desired by the user as shown in table 10.1. They are of primary importance for many activities of daily living. Bilateral amputations often cause more severe manipulation impairment than does a single arm amputation. Upper-extremity prostheses are not nearly as dextrous or as capable of fine motor control as a human limb. To achieve greatest functional capacity, prostheses must be designed for independent donning and doffing, for independent personal care, and for vocational activities. People with bilateral amputations may use a myriad of other assistive devices (e.g., specialty clothing, modified workbenches, computer access devices) in conjunction with their prostheses.

10.2.2. Terminal devices

Terminal devices may perform simple motions (e.g., prehension) or complex grasps (e.g., four-finger grasp). The more complex the function of the terminal device the more complex the control required, and hence the longer the time needed to learn to properly use the prosthesis. Many people prefer simple motions which are rapidly controlled. In many cases specialized terminal devices (e.g., screwdriver, pry bar) can be used to perform specific tasks. Terminal devices can be internally powered (i.e., powered by transferring motion of another part of the body to the terminal device) or externally powered (i.e., powered by batteries or other external power sources).

A split mechanical hook, made of aluminum or steel, often can perform tool-like activities and manipulations which provide a high degree of function. Two fingers can be opposed to one another with their tips forming a full contact area for gripping small objects. Most hooks incorporate one of two basic design concepts. One concept uses a movable finger which opens in a plane rotated with respect to the opposing finger along the wrist axis. The other concept uses a lyre shape designed to provide at least four contact points when grasping large cylindrical objects, and to provide flat grasping at the tip.

A mechanical hook or hand can be either voluntarily opening or closing. Voluntary opening terminal devices divert the force applied with the harness through a cable to pull on a spring-loaded lever. Relaxation of the harness force causes the finger to close with the spring force creating grasp. Voluntary closing devices divert force from the harness to open the movable finger against a weak opposing spring. This type of terminal device often incorporates a manual lock to reduce fatigue.

Some people prefer a hand and cosmetic gloves to a hook. Prosthetic hands may incorporate multiple degrees of freedom, and various grasps. Most prosthetic hands use cable tendons running through a composite (e.g., fiberglass, polyvinylchloride (PVC), carbon fiber) shell. The fingers are typically jointed, whereas the thumb may or may not be jointed. A prosthetic hand is often covered with a low modulus PVC glove for cosmesis. However, gloves have a tendency to stain.

10.2.3. Power sources for upper-extremity prostheses

The upper extremities are very flexible and quite strong for intact limbs. The energy efficiency of most prosthetic limbs is quite poor. With little energy available to perform the myriad of tasks desired by the prosthesis user, a functional hierarchy must be developed. Grip and elbow flexion and extension are considered by many to be the two most critical measures of prosthetic function. Terminal device pronation and supination, elbow rotation about the humeral axis, shoulder movement (e.g., extension/flexion, abduction/adduction) are also very important.

For people with single arm amputations, forces from the intact hand can be used to position and operate some functions of the prosthetic limb. In other cases, forces generated from muscles elsewhere in the body can be transferred to provide positioning and operation of the terminal device. A harness is used with many prosthetic devices to provide stability for the prosthesis, and to transmit forces and motions from another area of the body to the prosthesis. The harness should fit firmly, and distribute loads comfortably to allow maximum control of and force transfer to the prosthesis. Forces between 2 and 20 pounds are required to operate various functions of an above the elbow prosthesis. Typically opening and closing the terminal device requires the greatest force. The operating cable is connected to the terminal device via a variety of linkages. Many people prefer force amplification and proportional displacement. A method for accommodating a limited range of continuous motion, and still allowing large and small objects to be handled is to use a two-position opposing thumb. The thumb can be locked in two or more positions by releasing a latch and moving the thumb to the desired position and locking the latch. Other prostheses use a rack and dog (i.e., ratchet) mechanism to provide high mechanical advantage and large displacement. The terminal device closes proportionally until a reaction force develops which engages the dog in the rack. As further tension in the cable is applied, the rack ratchets the terminal device closed. This allows considerable grip force to be generated. However, the process may be slow.

Pneumatic actuators and power sources have been used with moderate success for upper-extremity prosthetics. Compressed gas offers the potential for a lightweight abundant energy source. Cylinders of compressed carbon dioxide are used as a pneumatic power source. Compressed carbon dioxide is commonly available (it is used in soda machines and fountains) and has desirable thermodynamic properties. Gas cylinders and associated valves present a potential risk of significant injury to the prosthesis user. High pressures and low temperatures are often required to store sufficient energy for practical applications. Most devices and systems are inefficient in converting the potential energy of the compressed gas to mechanical kinetic energy. This, coupled with the need to regulate the gas to lower operating pressures, reduces the operating time of the prosthesis dramatically.

Electric battery power is readily available and can be efficiently applied to upper-extremity prostheses. Nickel–cadmium or lithium batteries are used because they can be sealed and operated in any position. Lithium batteries have a high energy density. Standard alkali batteries can be used in emergency situations. Batteries are typically connected to highly efficient switching regulators to provide power to control electronics as well as actuator systems.

Multiple motors may be used in a single prosthesis. Therefore, weight and size are important considerations. Direct current motors provide high static torque, and simple proportional control. Gear-head or torques motors are used for actuation in prosthetic limbs. Gear-head motors operate at low voltages, but have a tendency to produce undesirable noise. Direct current torque motors can be designed to operate at low speeds. However, they are often heavy and bulky due to the additional windings. People generally prefer a fast response to control inputs, because most prostheses do not anticipate grasp configurations like a human hand. Some prostheses also include motorized wrist control.

10.2.4. Control sources for upper-extremity prostheses

The most common method of prosthesis control is through direct neuromuscular control with a harness. Control of the residual limb or another location of the body is transferred via the harness to control the prosthetic limb. Arm flexion, arm extension, and shoulder shrug are commonly used to control terminal devices. The harness also provides force and proprioceptive feedback to the person. In some cases, the residual limb or shoulder can be used to activate switches or locking mechanisms.

Direct neuromuscular control is improved by feedback received from a number of sources. Visual observation of the arm and terminal device provides important position information. Force, pressure and shear are transmitted through the socket and body harness which are useful in determining the forces applied to the prosthesis. The body and residual limb proprioceptive senses are important in prosthesis control. Prosthesis control can be enhanced by directly transferring motions in one area of the body to create motion of the prosthesis, for example, using elbow flexion and extension for direct mechanical control of closing and opening a hook. In this case, position and force information at the grip are transferred to the arm.

Electronic transducers can be used to provide indirect feedback from the prosthesis to the body. Force or pressure transducers mounted on or in the terminal device can be used to provide some degree of sensory feedback. Accelerometers can be used to provide vibration of the limb. Multiple sensors can be used to provide limited sensory feedback related to grip force or when an object slips. Transducers must be interfaced to the person in a meaningful way. Bladders, plungers, and vibrotactile actuators can be used to transfer the sense from the terminal device to the residual limb or other intact area of the body.

Myoelectric control of prosthetic limbs has demonstrated moderate success (figure 10.4). Gold-plated electrodes are implanted in the socket of the prosthesis to acquire electromyographic (EMG) signals produced by active muscle. The signals can be used to discriminate between various motions desired by the user. Myoelectric control can be used for elbow flexion/extension, wrist rotation, and terminal device prehension.

Several approaches have been applied to interpret myoelectric signals for prosthesis control. A simple approach is to place an electrode on each set of agonist/antagonist muscle. The relative level of activity is used to determine direction and force of the movement. An envelope detector is often used to condition the EMG signals. Typically, only a few EMG envelope thresholds are used to ensure reliable control.

More sophisticated control schemes can be used to reduce the number of electrodes and to reduce sensitivity to electrode placement. Autoregressive moving average (ARMA) models, and autoregressive integrated moving average (ARIMA) models have been used with single electrode pairs placed

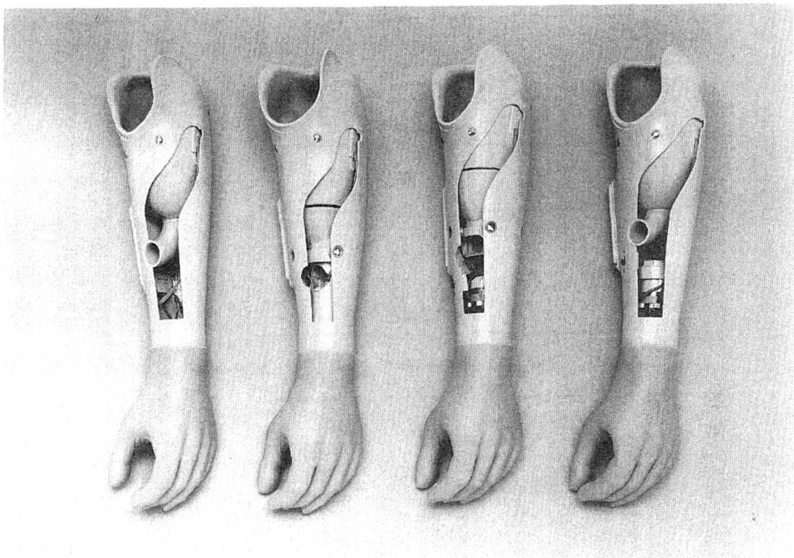

Figure 10.4. *Myoelectric prosthetic forearm with electronic hand. (Courtesy OTTO Bock USA.)*

over a set of muscles to condition EMG signals to produce multiple control signals. The concept is to make use of the prosthesis more coordinated and natural for the user.

An ARMA model uses a transfer function with poles and zeros (figure 10.5). The measured EMG signals are modeled as a function of current and past voltage values ($y(n)$), and an error signal ($e(n)$). Each sample instant is represented by n, and the time delay is represented by M for $y(n)$ and k for $e(n)$. The difference equation representation of a general ARMA model is described as follows

$$\begin{aligned} y(n) - a_1 y(n-1) - \ldots - a_M y(n-M) \\ = e(n) + b_1 e(n-1) + \ldots + b_k(n-k) \\ Y(z) = \frac{G(z)}{H(z)} E(z). \end{aligned} \tag{10.1}$$

The difference equation (10.1) can be represented as a transfer function in the z-transform domain between $Y(z)$ and $E(z)$. $Y(z)$ and $E(z)$ are the z-transforms of $y(n)$ and $e(n)$, respectively.

In equation (10.1) the $a_1 \ldots a_M$, and $b_1 \ldots b_k$ are called the ARMA parameters. The order of the ARMA process is (M, k). The output of the model is $y(n)$, and the $e(n)$ represents a white noise process of zero mean and variance σ_e^2. An ARMA process can be represented by an inner product

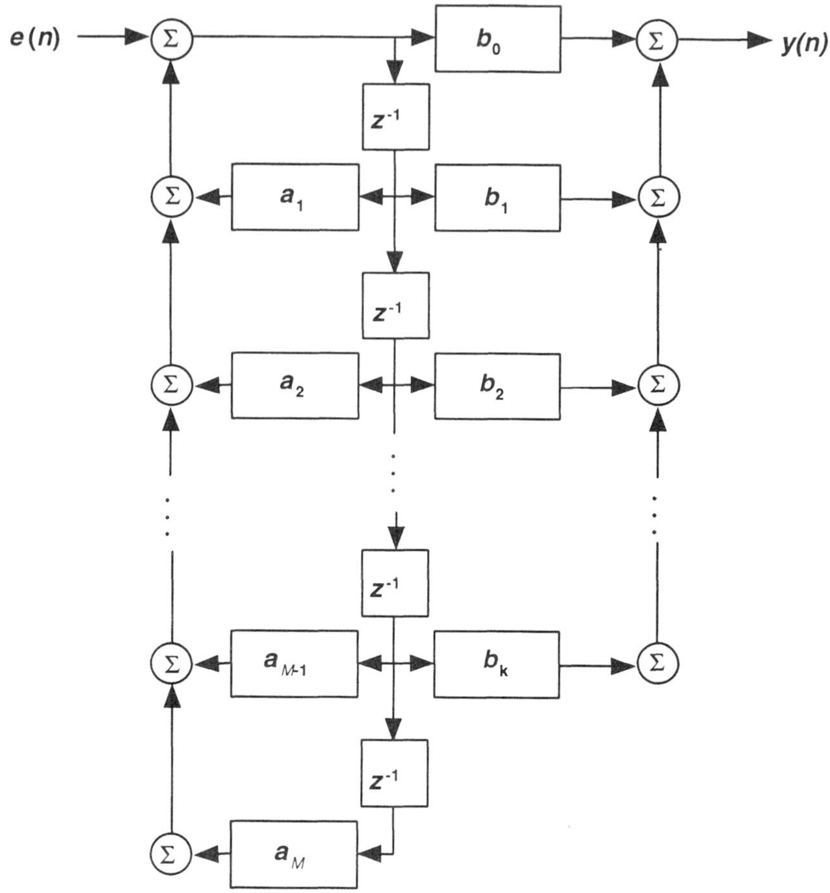

Figure 10.5. *Block diagram for order (M, k) ARMA model.*

which makes it simpler to estimate the model parameters.

$$\begin{aligned}
y(n) &= \theta^T \varphi(n) + e(n) \\
\theta^T &= [a_1, a_2, \ldots, a_M, b_1, b_2, \ldots, b_k] \\
\varphi^T(n) &= [y(n-1), y(n-2), \ldots, y(n-M), e(n-1), \\
&\qquad e(n-2), \ldots, e(n-k)].
\end{aligned} \tag{10.2}$$

The difficulty with determining the ARMA parameters is that the noise process is unknown *a priori*. Therefore, an estimate of the noise must be generated. The noise estimate is used to initialize the ARMA model parameter algorithm. The ARMA model algorithm is run until the model parameters converge. This method is outlined in table 10.2.

An ARMA model by definition implies the noise is colored (i.e., the amplitude of the noise is not constant across the frequency spectrum). An

Table 10.2. Outline of a simple algorithm for estimating ARMA model parameters. The initial error estimate is based upon an autoregressive model determined using the least-squares error criteria.

$$y(n) = \hat{\theta}^T \hat{\varphi}(n) + e(n)$$

where $\hat{\theta}^T = [a_1, a_2, \ldots, a_M]$ and $\hat{\varphi}^T(n) = [y(n-1), y(n-2), \ldots, y(n-M)]$.hfill
Using the least-squares error criteria for linear prediction

$$E[e^2] = \frac{1}{N}\sum_{n=1}^{N}\frac{1}{2}[y(n) - \hat{\theta}^T\hat{\varphi}(n)]^2$$

$$\hat{\theta} = \arg\min E[e^2] = \left[\frac{1}{N}\sum_{n=1}^{N}\hat{\varphi}(n)\hat{\varphi}^T(n)\right]^{-1}\frac{1}{N}\sum_{n=1}^{N}\hat{\varphi}(n)y(n).$$

The AR parameters are the product of the inverse autocorrelation matrix $(R(N))$, and the cross-correlation vector $(f(N))$.

$$R(N) = \frac{1}{N}\sum_{n=1}^{N}\hat{\varphi}(n)\hat{\varphi}^T(n) \quad \text{and } f(N) = \frac{1}{N}\sum_{n=1}^{N}\hat{\varphi}(n)y(n)$$

$$\hat{\theta} = R^{-1}(N)f(N).$$

The ARMA model parameters may be estimated by using the AR estimated model error as a starting point for the ARMA model error.

$\hat{\varepsilon}(n, \hat{\theta}) = y(n) - \hat{\theta}^T\hat{\varphi}(n)$

$i = 1$

while $[\theta(i) - \theta(i-1)] \leqslant$ threshold

$e(n, 1) = \hat{\varepsilon}(n, \hat{\theta})$

$\theta^T = [a_1, a_2, \ldots, a_M, b_1, b_2, \ldots, b_k]$

and

$\varphi^T(n) = [y(n-1), y(n-2), \ldots, y(n-M), e(n-1), e(n-2), \ldots, e(n-k)]$

$$\theta(i) = \arg\min E[e^2(i)] = \left[\frac{1}{N}\sum_{n=1}^{N}\varphi(n,i)\varphi^T(n,i)\right]^{-1}\frac{1}{N}\sum_{n=1}^{N}\varphi(n,i)y(n)$$

$e(n, i+1) = y(n) - \theta^T(i)\varphi(n, i)$

$i = i + 1$

end

variety of ARMA model algorithms exist. It is not uncommon to use a weighted algorithm, or prefiltering. Algorithms have been developed which estimate model parameters using frequency domain methods instead of time domain methods. Some data may contain low-frequency disturbances, offsets, trends, drift, and periodic variations which are undesirable. These sources often stem from external sources (e.g., motion artifacts, electrode placement, electrode variation) that we may not wish to contribute to the model development. Disturbances can be removed by pretreatment of the data, and by incorporating a noise component in our model parameters. The first approach involves direct removal of offsets and trends. The second method relies on noise models with poles on or close to the unit circle, such as the ARIMA models used in the Box–Jenkin approach. The ARIMA model is equivalent to using a high-pass filter on the data. ARIMA models incorporate an integrator in the noise model.

$$Y(z) = \left[\frac{1}{1-z^{-1}}\right]\frac{G(z)}{H(z)}E(z) \quad \text{or} \quad Y(z) = \left[\frac{1}{1-z^{-1}}\right]\frac{1}{A(z)}E(z). \tag{10.3}$$

The noise integrator is implemented starting at time zero (i.e., the integrator is multiplied by a unit impulse). The noise model could be given the flexibility to find the integrator parameters and/or to find a pair of complex poles on the unit circle to account for periodic variations.

Reliable myoelectric control requires consistent data with sufficient information content to provide the necessary degrees of freedom for prosthesis control. This requires well designed control systems, and reasonably accurate electrode placement. When designing myoelectric controllers, experimental and training data must be collected in conditions which resemble situations in which the prosthesis controller will be used. Most myoelectric controllers use low-order models (e.g., second-order models). However, if model accuracy is desired then a persistently exciting input (i.e., an input rich enough to excite all important modes of the system) and not too simple a feedback mechanism is required. Appropriate model order and structure are important for the design of a successful controller. Parameter variance is minimized with respect to variations in changes to muscle, electrodes, and the prosthesis by selecting parameters which have a clear effect on the output predictions. In order to achieve suitable sampling, the sampling frequency should be about ten times the estimated bandwidth of the system. Offsets, drift, trends, and other biopotentials must be treated with care when designing myoelectric controls.

10.3. UPPER-EXTREMITY ORTHOSES

The primary goal of most upper-extremity orthoses is to restore dexterity and versatility for manipulation of objects and the environment. Although strength is important when performing some tasks, it is not a primary consideration when designing upper-extremity orthoses. The upper extremities work as coordinated open kinematic chains with substantial degrees of freedom. The hand alone has numerous degrees of freedom. Each interphalangeal joint has one degree of freedom (i.e., flexion and extension). Each of the metacarpophalangeal joints has two degrees of freedom (i.e., flexion/extension and abduction/adduction). The wrist has two degrees of freedom (i.e., flexion/extension and ulnar/radial deviation), excluding migration. The elbow has two degrees of freedom (i.e., flexion/extension and axial rotation). The shoulder has three degrees of freedom as an amphiatrotic joint. The large number of muscles in the upper-extremity provide wondrous combinations of motion and force.

In the lower extremity, gravity can be manipulated to provide joint stability and aid in ambulation. In the upper extremity, only gravitational forces of the limbs themselves can be used. Typically, these forces are too small to be useful. Forces and motion of the upper-extremity orthosis may be generated from a number of other sources. The simplest method redirects forces and motions at the closest more proximal joint to the distal orthosis. Force and motion of another bodily motion can also be redirected to operate an orthosis. Elastic components (e.g., springs, elastic bands) can be used to apply forces or generate motions. Pneumatic actuators powered by compressed gas can be controlled to provide an effective orthosis. Battery powered electric motors can also be used with an orthosis.

Functional splints must closely follow the movements of the supported joints to provide smooth and coordinated function. The terminal device or hand is the most critical component of any upper-extremity orthosis. If the hand cannot be substantially more functional with an orthosis, then the orthosis will be discarded. Cosmesis is of utmost importance for many people, as orthoses are often readily visible. Size and weight are also very important. Upper-extremity orthoses must be durable, but must be light for efficiency and effectiveness.

10.3.1. Wrist-hand orthoses

The functions of the hands are to provide fine sensory feedback, fine motor control, and varying degrees of power. High grip strength usually involves use of all of the fingers, whereas precision grip uses a pinch-type grip with the thumb and an opposing finger tip. A fully functional orthosis must allow for full hand grip, three-fingered pinch, and thumb-to-side-of-index-finger key grip. These three grips provide the majority of functions provided by the unimpaired hand.

When muscles of the upper extremity are weakened so that gravity may influence the position of the limb or when spasticity is present to cause an imbalance, orthoses are used to prevent deformity due to contractures. Rest orthoses keep the joint in position of function during rest periods. Dynamic orthoses, which include elastic bands or springs to provide a deformation force, may also be used. A hand orthosis (HO) uses a three-point pressure

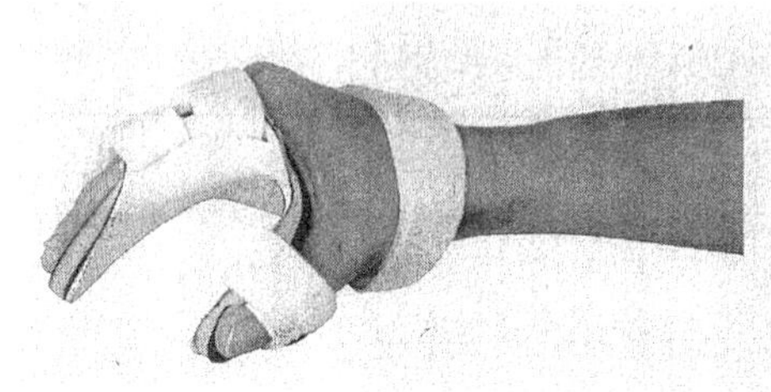

Figure 10.6. *Opponent hand orthosis which provides three-point pressure over the hand to prevent flexion deformity. (Photograph by Highland Drive VAMC Media Services.)*

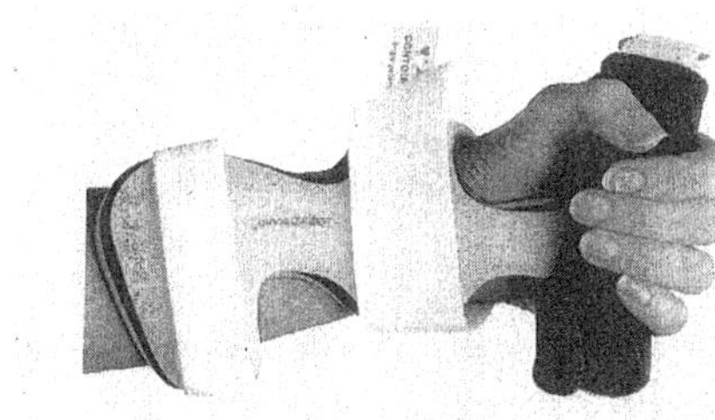

Figure 10.7. *Contoured hand and wrist orthosis provides three-point pressure over the hand and wrist to extend the wrist and fingers. (Photograph by Highland Drive VAMC Media Services.)*

system over the dorsum of the joint to provide interphalangeal extension to prevent flexion deformity of the finger joint (figure 10.6).

A wrist–hand orthosis (WHO) can be used to provide three-point pressure over the dorsum of the wrist, causing a bending moment to extend the wrist and fingers (figure 10.7). This type of orthosis helps the individual to use the functioning muscles to flex the fingers and wrist effectively.

Finger and thumb flexion for a precise grip can be created by diverting the force of wrist extension. The tenodesis splint is a popular orthoses for using wrist extension to create functional grip. Most tenodesis splints use a three-jaw-chuck pinch grip. Lack of tactile sensation makes use of the tenodesis splint more difficult. Other locations on the body can be used to power an orthosis when the individual does not have wrist function. Shoulder flexion, biscapular abduction, or chest expansion can be diverted through cables to provide wrist extension for a tenodesis splint (figure 10.8).

Compressed gas or electric motors can be used to provide grip. Gas can be permitted to flow from a pressurized cylinder into an expandable chamber (e.g., McKibben muscle). When the chamber expands, its length shortens. The shortening of the chamber pulls the wrist back and the flexor tendons pull the fingers together.

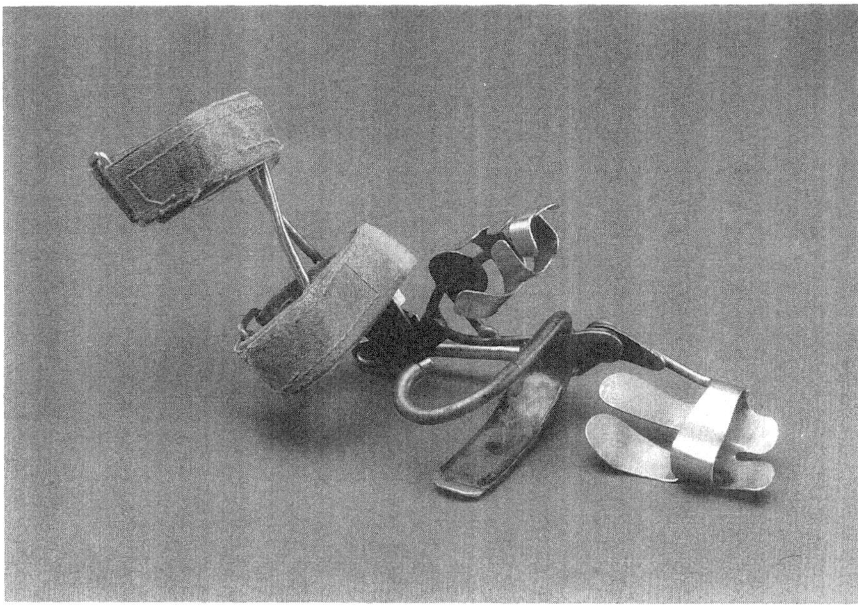

Figure 10.8. *Tenodesis splint designed to improve functional grip for people with quadriplegia. (Photograph by S Parsons, CSUS Media Services.)*

External electric power systems can also be used. Motor and gearing efficiency must be very high to provide a reasonable battery charge life. High-power battery sources, and high-efficiency motor controllers make electric power attractive for the development of externally powered orthoses.

10.4. LOWER-EXTREMITY PROSTHESES

The major concern of people of all ages who have undergone a lower-limb amputation has been whether or not they will be able to regain their previous level of physical activity. Prosthetic technology will allow many individuals to attain function nearly equal to pre-amputation levels. People with amputations cannot only participate in life's activities, but in many cases participate with non-disabled people.

10.4.1. Prosthetic feet

Partial foot amputation can often be treated with custom padding or inserts within a conventional shoe. A flexible shank may be added to shoes to prevent the shoes from bending sharply at the end of the foot. For people with Syme amputations a plastic laminate prosthesis provides functional ambulation.

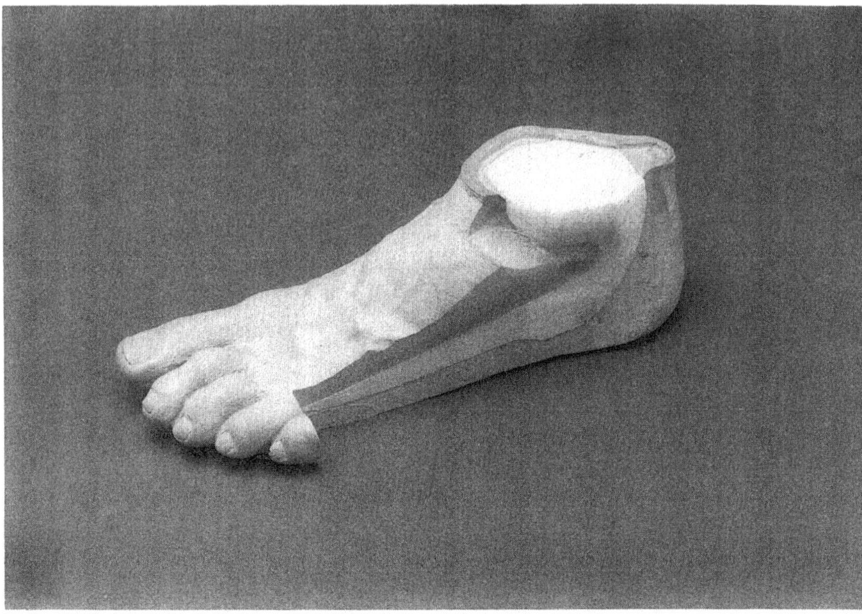

Figure 10.9. *Seattle Department of Veterans Affairs Rehabilitation Research and Development Center energy storing prosthetic foot. (Photograph by S Parsons, CSUS Media Services.)*

The development of energy-storing feet is possibly the greatest advance in prosthetic design. With the aid of energy-storing feet, people with lower-limb amputation are able to move quickly and to run with a foot-over-foot gait. Energy storage is accomplished via a flexible keel, which provides non-linear spring action similar to the push-off phase of walking or running (figure 10.9).

The stationary attachment flexible endoskeletal (SAFE) foot is considered to be the first modern energy-storing prosthetic foot design. Since then, a number of other energy-storing feet have been made commercially available. Energy-storing feet may be customized for an individual, or may be semi-custom to improve the ability to participate in certain activities. There are lightweight prosthetic feet with totally flexible keels to provide the appropriate dynamic elastic response for walking on uneven terrain. Lightweight graphite feet are available which provide a strong push at toe-off for activities like running. Feet which provide a strong push-off are referred to as 'super'-dynamic elastic response prosthetic feet. Proper alignment is critical for 'super'-dynamic elastic response prosthetic feet for the leg to roll smoothly from heel-strike to toe-off. This type of foot requires some accommodation time to adjust to the spring action.

For some people an energy-storing foot is not essential, and they are comfortable with a single-axis solid ankle cushioned heel (SACH) prosthetic foot. A SACH foot provides plantar flexion after heel-strike and a smooth transition to mid-stance, thus increasing knee stability. A SACH foot is helpful for people with AK, BK with short residual limb, or BK with a knee with flexion contracture. A multi-axis foot provides motion in dorsiflexion, inversion, and eversion as well as plantar flexion (figure 10.10). These features help to accommodate uneven terrain by having the foot conform to the terrain. Multi-axis feet can also be energy storing.

Swimming is a popular and effective form of physical conditioning. Swimming for a person with an amputation is important because it is a form of exercise that is not traumatic to the residual limb. Although most people can enjoy swimming without a prosthesis, some people choose to swim with one. A swim prosthesis provides balance in the water through even weight distribution. Both exoskeletal and endoskeletal prostheses can be used for swimming. Some prostheses are buoyant and may require the addition of weights. There are several types of swim legs: waterproofed walking prosthesis, peg leg, stubbies, and hollow-chambered legs. Standard feet that are molded without an external heel cushion are less susceptible to becoming water-soaked. Any foot or standard prosthesis used in the water should be treated with a waterproof coating (e.g., New Skin, foot paint). Fins can be added to stubbies or feet for added speed and control.

Many people in third-world areas require inexpensive, durable prosthetic feet. These people often have other needs than people in other areas. The Jaipur foot is widely used in India and other third-world countries (figure 10.11). It is inexpensive to produce, and functional for barefoot walking. The Jaipur foot is made of vulcanized rubber and fabric with a wooden keel. It uses three inserts (i.e., fore-foot and heel of microcellular rubber, and an ankle of laminated wood) covered with a layer of rayon or nylon cord-reinforced unvulcanized rubber, which is surrounded by a layer of vulcanized rubber. The foot is quite flexible in comparison with other prostheses. Many people use the prosthetic foot bare or with a loosely fitting sandal. The flexibility of the Jaipur foot in three planes is well suited for walking over uneven terrain, climbing trees, working in irrigated fields, and while squatting.

The complex articulations of the human ankle–foot complex must be precisely duplicated in the prosthetic foot for normal gait to occur. Kabra proposed using a processed human cadaver or amputated foot as a possible means of creating a fully articulated prosthetic foot (figure 10.12). His group tested three ankle–foot prostheses fabricated using a naturally articulated, fresh cadaveric human bone endoskeleton set in a fiber reinforced rubber shell. Radiographic and kinetic analyses before and after cyclic dorsiflexion for 5000, 100 000, and 3 million cycles revealed no structural or functional changes. Further research is required in this area. Living tissue has sensory receptors and remodeling capacity, whereas the processed foot does not. Human bones and joints, even with remodeling capability, have the tendency

(a)

(b)

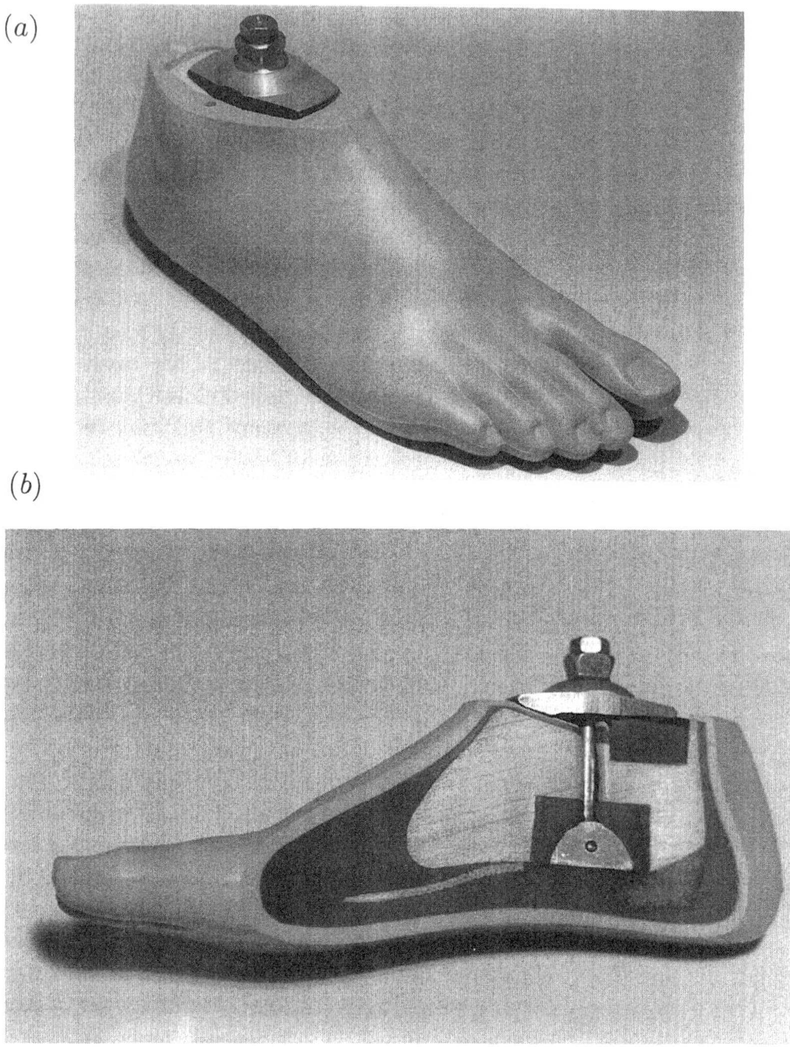

Figure 10.10. *(a) Multi-axial prosthetic foot. (Courtesy OTTO Bock USA.) (b) Cut-away view of multi-axial prosthetic foot made of a composite of materials to provide desirable cushioning and flexibility. (Courtesy OTTO Bock USA.)*

to fatigue and fail, especially when devoid of dynamic muscle stabilization and sensory input. Sources of suitable human cadaver feet are limited. The preparation of the human foot could be very labor intensive. Further work is required to determine the efficacy of using human feet as the basis for prosthetic feet. However, the knowledge gained by further study of fully articulated prosthetic feet is important.

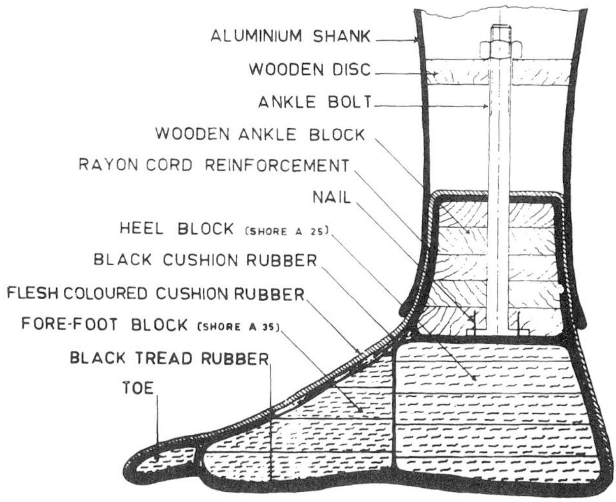

Figure 10.11. *Design schematic for the multi-axial Jaipur Foot. The laminated wood ankle block constitutes the stationary ankle. The flexible endoskeleton is composed of a microcellular rubber heel block and fore foot block encased in a cord-reinforced rubber exterior. The rubber exterior helps to modulate flexibility and to provide protection in the event of component failure. A carriage bolt attached to a wooden block inside the limb is used to hold the prosthetic foot to an aluminum shank. The lower end of the shank mold covers the ankle of the prosthetic foot for additional protection. (Courtesy S G Kabra, SMS Medical College, Jaipur, India.)*

10.4.2. Below-the-knee prostheses

The conventional patellar tendon-bearing (PTB) socket is adequate for activities of daily living, but is not suitable for active sports participation. The PTB does not evenly distribute the impact impulse generated by active people. It is also uncomfortable to run with a prominent patellar bar. A total surface-bearing socket provides for full contact, and damps impact loading. Side joints and a thigh lacer can also reduce the forces on the residual limb by transferring some weight to the thigh. The PTB does provide full contact between the distal end of the limb and the socket bottom. A PTB eliminates many problems associated with knee joint alignment, and reduces the hardware associated with the prosthesis.

Latex rubber suspension sleeves provide suction-cup-type suspension, and help reduce pistoning for people with BK amputation. They may be used without additional support. A waistbelt attached to a cuff suspension socket can provide additional suspension within the socket to prevent pistoning. Neoprene suspension sleeves can also provide atmospheric suspension. A silicone-suction-socket (3S) can provide good suspension properties and a socket interface that reduces shear forces. A simple 'Muley' strap just above the femoral condyles connected to smaller flexible straps pointing

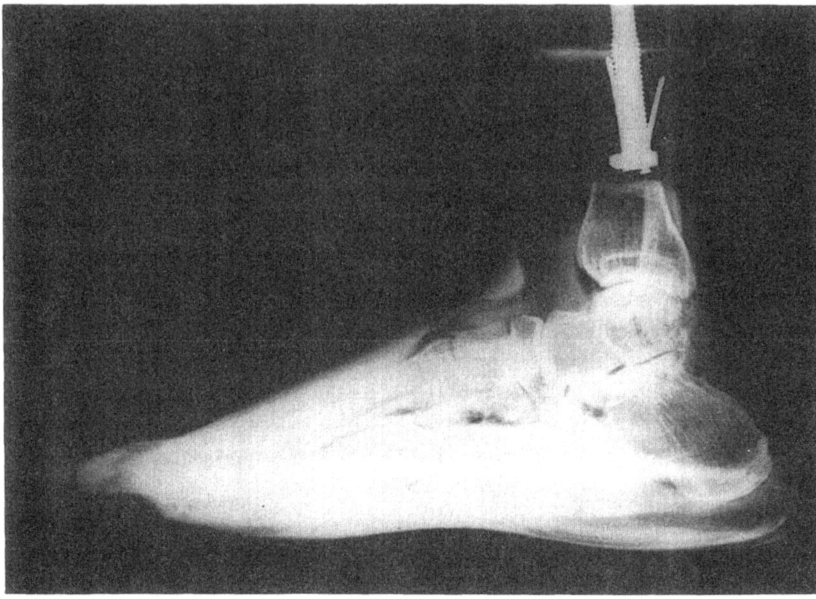

Figure 10.12. *Lateral radiograph of an articulated human cadaveric bone endoskeleton ankle–foot prosthesis. The fresh formalin fixed human cadaveric bones, with their articulations and major ligaments intact, are encased inside a cord reinforced rubber exterior that forms a hermetic seal. The foot articulations are preserved in their natural position by compressive forces from the elastic exterior. A carriage bolt is attached to the upper ends of the tibia and fibula by a wooden block formed over the tibio-fibular end. (Courtesy S G Kabra, SMS Medical College, Jaipur, India.)*

downwards and backwards along the medial and lateral aspects connected via a pivot to the shank is effective for some people.

Ankles can provide a wide range of motion and increased flexibility. An energy-storing foot that is not designed to provide multi-axis motion can be coupled with an ankle. Often, the ankle is mounted inside the prosthetic foot in a position that closely resembles that of a human foot.

Computer-aided design and manufacturing (CAD-CAM) systems in prosthetics provide an alternative to traditional methods of producing a positive mold used to make a socket. Several systems have been developed for using computer-aided design and manufacturing of prosthetic sockets. One method used for below the knee amputations is to load the residual limb with a casting jig. While loaded a prosthetist uses calipers to measure the anterio-posterior diameter at the mid-patellar tendon and the medio-lateral diameter at the tibial plateau. The length of the residual limb is measured with a fiberglass tape. The cross-sectional area along the residual limb can be measured by hand using custom equipment, or by using a cast and probe as described earlier. Measurement intervals of 1 to 2.5 cm are satisfactory.

Below-the-knee sockets come in three basic shapes: tapered, cylindrical, and bulbous. The measurements of the residual limb can be matched to the appropriate basic shape to optimize the smoothing process. The wire-frame or solid model representation of the socket can be viewed by the prosthetist, and modified as necessary. Common modifications include moving nodes on the wire frame model to effectively add or subtract material in a specific area from the male mold. This is sometimes referred to as the patch method. The scale of the wire-frame model can be changed about the vertical, horizontal, or both axes. This technique can be used to lengthen or shorten a socket, or to make the socket fit tighter or looser. Scaling the socket about both axes is sometimes called changing the 'ply'. CAD-CAM socket design tools can yield satisfactory sockets for the below-the-knee amputee.

10.4.3. Above-the-knee prostheses

Ischemia containment sockets provide alignment stability and control of the femur by using a high anterior brim to retain the ischial tuberosity within a lower-posterior ischial gluteal seat. Ischemia containment sockets are also known as narrow M/L, normal shape normal alignment (NSNA), or contoured adducted trochanteric-contained alignment method (CAT-CAM). A Silesian belt or Neoprene total elastic suspension (TES) belt can be used to provide suspension to an above knee (AK) suction socket. A flexible brim AK socket helps reduce shock and pressure against the ischial tuberosities.

A total contact socket (TCS) prosthesis provides contact over the surface of the distal end of the amputated limb. A TCS can reduce edema often associated with prostheses, and provides better sensory feedback. TCS prostheses are made from composites or plastic laminates formed over a mold made from the residual limb.

The knee joint should allow for stable and safe operation. Knees should be lightweight and provide ease of extension. A simple mechanical device, which can be manually released for sitting, can be used to lock the knee during ambulation. However, this is not very functional. It is desirable to allow some bending during the swing phase while providing adequate stability during the stance phase. Mechanical devices which allow the knee to flex when unweighted and which lock when a weight threshold is exceeded provide a means of knee flexion during the swing phase and stability during the stance phase. Pressure on the heel can be used to actuate the knee lock. Plantar flexion of the ankle can be used to cause extension of the knee during stance, providing anatomical stability without locking the knee.

Hydraulic knees allow stance and swing control, which provide stability during single-limb support. Hyperextension of the hydraulic knee allows it to lock, otherwise the knee provides hydraulic resistance to flexion. Force, above a threshold, at the prosthesis forefoot can be used to unlock the knee.

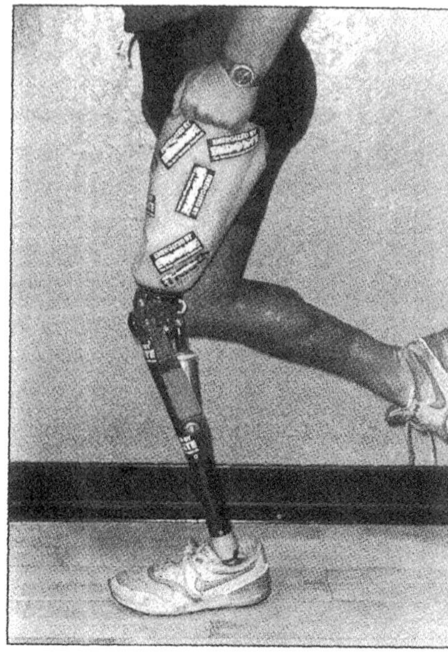

Figure 10.13. *The Endolite Stabilized StanceFlex (Bouncy) Knee is attached between the above-knee socket and the knee unit. The bouncy knee supports weight at mid stance and provides impact cushioning. Also shown is the Mauch Swing Phase Hydraulic Knee Unit, which is used to assist with cadence control while running. (Courtesy US Department of Veterans Affairs, photograph by Dale Tilly, Seattle VAMC, WA.)*

Many hydraulic knees incorporate a manual lock for activities requiring maximum stability (i.e., driving an automobile, standing on a bus, standing at a work cell), and a release which allows maximum flexibility. Hydraulic knees typically incorporate a piston in a cylinder which is perforated with roughened holes to allow fluid to flow from one side of the cylinder to the opposite side as the piston moves. The distribution of the holes within the cylinder determines the amount of damping. Fewer holes are placed near the ends of the cylinder to provide terminal deceleration. The distribution of holes is asymmetrical. A consequence of using hydraulic knees is that some of the body's energy is dissipated through the hydraulic piston and cylinder.

Many knees incorporate several degrees of freedom. Flexible or 'bouncy' knees are available to reduce impact and to provide knee stability (figure 10.13). Friction and elastic components can be adjusted for an individual. However, fixed friction and elastic components of the knee are only optimal for the walking speed at which they were adjusted. Increases in gait speed may cause excessive heel rise and premature locking of the knee. Decreases in speed may cause toe stubbing due to inadequate knee flexion. Allowing friction adjustment by the user helps to alleviate these problems

Figure 10.14. *A person with bi-lateral above-knee amputations running with energy-storing Sabolich Feet and CustomFlex Sockets. The efficient and ergonomic design of these prostheses make activities like running possible. (Courtesy of Sabolich P&O Center, Oklahoma City, OK.)*

for some people. The proper selection of the socket, knee, and foot can make activities such as heel–toe running possible for people with bilateral below knee amputations (figure 10.14).

Socket design for AK amputations requires several critical measurements. These measurements can be divided into two basic groups: linear distances from the appendicular skeleton (pelvic girdle) for design of the proximal (brim) portion of the socket; measurement of the residuum length and the transverse and longitudinal tissue distribution of the residuum for design of the distal (body) portion of the socket. An accurate representation of an AK socket consists of at least the following measurements: mediolateral linear distance from the anterior aspect of the adductor longus tendon (near the symphysis pubis) to the greater trochanter; anterior–posterior linear distance from the anterior aspect of the adductor longus tendon to the ischial tuberosity; pelvic vertical distance from the ischial tuberosity to the anterior superior iliac spine; distance from the ischial tuberosity to the lateral condyle of the intact femur; vertical distance from the ischial tuberosity to the end of the residuum (this position should be made with the person standing); transverse measurements along the residuum using a series of anthropometric rings; circumference of the thigh in the transverse plane, taken at the level of the gluteal fold just below the ischial tuberosity. These measurements are based on conventional practice and ease of identification.

10.4.4. Prosthetic gait analysis and assessment

Prosthetics designers must be familiar with all aspects of gait by people with lower-limb amputations. Walking with a prosthetic limb varies substantially with surface inclination, camber, and roughness. Designers must also be familiar with the biomechanics of periodic events like stumbling, slipping, and falling with which people with lower-limb amputations must live. Prostheses should be designed to minimize the occurrence and risk of injury associated with these events while providing functional and efficient gait.

Prosthetic limbs do not provide direct proprioceptive or kinesthetic feedback. However, some information about force and position is transferred to the user via the socket. Prosthetic limbs often produce different sounds, vibrations, and impulses with changes in gait. Once understood, this information can be used to adapt and improve gait. Socket design, limb–socket relations, and distal components are all important for successful prosthetic ambulation. The physical characteristics and activity objectives of the user vary the requirements for the prosthesis. People with lower-limb amputations use a higher than normal oxygen consumption during walking, with energy cost increasing with higher level or bilateral amputations. Dynamic elastic response or energy-storing feet have been developed to improve energy efficiency and provide damping. For level walking, improved energy efficiency of dynamic elastic response feet has not been demonstrated. However, prosthetic users report differences between dynamic elastic response and SACH feet.

Biomechanical assessment techniques have not become standard tools in most clinical settings. Simple and easily measurable parameters that are related to various pathologies or misalignments are much needed. Most research has concentrated in level, steady-state walking. Normal prosthetic gait kinetics and kinematics can be achieved with a large variety of active muscles and neuromuscular control schemes. Therefore, many pathologies, prosthetic misalignments, or discomforts might be compensated for, hidden, and/or missed in level, steady-state gait analysis. Electromyographic analysis has been able to provide little additional information applicable by clinicians.

Examining gait periods and functions other than level, steady-state walking may lead to better assessment tools for prosthetic fit and design. Study of the initiation of gait, stopping, approach to steps, or reactions to falls may lead to greater information about prosthetic gait. During periods of transition, greater neuromuscular control is required, hence more of the neuromuscular system is active. Often it is difficult to accurately record kinematic data during transitional phases. However, kinetic data may provide the relevant information. Kinetic analysis of a variety of transitional phases (i.e., starting, stopping, steps, stumbles) may lead to identification of various pathologies and misalignments.

Clinicians and people who use prosthetics often use a few simple variables for assessing fit and design. The age and ratio of actual weight to ideal weight of the person are helpful when selecting a prosthesis. The length of the residual limb, and percent length of the intact limb segment provide useful selection and evaluation criteria. The ratio of stride frequency to stride length is a simple evaluative measure. Many people prefer prostheses which provide the greatest damping and least shock.

10.5. LOWER-EXTREMITY ORTHOSES

An understanding of the biomechanics of the lower-extremity is important to the design of appropriate orthoses. The function of lower-extremity joints, and the effects of forces and moments induced must be well understood. The ankle, knee, and hip are the joints of primary concern during lower-extremity orthosis design. The ankle is often treated as a single degree of freedom joint (i.e., flexion and extension in the sagittal plane). Two degrees of knee joint motion are considered during orthosis design (i.e., flexion and extension in the sagittal plane, and rotation in the transverse plane). The hip joint is modeled with three degrees of freedom (i.e., abduction/adduction in the frontal plane, flexion/extension in the sagittal plane, and internal/external rotation in the transverse plane). Orthosis design must consider the degrees of freedom of the various joints involved, and determine the extent to which these joints will be restricted or fixed. Orthosis designers must also be aware of the interactions between joints. For example, plantar flexion at the ankle can be used to create hyperextension of the knee during some periods of the stance phase.

Lower-extremity orthoses are further complicated by weight bearing. Orthoses must be designed to support the limbs and joints under the weight of the body during standing and during ambulation. During ambulation the position of the center of gravity changes, and hence the loading on the orthosis changes. The orthosis design also needs to consider energy expenditure during ambulation. Large variations in the center of gravity position produce inefficient motion, and large stresses on the orthosis. Restrictions in the degrees of freedom at a joint can have a dramatic effect on the forces developed at each joint of the lower limb encompassed by the orthosis.

Sensory feedback, and comfort are important factors related to effective ambulation with an orthosis. Proprioceptive perception can greatly enhance the ability to ambulate, and the orthosis design should not reduce perception to joint position. If the person is insensate, greater joint stability may be appropriate. Care must be taken to avoid causing injury or discomfort to the end user.

Orthoses to enhance or provide stability of paralyzed limbs are often among the most difficult to design, develop, and apply. The desired degrees of freedom at each joint must be determined based upon the individual's impairment, physical condition, and desires. A thorough biomechanical analysis is required to determine the extent of functional performance of the limbs and orthoses. From the assessment, functional goals for the orthoses are developed, design and material selection is determined, fabrication proceeds, and eventually fitting and training is provided.

10.5.1. Ankle–foot orthoses (AFO)

Many people with lower-limb impairments can ambulate effectively with rigid AFOs. Typically a rigid AFO is constructed of molded plastic (e.g., translucent polypropylene). If the foot is rigidly placed in plantar flexion, the foot becomes flat on the floor earlier, creating a strong and long-lasting force at the knee which can be used to compensate for quadricep or knee extensor weakness. If the AFO is shaped to place the foot in dorsiflexion, the foot will take longer to be flat on the floor and the extension force magnitude and duration will be reduced. This makes it easier to bend the knee, which could create a less stable gait. Adding heels pads, or altering the stiffness of the orthosis will change the forces at the knee and will alter gait. A properly designed and appropriately applied AFO can significantly reduce the energy costs associated with ambulation.

Paralysis of the lower limbs requires extensive training and a suitable orthoses to accomplish reciprocal ambulation. Many mechanical orthoses (e.g., metal long and short leg braces) are considered heavy, bulky and cosmetically unappealing (figure 10.15). Some people persist in using orthoses despite their limitations. The ability to ambulate is considered to be important to self-esteem for some individuals. Standing and ambulating does provide some physiological conditioning and remodeling.

The Vannini–Rizzoli Stabilizing Limb Orthosis (VRSLO) is a polypropylene orthosis configured to position the lower portion of the leg for ambulation (figure 10.16). There are three models of the VRSLO (e.g., winter, summer, and elastic). The VRSLO was designed as an alternative to the standard mechanical orthosis. The polypropylene orthosis is inserted into a specially designed leather boot which extends to just below the knee. The VRSLO is targeted for people with lower-limb paralysis. It can be used for reciprocating gait ambulation.

Basically there are three approaches to providing reciprocating ambulation to people with lower-limb paralysis: purely mechanical orthoses (e.g.,

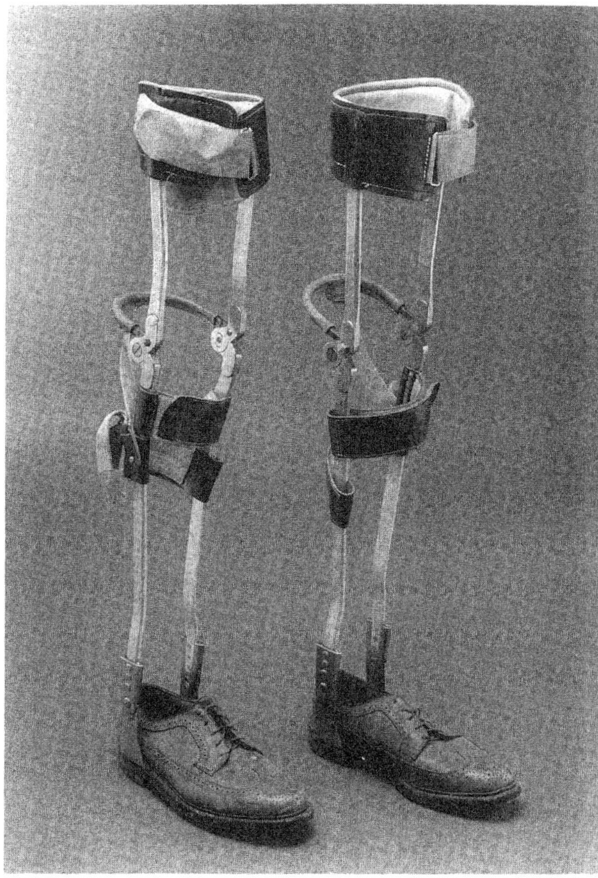

Figure 10.15. *Metal frame long leg braces with pressure activated knee lock. (Photograph by S Parsons, CSUS Media Services.)*

VRSLO, Lousiana State University Reciprocating Gait Orthosis); hybrid devices using functional neuromuscular stimulation in conjunction with mechanical bracing; and functional neuromuscular stimulation alone. All of these methods require substantial training, and are accompanied by high metabolic energy costs. Most mechanical orthoses are rejected by consumers. Functional neuromuscular stimulation systems, despite vigorous research, remain primarily experimental.

The VRSLO represents an attempt to develop a lightweight, simple and easy to use mechanical reciprocating gait orthosis. The result is an orthosis, which requires the assistance of an ambulation aid, that is simple to use, and requires reduced energy expenditure. The orthosis is easier to use with wheelchairs and personal transportation than other orthoses, and it is more cosmetically appealing. The approximate weight of a winter version is 4 pounds.

Figure 10.16. *Vannini–Rizzoli Stabilizing Orthoses made of molded plastic core and leather exterior. This orthosis is designed to be a lightweight ambulation and standing aid. (Photograph by S Parsons, CSUS Media Services.)*

The skeletal structure of the VRSLO comprises a flat, rigid sole with a posterior, polypropylene half-shell designed to partially enclose the lower leg from approximately 2 cm below the distal portion of the patella to the distal end of the toes. The skeletal structure of the boot is internally padded or lined for skin protection. The boot is fastened by a central lace and two zippers (one either side of the lace). Once the lace is properly adjusted, the zippers can be used while donning and doffing the orthosis. A Velcro closure system can be used as well.

The insole of the orthosis is angled to provide 10 to 15 degrees of plantar flexion. This places the center of gravity of the wearer anterior to the ankle joint. Holding the feet in this position stabilizes the knees when standing.

The individual can control static equilibrium by holding the head upright, and keeping the hips and knees extended. Static equilibrium is maintained during the fitting phase by placing wedges under the skeletal structure to obtain different degrees of plantar/dorsiflexion and pronation/supination. Ambulation is achieved by shifting the upper body slightly to the left or right and forwards. This causes the whole body center of gravity to move forwards. The person is supported by parallel bars, a walker, or platform canes then moves the unweighted foot forwards in a pendulum fashion.

10.5.2. Knee–ankle–foot orthoses (KAFO)

For many people with paralysis, long-term use of a KAFO for ambulation is typically not recommended. Energy costs associated with ambulation are heavily dependent upon the degree of impairment. If the individual has paralysis of the lower extremities and abdominal muscles energy cost during ambulation will be three to five times normal. If hip flexors/extensors and abdominal muscles are functional then the energy cost is dramatically reduced. For individuals who have some leg function, a KAFO may be an effective form of ambulation.

Stabilization of the hip, knee, and ankle during all phases of gait must be considered. A standard KAFO places the ankle in plantar flexion with very limited flexion and extensions in the sagittal plane, the knee is typically immobilized, and the hip is unrestricted. Some orthoses incorporate shoes, while others use a polypropylene insert which allows the user to change shoes. In some cases the hips are restricted to flexion/extension in the sagittal plane with a pelvic band or corset. With a standard KAFO, reciprocating gait is often unstable, and inefficient. Hence, people use a swing-through gait.

10.5.3. Biodynamics of orthosis ambulation

The gait cycle during bipedal locomotion can be divided into several categories. The start of the gait cycle is often defined by foot strike (i.e., when the foot strikes the floor). During normal walking the stance phase (i.e. the period where the lead foot is in contact with the floor) accounts for about 60% of the time for a complete gait cycle. For the remainder of the time the leg is in the swing phase. In many cases there is some time (about 25% of the total time in some cases) where both legs are in contact with the floor. This period is defined as the double support phase.

The energy cost and physical difficulty of using ambulation aids and orthotic devices are great for most individuals. People with paraplegia are sometimes effective ambulation aid and orthosis users. The swing through gait is preferred by many lower-limb impaired orthosis users. Most KAFOs needed by people with upper lumbar and thoracic level lesions are particularly cumbersome. Upright ambulation may help overcome some architectural and environmental barriers. However, ambulation is energy demanding, and may cause excessive wear on joint cartilage and joint capsules. The user may also be at risk for falls.

The metabolic cost for orthosis ambulation is near maximal (i.e., 80% of maximal aerobic capacity) for some users. The cause for the high-energy cost is not well understood. The high cost may be due to the mechanical work required to perform the activity or due to the poor physiological and biomechanical efficiency of the activity. Ambulation aid assisted gait may result in higher vertical fluctuations in the body center of mass than other activities. Excessive lateral crutch excursions, and the shock of crutch strike can also lead to higher energy costs. Using the arms to support the body, and the discontinuous nature of orthosis ambulation accounted for increased difficulty and energy demand relative to normal biped walking.

During the first phase of swing-through gait, muscular control of shoulder joint rotation contributes to increased ground clearance, control of body orientation, and timing control of the body-swing phase. Relatively high energy demands are required to stabilize the joints of the arms and shoulders during the swing phase, and for the active lifting of the toes to initiate the body-swing phase. Energy cost may be reduced by designing orthoses which address the factors of the body-swing phase.

10.6. FUNCTIONAL NEUROMUSCULAR STIMULATION (FNS)

Muscles can be stimulated either by electrodes placed on the skin surface (transcutaneous electrodes) or by electrodes positioned within the body near nerves inside muscles (i.e., intramuscular electrodes or percutaneous electrodes). Intramuscular stimulation is potentially more selective and can lead to finer control of muscle contraction. The use of intramuscular electrodes can be accomplished with an external or implanted stimulator. Surgery is required to implant intramuscular electrodes.

Electrical stimulation of paralyzed muscles using surface electrodes inherently results in a physiologically unnatural muscle activation. Surface activation is characterized by synchronous activation and inverse recruitment of motor units of the muscle. Stimulation by surface electrodes is physically non-invasive and is accepted in clinical practice—it will probably remain viable even when implantable electrodes are widely available. Due to the poor selectivity and reachability of muscle with transcutaneous electrodes, supramaximal stimulation is often necessary to obtain functional movement. Disturbances often cannot be compensated for due to the lack of recruitment modulation. The fast rate of fatigue reduces the effective functional output time with stimulation. This must be compensated for by controlling the burst time increment (figure 10.17).

Nerves and muscles can be stimulated by applying current to the target tissue. A sufficient amount of charge per unit area must be displaced across

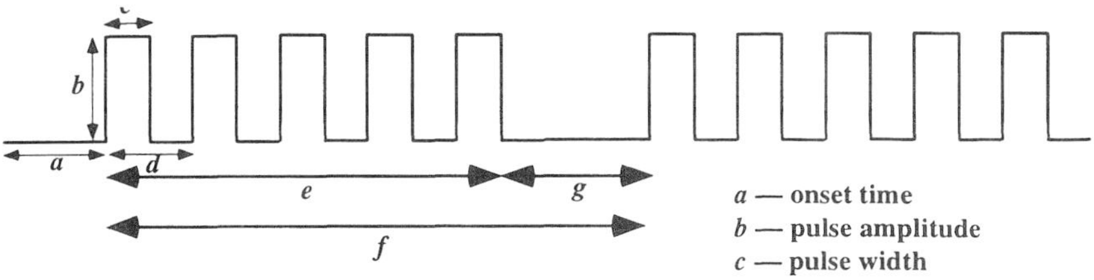

Figure 10.17. *Description of FNS signal.*

the cell membrane in order to raise the transmembrane potential to its firing threshold. The proximity of the electrodes to the target tissue determines the amount of current that will spread to the surrounding tissue and be reduced in current density. Typically 1–50 mA must be supplied to the current generator. Excitable tissue will respond to current pulses of different amplitude and time duration. Stimulus strength is related to current density. The stimulation current has a pulse repetition rate (frequency), pulse width (duration), and an amplitude. Stimulus strength is related to charge density (current $\times$ time/area) at the point of excitation. Charge densities must be within the range of 10–300 $\mathrm{mC\,cm^{-2}}$ to be effective. The pulsed waveform may be monophasic or biphasic. Biphasic pulses are used when charge build-up in the tissue is not negligible. Stimulation pulses commonly consist of a burst of a carrier signal to reduce potential pain experienced by the stimulated person. Nerve cell stimulation requires the establishment of sufficient current density in the target area for a length of time (100–600 μs) dictated by the tissue chronaxie. The pulse amplitude required to deliver the necessary charge depends on the size of the electrode and the degree of current spread in the tissue between electrodes and targets. Current generators must have the ability to output a regulated current between 0 and 60 mA, at a voltage of 0 to 180 V. Muscle stimulation is limited to 50 pps (pulses per second) to prevent tetany and rapid fatigue. Low duty-cycle waveforms help to maintain charge balance in the tissue and minimize risk of damage. A 10% duty cycle permits safe operation of a 100 pps for a duration of 1 ms.

The optimal stimulated muscle activation for generation of functional movements is critical for developing practical systems. Successful functional neuromuscular stimulation requires knowledge of each muscle's response to stimulation and muscle fatigue during electrical stimulation. External control is needed to perform the motor tasks desired by the user. Some FNS systems apply open loop control using event or time triggered stimulation patterns. External disturbances and muscle fatigue make this approach ineffective. Powerful automatic control systems which respond to simple user commands are required to provide the user maximal flexibility and

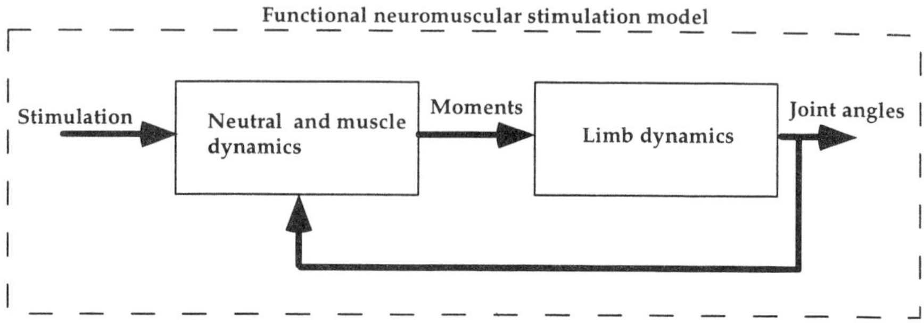

Figure 10.18. *Block diagram of FNS model.*

function. Non-linear and adaptive control systems are employed to yield smooth and reliable control. A problem which persists for FNS researchers is the difficulty in obtaining meaningful physiological feedback from the stimulated joint–muscle system. For motion control via FNS, models are required of the segments and joints, and the muscles. The relation between joint torques and artificially stimulated muscles is required. The effects of time delay between stimulation and force production, force–length relations, muscle fiber recruitment, and point of origin and insertion are important. Magnetic resonance imaging (MRI) is used to determine the geometry of the muscle–joint system. Models are used to predict muscle forces given stimulation profiles, and movements. Model-based control strategies prefilter the stimulation pattern to account for off-line estimated non-linearities in the muscle–limb system. Muscle fatigue may be compensated for by adjusting the stimulation from cycle to cycle on the basis of errors between the desired and identified model parameters for the motion dynamics. Increasing burst time can be used to compensate for fatigue.

The time delay between muscle stimulation and muscle activation is called the neutral dynamics. The muscle dynamics contain the non-linear relationship between input activation and generated joint torque. The joint torque also depends upon the current joint angle, and the joint angular velocity and acceleration. The limb dynamics contain all passive characteristics of the freely swing limb. Passive characteristics consist of mass, inertia, Coriolis properties of connected limbs, and damping, elasticity, and stiffness of involved joints (figure 10.18).

10.6.1. Upper-extremity FNS

Paralyzed hand and forearm muscles may be stimulated electrically via percutaneous or subcutaneous electrodes. Neuroprostheses enhance independence in activities of daily living (ADL) by giving users unimanual control of lateral (key grip) and palmar (three-jaw chuck) grasp and release. People without neurological impairments use palmar and lateral prehension to grasp 34 and 58% of objects, and 64 and 34% to hold objects, respectively. Assessing function is essential in determining the potential benefit of the neuroprosthesis. FNS has been used to provide grasp and release in people with quadriplegia. One grasp pattern can be achieved with a few percutaneous electrodes. To provide palmar and lateral grasp patterns more electrodes are required (i.e., up to 16 electrodes have been used). Proportional control has been obtained using residual voluntary movement. Shoulder and wrist motion have been used to provide control signals. Control is often obtained by using movement of the contralateral shoulder, using an external angle transducer and associated electronics. For example, elevating the shoulder can be used to close the hand by increasing flexor stimulation, and depressing the shoulder stimulates extensors to open the hand. Switching can also be used to activate the FNS system, and to select grasp patterns. An orthosis with a locking mechanism can be used to provide joint stability and maintain the grasp.

Implantable FNS neuroprostheses can be used to reduce maintenance demands, improve reliability, and improve function. The percutaneous electrode–skin interface is a source of considerable variability and maintenance. Elimination of external wiring reduces the risk of snagging on clothing or other external sources. Function is improved through improved electrode placement. The primary muscles for hand function are stimulated to provide functional grasping (see table 10.3). The intrinsic muscles of the fingers are excluded because precise electrode placement within these muscles is difficult, and a considerable number of electrodes would be required. Muscles must be assessed for peripheral nerve damage before stimulation. Muscles without sufficient peripheral nerve activity are excluded for FNS.

Several steps must be completed for each person to successfully implement an FNS hand grasp system. The electrode profile for each electrode–muscle combination must be determined. The electrode profile is described by the threshold, primary muscle force, gain, selectivity, direction, and length dependency. The threshold is the minimum stimulus required to elicit a visible muscle contraction. The threshold is a function of pulse duration, amplitude, and stimulus period. The muscle recruited at the lowest stimulus level is considered the primary muscle. Muscles recruited above this point are referred to as secondary muscles. The threshold for each digit to be stimulated must be determined. The primary muscle force is the maximum force generated by the primary muscle just below the activation level of the secondary muscles. The system gain for each digit is defined by the ratio of the change in force, or in some cases the change in position for the primary muscle, per change in stimulus. The percentage of the total force recruited by the primary muscle is referred to as the selectivity. The force of the primary muscle is determined at the maximum stimulation level just below activation of any secondary muscles. Selectivity is a measure of the isolation

Table 10.3. *Primary muscles for hand grasp function.*

Finger flexors	Flexor digitorum profundus Flexor digitorum superficialis
Finger extensors	Extensor digitorum communis Extensor indicus proprius
Thumb flexors	Adductor pollicis Flexor pollicis brevis Flexor pollicis longus
Thumb extensors	Extensor pollicis longus Extensor pollicis brevis
Thumb abductors	Abductor pollicis brevis Opponens pollicis

of a single anatomically defined muscle. Since digits are capable of moving in more than one plane, the direction of force application or movement is recorded. Muscles change length during contraction. Therefore, the change in muscle force, with a constant stimulus, as a function of forearm, wrist, and digit position is defined as the length dependency.

Restoration of functional control to the upper extremities of individuals with neurologically impaired upper extremities is an important goal for comprehensive rehabilitation. The synthesis of coordinated functional grasp–release patterns must provide reliable and repeatable movement, and be achieved using methods natural for the user. Hand function requires high selectivity, and the sensitivity of skin surface electrodes to placement may require the use of implanted electrodes. The number of electrodes must be optimized. Increasing the number of electrodes may provide finer control, but at the expense of higher complexity and decreased system reliability. Control algorithms and electrode placement must account for individual anatomy and physiology. With prolonged neurological impairment, some people develop hand deformities (e.g., limitations in range of motion, joint laxity, denervation of primary muscle groups, claw hand deformity) which require special consideration when developing hand grasp systems.

Surgical alterations can be performed to enhance the performance of FNS grasping. Tendon transfers, tenodesis of the long tendons of the fingers, Zancolli-lasso intrinsicplasty, and fusion of the interphalangeal joint of the thumb have been used to enhance grasping with FNS. Tendon transfers can be used to replace function lost due to lower motor neuron lesion of muscles required for grasping. The tendon of an unused intact muscle is transferred to the tendon of the denervated muscle required for grasping. Surgical procedures are used to reduce the number of actuator muscles and degrees of freedom required to control hand grasp.

10.6.2. Ambulation via FNS

Gait restoration in people with paraplegia or quadriplegia by means of electrical stimulation of paralyzed muscles is being investigated by many groups worldwide. Some of the most promising systems use FNS in combination with an orthosis, forming a hybrid system. Research activities in FNS-assisted gait include the areas of neurophysiology, orthosis design, automatic control systems, gait analysis, biomechanical modeling, and clinical implications. The electrical stimulation of paralyzed muscles can yield active generation of movement, and an orthosis supplies adequate support and restricts the number of degrees of freedom.

Muscle fatigue plays a significant role in the efficiency of FNS-assisted gait. Muscle fatigue increases with joint angular velocity and stimulation duty cycle. A mechanical orthosis can provide static support which is not easily accomplished by electrical stimulation because of muscle fatigue. An orthosis limits the number of degrees of freedom which can be controlled by active muscle contractions induced by electrical stimulation. A reciprocating gait orthosis can be used to restrict movement to hip movement in the sagittal plane. Hip extension in one leg can be coupled to hip flexion in the other leg. A reciprocating gait orthosis allows no knee or ankle flexion, which results in a rigid unnatural walking pattern. An integrated orthosis-stimulation system has the potential of improving gait in people with paralyzed limbs. A hybrid system which produces a smooth 'natural' gait has the potential to reduce the high energy costs and muscle fatigue associated with pure FNS or pure mechanical orthosis systems. The knee plays a pivotal role in the success of hybrid systems. Efficiency can be improved by allowing for knee flexion during the gait swing phase, and mechanically locking the knee during the stance phase. Limited ankle dorsi and plantar flexion help to smooth the gait as well.

External control is required to provide efficient gait. Stimulated muscle properties change over time and may also change state or phase. Fatigue is also an important consideration in control systems design. Stimulated muscles can produce limited contractile force, which is rate limited and has a finite effective time before fatigue sets in, making the contractile force insufficient to produce reliable motion. Lower extremity FNS provides for two basic tasks: standing and walking. Walking is a cyclical task with time-varying actuator properties; this alludes to possible control via adaptive control. In an adaptive control system the coefficients of the controller adapt to changes in the parameters of the system being controlled. As the stimulated muscles change their characteristics due to the stimulation pulse width, fatigue, contractile length, or contractile force, the controller would change as well. Adaptive optimal control requires knowledge of the system to be incorporated into the algorithm used to adapt the coefficients of the controller. Optimal control can be used to increase gait efficiency, as well as to reduce the risk of injury. Stimulated muscle may exhibit behavioral changes which require the controller to change form. In addition, the desired task (e.g., standing, walking, negotiating stairs) may require a different controller. A high-level finite-state controller can be used to compensate for these changes. Crude cyclic walking can be achieved using stimulation patterns which are event or time triggered in an open loop controller. However, they are sensitive to external disturbances and fatigue.

A hybrid system can be viewed as a system with active and passive components. The artificially stimulated muscles generate joint torques that result in movement. The orthotic components are passive and stabilize the joints. These two components must interact properly to produce smooth and efficient motion. Computer models can be used to study the interaction between the stimulated muscle, the orthosis, and the human limb. The stimulated muscle and the corresponding joint can be modelled using forward dynamics methods by specifying a joint torque and estimating the movement from the geometry and dynamics of the limb. The influence of the orthosis can be modeled as a set of constraints placed upon the joints. This method can be extended to determine the optimal controller for the system. For example, the stride length, and walking velocity could be specified, and a criterion such as energy cost could be minimized. With a cost function relating energy, stride length, walking velocity, joint force, and joint torque, the appropriate joint forces and torques could be determined which would minimize energy cost for the specified stride length and walking speed. The individual properties and abilities of the person (i.e., body mass, size, residual function) must be accounted for in the models.

Gait analysis (kinematics and kinetic) is used to evaluate the performance of walking with FNS systems. Standard kinematic and kinetic analysis tools (e.g., video taping and force plates) can be used with an inverse dynamics model to determine the biomechanical output of the joints (e.g., net joint moments, net joint forces). The inverse dynamic model must incorporate the effects of ambulation aids which are required for FNS ambulation. The net joint moment and net joint force values are compared with the simulations originally prepared to produce the motion. This information can be used to determine how the energy used in ambulation is distributed between the arms and legs. Currently, most of the energy used for forward propulsion comes from the arms and shoulders.

FNS to restore ambulation to people with paralyzed limbs lacks clinical acceptance due to poor reproducibility and reliability of position control, and high energy cost. Muscle fatigue, spasticity, inadequate timing of stimulation, variability in response of stimulated muscle, poor selectivity (especially with transcutaneous electrodes) and external disturbances contribute to the overall poor clinical performance of FNS ambulation systems.

10.7. AMBULATION AIDS

People with lower-limb pain, sensation, coordination, or weightbearing impairments may require an assistive device for safe walking and independent function (figure 10.19). Ambulation aids are used to reduce weightbearing on an impaired limb or to ameliorate limitations due to poor balance or motor control. Selection of the proper assistive device facilitates an efficient and functional gait. Assessment of the ambulation aid should begin with a functional assessment of the individual's upper- and lower-limb strength, function, and coordination. Clinicians should account for each patient's general physical and mental condition. A thorough assessment of the patient's aided and unaided gait is best for optimizing function. Force-plate (kinetic) and motion (kinematic) analyses can aid in the design and prescription of ambulation aids. However, several simple clinical assessment tools exist which are appropriate for many people. The assistive device (e.g., cane, crutch, walker) must be matched to the user's abilities, preferences, and lifestyle. Canes, crutches, and walkers can support between 25 and 50% of the individual's body weight. Canes, typically, should not be used to support in excess of 25% of a person's body weight. Construction durability and price are also important considerations.

Ambulation aids should be designed to require minimum maintenance. Fasteners and components should be kept to a minimum. Materials should be durable, and resist burning. Many ambulation aids are constructed of wood or aluminum. Handles on aluminum canes, crutches, and walkers should be made of a pliable thermal insulator, because aluminum is a good conductor of heat. Tips, handles, and other parts that wear must be replaced regularly to ensure safe use. Worn, damaged or missing tips are related to most incidents involving ambulation aids. Folding parts and locking mechanisms should operate smoothly and with little force. Ambulation aids are typically simple mechanical devices.

10.7.1. Canes

Canes consist of a handle, shaft, base, and tip. Canes are designed to provide additional contact with the ground, and to alter walking biomechanics in a way which relieves painful joints or provides additional balance. There are several types of canes. Single-point or straight canes have a single point in contact with the ground and may be adjustable or cut to length. Large or broad base canes, also known as tripod or quad canes, have three or four feet attached near the end of the canes which provide a base of support. Side walkers are folding-frame cane devices designed to be held by one hand. Canes handles may be either straight, molded, or curved.

Single-point canes are not designed to relieve weightbearing from the legs. Canes may be used singly or in pairs depending upon the person's needs for

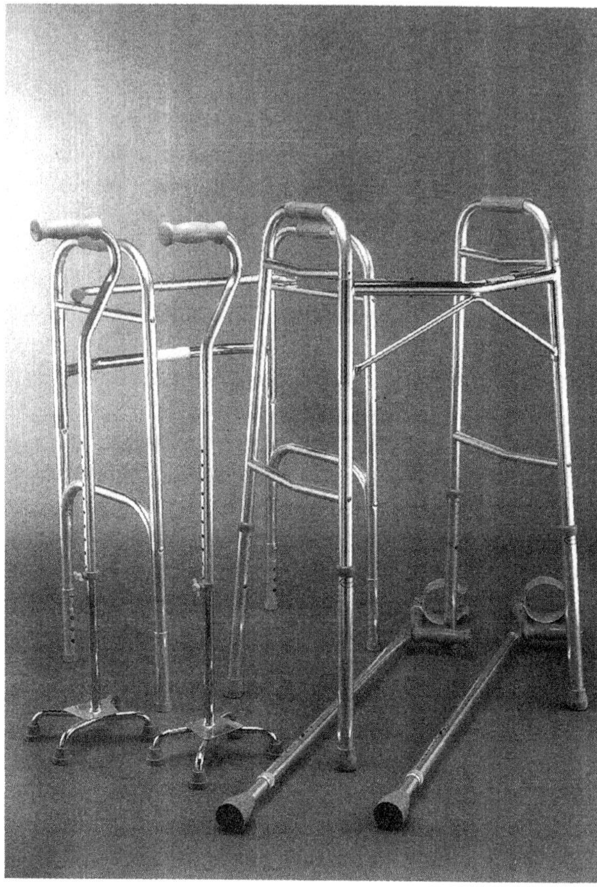

Figure 10.19. *Ambulation aids used in conjunction with prosthetics and orthotics. (Photograph by S Parsons, CSUS Media Center.)*

stabilization and support. Single canes are used when balance or strength are mildly impaired, or to relieve weight on one leg to alleviate pain. If a single cane is used it should be held opposite of the impaired leg. Single-point canes provide less support, but tend to interfere less with the feet. Side walkers or broad base canes may be used for weightbearing support. Many people with hemiparesis choose these types of canes. Interference with the feet can be minimized by angling the canes legs away from the person's body.

A problem which sometimes occurs when using canes is that people tend to lean towards the cane which alters their upright posture. The grip of the canes should be designed to place the hand and wrist in a near neutral position during the gait cycle. The elbow should be flexed at 20 to 30 degrees while standing with the shoulders level.

10.7.2. Crutches

Crutches are more suitable for active people. They provide good support on uneven surfaces, or while climbing stairs. Crutches come in various lengths and configurations (e.g., tips, pads). Crutches are either axillary, platform, or forearm (also known as Lofstrand crutches). Canadian crutches are a type of forearm crutch with a hinged wrist cuff placed directly above the handgrip.

Axillary crutches use an axillary pad placed between the chest wall and the arm. They work well for people who have limited arm strength. However, caution should be observed not to place excessive weight on the axillary pad as this may occlude circulation. Typically, the axillary pad should be 50 to 100 mm below the armpit. Forearm crutches use a wrist cuff angled with respect to the hand grip. The wrist cuff provides a weightbearing surface, and helps to hold the crutch when the hands are off the grips. Forearm crutches can be made lightweight and streamline. Some forearm crutches use a telescoping shaft. Additional stability, at the cost of weight bearing, can be achieved by adding a four-legged base via a hinge to the end of the forearm crutch. Platform crutches use a padded platform or trough to support the forearms. Leather straps are used to provide additional support and hold the crutch to the person's arm. Platform crutches effectively provide stability and weight while minimizing the stress on the individual's hands and wrists. Donning and doffing platform crutches is more difficult than with other types of crutches.

10.7.3. Walkers

Walkers provide a more stable base of support than other ambulation aids, but they are less versatile. Walkers work well on flat, level surfaces, but may present a problem on uneven surfaces or in narrow passages (e.g., doorways). Walkers can remain upright without the assistance of the user, and can support greater weight than other devices. Another advantage of walkers is that there are several attachments available for supporting or carrying items. Seats may be added to provide a resting place for the person, and bags or trays can added to carry personal belongings (e.g., eyeglasses, purse, keys).

Walkers may be equipped with wheels so that they may be rolled instead of lifted. Most walkers with wheels incorporate either wheel locks (i.e., parking brakes which are either fully locked or fully unlocked) or brakes (i.e., a device which provides resistance to rolling that is usually operated by a lever through a cable). Brakes may require good grip strength to be operated effectively. Articulated walkers are designed to allow each side of the walker to be moved forward independently thus permitting reciprocating gait. Reciprocating gait (articulated) walkers yield a more normal gait pattern but do not provide as much stability or support as much weight as other types of walkers.

10.8. AIDS TO DAILY LIVING

Most aids to daily living address specific problems of restricted manipulation or mobility related to a single task. Assistive devices have been developed for assisting with performance of many activities of daily living (ADL). For people with upper-extremity impairments most aids to daily living require specialized handles. Several handle modifications are used (e.g., extended, expanded, slip-on grip, swivel, suction, offset, strapped). Residual function and activity determine the appropriate handle.

Eating utensils may be modified or purchased to include specialized handles. Utensils are modified for limited finger dexterity, lack of reach, and for single-handed use. Swivel or angled handles may help people with poor fine motor control. Combination knife–fork or fork–spoon utensils can help with single-handed eating. Food preparation can also be simplified with some simple tools. Rotary or electric knives make cutting food easier. Plates and cutting boards can be modified or purchased to ease food holding. Electric appliances can often be easily modified to accommodate people with limited hand function.

Dressing can be simplified through the use of devices to extend reach and improve grasp. Many varieties of buttoners can be used for one-handed shirt or blouse buttoning. A buttoner basically passes through the buttonhole and loops around the button. Once the button is in its grasp the individual twists or pulls it through the hole. Once in place, pushing on the handle causes the buttoner to release the button. Many simple devices can be purchased or made to aid in donning/doffing socks, shoes, and zippered clothing items.

A person with a disability can gain considerable independence with simple grooming aids. Many devices (e.g., tooth brushes, hair brushes, combs, electric razors) can be purchased or simply modified to require limited or no hand dexterity. Most of these modifications are inexpensive, and often use commonly available items.

FURTHER READING AND REFERENCES

Abbas J J and Chizeck H J 1991 Feedback control of coronal plane hip angle paraplegic subjects using functional neuromuscular stimulation *IEEE Trans. Biomed. Eng.* **BME-38** 687–98

Abul-Haj C J and Hogan N 1990 Functional assessment of control systems for cybernetic elbow prostheses - part I: description of the technique *IEEE Trans. Biomed. Eng.* **BME-37** 1025–36

Abul-Haj C J and Hogan N 1990 Functional assessment of control systems for cybernetic elbow prostheses - part II: application of the technique *IEEE Trans. Biomed. Eng.* **BME-37** 1037–47

Andrews B 1993 Hybrid FES orthosis for paraplegic locomotion *Proc. RESNA 16th Ann. Conf. (Las Vegas, NV, 1993)* pp 401–3

Anglin C R, Gosine R G and Hershler C 1992 Kinematic analysis of powered upper-extremity orthoses *Proc. RESNA Int. '92 (Toronto, Ont., 1992)* pp 105–7

Burgess E M and Rappoport A 1993 *Physical Fitness: A Guide for Individuals with Lower Limb Loss. A Clincal Guide* (Washington DC: Rehabilitation Research and Development Service, US Department of Veterans Affairs)

Butler P B, Farmer I R, Abist R P and Patrick J H 1982 Use of the Orlau swivel walker for the severely handicapped patient *Physiotherapy* **68** 324–6

Caldwell R R, Buck C S, Lovely D F and Scott R N 1985 A myoelectric b/e prosthetic system for young children *Proc. RESNA 8th Ann. Conf. (Memphis, TN, 1985)* pp 69–70

Chia T L, Chow P C and Chizeck H J 1991 Recursive parameter identification of constrained systems: an application to electrically stimulated muscle *IEEE Trans. Biomed. Eng.* **BME-38** 429–42

Chizeck H J, Lalonde R, Chang C W, Rosenthal J A and Marsolais E B 1985 Performance of a closed loop controller for electrical-stimulated standing in paralyzed patients *Proc. RESNA 8th Ann. Conf. (Memphis, TN, 1985)* pp 231–3

Cook A M and Webster J G 1982 *Therapeutic Medical Devices: Application and Design* (Englewood Cliffs, NJ: Prentice-Hall)

Czerniecki J M, Gotter A and Munro C 1991 Joint moment and muscle power output characteristics of below knee amputees during running: the influence of energy storing prothetic feet *J. Biomech.* **24** 63–75

Douglas R, Larson P F, D'Ambrosia R and McCall R E 1983 The LSU reciprocation-gait orthoses *Orthopedics* **6** pp 834–9

Durfee W K, Chiou J T and Beck H L 1993 Dynamic musculo-skeletal model for FES-aided gait applications *Proc. IEEE-EMBS 14th Int. Conf. (Paris, 1993)* vol 14 pp 1335–6

Flowers W, Cullen C and Tyra K P 1990 A preliminary report on the use of a practical biofeedback device for gait training of above-knee amputees *J. Rehab. Res. Dev.* **23** 7–18

Franken H, Vletnik P, Baardman G, van Alste J and Boom H B K 1993 Experimental control of the swing phase of paraplegic gait induced by surface electrical stimulation *Proc. 2nd Eur. Conf. on the Advancement of Rehabilitation Technology (Stockholm, 1993)* pp 16.2

Gage J R and Ounpuu S 1989 Gait analysis in clinical practice *Seminars in Orthopedics* **4** 72–87

Gottschalk F, McClellan B, Carlton A and Mooney V 1985 Early fitting of the amputee with a plastic temporary adjustable below-knee prosthesis *Proc. RESNA 8th Ann. Conf. (Memphis, TN, 1983)* pp 373–5

Hall J, Clarke A K and Harrison R 1990 Guidelines for prescription of walking frames *Physiotherapy* **76** 118–20

Hausdorff J M and Durfee W K 1991 Open loop position control of the knee joint using electrical stimulation of the quadriceps and hamstrings *Med. Biol. Eng. Comput.* **BME-29** 269–80

Haykin S 1986 *Adaptive Filter Theory* (Englewood Cliffs, NJ: Prentice-Hall)

Heckathorne C W and Uellendahl J 1990 Socket design for a prototype EPP shoulder disarticulation prosthesis *Proc. RESNA 13th Ann. Conf. (Washington, DC, 1990)* pp 232–3

Hillstrom H J and Moskowitz G D 1993 Robust intent recognition for prosthesis control *Proc. IEEE-EMBS 14th Int. Conf. (Paris, 1993)* vol 14 No 4 pp 1448–9

Hortensius P, Quanbury A and Onyshko S 1985 A microcomputer based myoelectric limb controller *Proc. RESNA 8th Ann. Conf., (Memphis, TN, 1985)* pp 353–5

Irvine B 1990 A physical therapist's perspective on canes, crutches, and walkers *REquest Product Comparison and Evaluation: Canes, Crutches, and Walkers* (The Rehabilitation Engineering Center at The National Rehabilitation Hospital) pp 2–4

Isakov E, Mizrahi J and Najenson T 1986 Biomechanical and physiological evaluation of FES-activated paraplegic patients *J. Rehab. Res. Dev.* **23** 9–19

Ito K, Tsuji T, Kato A and Ito M 1993 An EMG controlled prosthetic forearm in three degrees of freedom using ultrasonic motors *Proc. IEEE-EMBS 14th Int. Conf. (Paris, 1993)* vol 14 No 4 pp 1487–1488

James W V 1991 Principles of limb fitting and prostheses *Ann. R. Coll. Surg. England* **73** 158–62

Jin Z and Chizeck H J 1992 Instrumentation parallel bars for three-dimensional force measurement *J. Rehab. Res. Dev.* **29** 31–8

Johnson M W and Peckham H. 1990 Evaluation of shoulder movement as a command control source *IEEE Trans. Biomed. Eng.* **BME-37** 876–85

Kabra S G 1990 Articulated cadaveric bones as a structural endoskeleton in an ankle–foot prosthesis: a preliminary report *J. Rehab. Res. Dev.* **27** 43–52

Kabra S G and Narayanan R 1991 Ankle-foot prosthesis with articulated human bone endoskeleton: force-deflection and fatigue study *J. Rehab. Res. Dev.* **28** 13–22

Keller A D, Taylor C L and Zahn V 1947 Studies to determine the functional requirements for hand and arm prosthesis *Final Report National Academy Science* Contract No VAm-21223

Kilgore K L, Peckham P H, Keith M W and Thorpe G B 1993 An implanted upper-extremity neuroprosthesis *Proc. RESNA 16th Ann. Conf. (Las Vegas, NV, 1993)* pp 404–6

Kilgore K L, Peckham P H, Thorpe G B, Keith M W and Gallaher-Stone K A 1989 Synthesis of hand grasp using functional neuromuscular stimulation *IEEE Trans. Biomed. Eng.* **BME-36** 761–70

Kostov A, Stein R B, Armstrong W W and Thomas M 1993 Evaluation of adaptive logic networks for control of walking in paralyzed patients *Proc. IEEE-EMBS 14th Int. Conf. (Paris, 1993)* vol 14 No 4 pp 1332–4

Krouskop T A, Goode B L, Doughtery D R and Hemmen E H 1985 Predicting the loaded shape of an amputee's residual limb *Proc. RESNA 8th Ann. Conf. (Memphis, TN, 1985)* pp 225–227

Krouskop T A, Muilenberg A L, Doughtery D R and Winningham D J 1987 Computer-aided design of a prosthetic socket for an above-knee amputee *J. Rehab. Res. Dev.* **24** 31–8

LeBlanc M A 1985 Evaluation of cable vs. hydraulic transmission of forces for body-powered arm prostheses *Proc. RESNA 8th Ann. Conf. (Memphis, TN, 1985)* pp 71–75

LeBlanc M A, Carlson L E and Nauenberg T 1992 Design and testing of four underarm crutches *Proc. RESNA Int. '92 (Toronto, Ont., 1992)* pp 457–459

Liang D H, Kovacs G T A, Storment C W and White R L 1991 A method for evaluating the selectivity of electrodes implanted for nerve simulation *IEEE Trans. Biomed. Eng.* **BME-38** 443–9

Lin D C and Rymer W Z 1991 A quantitative analysis of pendular motion of the lower leg in spastic human subjects *IEEE Trans. Biomed. Eng.* **BME-38** 906–18

Ljung L 1987 *System Identification: Theory for the User* (Englewood Cliffs, NJ: Prentice-Hall)

Lyles M and Munday J 1992 Report on the evaluation of the Vannini–Rizzoli Stabilizing Limb Orthosis *J. Rehab. Res. Dev.* **29** 77–104

Marsolais E B and Kobetic R 1985 Kinematics of paraplegic gait produced by electrical stimulation *Proc. RESNA 8th Ann. Conf. (Memphis, TN, 1985)* pp 108–10

McDonnell P M, Scott R N, Dickison J, Theriault R A and Wood B 1989 Do artificial limbs become part of the user? New evidence *J. Rehab. Res. Dev.* **26** 17–24

Meadows P M, and Campbell J M 1993 Functional neuromuscular stimulation: at the crossroads *Proc. RESNA 16th Ann. Conf. (Las Vegas, NV, 1993)* pp 386–388

Michael J W 1989 Reflections on CAD-CAM in prosthetics and orthotics *J. Prosthet. Orthot.* **1** 116–21

Mohan D, Sethi P K and Ravi R 1992 Mathematical modelling and field trials of an inexpensive endoskeletal above-knee prosthesis *J. Prosthet. Orthot.* **16** 118–23

Moleshi K A, Solomonidis S E and Paul J P 1992 Development of a 3-dimensional measurement system for measuring alignment of artificial limbs *Proc. RESNA Int. '92 (Toronto, Ont., 1992)* p 347

Moran P, Khomeshwari A, Zhu H, Wertsch J J and G F Harris 1993 A microprocessor-based biofeedback cane system *Proc. IEEE-EMBS 14th Int. Conf. (Paris, 1993)* vol 14 No 4 pp 1542–1543

Morin E, Parker P A and Scott R N 1993 Operator error in a level coded myoelectric control channel *IEEE Trans. Biomed. Eng.* **BME-40** 558–62

Nannini N and Horch K 1991 Muscle recruitment with intrafascicular electrodes *IEEE Trans. Biomed. Eng.* **BME-38** 769–76

Nava L C 1985 Design and use of improved walking aids *J. Biomed. Eng.* **BME-7** 329–33

Nissan M 1991 The initiation of gait in lower-limb amputees: some related data *J. Rehab. Res. Dev.* **28** 1–12

Oshima T and Saito Y 1985 CAD/CAM System applied to the foot shape for the prosthetic device *Proc. RESNA 8th Ann. Conf. (Memphis, TN, 1985)* pp 222–4

Patterson P and Katz J A 1992 Design and evaluation of a sensory feedback system that provides grasping pressure in a myoelectric hand *J. Rehab. Res. Dev.* **29** 1–8

Peckham P H, Keith M W and Freehafer A A 1988 Restoration of functional control by electrical stimulation in the upper extremity of the quadriplegic patient *J. Bone Joint Surg.* **70A** 144–8

Pettit D J, Robertson R N and Cooper R A 1992 Development of a dynamometer for measuring dynamic hand function *Proc. NACOB II: The Second North American Congress on Biomechanics (Chicago, Il, 1992)*

Pinzur M S, Perona P, Patwardhan A and Havey R 1991 Loading of the contralateral foot in peripheral vascular insufficiency below-knee amputees *Foot Ankle* **11** 368–71

Popovic D, Oguztoreli M N and Stein R B 1991 Optimal control for the active above-knee prosthesis *Ann. Biomed. Eng.* **BME-19** 131–50

Popovic D, Tomovic R and Schwirtlich L 1989 Hybrid assistive system—the motor neuroprosthesis *IEEE Trans. Biomed. Eng.* **BME-36** 729–37

Popovic D B and Kalanovic V D 1993 Output space tracking control for above-knee prosthesis *IEEE Trans. Biomed. Eng.* **40** 549–57

Prochazka A 1993 Comparison of natural and artificial control of movement *IEEE Trans. Rehab. Eng.* **BME-1** 7–17

Prosthetic-Orthotic Resource Directory 1993 (Resource Unit for Information and Education, Northwestern University Rehabilitation Engineering Program)

REquest 1990 Canes, Crutches and Walkers Product comparison and evaluation (REquest, Rehabilitation Engineering Center, National Rehabilitation Hospital)

Rovick J S and Childress D S 1988 Pendular model of paraplegic swing-through crutch ambulation *J. Rehab. Res. Dev.* **25** 1–16

Sanders J E, Boone D A and Daly C H 1990 The residual limb/prosthetic socket interface: normal stress and shear stress *Proc. RESNA 13th Ann. Conf. (Washington, DC, 1990)* pp 234–235

Sauter W F 1991 The use of electric elbows in the rehabilitation of children with upper limb deficiencies *Prosthetics and Orthotics Int.* **15** 93–5

Schneider K, Hart T, Zernicke R F, Setoguchsi Y and Oppenheim W 1993 Dynamics of below-knee child amputee gait: SACH foot versus flex foot *J. Biomech.* **26** 1191–204

Shiavi R, Bugle H J and Limbird T 1987 Electromyographic gait assessment, part 1: adult EMG profiles and walking speed *J. Rehab. Res. Dev.* **24** 13–23

Siegel J D 1990 A physician's perspective on prescription of ambulation aids *REquest Product Comparison and Evaluation: Canes, Crutches, and Walkers* (The Rehabilitation Engineering Center at The National Rehabilitation Hospital, Washington, DC) pp 5–7

Skinner H B and Effeney D J 1985 Gait analysis in amputees *Am. J. Phys. Med.* **64** 82–9

Solomonow M, Baratta R, Harris M and D'Ambrosia R 1993 Evaluation of adaptive logic networks for control of walking in paralyzed patients *Proc. IEEE-EMBS 14th Int. Conf. (Paris, 1993)* vol 14 No 4 pp 1347–8

Snell F 1992 Bringing prosthetics into the 21st century *J. Arkansas Med. Soc.* **89** 337–8

Stallard J, Major R E and Patrick J H 1989 A review of the fundamental design problems of providing ambulation for paraplegic patients *Paraplegia* **27** 70–5

Suzuki K 1972 Force plate study on the artificial limb gait *J. Japan. Orthop. Assoc.* **46** 503–16

Thorpe G B, Peckham P H and Crago P E 1985 A computer-controlled multichannel stimulation system for laboratory use in functional neuromuscular stimulation *IEEE Trans. Biomed. Eng.* **BME-32** 363–70

Topper A K and Fernie G R 1990 An evaluation of computer-aided design of below-knee prosthetic sockets *Prosthet. Orthot. Int.* **14** 136–42

Torburn L, Perry J, Ayyappa E and Shanfield S 1990 Below-knee amputee gait with dynamic elastic response prosthetic feet: a pilot study *J. Rehab. Res. Dev.* **27** 369–84

Torres-Moreno R, Morrison J B, Cooper D, Saunders C G and Foort J 1992 A computer-aided socket design procedure for above-knee prostheses *J. Rehab. Res. Dev.* **29** 35–44

Uellendahl J, Heckathorne C W, Krick H, Duff S 1990 Four-function hybrid arm prosthesis incorporating an electric wrist rotator and prototype electric prehensor *Proc. RESNA 13th Ann. Conf. (Washington, DC, 1990)* pp 167–8

van Alste J A, Boom H B K, Grootenboer H J, Hermens H J, Koopman H F J M, Veltnik P H and Zilvold G 1993 A hybrid orthosis approach in walking restoration for paraplegics *Proc. 2nd Eur. Conf. Advancement Rehabilitation Technol. (Stockholm, 1993)* pp 16.1

van de Belt D, Grootenboer H J and Koopman H F J M 1993 Simulation and evaluation of FES induced walking *Proc. IEEE-EMBS 14th Int. Conf. (Paris, 1993)* vol 14 No 4 pp 1345–6

Walker P S, Kurosawa H, Rovick J S and Zimmerman R A 1985 External knee joint design based on normal motion *J. Rehab. Res. Dev.* **22** 9–22

Wilson J F and Gilbert J A 1982 Dynamic body forces on axillary crutch walkers during swing-through gait *Am. J. Phys. Med.* **61** 85–92

Wirta R, Goldbranson F, Mason R and Calvo K 1990 Analysis of below-knee suspension systems: effect on gait *J. Rehab. Res. Dev.* **27** 385–96

Wirta R, Mason R, Calvo K and Goldbranson F L 1991 Effect on gait using various prosthetic ankle-foot devices *J. Rehab. Res. Dev.* **28** 13–24

Xu Y, Hunter I W, Hollerbach J and Bennett D J 1991 An airjet actuator system for identification of the human arm joint mechanical properties *IEEE Trans. Biomed. Eng.* **BME-38** 1111–22

Yamaguchi G T and Zajac F E 1990 Restoring unassisted natural gait to paraplegics via functional neuromuscular stimulation: a computer simulation study *IEEE Trans. Biomed. Eng.* **37** 886–902

Zhang Y T, Parker P A, Herzog W, Hulliger M and Guimaraes A 1992 Distributed random electrical neuro-muscular stimulation: effects of inter-stimulus interval statistics on EMG spectrum *Proc. RESNA Int. '92 (Toronto, Ont., 1992)* pp 235–7

CHAPTER 11

RECREATIONAL DEVICES AND VEHICLES

11.1. INTRODUCTION

People with disabilities participate in nearly every recreational activity that exists. Some activities require specialized adaptions to equipment, whereas others do not. Recreation is an important aspect of people's lives. Good recreation habits can lead to a fuller and healthier life. People participate in a variety of recreational activities from gardening to sky-diving. Recreation activities change over a person's life span as well. It is important for people with disabilities to learn healthy recreation activities during rehabilitation.

Shortly after World War II, Sir Ludwig Guttmann and his colleagues originated wheelchair sports as a rehabilitation tool at Stoke Mandeville Hospital in England. This developed out of the need to provide exercise and recreational outlets for the large number of young people recently injured in the war. News of Dr Guttmann's success with the rehabilitation of his patients through the use of sports soon spread through Europe and to the United States. In 1948, he organized 'Games' for British veterans with disabilities. In 1952, the Games developed into the first international wheelchair sporting competition for people with physical disabilities, with participants from the Netherlands, The Federal Republic of Germany, Sweden, Norway, and Israel. During this event, the International Stoke Mandeville Games Federation (ISMGF) was formed to govern and develop wheelchair sporting competitions. The ISMGF established ties to the International Olympic Committee (IOC), thus expanding the scope of wheelchair sports. As the international sports movement for people with disabilities grew and the number of international multi-disability events increased, the ISMGF was expanded to include all wheelchair sporting events. After the reorganization, the ISMGF was renamed the International Stoke Mandeville Wheelchair Sports Federation (ISMWSF).

The first international games for the disabled held in conjunction with the Olympic Games took place in 1960 in Rome, Italy. The name 'Paralympics' was coined during the 1964 Tokyo games and, as such, was subsequently held every four years.

11.2. RACING WHEELCHAIRS

In the early years of wheelchair racing, participants used bulky standard (depot type) wheelchairs and did not compete in events with distances over 200 m. In the 1970s, athletes started to modify their wheelchairs for specific sports, and began to take an interest in road racing.

In 1975, a young man with paraplegia became the first person to compete in the Boston Athletic Association Marathon in a wheelchair. This opened the door for many future road racers, prompting Dr Caibre McCann, a leading physician for the ISMGF, to say: 'Running is natural, but propelling yourself in a wheelchair is an unnatural phenomenon. People never realize what a wheelchair athlete is capable of. This is a breakthrough in man's limits.' Within a few years, several recognized US road races initiated wheelchair divisions and more people with disabilities began to train for these races than had ever been anticipated. In 1976, the ISMGF started to coordinate with other international sports organizations to launch a unified international sports movement for people with disabilities.

Racing wheelchairs began to evolve as special-purpose pieces of equipment easily distinguishable from everyday wheelchairs. Distances on the track were extended to include races up to 1500 m, and during this time the mile record was dropped to below 5 minutes.

The early 1980s saw the development of more sophisticated racing wheelchairs and training techniques. By 1985, most racing wheelchairs no longer had any components in common with everyday wheelchairs (which had also improved dramatically), and George Murray became the first wheelchair racer to break the 4 minute mile. In the years that followed, wheelchair racing continued to progress with improved equipment, training, and nutrition; consequently, world records were continuously being broken. Wheelchair racing began the path towards recognition as a legitimate Olympic sport in 1984 when the men's 1500 m and the women's 800 m wheelchair races were included as demonstration events in the Olympic Games held in Los Angeles, CA.

The general wheelchair user has also benefited from the advances brought about in wheelchair design by racing. The lightweight and superior maneuverability of the sports wheelchair, once only used by athletes, is rapidly making it the wheelchair of choice for nearly all wheelchair users.

Some personal considerations must be made when specifying or designing a racing wheelchair. Athletes with paraplegia, quadriplegia, or amputations have different needs, and each person has his or her own anatomical structure.

Many races require brakes, and they make training safer. Brakes can be easily accessible from a comfortable position. The brake levers must be long enough and at the proper angle so that sufficient leverage can be applied to stop the racing chair. This is particularly important for people with quadriplegia.

It is important that a racing chair fits the user properly. Racing chairs are similar to shoes, in that a poorly fitted chair can be uncomfortable and awkward. A racing chair should fit as closely as possible without causing discomfort or pressure sores. Most manufacturers ask for a number of anatomical measurements when a chair is ordered. The most common measurements requested are hip width, chest width, upper-leg length, arm length, trunk length, height, and weight.

11.2.1. Racing wheelchair design restrictions

Wheelchair racing, like any other sport, has rules to define competition. The purpose of wheelchair racing is to cover a predesignated course in a minimum amount of time, and to finish ahead of the other competitors in the process. There are rules governing the conduct of the competitors and design of the equipment.

The International Stoke Mandeville Wheelchair Sports Federation specifies constraints on equipment:

- The wheelchair shall have at least three wheels.
- The diameter of the two larger wheels, including the inflated tire, shall not exceed 70 cm. The other wheels shall not exceed 50 cm in diameter including the inflated tire.
- Only one pushrim shall be attached to each of the drive wheels.
- The widest measurement of the wheelchair shall be a straight line measured from the outside edge of the pushrim or the tire, whichever is greater at the widest point. No part of the wheelchair may protrude beyond the width of this measurement.
- Propulsion and direct steering shall be made by the hands and arms only.
- No gears, levers, chains or any other mechanical devices may be used in propelling the wheelchair.
- There shall be no device as part of the wheelchair or the athlete that functions for the primary purpose of reducing air resistance.
- Athletes must ensure that no part of their lower limbs can fall to the ground during an event.
- An athlete's wheelchair must be structurally sound.
- An athlete must be securely seated in the wheelchair and will remain so while racing.

These rules have been developed to make competition fair and equitable. The rules allow a wide variety of designs to be used and thus are not overly prohibitive. The racing wheelchair must be designed to meet the specifications of the rules and yet assist the racer in completing the course as rapidly as possible in a safe manner.

11.2.2. Characteristics of racing wheelchairs

There are a variety of different designs for racing wheelchairs; however, most have some common characteristics. Racing wheelchair frames are constructed of aluminum (6061T6), chromolly steel (4130 or 4140), or titanium. Racing wheelchairs may be equipped with three or four wheels. In the infancy of wheelchair racing, most racing wheelchairs were designed with four wheels. However, most wheelchairs today use three wheels because of changes in rules and design innovations.

Athletes sit near rear axle height. Four-wheeled wheelchairs commonly employ a rectangular main frame with a cross-member immediately fore and aft of the seat. Three-wheeled wheelchairs are similarly constructed using a triangular (as viewed from above) main frame. The rear wheels are mounted rigidly to the frame by a threaded insert welded either through the side frame tubes or into a cross-member. The rear wheel inserts are aligned such that there is no toe in/out and that there is between 2 and 15 degrees of wheel camber. This helps to make the chair more stable and allows the athlete to reach the bottom of the pushrims without hitting the top of the wheels or pushrim. Toe in/out causes problems for wheelchair racers as it can cause a significant increase in rolling resistance if not aligned properly and toe in/out may change with use, therefore some manufacturers incorporate an alignment mechanism into the frame. Three methods are commonly used: a cross-member is fixed to one side of the side frame via a bolt and spacers are used to lengthen or shorten the cross-member; hence aligning the rear wheels; a similar method employs a 'turn-buckle' mechanism which can be expanded or contracted to align the rear wheels; another method uses an angled cross-member with the rear axle insert pressed into it. Rotating the cross-member is used to align the rear wheels.

The seat cage is welded on top of the main frame, and serves the purpose of supporting the athlete. The seat cage should fit snugly around the athlete. Some elite racers use racing wheelchairs which incorporate seat and leg support as an integral unit. Elite athletes usually prefer a tighter fitting seat cage than do novice athletes because they are more accustomed to pushing a racing wheelchair. The seat and leg support should hold the athlete solidly, so that his or her energy can be focused upon propelling the wheelchair, and not on maintaining balance. A seat cage also offers the athlete greater control over the racing chair than does a conventional seat and provides some protection in the event of an upset. Seat cages have side panels to provide support and to prevent the athlete from rubbing against the wheels. Well designed side panels follow the curvature of the wheel, and allow the athlete a large range of motion fore and aft. Most seat upholstery is made

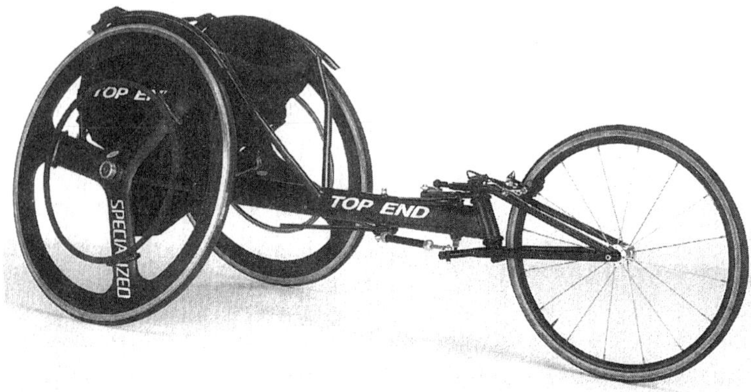

Figure 11.1. *Aluminum racing wheelchair with tiller steering front fork and spring compensator mechanism. The seat is designed for rider to be positioned in a kneeling posture. Rear wheels are tri-spoke design constructed with composite materials to reduce wind resistance. (Photograph courtesy of TOP END by action.)*

of nylon or cotton canvas slung from the seat cage. Some racing chairs use plastic or fiberglass seats. Athletes typically use low-profile foam or air-flotation cushions.

Racing wheelchair frames are designed to be stiff and lightweight. The typical racing wheelchair frame weighs less than five pounds, excluding the wheels. The frame and seat cage are made to fit each individual, and cater for different disability etiologies and levels. The location of the rear axles with respect to the seat cage is dependent upon experience and disability etiology. Experienced athletes with paraplegia prefer between 15 to 25 cm from the seat back to the rear axles inserts. Experienced athletes with quadriplegia prefer between 5 and 20 cm from the seat back to the rear axle inserts. More stable configurations are recommended for novice athletes. The seat cage upholstery adjustment and rear axle's position must be such that the athlete can position his/her shoulders over the front edge of the pushrims and be able to reach the bottom of the pushrims with both arms.

Many athletes use three-wheeled racing wheelchairs. Advances in steering geometry and front wheel assemblies have contributed to improvements in race performance and safety (figure 11.1). Front wheel assemblies are designed to follow the path selected by the athlete without requiring corrections from the pushrims. Two methods are commonly used to attach the front wheels to racing wheelchairs: trailing arm or fork. Trailing arm racing chairs have the front wheel(s) mounted on a cantilever axle. The frame is swept down to axle height, minimizing the moment arm of the ground reaction forces with respect to the front housing. This design also makes the front tires easily accessible in the event of a puncture. On four-wheeled trailing arm racing chairs the front wheels are mounted in line with the rear wheels yielding the maximum allowable width, and the most stable configuration. Three-wheeled racing chairs often use a leading fork, similar to those used on bicycles; this minimizes frame length for a given wheelbase, helps to improve high-speed handling, and makes the frame lightweight.

Racing wheelchairs must be propelled by arm power only, and steered by hands and arms only. This leaves the uncompensated front wheel assembly susceptible to downhill turning moments due to road crown and to caster flutter. These factors affect the energy required to propel the racing chair and thus the efficiency. Crown compensators on four-wheeled chairs and tie-rod linkages with Ackerman steering geometry effectively minimize the effects of crown and caster flutter. Racing chairs are steered by using a lever attached to the front fork(s) or trailing arm(s) (figure 11.2). Reduction or increase in the steering is governed by the length of the lever.

Lightweight wheels have been designed and are commonly available for wheelchair racing. Racing wheelchair rear hubs have cantilever threaded axles ($\frac{7}{16}$ and $\frac{1}{2}$ inch thread sizes are most common). The hub shells are made of aluminum alloy and are either standard or high flange. Spoke patterns are from 24 to 36 spokes per wheel. Tri-spoke composite wheels are also used on some racing chairs. Many athletes use Mylar wheel covers on their rear wheels. Front hubs are either fork mount or cantilever ($\frac{5}{16}$ or $\frac{7}{16}$ axle). Front hubs are made for 10 to 20 spokes. Some athletes use composite disk wheels on their racing chairs. Front and rear wheels use aluminum alloy rims (18 mm wide). Rear wheels are typically 60 to 70 cm, whereas front wheels are typically 35 to 50 cm in diameter. Athletes use both clincher and sew-up tires. Clincher tires are usually used with latex tubes. Rear tires may weigh as little as 150 g, while front tires may weigh as little as 75 g. Racing wheels are radially spoked to reduce weight.

Racing wheelchairs are propelled by hitting the pushrims as rapidly and with as much force as can be sustained for the distance required. Pushrims used on racing chairs are smaller than those used on chairs used for mobility. Pushrims are tubular rings of aluminum attached to the rear spokes. The attachment is made via threaded stand-offs with two spacers sandwiched around a spoke and clamped to the stand-off via a small bolt. A pushrim will have several attachments to the wheel. Pushrims may be coated to achieve higher friction between it and the athletes gloves (a tire or a high density foam coating are most common).

Many racing wheelchairs have features that aid the athlete's safety. Some of these features are a cycle computer which aids in pacing and training;

Figure 11.2. *Titanium racing wheelchair with feet forward seating, and classical fork steering. Wheels use alloy rims and hubs with stainless steel spokes. (Photograph courtesy of Halls Wheels, Inc.)*

a water bottle for fluid replacement; and caliber brakes (a safety feature to help control the tremendous speeds attainable on some downhill sections).

11.2.3. Racing wheelchair design issues

Racing wheelchairs are very user specific, and athletes vary greatly in their abilities. Thus, it is difficult if not impossible to design racing wheelchairs that are effective for every athlete. This is one of the primary reasons why many racing wheelchairs are hand-crafted for each individual based upon his or her specific needs.

A racing wheelchair must be functional, yet lightweight, stiff, and safe. These are classical design trade-offs for most racing applications, and they must be made to achieve a practical design. It is not acceptable to compromise safety. A properly designed racing wheelchair provides optimal interaction between the chair and the athlete. This requires an understanding of the mechanical design, and of the athlete's needs.

The seat of the racing wheelchair should fit tightly to the body. Getting in a racing wheelchair should be like slipping on a glove. Many new athletes

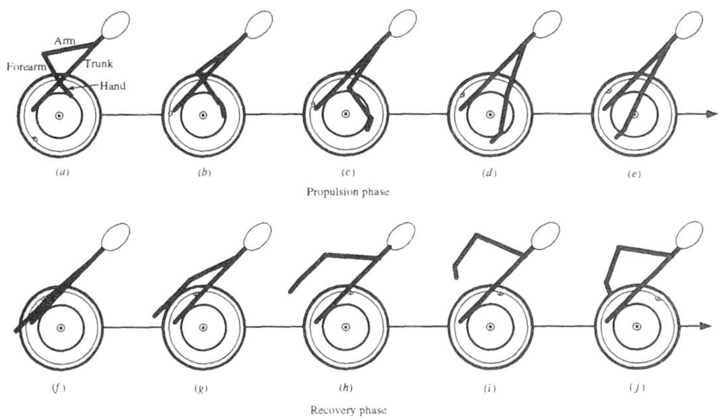

Figure 11.3. *Schematic diagram illustrating the phases of an efficient racing wheelchair stroke.*

make the mistake of getting a loose fitting wheelchair. Top athletes can only fit into their racers when wearing racing/training tights. To push properly the athlete has to sit properly. If the athlete is flexible enough and feels comfortable leaning on his/her knees, then the kneeling position is probably going to be the fastest. If this is uncomfortable, after some trial period, or if the athlete has very good trunk control, then a more upright posture is best. When seated in the chair and kneeling or lying upon his/her knees the athlete should be able to touch the ground with both hands, and be able to reach all the way around both push-rings. The center of each shoulder should line up with the front of each push-ring in the fully down position. The athlete should move his/her arms simulating stroking and test how difficult it is to breathe. Breathing must be synchronized with stroking, but if it seems difficult to breath the knees should be raised.

The wheelchair racing stroke can be divided into two primary phases: propulsion and recovery. Propulsion is when the arms are applying force, and recovery is when they are off the push-rings getting ready for the next stroke. In order to effectively push a racing wheelchair, the upper body must be used to rapidly apply a large force to the push-rings. Theoretically, the greatest velocity can be achieved by applying a large force over a long time period. However, it is difficult for an athlete to effectively apply a large force during the entire propulsion phase. The use of larger muscles and more of the upper body can assist in developing greater force over the propulsion cycle. The result is a larger force value and longer time of application which creates a larger change in the athlete's and wheelchair's momentum. However, the price for this higher mechanical energy is a higher metabolic energy cost (i.e., the amount energy used by the body).

A racing wheelchair stroke can be divided into five phases (figure 11.3). The five phases are defined as follows: (1) pushrim contact (*a*); (2) pushing through to the bottom of the pushrims (*b–d*); (3) push-off or follow through (*e*); (4) elbow drive to the top (*f–h*); and drive forward and downward (*i–j*). During the entire stroke the head should remain in-line with the trunk. Head movement should be kept to a minimum. Elbow height before the drive forward and downward is critical to generating propulsion force. Maximal elbow height requires strength and flexibility. Optimal force transference from the body to the push-ring requires properly fitted and designed gloves, and a non-slip push-ring coating. The hand should be in-line with the forearm at contact (i.e., wrist in the neutral position) to transfer maximum energy. The athlete must push continuously from contact through to the bottom of the push-rings. If the propulsion phase is done properly, the hand will push-off the push-ring with a flick. Then the momentum of the arms is used to carry the elbows upwards, finally the elbows are lifted to their peak. When the elbows are at their peak, the athlete should contract chest, shoulder and arm muscles to punch the push-rings.

Some personal considerations must be made when specifying a racing wheelchair. Paraplegic, quadriplegic, and amputee athletes have different preferences, and each person has his/her own abilities and anatomical structure. There are three basic seats to consider: kneeling bucket; kneeling cage; and upright cage. The kneeling position is very aerodynamic and has allowed some athletes to make tremendous improvements in their performance. Many athletes with paraplegia or quadriplegia use the kneeling position quite effectively. Athletes inexperienced with the kneeling position should use a kneeling cage. This affords them the option of sitting upright or kneeling and permits some adjustment of body position. Experienced athletes prefer a kneeling bucket because of the reduced weight and improved aerodynamics. Upright cage seats work well for athletes with lower-limb amputations, and for athletes with low levels of paraplegia. These athletes have the trunk control to adjust their body position while racing. Athletes with contractures should order a cage seat, unless they have substantial experience. Athletes are held in their racing chairs with straps and webbing. Additional padding or extra width can be requested to reduce the risk of developing pressure sores for people who are prone to breakdown. Although many competitors do not use a seat cushion in their racing chairs, the danger of skin breakdown makes use of a cushion advisable.

Most road and track races require helmets. Helmets have improved the safety of wheelchair racing, and every athlete should purchase a helmet when they purchase their racing wheelchair. The helmet must meet national safety standards, and fit the athlete snugly. Many helmets provide some adjustment. It also important that the helmet provides good airflow over the head. The head is a primary area for cooling the body of many wheelchair athletes. Aerodynamic helmets can enhance speed over no-helmet.

Most racing wheelchairs come equipped with brakes, steering levers, and compensators. These items all make racing safer and more fun. However,

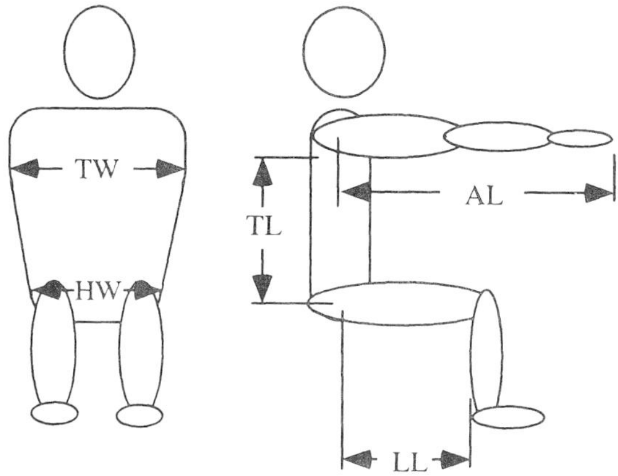

Figure 11.4. *Anatomical measurements for specification of a racing wheelchair. Arm length (AL), hip width (HW), leg length (LL), trunk length (TL), and trunk width (TW) are all critical measurements.*

all people are not alike. When ordering a racing wheelchair customers and manufacturers must ensure that all brake, compensator, and steering levers are easily reached and operated by the athlete. The levers must be of sufficient length and in a position where the athlete can effectively apply the necessary force and movement to operate these controls. Many compensator mechanisms come with stiff springs to ensure the wheelchair tracks are straight. To steer the wheelchair, adequate strength must be used to overpower the compensator. Softer springs or a longer steering lever can be requested to reduce the force required to steer the racing chair.

It is important that a racing wheelchair fit the user like a glove. Racing chairs are like shoes or gloves, in that a poorly fitting racing chair can be uncomfortable and awkward. A racing chair should fit as closely as possible without causing discomfort or pressure sores. Most manufacturers ask for a number of anatomical measurements when a racing chair is ordered. The most common measurements are hip width, chest width, upper-leg length, arm length, trunk length, height, and weight (figure 11.4, table 11.1).

It is important for the manufacturer to know of the athlete's special needs, such as asymmetry or limited range of motion (e.g., contractures). The manufacturer should be informed of any pelvic or spinal curvatures (e.g., lordosis, kyphosis), and whether the athlete is prone to fractures or skin breakdown. Type of disability and racing ability are also important factors that should be considered when specifying a racing wheelchair. A properly fitting racing wheelchair translates into maximum performance with minimum risk of injury.

Wheelchair racing is an important part of the rehabilitation of many people with disabilities, and is indeed a significant activity in the daily lives of many people. Exercise is an important modality in the well being of all

Table 11.1. *Critical body measurements for design and selection of a racing wheelchair.*

Hip width measurements:	This is the distance between hip joints. To measure this distance, the hands should be used to distinguish where the hip joint is located, and the distance between the hands should be measured.
Chest width measurements:	When exercising, muscles fill with blood and expand. Consequently, this measurement should be taken with the chest expanded. In order for the top of the seat to be comfortable but snug, the chest measurement should be made after taking a deep breath and while flexing the chest and upper-body muscles. The measurement is made across the back between the underarms.
Upper-leg length:	This measurement is taken from the hip joint to the inside of the knee. If amputated above the knee, the residual limb length should be used.
Arm length:	This measurement is taken from the underarm to the cleft between the thumb and index finger. Arm length is used to recommend the correct rear wheel and push-ring sizes.
Trunk length:	This measurement is taken from the hip joints to the underarms.

people including those with mobility impairments, and exercise should be encouraged and methods developed.

11.3. ARM-POWERED BICYCLES AND TRICYCLES

People with mobility impairments want alternatives when choosing recreational activities. Some people are interested in physically demanding sports such as marathons or triathlons. Other people prefer touring. Alternatives to racing wheelchairs, that offer the increased efficiency of levers and gears are attractive to some people. Alternatives to wheelchair locomotion have been in development for a number of years. The commercial availability of arm-crank recreational equipment was delayed because of: the lack of awareness among people with mobility impairments of the equipment's existence; the ineligibility of this type of product for purchase by insurance carriers; the high risk associated with production of low-volume adaptive recreation equipment, since liability insurance is difficult to obtain at a reasonable price; the complexity of home-building an arm-powered vehicle. At present there are only a few restrictions on arm-powered vehicles for use in competitions. The vehicle can use no motors or external energy sources and may have no structures that serve the sole purpose of reducing air/wind resistance.

11.3.1. Riding an arm-powered vehicle

For people with mobility impairments, hand cycles are their equivalents of bicycles. The present generation of hand cycles are geared much like mountain bikes with 15 to 21 gears. The wide range of gears gives riders the ability to select the most efficient gear for the grade of the roadway. Whether going downhill or uphill, the rider searches for the optimal gear ratio. The goal is to keep pedal turnover high and effort low. A cadence of between 80 to 100 revolutions per minute minimizes energy consumption and risk of musculoskeletal injury. Before a person rides independently several factors must be considered: where will most of the riding be done (beach, neighborhood, country, city); what kind of riding will be done (recreational, racing, touring); whether the hand cycle is appropriate based upon the abilities of the rider (disability etiology, physical function). When the pedal(s) are fully extended (hands on the pedals, back against the backrest) the elbows should be slightly bent.

11.3.2. Arm-powered vehicle designs issues

The motivation for designing arm-powered vehicles is to create a vehicle which converts the user's energy efficiently to motion while maintaining control on the vehicle over normal surface streets. Ideally, arm-powered vehicles would be designed to be transported in a passenger automobile without requiring disassembly. Standard bicycle components should be used whenever possible. Arm-powered vehicle frames are typically constructed from steel alloy (SAE 4130) tubing. Commonly, tubing ends are milled for best fit and stronger joints. Frames may be either brass welded (simple and inexpensive) or inert gas welded.

A limitation of most arm-powered vehicles is the balance of the pilot with the mobility impairment. Accustomed to using a wheelchair for mobility, many people have unlearned their ability to balance. The loss of some kinesthesia and neuromuscular control of the trunk and lower limbs make balancing a bicycle more difficult. A bucket-type seat as used on a racing wheelchair is effective in ameliorating these challenges for some people. Care should be taken when placing the rider's legs, as some positions may require placing the crank center further away from the rider so that the cranks clear the lower legs. This may result in a suboptimal center position and make steering more difficult at the front end of the crank cycle. Some bicycles position the rider low enough to the ground that the rider may place his/her hand on the ground when starting or stopping. Some people may prefer fully retractable side wheels (figure 11.5). Arm-powered bicycles are turned by a combination of leaning and turning the crank arms about the steering axis. Using side wheels may limit the amount the rider can lean the bicycle. When side wheels are used they must be designed not to interfere with normal lean steering. Height of the center of gravity, the balance of

(a)

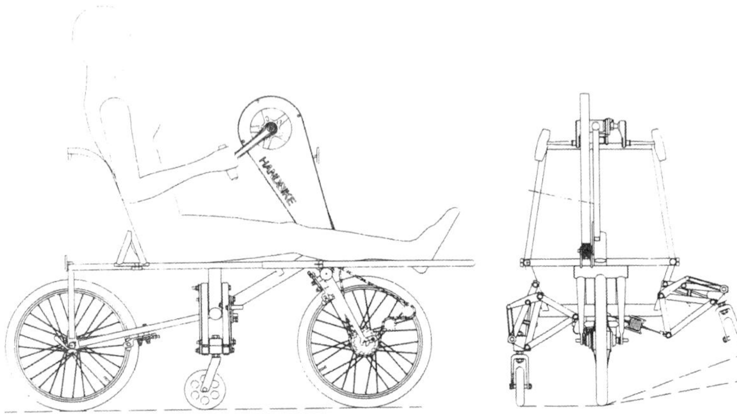

(b)

Figure 11.5. *Handbike arm-crank bicycle with retractable outriggers. This bicycle is designed primarily for touring, and has a small turning radius for close quarter maneuvering. (Figure courtesy Palo Alto Department of Veterans Affairs Rehabilitation Research and Development Center.)*

the user, and the vehicle's steering geometry interact. These factors must be considered when designing arm-powered vehicles to make them safe and effective.

With many arm-powered vehicles the steering and drive train are interconnected. One of the primary design considerations of arm-powered vehicles is: how does one power the vehicle and maintain directional control? Some designs have decoupled the steering from the drive train with

Figure 11.6. *Arm-crank tricycle with composite seat and (skate-board truck) steering. This type of tricylce is popular for racing and touring. (Photograph courtesy of Brike International, Ltd.)*

moderate success by using the tilt (from side-to-side) of the seat for steering (figure 11.6). These vehicles have a large turning radius for their size. Arm-powered bicycles require finer steering control to maintain balance, hence direct arm steering is used.

Crank-arms may be positioned adjacent or opposed. Using the cranks opposed provides for greater mechanical efficiency. However, damping must be used to minimize the moment about the head-set created by the opposed crank-arms. This moment may cause the front wheel(s) to turn from one side to the other as the rider cranks the vehicle, which results in the rider and vehicle moving side-to-side when uncompensated. The undesired turning moment can be minimized with a friction damped (nylon bushes in the head-set) or a viscous damper. Steering a bicycle is different from a tricycle; the former is accomplished primarily by leaning with the front wheels turned opposite to the desired direction to initiate the turn. A tricycle, in contrast, relies on turning the front wheels in the direction of the turn without leaning (figure 11.7).

It is important to have a wide range of gear ratios available in arm-powered vehicles, since the power output of the arms can be quite limited and fine increments are required to achieve optimal pedaling rates. A chain guard must be used to help keep fingers out of the sprockets, and to protect the rider in the event of an accident. Easily reachable indexing-type shift levers

Figure 11.7. *Touring arm-crank tricycle with sling seat and backrest upholstery designed to allow the riders to sit in posture similar to their wheelchair. The crank handles are ergonomically designed for rider comfort. (Photograph courtesy TOP END by Action.)*

should be used.

Braking may be accomplished via standard bicycle brakes. Levers must be positioned so that they can be grasped with one hand. Typically, the rider will steer with one hand while braking with the other. Some arm-powered vehicles use a cam-activated brake, which is engaged by reversing the crank direction. This simplifies braking, but prohibits back pedaling for balance at low speeds and when attempting to maximize one's leverage on inclines.

11.3.3. Add-on units and tandems

An arm-powered vehicle does not have to be separate from the wheelchair. Add-on units have been designed to convert a wheelchair into an arm-crank vehicle. Some of these devices incorporate quick-release mechanisms which are used to attach a self-contained front wheelgear-crank system. Another approach is to attach cranks to each of the rear drive wheels. This type of device may be quick-release as well. Quick-release add-on units are useful for people who desire a device which allows them to use their wheelchair as an arm-crank vehicle without transferring. When desired the user may remove the add-on unit and use their wheelchair normally. Add-on units are often easier to transport in an automobile. Usually performance is traded-off for convenience with add-on units. However, add-on units are often less expensive to purchase and quite functional for recreational riders.

Power output of human-powered vehicles may be enhanced by using a tandem where a leg crank is used by an ambulatory rider and an arm crank is used by the other rider (figure 11.8). Tandems may be two- or three-wheeled. The combined power output of the arm–leg-powered tandem is greater than the power output of the arm-powered vehicle alone. The ambulatory rider may either sit in front of or behind the other rider. When the ambulatory rider sits in the rear s/he can sit higher and see over the other rider providing both a clear view. When the ambulatory rider sits forward, a carriage can be used to allow the wheelchair rider to roll into the carriage and crank from the wheelchair. Steering is often performed by the ambulatory rider so that the cranking is decoupled from the steering. Shifting may also be done by the ambulatory rider so that pedaling does not have to be interrupted to change gears. The drive train of tandems is most efficient when each rider can pedal at his/her own pace and coast independently.

11.4. OFF-ROAD VEHICLES

Mountain biking has become very popular and some wheelchair users are joining the movement. Some wheelers preferring not to wait for all trails to become accessible have decided to change their circumstances by developing new or modifying existing equipment. Adaptive off-road vehicles open-up trails that would be impassable in a conventional wheelchair. There are several versions of adaptive off-road vehicles.

The off-road wheelchair looks similar to a four-wheeled racing wheelchair from a distance, with large front wheels and a lot of rear-wheel camber. These chairs use wide tires with a deep thread for added traction and resistance to puncture. Dual pushrims (two per rear wheel) help to find the proper gear. Handlebars and four-wheel brakes help to control the chair on steep rough descents. The center of gravity is moved more forward of the rear wheels than in other chairs to provide additional stability. The steering geometry must also be designed to accommodate high speeds over rough terrain. The frame and wheels have to be made to withstand much higher forces than racing or conventional wheelchairs. The extra equipment and strength required for off-road wheelchairs makes them heavier (typically around 40 pounds for manual chairs).

Mountain-bike and hiking trails tend to be steeper than paved roads or pathways. Off-road wheelchairs must be able to climb steep hills. This is accomplished by lowering the seat, moving the center of gravity forward, and lengthening the wheelbase. Two pushrim sizes make it easier to push on hills and flat ground.

Conventional steering from the pushrims is insufficient for off-road use. The larger front wheels of an off-road wheelchair are steered using

(*a*)

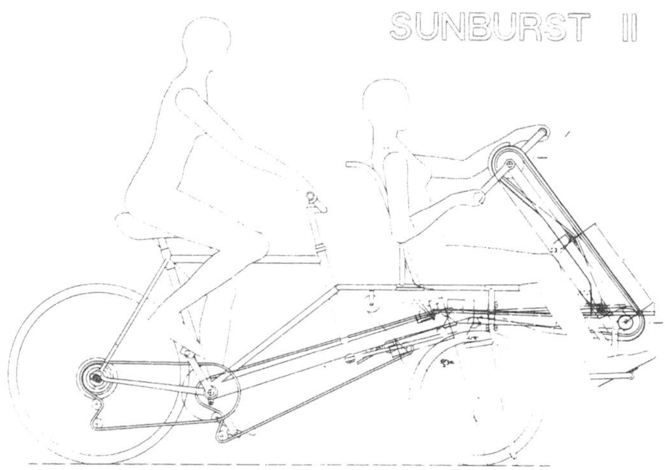

(*b*)

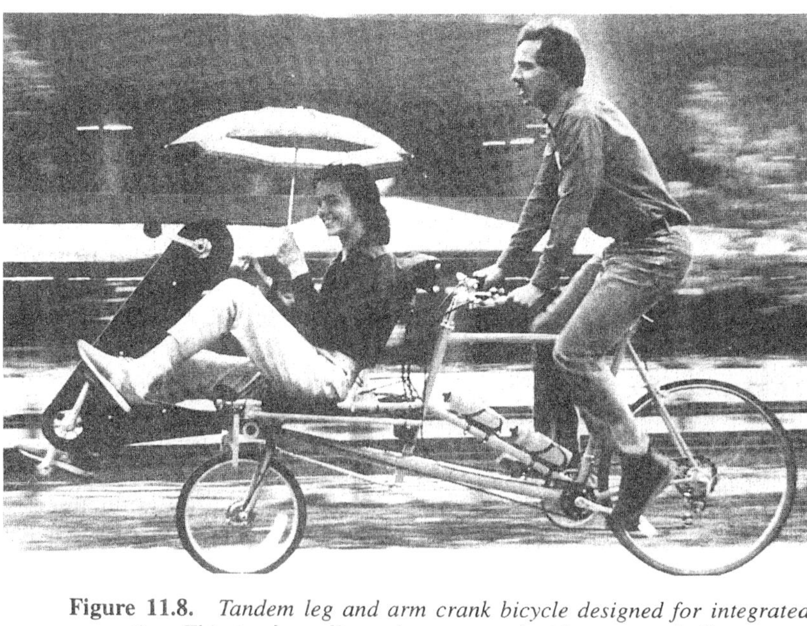

Figure 11.8. *Tandem leg and arm crank bicycle designed for integrated recreation. This tandem allows the arm-crank rider to travel distances at greater speeds than is possible with arm-crank alone. Front rider may use either arms or legs, or both arms and legs. (Figure courtesy Palo Alto Department of Veterans Affairs Rehabilitation Research and Development Center.)*

handlebars. The front wheels are connected via tie-rod linkage and a spring compensator is used to center the front wheels when they are not being turned by the rider.

Dynamic braking is required on off-road wheelchairs. The brakes are typically located on the handle bars to allow simultaneous braking and directional control. Conventional wheelchairs use wheel locks which only restrict motion when the wheelchair is stationary.

Powered off-road vehicles are also available (Ontario Drive & Gear Limited, New Hamburg, Ontario, Canada, Recreatives Industries, Inc., Buffalo, New York, USA). Several off-road, amphibious vehicles can be totally hand controlled. These vehicles incorporate roll protection guards, a seat belt harness, and an on-board fire extinguisher. A swing-arm and crane may be incorporated for some users. Many of these vehicles use six- or more-wheel drive and a vacuum-formed high-density polyethylene body. Two hand-held levers are used to turn and activate the brakes. The vehicles commonly use twist grip throttles. These vehicles may climb hills as steep as 45 degrees. Vehicles use either internal combustion engines or electric motors.

11.5. WATER SPORTS

11.5.1. Sailing

Wheelchair athletics has yet to be completely integrated with non-disabled sport. Many wheelchair sports focus on the challenges desired by a narrow segment of the population. The wheelchair experience and acclimatization process represents a long continuum of development and maturation (educationally, socially, and economically). Recreation is important to socialization. Sailing is a recreational activity which can be enjoyed alone or in the company of others. Sailing covers the broad spectrum from pleasure cruising to international competition. Sailing is a life-long sport which is enjoyed by many.

Prosthetics, wheelchairs and disability classification systems are the not the primary issues in sailing; mental and sailing ability are. Sailing allows people to participate in a fully integrated sports and recreational activity. Wheelchair sports only permit the opportunity to be the best wheelchair athlete one can be.

Many boats can be adapted to meet the needs of a wide variety of sailors. The *Freedom Independence 20* is a boat designed from the keel up for disabled sailors. The boat includes a wider hull to provide greater stability for boarding. The keel is heavily ballasted to help keep the boat upright. Full flotation supports the hull and keel along with a foam-filled mast to prevent the boat from over-turning. All sail lines lead back to the cockpit giving the sailor greater freedom and independence.

The *Freedom Independence* incorporates two counter-weighted seats, one for crew and helmsman, that tack from one side of the boat to the other. Below deck weights are used to counter the weight of the sailor. Simply releasing a line allows the sailor to move to the high side of the boat. The molded fiberglass seat has adjustable seating angles and backrest heights (headrests are optional). Benches with hinged gangplanks resting on the boat's gunwale provide a sliding surface from the dock to the cockpit.

Trapseat catamaran sailing is also popular among sailors with disabilities. The popular *Hobie 16* can be fitted with a trapseat: a seat with an aluminum frame and mesh or padded upholstery mounted outside each hull. At times an ambulatory crew member must be used to assist with steering the boat. The *Hobie* and trapseat make a practical and affordable combination.

Mini-12 m sailboats are compatible with most sailors with disabilities. Most have heavily ballasted keels for single-person sailing. The sailor sits low in the hull, usually with just their head above the deck. A full complement of controls and on some boats a spinnaker lead back to the sailor. These boats are easily adapted by removing the foot rudder pedals and replacing them with a tiller or PVC pushrods.

Many other boats can be modified for use by people with disabilities. Sailing offers a unique integrated recreational activity for people of all ages.

11.5.2. Water skiing

Adaptive water skiing equipment has been developed for people who are interested in an active water sport. Water skiing can be enjoyed using knee skis, sit skis, and variations of the Kan Ski. Novices use larger skis, whereas experienced skiers use shorter six foot boards approximately 13 inches wide. Skiers sit about six inches off the board in an aluminum alloy frame which holds them snugly in place but allows them to easily escape in a spill. The seat frame is typically open in the back so that when the rope tension is released the skier easily floats out the back. The skis also incorporate foot straps. The seat cage and the foot straps must be well padded.

11.5.3. Kayaking and canoeing

Sea and river kayaking and canoeing can be fully integrated and accessible recreational activities for many people. People should enter in instructional courses, and begin boating on flat water. Once the necessary skills are developed then people can move to more difficult rivers and seas. The level of difficulty in navigating the water is rated on a scale from one to six, with six being most difficult. Neophytes should start on level-one waters, and if they desire move to more difficult waters as their skills progress. People with lower-limb mobility impairments can typically progress to class-three waters at a normal pace. Kayaking and canoeing are great fitness and recreational activities which do not use a wheelchair; this makes these activities attractive to some. Another appealing aspect of boating is that people can explore wilderness areas that they may not otherwise be able to reach.

Kayak and canoe adaptations are necessary for some people because control of the boat is, in part, accomplished by the boater shifting his/her weight. This is usually accomplished by bracing with feet or knees. It is also important to retain balance and stability. Greater control is achieved by having the boat fit the boater, also known as 'wearing' the boat. The extent of one's physical limitations determines how much adaptation is required. The adaptations required vary from individual to individual. People should contact their local boating chapter for more information.

Most boats require modification to the seat for better trunk support, and provide a suitable seating base. A snug fitting seat which can be exited in emergency improves control and safety. Some people use Velcro straps. However, contoured foam with vacuum-formed vinyl upholstery can provide good postural support. Molded fiberglass seating systems are preferred by some athletes. People with quadriplegia may require greater postural support and adaptive paddle grips.

Selecting a suitable boat is perhaps the most important factor related to enjoyable boating. Typically two types of kayaks are used: sea kayaks and river kayaks. Sea kayaks range from 16 to 20 feet in length. Sea kayaks are built for stability, tracking, and speed of long distances. The cockpit tends to be larger than for a river kayak, and stowage space is typically included. Double-bladed paddles are used for kayaking. Sea kayaks come in single and double person versions. A river kayak is usually between 10 to 13 feet long. They are built for maneuverability. Cockpit sizes vary considerably with manufacturer and model. River kayaks are generally intended for shorter trips and do not customarily include storage space. People with lower degrees of impairment can learn many maneuvers in river kayaks including the 'Eskimo Roll'.

A canoe can carry either one or two people depending upon design. The length of canoes varies from 13 to 18 feet. A canoe is an open boat which offers ready storage of gear and easier transfer than a kayak. Specialized seating must often be designed which provides the impaired person sufficient support and adequate cushioning. When canoeing a single-bladed paddle is used. Stability of canoes and kayaks varies with design and purpose. Often it is better to complete a course or rent a boat before purchasing one. People should take lessons from a certified instructor before boating on their own.

11.6. ADAPTIVE SKI EQUIPMENT

Many athletes enjoy the exhilaration of moving through snow, powered by gravity and controlled by skill. With the right adaptations, many people with disabilities can enjoy moving freely through the snow in alpine skiing or the freedom of the wilderness in cross-country skiing. Many beautiful ski areas have been developed with wide trails and lift facilities to assist skiers in getting to the top of the mountain. There are many means of getting to the top of the mountain: rope and cable tows, poma lifts, T-bars, chair lifts, gondolas, trams. Ski facilities are usually located in remote areas which require snow tires or chains on automobiles.

Many ski facilities offer instruction to help make skiing safer and more enjoyable. All inexperienced skiers and those wishing to improve their skills should participate in organized instruction. Skiers with disabilities must learn to adapt to the winter sports environment and to educate ski personnel about their needs and requirements. People should learn as much as possible before hitting the slopes; this will make the ski experience more enjoyable and successful. Skiers must be sensitive to the concerns of ski area managers: potential injury to ski lift personnel, slowing the ski lift, liability concerns, congestion of slope, and accessibility.

11.6.1. Clothing and personal preparation

Winter sports athletes should dress for extreme weather conditions due to the variability of weather conditions throughout the ski season. Layering clothing is best. Wear a T-shirt or long underwear, a turtleneck, a sweater, and a waterproof windbreaker or down vest or ski parka, depending on weather conditions. People with lower-limb impairments should pay particular attention to keeping their legs and feet warm. Heavy socks and moon boots or hiking boots not only provide warmth, but protection as well. People with loss of sensation should periodically check their lower limbs for cooling. Waterproof mittens and gloves, and goggles should be worn at all times to protect the skier. Downhill skiers should wear helmets. Goggles should be worn whenever possible. Sunglasses should be avoided because they can break and cause injury.

To prevent injury it is important to strengthen and increase the flexibility of the arms and shoulders. Unlike pushing, skiing requires the skier to resist the upward and backward force from the poles or outriggers. Weightlifting and stretching routines should be employed to minimize risk of injury.

11.6.2. Sit-skiing

The Norwegian cross-country pulk (sled) was the forebearer to the sit-ski for downhill skiing. Winter Park Ski Area in Colorado experimented with devices for downhill skiing during the mid-1970s. Many people found the sensation of moving down the ski slopes remarkable, but the control of the pulk's speed and direction was difficult. The development of the Arroya series of sit-skis by Peter Axelson and the US Veterans Administration.

Figure 11.9. *Arroya sit-ski with its patented four control surfaces allows good control over speed and direction. This device has helped to popularize skiing for people with lower limb impairments. Now thousands of people with disabilities enjoy skiing. (Courtesy of Beneficial Designs Inc, Santa Cruz, CA.)*

The Arroya used a patented four-control surface design which allowed the skier greater control over speed and direction (figure 11.9). Other companies developed sit-skis as well, many of which use runners or blades along the bottom to aid in control.

A sit-ski must function effectively as a boot, binding, and ski. Sufficient padding and adequate cushioning to prevent pressure sores must be provided by the sit-ski. The sit skier must be held in place securely via straps or other hardware. A cover should be used to keep the skier dry and warm. The height of the backrest should provide the skier adequate support based upon his/her level of physical impairment. Sit-skis intended for downhill skiing should contain an evacuation harness and securing mechanism for chair lifts.

A variety of implements are used to control speed and direction when sit-skiing: kayak-type pole, ski grip and spike or disk, brass knuckles, shortened ski poles, and outriggers (Canadian crutches with short skis attached). Long poles are sometimes used by skiers to propel themselves across flat areas.

Sit-skis and pulks may also be used for cross-country skiing. For cross-country skiing the weight of the sit-ski or pulk must be kept to a minimum as the skier must propel the ski up and down moderate grades. Longer ski poles are used when cross-country skiing in order to get greater leverage and a longer stroke for polling.

11.6.3. Mono- and bi-skiing

Great progress was made in the sport of adaptive skiing when mono- and bi-skis were developed. A mono-ski is a device which uses a molded plastic or composite seat attached via a suspension system to a standard ski (figure 11.10). A mono-ski may be viewed as a dynamic ski orthosis. The seat of the mono-ski is raised from 250 to 500 mm above the snow. This permits the skier to lean and take advantage of the ski's edges. Suspension helps to cushion impacts and to provide greater control by keeping the ski in contact with the snow. The suspension acts much like the legs and knees of a standing skier. A bi-ski uses two skis attached to a molded seat via suspension incorporated into a four-bar linkage, all of which helps to keep the skis on the snow. Bi-skis provide two ski edges to be used simultaneously giving the bi-ski greater control. Mono-skis and bi-skis are often the preferable adaptive ski devices for many people. Air shocks are used so that the suspension can be adjusted for the terrain and preference of the skier. Often the travel of the suspension can be adjusted as well. If the springs are removed from a shock absorber, the damper which controls the rate of compression will remain. Some dampers allow adjustment via a flow valve.

Devices using skis are faster and more maneuverable than sit-ski devices. Mono/bi-skis can be skied as fast as standing skiers ski. These devices are easily integrated on common ski slopes. Mono/bi-skiers use outriggers for control. Often the back edge of the shortened ski has serrated edges. A cable and motorcycle brake lever mounted to the outrigger handle are used to flip the ski on end so that the serrated edge can be used like a ski pole tip. The skier turns a mono/bi-ski by shifting the center of gravity of the ski and skier. This allows the skier to adjust the force on the edge of the ski(s) to control the turn. Outriggers are used to assist with control and balance.

A seat bucket is used to provide support and aid in control over the ski. A mono/bi-ski must fit like the standing skier's boots. Sufficient padding is required to prevent pressure sores, while a snug fit is required to give the skier optimal control. Often a custom contoured plastic or composite seat with straps to secure the skier into the seat are used. When the seat is designed properly the ski and skier will act as one allowing the skier to control direction and speed by moving and leaning. The seat must be designed to accommodate each skier's physical abilities. Fore and aft adjustment is one of the most important features of a mono-ski. Fore and aft adjustment must be matched to the ability of the skier to move their center of mass. Lateral adjustment is also important so that the mass of the skier can be centered over the ski. The legs of the skier must also be secured into

Figure 11.10. *The mono-ski greatly increases the speed and manuever-ability of skiers with lower limb impairments. The mono-ski also provides the ability to independently use a chair lift. Mono-skis use standard skis and incorporate suspension systems. A mono-ski is a dynamic ski orthosis. (Courtesy of Beneficial Designs Inc, Santa Cruz, CA.)*

the mono/bi-ski to protect the skier from injury, and to help in controlling the ski. Some skis incorporate the legrests into the seat design, whereas other designs use a separate legrest and footplate. Commonly the legs are strapped together and to the legrests and footplates. The angle of the knees with the thighs must be set to accommodate the functional balance of the skier. The less upper-body stability the greater the knee angle. The knees should always have some bend to resist hyperextension.

Mono/bi-skis are taller than sit-skis and are nearly high enough to slide onto a common chair lift. This led to the development of chair lift mechanisms which elevate the seat on a mono/bi-ski. Thus when the chair lift mechanism is engaged the seat on the ski is high enough that the seat of the chair lift slides underneath it. This allows the skier to use the chair lift with minimal or no assistance, contrary to a sit skier who must be lifted onto the chair lift. Chair lift mechanisms often consist of a shock absorber release and a cam which lifts the chair when engaged with a handle operated by the skier. Mono-skiers should always be aware of the chairlift clearance height. This can simply be measured by using a skiing partner's pole. Many mono-skiers are unable to use equipment designed for evacuating ambulatory skiers in the event of a lift failure. Mono-skiers should carry a personal evacuation harness.

Skis can also be changed for the type of skiing and snow. The use of a ski also enables selecting wax for snow conditions. This innovation opened many opportunities for personalizing the ski device to an individual's preferences and abilities. It also helped to improve speed and maneuverability. The flex of ski may be hard or soft and non-linear or linear. It is important that the mounting of the frame onto the ski does not interfere with the intended flex of the ski. The mounting between the frame and ski must allow some relative motion so that the ski can flex otherwise the ski or mounting hardware will prematurely fatigue. The mono-ski center of mass in the skier's neutral position should be over the midcord of the ski. The mounting points for the mono-ski frame should also be located at the midcord as this is where the ski is designed to be loaded.

Several factors are critical to the design and selection of an appropriate mono-ski for each skier. The design of the mono-ski and the actions of the skier must work in harmony. Skiers must shift their body mass to put the ski on edge. This is called angulation. If the skier sits high he/she will only be able to make small adjustments in the center of mass position. If the center of mass is too low, the skier will be unable to lean because of interference between the seat and the snow. The center of mass should be as low as possible without having any component of the mono-ski, other than the ski itself, come into contact with the snow while turning. Skiers with skeletal asymmetries or spinal deformities may require canting. Cants are wedges placed between the ski and the frame.

11.6.4. Three- and four-track skiing

People with lower-limb disabilities can often ski using standard skis and boots fitted to their prostheses. Three-track skiing involves using a ski on one leg while using two outriggers with short skis. Three-track skiing is usually done by people who have one unimpaired leg while the other is amputated above the knee. People who have one or both legs amputated below the knee often four-track ski. Four-track skiing involves using two skis with boots fitted to the prostheses and two outriggers. It is not unusual for three- and four-track skiers to blend with other skiers on ski slopes. Some types of prosthesis are not well suited for three- or four-track skiing. The prosthesis and boot must fit well to give the skier adequate control. Also, some prosthesis are not designed to withstand the stress imposed by skiing. People should consult with the prosthesis designer/manufacturer or a certified prosthetist on whether the prosthesis is suitable for skiing.

11.7. RECREATIONAL VEHICLES

Extensive improvements have produced easy to drive, comfortable to live in, and accessible recreational vehicles (RVs). Many camp grounds provide clean, modern, and accessible facilities. Sometimes resorts include golf courses, pools, and health spas. Full hook-ups (e.g., electricity, water, sewer) are standard.

Recreational vehicles include living space, storage space, kitchens, and bathrooms. Many RVs are equipped with a microwave, stove, conventional oven, refrigerator, dual sinks, and coffeemaker. Larger RVs may include a TV/VCR, a sound system, a queen/king size bed, and a living room. Many motor-homes or mini-homes are built with aisles wide enough to accommodate a wheelchair. Most RV manufacturers offer a number of options which can make their products more accessible to people with disabilities. Some RV manufacturers or dealerships will perform custom modification (e.g., lift installation).

RVs allow travelers to be independent of hotels and restaurants. With full kitchen facilities, families can control what and when they eat, making it easier to accommodate special dietary needs. RV clubs have been organized in many parts of the world allowing enthusiasts to meet and enjoy the company of others who share the same interests in traveling.

Many people use their RVs as a guest home, or as a second home at a favorite hideaway. The range of RVs, from folding camping trailers to large size motor-homes, offers something for nearly everyone. Some RVs are more accessible than others, and modifications vary with model and manufacturer, therefore people should view, and if possible try, several RVs to determine which best meets their needs.

FURTHER READING AND REFERENCES

Asato K T, Cooper R A, Baldini F D and Robertson R N 1992 Training practices of athletes who participated in the National Wheelchair Athletic Association Training Camps *Adapted Phys. Act. Q.* **9** 249–60

Axelson P 1988 A guide to mono-skis and mono-skiing *Sports 'N Spokes* **14** 23–34

Beck K 1992 Evergreen folding camper *Paraplegia News* **46** 31–2

Charles D, James K B and Stein R B 1988 Rehabilitation of musicians with upper limb amputations *J. Rehab. Res. Dev.* **25** 25–32

Cooper R A, Bedi J, Horvath S M, Drechsler-Parks D and Williams R 1992 Maximal exercise responses of paraplegic wheelchair road racers *Paraplegia* **30**

Cooper R A 1992 Contributions of selected anthropometric and metabolic parameters to 10K performance—a preliminary study *J. Rehab. Res. Dev.* **29** 29–34

Cooper R A 1989 Racing wheelchair crown compensation *J. Rehab. Res. Dev.* **26** 25–32

Cooper R A 1989 Racing wheelchair rear wheel alignment *J. Rehab. Res. Dev.* **26** 47–50

Cooper R A 1992 The contribution of selected anthropometric and physiological variables to 10K performance of wheelchair racers: a preliminary study *J. Rehab. Res. Dev.* **29** 29–34

Cooper R A 1989 An international track wheelchair with a center of gravity directional controller *J. Rehab. Res. Dev.* **26** 63–70

Cooper R A 1990 Wheelchair racing sports science: a review *J. Rehab. Res. Dev.* **27** 295–312

Cooper R A 1992 Racing wheelchair roll stability while turning: a simple model *J. Rehab. Res. Dev.* **29** 23–30

Coutts K D 1991 Dynamic characteristics of a sport wheelchair *J. Rehab. Res. Dev.* **28** 45–50

Engel P and Seeliger K 1986 Technological and physiological characteristics of a newly developed hand-lever drive system for wheelchairs *J. Rehab. Res. Dev.* **23** 37–40

Golbranson F L and Wirta R W 1982 *Wheelchair III: Report of a Workshop on Specially Adapted Wheelchairs and Sports Wheelchairs* (Washington, DC: RESNA)

Howell G H, Brown D R, Bloswick D S, Bean J and Gooch JL 1993 Design of a device to exercise hip extensor muscles in children with cerebral palsy *J. Assistive Technol.* **5** 119–29

Loverock P 1989 The athlete of the future *Los Angeles Times Mag.* March 12

MacLeish M S, Cooper R A, Harralson J and Ster J S 1993 Design of a composite monocoque frame racing wheelchair *J. Rehab. Res. Dev.* **30**

Maarlewski-Probert B 1992 The RV industry is meeting the challenge *Paraplegia News* **46** 21–3

Rick Hansen Centre 1988 *Proc. Natl Symp. Wheelchair Track And Roadracing, Department of Physical Education and Sport Studies (University of Alberta, 1988)*
Wade J 1993 A league of its own *REHAB Management* **6** 44-51

CHAPTER 12

REHABILITATION ROBOTICS

12.1. INTRODUCTION

The concept of using robots as manipulation devices for people with upper-limb impairments was first proposed nearly 30 years ago. Since that time substantial progress has been made in developing robotic systems for rehabilitation applications. Robotic systems may promote a greater feeling of independence on the part of the user and reduce the sense of dependence on a personal assistant. Advances in robotic systems indicate that for some applications such systems may be cost effective.

The term 'robot' has its roots from the Czech term, *robota*, meaning labor or work. The term, robot, began to acquire its modern meaning in 1921 when it was used in Karel Čapek's play 'Rossum's Universal Robot', which depicted the story of a society of artificial men and women. In 1941, Isaac Asimov introduced the 'Three Laws of Robotics' in his book 'I, Robot': (1) a robot may not injure a human being, or through inaction allow a human being to come to harm; (2) a robot must obey the order given it by human beings except where such orders would conflict with the First Law; (3) a robot must protect its own existence as long as such protection does not conflict with the First or Second Laws (2). Although these laws were developed for a fictional story, they are of interest in rehabilitation robotics as well.

The Robot Institute of America (RIA) defines a robot as a reprogrammable multifunctional manipulator designed to move material, parts, tools, or specialized devices through variable programmed motions for the performance of a variety of tasks. Reprogrammability is a key component in the definition of robots.

Presently, several robotics systems are being developed in private, government and university research laboratories to assist people with upper-limb impairments. Two basic approaches have received substantial attention: (1) robotic workstations and (2) mobile robotic systems. Robotic workstations are systems optimized to operate within a semi-fixed environment. Such systems are being developed for kitchens and for a personal office environment. Workstation robotic systems may be mounted to a work surface or on an overhead mounting system. Mobile robotic systems present a greater number of technical challenges because the operating environment is expanded as well as the number of tasks performed by the system. Studies have been performed to explore the feasibility of low-end mobile robotic systems. The HERO 2000 (Heathkit Inc.) has been used as an experimental base with limited success.

Rehabilitation engineering has discovered the human-substitute concept and has employed robotics systems to serve as external assistants or artificial extensions of missing or impaired limbs. Some robotic systems have been accepted by people with upper-extremity disabilities, primarily quadriplegia. Functions of rehabilitation robotics systems range from pick and place operations to performing activities of daily living (ADL). Activities such as preparing simple meals, feeding, washing and brushing teeth have been performed successfully. Rehabilitation robotics should strive to develop systems which augment or support functions required for independence. Industrial robots have been well developed and can reduce product costs while improving product quality. Service robotic systems have been developed to a much smaller extent. Service robotic systems will help people with disabilities, but perhaps of greater economic importance they will be attractive to a broader range of the public. If designed properly service robotic systems will be used by people who want them, and not just by people who need them. Japan and the United States have entered into a joint venture to develop a useful, two-armed, mobile, sensate, servant robot for clinical study. It was proposed that the large aging populations in both countries would benefit. Robotic systems demonstrate substantial potential to assist people with several tasks. Robotic systems may help with household tasks (e.g., cleaning, food preparation) and work (i.e., operating computer peripherals, commercial cleaning, labor tasks).

The user of a robotic system must be able to control the robot by communicating what he/she wants performed. User friendly communication is the key to successful service robotic applications. The user with a disability should be able to direct a robotic system in much the same way as a human assistant is directed. The robotic system must incorporate the artificial intelligence, sensing, voice recognition and manipulation required to perform the required tasks.

12.1.1. Basic concepts

There are three basic categories of robotic systems. The workstation (workcell) robotic systems are designed to work in a fixed or static environment. Workstation robots have limited or no mobility. Telemanipulators or teletheses are robotic manipulators that neither replace (i.e., like a prosthesis) or assist (i.e., like an orthosis) a limb, but instead consists of a remotely controlled tool not attached to the body. Mobile robots are intelligent machines which have the ability to move under their own power and to operate in a flexible environment.

Robotic systems must be capable of operating in unstructured free space with obstacle avoidance. Either the user must be able to specify the motion of the end effector, or the robot must be able to interpret commands from the user and perform task planning. Point control or path control must be performed within the six degrees of freedom (e.g., three translation, three rotation) of the end effector. Some methods use preprogrammed motions, but this does not allow the user to perform new motions. However, requiring the operator to control each of the robot's joints is often too great a burden. A more desirable method is to allow the user to control motions that are not motion specific, but are task specific. This allows the user to control a single degree of freedom in task space, while the robot controller determines the motions in joint space. This method requires a sophisticated controller.

The objective of most rehabilitation robotic systems research is to assist people with impaired arm and hand function to operate with maximum autonomy in a minimally structured environment. This goal can be approached in many ways. The robotic system can be autonomous in executing a specific task, or autonomous at a specific workstation, or can provide assistance with activities of daily living, or provide general assistance to reduce the number of hours the user is dependent upon a human assistant. System performance must be determined by the performance of the user and the robotic system in the environment. Optimal performance will likely come from the integration of a variety of technologies to create a 'Smarthouse' or 'Smartoffice' system. Robotic devices, mobility aids, environmental controls, computer access devices, communication devices and security systems could be integrated to provide maximum autonomy within the environment. Human assistance may be required to perform specific tasks or for emergency situations. The integration of multiple technologies and human assistance needs to be investigated.

With many complex rehabilitation products, the ultimate success for the user and in some cases purchaser is determined by the subtle interplay between the user interface design, functional capabilities, aesthetics, geometry, software, reliability and cost. The quality of the user interface can determine the acceptance of a product even at substantial cost penalty. Most users prefer transparent directness between task request and task performance. User interface design must be flexible to user preferences and future system enhancements. Function keys and voice vocabulary can be customized and alternate input systems can be used. The perspective of the system developer influences the robotic system interface. For example, the MANUS project viewed the robotic arm as a prosthesis and used a track ball or joystick as the primary user interface, whereas the Palo Alto VA and Stanford University (VASU) team viewed the robotic system as

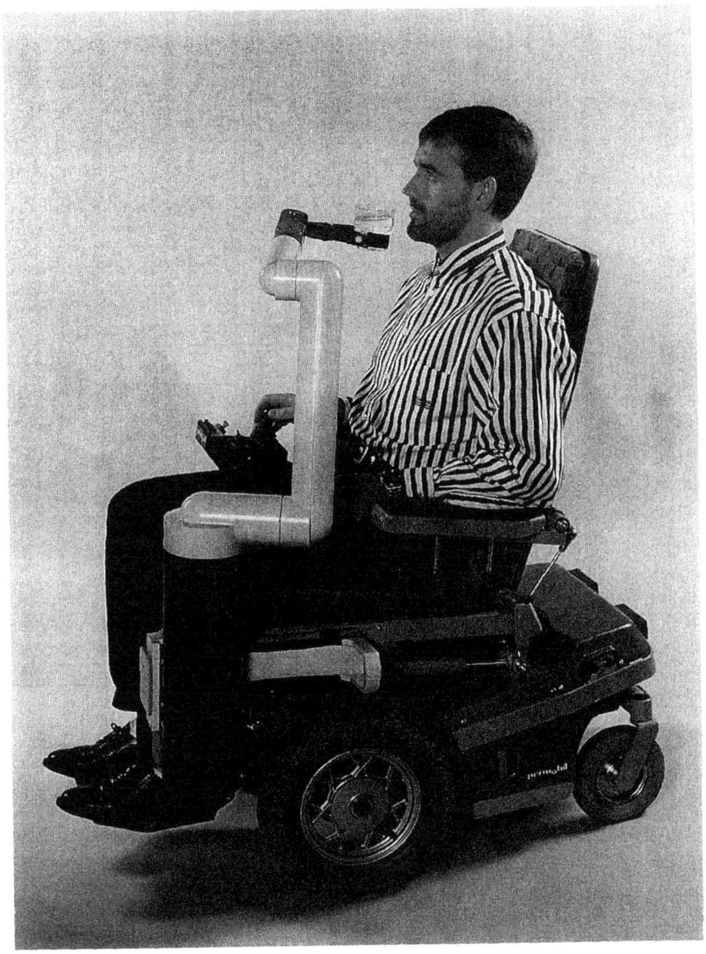

Figure 12.1. *MANUS robotic manipulation arm mounted to a power wheelchair. This type of system can give people with severe physical impairments greater ability to manipulate the environment. (Photograph courtesy PERMOBIL Corporation of Sweden.)*

a tool under the user's control and chose a voice-activated interface. The user's relationship to the robotic system must be clearly understood. The MANUS system is geared towards direct manipulation in an unstructured environment (figure 12.1) and the VASU system is directed toward repetitive task execution in a structured environment (figure 12.2). Flexible interfaces may accommodate both requirements.

User interfaces for robotic systems must be readily adapted from one person to another. The interface must also be able to accommodate changes in the user's abilities over time. One method of doing this is to have

Figure 12.2. *DEVAR robotic workstation being used to assist programmer with quadriplegia in his work area. This type of robotic system can provide people with greater independence in their work and home environments. (Photograph courtesy Palo Alto Department of Veterans Affairs Rehabilitation Research and Development Center.)*

the software sense the traits of the user through recording the sequences most often used. Algorithms similar to those used for word prediction in word-processing software could be helpful. The number of modes in which the system operates should be minimized to match each operator's needs. Robotic systems should be easily integrated with other assistive devices (e.g., environmental control units, communication devices). Alternative input devices must be available. Knowledge gained through developing computer access devices should be applied.

12.1.2. Robots as stationary manipulation aids

To be an effective manipulator a robotic system must be capable of performing many tasks that an operator with a disability may wish to perform. To use a robotic system to brush teeth, the robot must be capable of negotiating around obstacles, making contact with the surface of the teeth in various orientations, and applying the correct amount of force in the proper directions to follow the shape of the mouth and placement of the teeth. Manual operation of a telethesis can be accomplished by specifying the position of the end-effector at each point (i.e., end-point control) or by specifying direction and velocity at each instant (i.e., path control). Point control requires deriving the position solution at each point, whereas path control requires deriving the velocity solution at each point. Path control can result in smoother movement.

The mechanical structure, electronics and software implementation determine the tasks that may be performed. Viewing mechanical structure, electronics and software as subsystems of a single mechatronics system may reduce monetary cost and produce a more compact and aesthetic design. The MANUS system is an example of this approach. By using a mechatronics systems design approach, the MANUS designers were able to minimize the mechanical and electronic hardware required for the manipulator. Designers must decide which tasks are to be performed by mechanical or electrical hardware, or computer software. The choice of cabling and connectors can affect appearance and monetary cost of robotic systems. The MANUS system uses a proprietary two-wire communication bus.

12.1.3. Mobile manipulators and autonomously mobile robots

Mobility of manipulators and autonomously mobile robotic systems have an impact on functional and hardware design requirements. Mobile manipulators and autonomously mobile robots often cost more than workstation (fixed) manipulators. Mobile robots must function in dynamic, potentially unstructured, environments. Wheelchair-mounted manipulators have been used to minimize the complications associated with autonomous robots. Wheelchair-mounted manipulators may not require the same reach as other robotic systems, because of the ability to align the base. However, wheelchairs are not designed for mounting manipulators. Therefore, only certain locations may be suitable for mounting a manipulator which can limit the range of motion (i.e., ability to reach the floor, ability to use lap tray) and limit the size and weight of objects manipulated. Many users are strongly opposed to manipulators that increase the size of the wheelchair as this may limit their mobility. Cosmesis is important for wheelchair-mounted systems because of their close proximity to the user. Power consumption and sources for wheelchair-mounted robots must also be carefully considered.

Autonomously mobile robots may be used to perform a function while the user remains stationary. This may permit the user to perform some function while unable to move, while the autonomous robot performs its function. Autonomous robots require more sophisticated control and onboard intelligence. They also require an array of sensors and navigation systems. Autonomous rehabilitation robots share many features in common with service robots and are excellent candidates for artificial intelligence. Teleoperation of mobile robotic systems has been applied, but has required rather cumbersome control panels.

Mobile robots must have the ability to move independently in their operating environment while avoiding fixed and random obstacles. This quality is necessary to minimize the interactions required between the user and robot. It is also desirable to have the robot capable of following the operator. This frees the operator from the burden of controlling the robot when traveling from one location to another and permits navigation of terrain unfamiliar to the robot.

12.1.4. Cost/benefit analysis

Robotic systems as with other forms of assistive technology need to be developed with some consideration for the cost/benefit ratio. Physical, psychological, social and financial costs must be considered when developing complex technologies. Robotic systems may provide the user with greater control over the environment, independence in performing specific tasks, reduction in workload for personal assistants, greater autonomy in the home or workplace, and more time for interpersonal relationships. However, the user must accept the burden of: relying on a machine which may be slow and impersonal compared with a human assistant; receiving unwanted attention by using a robotic system; undertaking modifications to their living and working environment. Some robotic systems are mounted to the user's wheelchair to provide a mobile base. This makes the robot available for use nearly all of the time. However, attachment of a robot to the wheelchair adds additional weight and width which makes it more difficult to negotiate obstacles (e.g., doorways, hallways). It can also make it more difficult to transfer in and out of the wheelchair and to use desks or tables.

The goals and desires of the user determine, to a large extent, whether or not the robotic system will be accepted. People have varying degrees of desire to achieve functional independence and tolerance of technology. Some people need or desire the human contact provided by the personal assistant and develop a close relationship with their assistant. People also have varying reactions to technology, especially unfamiliar technology. Appearance is an important factor in the acceptance of robotics technology. Robotic systems gain greater acceptance if the user's human appearance and dignity are minimally affected. The robotic system should be as non-obtrusive as possible. People tend to accept robotic technology if it can be presented as an aesthetically appealing tool separate from the user.

The actual monetary cost of robotic systems is also very important. Monetary cost and human benefit and acceptance are often at odds with each other. Monetary cost and human benefit must be weighed in the design process to produce a positive cost/benefit ratio. This may be achievable by determining the minimal features required for acceptable task performance and appearance. The number of degrees of freedom of the robotic manipulator determines the end-effector orientation and accessible volume within the constraints of the mechanical ranges. Increasing the degrees of freedom makes the robot more dextrous at additional monetary cost. In some instances it is more cost effective to develop a special purpose

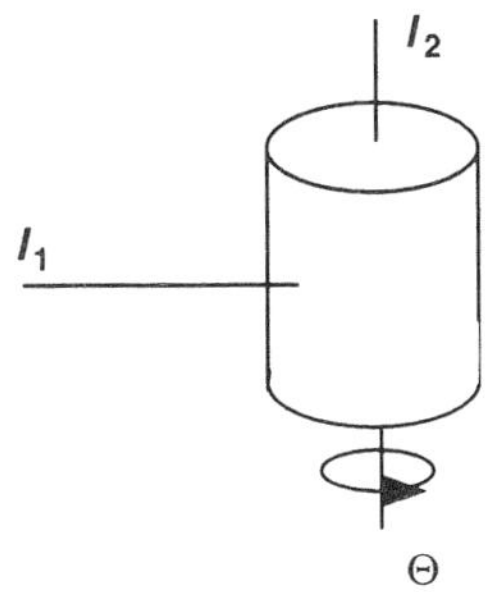

Figure 12.3. *Representation of a revolute joint.*

manipulator or a minimal robotic arm. Regardless of cost, a robotic system at the service of people with disabilities must attain the functionality and reliability required by the end-users, their employers, personal care assistants and sponsors.

12.2. COMPONENTS AND CONFIGURATIONS OF ROBOTS

Robot manipulators are composed of links connected by joints into an open kinematic chain. Joints are typically rotary (revolute) or linear (prismatic). A revolute joint allows relative rotation between two links. A prismatic joint allows relative translation between two links. Each joint represents the connection between two links. The joint variables for revolute joints are represented as the variable θ shown in figure 12.3. For prismatic joints, variables are represented by d as shown in figure 12.4. Joints of a robot have been controlled by a variety of actuators (e.g., electromechanical, hydraulic, pneumatic, memory metal). The joints of a robot determine the degrees of freedom. A minimum of six degrees of freedom (i.e., three for position and three for orientation) are required to reach an arbitrary point in a three-dimensional workspace. Some tasks may require greater than six degrees of freedom. However, the control complexity increases with additional degrees of freedom. The total volume which can be swept out by the end-effector is defined as the workspace. For mobile robots the workspace may be extended to the environment that the robot can access. The workspace of a robot may be partitioned into the reachable workspace and the dextrous workspace. The reachable workspace is defined as the volume in which the space can be reached in at least one orientation. The dextrous workspace defines the volume which can be reached in any arbitrary orientation.

12.2.1. Configurations

The number of joints determines the degrees of freedom (DOF) of the manipulator. Typically, a manipulator will possess at least six DOF: three

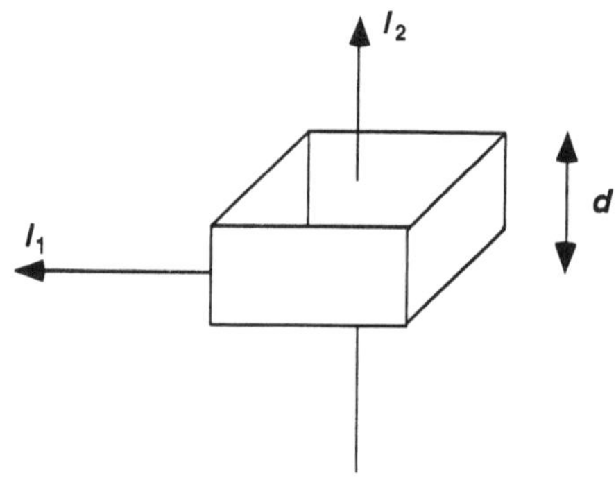

Figure 12.4. *Representation of a prismatic joint.*

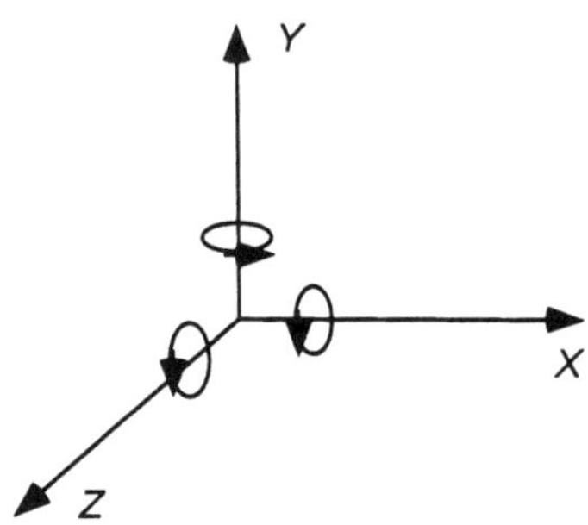

Figure 12.5. *Coordinate system in three-space.*

for position and three for orientation (figure 12.5). A manipulator having more than six degrees of freedom is referred to as kinematically redundant. Most robots are designed for a broad class of applications. A particular configuration can perform certain tasks better than others. For this reason, there are several robot configurations commonly in use. The primary application (e.g., materials handling, assembly, environmental control) often determines the kinematic structure. Robot manipulators can be classified by geometry, kinematic structure, application and control scheme. Manipulators are often described by their arm, with the wrist being described separately. There are five common manipulator geometries: articulated configuration, spherical configuration, SCARA (Selective Compliant Articulated Robot for Assembly) configuration, cylindrical configuration, Cartesian configuration.

An articulated manipulator, also known as a revolute or anthropomorphic manipulator, provides large freedom of movement in a compact space. Articulated manipulators use either elbow joints or parallelogram linkages. The Unimation PUMA (Programmable Universal Manipulator for Assembly) is an articulated manipulator used for rehabilitation applications. Parallelogram linkages are less dextrous than elbow joints, but provides for lighter links and simpler manipulator dynamics. Replacing the revolute joint in the elbow of an articulated manipulator by a prismatic joint creates a spherical configuration. A spherical manipulator has the z-axis of each joint mutually perpendicular and is simply described in terms of a spherical coordinate system. The SCARA is designed for assembly and is a revolute-prismatic manipulator. The SCARA configuration has the z-axis of each of the joints mutually parallel. The cylindrical configuration uses a revolute first joint and prismatic second and third joints. The joint variables for a cylindrical robot are the cylindrical coordinates of the end-effector. Cartesian manipulators use prismatic joints. The joint variables for a Cartesian manipulator are the Cartesian coordinates for the end-effector. Cartesian manipulators are useful as gantry robots to transport objects and materials.

The wrist and end-effector are critical to the performance of the manipulator. The joint between the arm and end-effector is referred to as the wrist. Wrists are often revolute or spherical and may possess several degrees of freedom. The spherical wrist simplifies kinematic analysis and provides maximal decoupling of object position and orientation. The PUMA robot uses a spherical wrist and possesses six degrees of freedom. The end-effector permits work to be done and is the interface to the world. Grippers are the simplest of the general purpose manipulators. Grippers are only capable of opening and closing which is adequate for a variety of operations. Special purpose end-effectors which can be easily and quickly exchanged to perform a specific task have been effective for some applications. Robotic anthropomorphic hands have been and are being developed as general purpose end-effectors. Anthropomorphic hands show potential for performing fine motor control.

12.2.2. Robotic systems

A robotic system should be viewed as more than a series of mechanical linkages. A robotic system consists of an arm, power source, end-effector, sensors, actuators, computer interface, and control computer (figure 12.6). The programming software is also an integral part of the system. A robotic system must possess some form of intelligence. A robotic arm is a computer-controlled manipulator consisting of several links connected in series by revolute or prismatic joints. One end of the kinematic chain is connected to a supporting base, while the other is equipped with an end-effector and is free to manipulate objects in the environment. The use of sensors allows the robot to interact more effectively and intelligently with its environment. Robot sensors may sense the internal or external state of the system. Internal-state sensors detect variables related to the internal operation of the robot such as joint position, motor current, grip force, etc. External-state sensors are used for object identification, obstacle avoidance and coordination. External

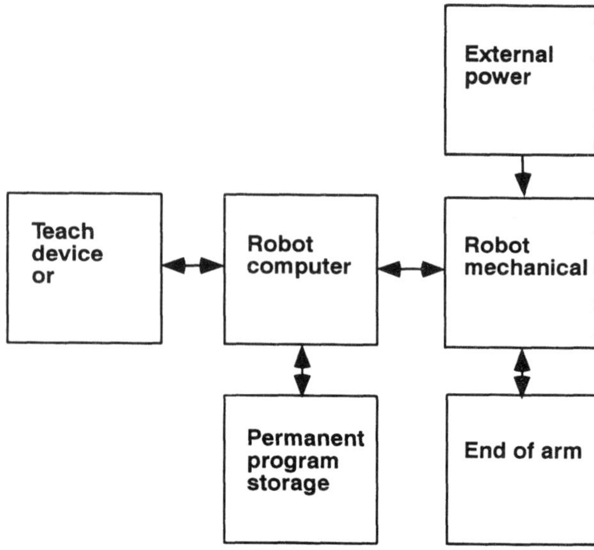

Figure 12.6. *Simple block diagram of a robotic system.*

states include range, proximity, touch, and temperature. Machine vision is possibly the most potentially powerful external sense. Robot vision can be defined as the process of acquiring, characterizing and interpreting visual information from the robot's environment.

A significant challenge in using robotic systems as assistive devices is finding suitable and efficient communication between the user, programmer and the robotic system so that the user can perform the desired tasks. The three basic approaches used to communicate with robotic systems are discrete word recognition, teach and playback and high-level programming languages. Speech recognition systems often require users to speak in discrete words, with pauses between words. Real-time word recognition is possible with high-speed computers and specialized hardware. Teach mode requires manually guiding the robot, at a reduced speed, through the entire task to be performed. The recorded motion is edited and then played at normal speed to control the robot. High-level programming presents a more general and flexible approach to robot programming. This approach permits using sensors to alter robot programming to adapt to the environment.

Sensors may provide information which can be used to make robots more flexible. Planning a robot's path and controlling the robot as it executes that path requires sophisticated techniques. Path planning requires solving the steps necessary to accomplish some stated goal, given the initial situation. Many robots have been programmed to use sensors to perform a primitive set of actions in some well structured environment. An intelligent program attempts to develop a path from the current state to some state specified by the user. The path typically consists of a sequence of primitive operations. Real-time path planning and control require powerful and efficient algorithms, operating on high-speed computer systems.

12.2.3. Characterization of robotic systems

The specification of a robotic system in terms of resolution, speed, programming features and end-effector capabilities will generally affect function and cost. Specifications of the system need to be scaled based upon the application. Many human service applications require relatively slow, compliant, high-resolution robots, with an exemplary programming environment and a full six degrees of freedom.

When a human performs a manipulative task, the activity may be decomposed into primitive movements such as grasping, rotation, translation, lifting and placement. Using primitive moves, people can repeat a sequence of moves controlling various parameters and achieving functionally different tasks. For example, washing and drying one's face are similar tasks, with different functions. Robotic systems can be designed to have similar attributes. Instructability is the term used to define the degree to which the robot can be programmed to follow a defined sequence of primitive moves. Repeatability is the measure of how close a manipulator can return to a previously taught point. Most robots exhibit high repeatability. When teaching a sequence, the operator guides the manipulator through a series of primitive steps that are saved under a symbolic task name. The sequence is executed upon issuance of the symbolic name associated with the desired task.

Adaptability is an extension of the workstation feature which permits the definition of new objects and locations in the workspace. In many cases, a fixed internal model of the environment is programmed into the computer. An adaptable robotic system permits updating the internal environmental model when objects are removed or introduced into the environment. Often only end-effector proximity sensors, joint encoders and the operator's vision are used to update the environmental model. Most robotic systems are inaccurate without external sensing of end-effector position. Accuracy is a measure of how close the manipulator can come to a given point within its workspace. Accuracy and repeatability can be improved by increasing controller resolution. Controller resolution is the smallest increment of motion that the controller can sense.

Modularity is important in the design of robotic systems. Modularity defines the number of similar components within the robotic system. Hardware and software can be modular. Hardware modularity refers to using similar units (modules) which perform related tasks (i.e., making all servo amplifiers the same, using the same type of motor for various joints). Software modularity means creating subroutines to perform commonly requested tasks. Primitives are constructed from these basic subroutines and in turn programs are created using sequences of primitives. Modular design makes maintenance and upgrading simpler.

12.3. ROBOT KINEMATICS

Kinematics is the study of the geometry and time-dependent motion of connected links, without consideration of the forces/moments that produce the motion. The motion of a body in free space can be uniquely described by two velocity vectors (i.e., one for translation and another for rotation). To move an end-effector along a straight line, the controller must specify the direction of the line then the velocity of the motion. Commonly a spherical coordinate system is used, but the appropriate coordinate system depends on the structure of the robot. Complex motions can be synthesized by concatenating path segments to accomplish the desired task as well as to avoid collision with obstacles, keep a liquid-filled container upright, or remain within a working envelope. If the transitions between motion segments are smooth, better task performance can be achieved. Getting the robot to act appropriately to task-based commands and to control the joints to provide the desired spatial path requires derivation of the inverse Jacobian for the particular robotic system.

The human hand uses over 50 muscles to perform various functions. Most robotic manipulators use seven actuators (i.e., six for position and one for grasp) to perform many of the same functions. Complex end-effector functions require interchangeable end-effectors (i.e., task specific end-effectors) or redundant degrees of freedom. Robots remain in their home position until given a command to perform some task. To perform most tasks the robot must move from its home position to a final position while maintaining a certain end-effector attitude and velocity profile. To successfully obtain this objective the position of the end-effector must be described with respect to a common coordinate system along the entire path from initial to final position. Internal sensors allow the robot to monitor the internal state of the system.

12.3.1. Forward kinematics

Forward kinematics are used to determine the position and orientation of the end-effector in terms of joint variables. Forward kinematics commonly involves the process of solving for the Cartesian configuration of a robot manipulator in terms of its links. The position of the end-effector is described with respect to a common coordinate system. Control is simplified by using a fixed coordinate system which is often called the world or base coordinate system. The base of the robot manipulator arm is defined as the origin or

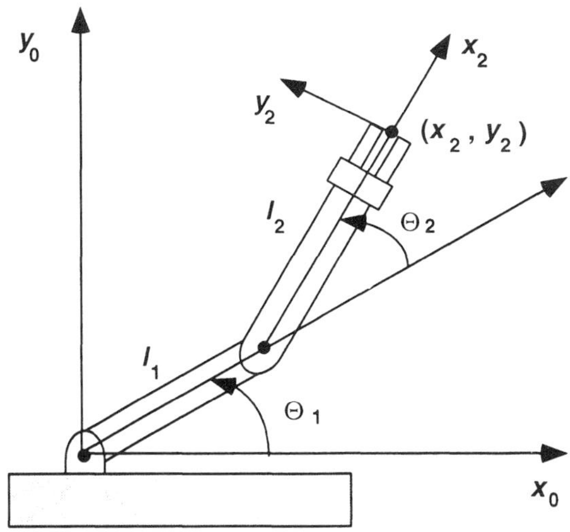

Figure 12.7. *Kinematics of two-link revolute robot.*

the world coordinate system (figure 12.7). A two-link articulate robot will be used to illustrate the use of forward kinematics in this section.

Optical encoders can be used to measure the joint angles θ_1 and θ_2. It is desired to express position (x_2, y_2) in terms of these joint angles. Position is relative to the world coordinate frame to which all objects, including the manipulator, are referenced. The origin of the reference frame is denoted $O_0x_0y_0$. Coordinates (x_2, y_2) of the end-effector are expressed in the world coordinate frame by

$$\begin{aligned} x_2 &= l_1 \cos\theta_1 + l_2 \cos(\theta_1 + \theta_2) \\ y_2 &= l_1 \sin\theta_1 + l_2 \sin(\theta_1 + \theta_2). \end{aligned} \tag{12.1}$$

The lengths of the links of the manipulator segments are defined by l_1 and l_2. The orientation of the end-effector can also be determined from the manipulator geometry.

Orientation of the end-effector reference frame relative to the world coordinate frame is given by the direction cosines of the x_2 and y_2 axes relative to the x_0 and y_0 axes

$$\begin{gathered} \boldsymbol{i}_2 \cdot \boldsymbol{i}_0 = \cos(\theta_1 + \theta_2) \qquad \boldsymbol{i}_2 \cdot \boldsymbol{j}_0 = \sin(\theta_1 + \theta_2) \\ \boldsymbol{j}_2 \cdot \boldsymbol{i}_0 = -\sin(\theta_1 + \theta_2) \qquad \boldsymbol{j}_2 \cdot \boldsymbol{j}_0 = \cos(\theta_1 + \theta_2) \\ \begin{bmatrix} \boldsymbol{i}_2 \cdot \boldsymbol{i}_0 & \boldsymbol{j}_2 \cdot \boldsymbol{i}_0 \\ \boldsymbol{i}_2 \cdot \boldsymbol{j}_0 & \boldsymbol{j}_2 \cdot \boldsymbol{j}_0 \end{bmatrix} = \begin{bmatrix} \cos(\theta_1 + \theta_2) & -\sin(\theta_1 + \theta_2) \\ \sin(\theta_1 + \theta_2) & \cos(\theta_1 + \theta_2) \end{bmatrix}. \end{gathered} \tag{12.2}$$

In the equation above, $\boldsymbol{i}_0$, $\boldsymbol{j}_0$ are the standard orthonormal unit vectors in the base frame and $\boldsymbol{i}_2$, $\boldsymbol{j}_2$ are the standard orthonormal unit vectors in the end-effector frame. These are the forward kinematic equations for the two-link

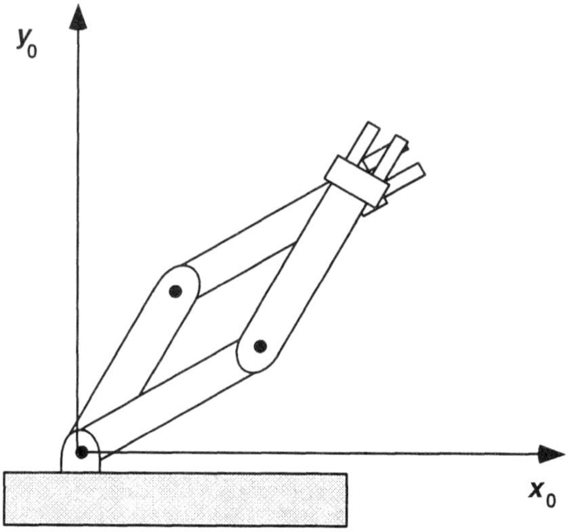

Figure 12.8. *Illustration of elbow up and elbow down configuration.*

robot described in figure 12.7. More complex robots follow a similar pattern, but are more difficult to write in closed form, therefore, homogeneous transformations are used to simplify transformation from one coordinate frame to another. Once the kinematic relationship has been determined, the inverse kinematic relationship can be sought. Solving for the inverse kinematic relationships from knowledge of the forward kinematics is neither simple nor straightforward.

12.3.2. Inverse kinematics

Inverse kinematics requires solving for the joint variable trajectories which will produce the desired Cartesian trajectory in the world coordinate frame. A relationship between the joint variables θ_1, θ_2 in terms of x and y coordinates is required. Both the forward and inverse kinematic relationships involve more than just the static positional relationships. The velocity and acceleration trajectories must also be considered. For relatively simple robots like the one described here, a little thought and some mathematical manipulations can lead to an inverse solution. Since the forward kinematics is non-linear, a solution may not be easy to find nor is there a unique solution in general. For example, with the 2-link planar robot described herein, there would be no solution if x and y were out of reach. If given the point is within reach of the manipulator there may be two or more solutions. This is referred to as the elbow up and elbow down configuration (figure 12.8).

For the robot control system to be effective it must be capable of coping with indeterminate inverse dynamics. A nominal trajectory must be selected too for path planning with or without the use of external sensors. The law of cosines can be used to determine the joint angles from end-effector position and length of the manipulator links

$$\cos\theta_2 = \frac{x^2 + y^2 - l_1^2 - l_2^2}{2l_1 l_2} \equiv D \quad \Rightarrow \quad \theta_2 = \cos^{-1} D. \tag{12.3}$$

The cosine function as described above gives the same solution for both elbow up and elbow down configurations. A better method is to use the definition in equation (12.3) to determine the sine of θ_2

$$\sin\theta_2 = \pm\sqrt{1 - D^2}. \tag{12.4}$$

This method yields both the elbow up and elbow down angles. The possible θ_2 angles can be found using the arctangent function

$$\theta_2 = \tan^{-1}\left(\frac{\pm\sqrt{1 - D^2}}{D}\right). \tag{12.5}$$

The sine and cosine functions are not unique and they are sensitive to numerical variations in certain regions. The two-argument arctangent function (ATAN2) is best used to determine joint angles. The two-argument arctangent function uses both the magnitudes and signs of the arguments so that the angles are uniquely defined in all four quadrants, a distinction lost with the single argument arctangent function. Furthermore, problems of division by zero are also avoided.

The angle θ_1 must depend upon θ_2. Physically one would expect to get a different value for θ_1 depending on the value of θ_2 selected. The value of θ_1 at each instant can be calculated from θ_2 and the manipulator configuration

$$\theta_1 = \tan^{-1}\left(\frac{y}{x}\right) - \tan^{-1}\left(\frac{l_2 \sin\theta_2}{l_1 + l_2 \cos\theta_2}\right). \tag{12.6}$$

These inverse kinematic relationships can be stored in the control computer and subsequently used to determine the desired joint variable trajectories corresponding to a desired end-effector path. Some consideration must be given to how the joint variables are physically measured and stored in the control computer. Accuracy problems can occur if joint variables are measured relative to external coordinate sensors. The mismatch between the physical Cartesian position and the position measured by the sensors can lead to significant deviations from the desired path. Furthermore, even if the home position were known exactly there is no guarantee that the robot's manipulator will follow the desired path without error. The proper controller is required to minimize path errors. Many robotic systems are programmed to return to the home position periodically to reduce the amount of error due to the variations in initial conditions.

12.3.3. Velocity kinematics

In order to follow a contour at constant velocity, or at any prescribed velocity, we must know the relationship between the velocity of the end-effector and the joint velocities. The can be accomplished by differentiating the end-effector Cartesian coordinates

$$\begin{aligned} \dot{x} &= -l_1 \sin\theta_1\dot{\theta}_1 - l_2\sin(\theta_1+\theta_2)(\dot{\theta}_1+\dot{\theta}_2) \\ \dot{y} &= l_1\cos\theta_1\dot{\theta}_1 - l_2\cos(\theta_1+\theta_2)(\dot{\theta}_1+\dot{\theta}_2). \end{aligned} \tag{12.7}$$

Equation (12.7) can be written as a matrix equation which is more appropriate for computer control

$$[\dot{\boldsymbol{x}}] = \begin{bmatrix} \dot{x} \\ \dot{y} \end{bmatrix} = \begin{bmatrix} -l_1\sin\theta_1 - l_2\sin(\theta_1+\theta_2) & -l_2\sin(\theta_1+\theta_2) \\ l_1\cos\theta_1 - l_2\cos(\theta_1+\theta_2) & l_2\cos(\theta_1+\theta_2) \end{bmatrix} \begin{bmatrix} \dot{\theta}_1 \\ \dot{\theta}_2 \end{bmatrix} = \mathbf{J}\dot{\boldsymbol{\theta}}. \tag{12.8}$$

The matrix **J** is called the Jacobian of the manipulator, and represents a fundamental relation. The determination of the joint velocities from the end-effector velocities is conceptually relatively simple since the velocity relationship is linear. Joint velocities are found from the end-effector velocities via the inverse Jacobian

$$\dot{\boldsymbol{\theta}} = \mathbf{J}^{-1}\dot{\boldsymbol{x}}. \tag{12.9}$$

For the two-link articulate robot example developed earlier, the .joint velocities can be found explicitly from the inverse Jacobian

$$\dot{\boldsymbol{\theta}} = \frac{1}{l_1 l_2 \sin\theta_2}\begin{bmatrix} l_2\cos(\theta_1+\theta_2) & l_2\sin(\theta_1+\theta_2) \\ -l_1\cos\theta_1 + l_2\cos(\theta_1+\theta_2) & -l_1\sin\theta_1 - l_2\sin(\theta_1+\theta_2) \end{bmatrix}\dot{\boldsymbol{x}}. \tag{12.10}$$

The determinant of the Jacobian for this example is $l_1 l_2 \sin\theta_2$. The Jacobian does not exist for all values. In such cases the manipulator is said to be in a singular configuration. At singular configurations fine control is unachievable (i.e., the manipulator cannot be moved in some directions). Singular configurations are indications of non-unique solutions of the inverse kinematics, and are best avoided during path planning.

12.3.4. Rotations

A large part of robot kinematics is concerned with the establishment of various coordinate systems to represent the positions and orientations of

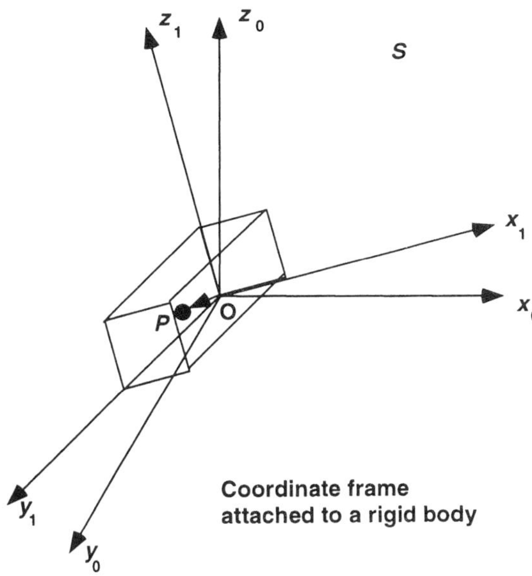

Figure 12.9. *Rotation of an object in space.*

rigid bodies and with transformations among these coordinate systems. Homogeneous transformations combine operations of rotation and translation into a single matrix multiplication. Homogeneous transformations are useful for planning motion from one point to another (figure 12.9). Rotations are an important aspect of robot motion.

Coordinates of point p within a rigid body in the coordinate system S in the $\mathrm{O}x_1y_1z_1$ frame can be related to coordinates of p in a fixed frame (non-rotated) $\mathrm{O}x_0y_0z_0$. The standard orthonormal basis for $\mathrm{O}x_0y_0z_0$ will be denoted $\boldsymbol{i}_0, \boldsymbol{j}_0, \boldsymbol{k}_0$; thus $\boldsymbol{i}_0, \boldsymbol{j}_0, \boldsymbol{k}_0$ are unit vectors along the x_0, y_0, z_0 axes respectively. Similarly, the standard orthonormal basis for $\mathrm{O}x_1y_1z_1$ will be defined as $\boldsymbol{i}_1, \boldsymbol{j}_1, \boldsymbol{k}_1$. Then the vector from the common origin to the point p on the object can be represented either with respect to to $\mathrm{O}x_0y_0z_0$ as

$$\boldsymbol{p}_0 = p_{0x}\boldsymbol{i}_0 + p_{0y}\boldsymbol{j}_0 + p_{0z}\boldsymbol{k}_0 \tag{12.11}$$

or with respect to $\mathrm{O}x_1y_1z_1$ as

$$\boldsymbol{p}_1 = p_{1x}\boldsymbol{i}_1 + p_{1y}\boldsymbol{j}_1 + p_{1z}\boldsymbol{k}_1. \tag{12.12}$$

Since both of the vectors are representations of the same vector p, the relationship between coordinate systems can be developed

$$\begin{aligned} p_{0x} &= \boldsymbol{p}_0\boldsymbol{i}_0 = \boldsymbol{p}_1\boldsymbol{i}_0 = \boldsymbol{p}_{1x}\boldsymbol{i}_1\boldsymbol{i}_0 + \boldsymbol{p}_{1y}\boldsymbol{j}_1\boldsymbol{i}_0 + \boldsymbol{p}_{1k}\boldsymbol{k}_1\boldsymbol{i}_0 \\ p_{0y} &= \boldsymbol{p}_0\boldsymbol{j}_0 = \boldsymbol{p}_1\boldsymbol{j}_0 = \boldsymbol{p}_{1x}\boldsymbol{i}_1\boldsymbol{j}_0 + \boldsymbol{p}_{1y}\boldsymbol{j}_1\boldsymbol{j}_0 + \boldsymbol{p}_{1k}\boldsymbol{k}_1\boldsymbol{j}_0 \\ p_{0z} &= \boldsymbol{p}_0\boldsymbol{k}_0 = \boldsymbol{p}_1\boldsymbol{k}_0 = \boldsymbol{p}_{1x}\boldsymbol{i}_1\boldsymbol{k}_0 + \boldsymbol{p}_{1y}\boldsymbol{j}_1\boldsymbol{k}_0 + \boldsymbol{p}_{1k}\boldsymbol{k}_1\boldsymbol{k}_0. \end{aligned} \tag{12.13}$$

The above three equations can be written together as one matrix equation.

$$p_0 = \mathbf{R}_0^1 p_1. \tag{12.14}$$

The rotation matrix between the two coordinate systems is defined as follows:

$$\mathbf{R}_0^1 = \begin{bmatrix} i_1 \sum i_0 & j_1 \sum i_0 & k_1 \sum i_0 \\ i_1 \sum j_0 & j_1 \sum j_0 & k_1 \sum j_0 \\ i_1 \sum k_0 & j_1 \sum k_0 & k_1 \sum k_0 \end{bmatrix}. \tag{12.15}$$

This matrix represents the transformation matrix from the coordinates of p with respect to the frame $Ox_0y_0z_0$. We can also write the inverse of this transformation

$$p_1 = \mathbf{R}_1^0 p_0$$

$$\mathbf{R}_1^0 = \begin{bmatrix} i_0 \sum i_1 & j_0 \sum i_1 & k_0 \sum i_1 \\ i_0 \sum j_1 & j_0 \sum j_1 & k_0 \sum j_1 \\ i_0 \sum k_1 & j_0 \sum k_1 & k_0 \sum k_1 \end{bmatrix}. \tag{12.16}$$

The matrix $\mathbf{R}_1^0$ represents the inverse of the transformation matrix $\mathbf{R}_0^1$. Since the inner product is commutative, i.e., $i_0 \cdot j_0 = j_0 \cdot i_0$. This leads to some useful properties for the rotation transformation matrix

$$\mathbf{R}_1^0 = \left(\mathbf{R}_0^1\right)^{-1} = \left(\mathbf{R}_0^1\right)^T. \tag{12.17}$$

A matrix whose inverse is its transpose is said to be orthonormal. It can be shown that the determinant of the transformation matrix is equal to ± 1. For example, suppose the frame $Ox_1y_1z_1$ is rotated through an angle θ about the z_0 axis. It is desired to find the resulting transformation matrix $\mathbf{R}_0^1$. Positive sense for the angle θ is given by the right-hand rule. In this case the transformation matrix is related by the sine and cosine of θ

$$\begin{gathered} i_0 \cdot i_1 = \cos\theta \quad j_1 \cdot i_0 = -\sin\theta \\ j_0 \cdot j_1 = \cos\theta \quad i_1 \cdot j_0 = \sin\theta \\ k_0 \cdot k_1 = 1. \end{gathered} \tag{12.18}$$

All other dot products are zero. Hence, the rotation matrix is defined. This is a basic rotation matrix

$$\mathbf{R}_0^1 = \begin{bmatrix} \cos\theta & -\sin\theta & 0 \\ \sin\theta & \cos\theta & 0 \\ 0 & 0 & 1 \end{bmatrix}. \tag{12.19}$$

In this case $\mathbf{R}_0^1$ is commonly called $\mathbf{R}_{z,\theta}$. Some useful properties of this basic rotational matrix are listed below.

$$\begin{aligned} &\mathbf{R}_{z,\theta} = \mathbf{R}_0^1 \Rightarrow \mathbf{R}_{z,0} = I \\ &\mathbf{R}_{z,\theta}\mathbf{R}_{z,\varphi} = \mathbf{R}_{z,\theta+\varphi} \\ &\mathbf{R}_{z,\theta}^{-1} = \mathbf{R}_{z,-\theta}. \end{aligned} \tag{12.20}$$

The other basic rotation matrices can be derived in a similar manner. This will be left as an exercise for the reader.

$$\mathbf{R}_{x,\theta} = \begin{bmatrix} 1 & 0 & 0 \\ 0 & \cos\theta & -\sin\theta \\ 0 & \sin\theta & \cos\theta \end{bmatrix}$$
$$\mathbf{R}_{y,\theta} = \begin{bmatrix} \cos\theta & 0 & \sin\theta \\ 0 & 1 & \\ -\sin\theta & 0 & \cos\theta \end{bmatrix}. \tag{12.21}$$

The matrices $\mathbf{R}_{x,\theta}$, $\mathbf{R}_{y,\theta}$ have similar properties as $\mathbf{R}_{z,\theta}$. The rotation matrix can also be interpreted as specifying the orientation of the coordinate frame $Ox_1y_1z_1$ relative to the frame $Ox_0y_0z_0$. The columns of a rotation matrix are the direction cosines of the coordinate axes $Ox_1y_1z_1$ relative to the coordinate axes $Ox_0y_0z_0$. For example, the first column of $\mathbf{R}_0^1$ specifies the direction of the x-axis relative to the $Ox_0y_0z_0$ frame. If the vector $\boldsymbol{p}_0 = (1, 1, 0)^T$ is rotated about the y_0 axis by $\pi/2$, then $\boldsymbol{p}_1$ the resulting vector is as follows

$$\boldsymbol{p}_1 = \mathbf{R}_{y,\pi/2}\boldsymbol{p}_0$$
$$\begin{bmatrix} 0 & 0 & 1 \\ 0 & 1 & 0 \\ -1 & 0 & 0 \end{bmatrix}\begin{bmatrix} 1 \\ 1 \\ 0 \end{bmatrix} = \begin{bmatrix} 0 \\ 1 \\ -1 \end{bmatrix}. \tag{12.22}$$

12.3.5. Composition of rotations

Recall the matrix $\mathbf{R}_0^1$ represents a rotational transformation between frames $Ox_0y_0z_0$ and $Ox_1y_1z_1$. Suppose we now add a third coordinate frame $Ox_2y_2z_2$ which is related to $Ox_0y_0z_0$ and $Ox_1y_1z_1$ by rotational transformations. A given point p can be represented in three ways: p_0, p_1, and p_2 in the three frames. The relationship between these three representations of p is related by their transformation matrices

$$\boldsymbol{p}_0 = \mathbf{R}_0^1\boldsymbol{p}_1 = \mathbf{R}_0^2\boldsymbol{p}_2 \quad \text{and} \quad \boldsymbol{p}_1 = \mathbf{R}_1^2\boldsymbol{p}_2. \tag{12.23}$$

Each $\mathbf{R}_i^j$ in equation (12.23) is a rotation matrix. The relationship between $\boldsymbol{p}_0$ and $\boldsymbol{p}_2$ can be found by substitution

$$\boldsymbol{p}_0 = \mathbf{R}_0^1\boldsymbol{p}_1 = \mathbf{R}_0^1\mathbf{R}_1^2\boldsymbol{p}_2 \quad \Rightarrow \quad \mathbf{R}_0^2 = \mathbf{R}_0^1\mathbf{R}_1^2. \tag{12.24}$$

The frame relative to which the rotation occurs is called the current frame. For example, suppose a rotation matrix $\mathbf{R}$ represents a rotation of θ degrees about the current y-axis, followed by a rotation of ϕ degrees about the current

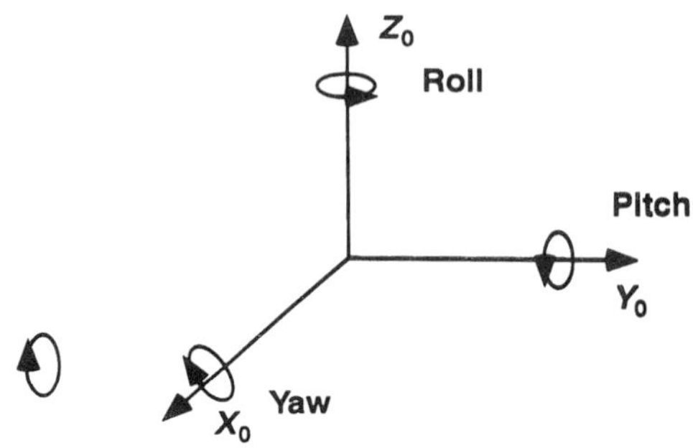

Figure 12.10. *Definition of roll, pitch, and yaw axes for robots.*

z-axis. Then the rotation matrix for both rotations can be found using the equation

$$\mathbf{R}_{y,\theta}\mathbf{R}_{z,\phi} = \begin{bmatrix} \cos\theta & 0 & \sin\theta \\ 0 & 1 & \\ -\sin\theta & 0 & \cos\theta \end{bmatrix} \begin{bmatrix} \cos\phi & -\sin\phi & 0 \\ \sin\phi & \cos\phi & 0 \\ 0 & 0 & 1 \end{bmatrix}$$
$$= \begin{bmatrix} c_\theta c_\phi & -c_\theta s_\phi & s_\theta \\ s_\phi & c_\phi & 0 \\ -s_\theta c_\phi & s_\theta s_\phi & c_\theta \end{bmatrix} \tag{12.25}$$

where $c_\theta = \cos\theta$ and $s_\theta = \sin\theta$.

It is important to remember that the order in which a sequence of rotations is carried out, and consequently the order in which the matrices are multiplied together, is crucial. Rotation, unlike position, is not a vector quantity and is therefore NOT subject to the laws of vector addition, hence rotation matrices do not commute in general.

Euler angles are commonly used to specify a rotation matrix. We can specify the orientation of the frame $Ox_1y_1z_1$ relative to the frame $Ox_0y_0z_0$ by three angles (ϕ, θ, ψ), known as Euler angles. A rotation matrix $\mathbf{R}$ can be described as a product of successive rotations about the principal coordinate axis $Ox_0y_0z_0$ taken in a specific order. The rotations define roll, pitch, and yaw angles (ϕ, θ, ψ), respectively (figure 12.10).

12.3.6. Homogeneous transformations

The order of rotation can be specified as x–y–z. The resulting transformation matrix for a fixed frame is found by multiplying the three basic rotation matrices

$$\mathbf{R}_0^1 = \mathbf{R}_{z,\phi}\mathbf{R}_{y,\theta}\mathbf{R}_{x,\psi}$$

$$\mathbf{R}_0^1 = \begin{bmatrix} \cos\phi & -\sin\phi & 0 \\ \sin\phi & \cos\phi & 0 \\ 0 & 0 & 1 \end{bmatrix} \begin{bmatrix} \cos\theta & 0 & \sin\theta \\ 0 & 1 & \\ -\sin\theta & 0 & \cos\theta \end{bmatrix} \begin{bmatrix} 1 & 0 & 0 \\ 0 & \cos\psi & -\sin\psi \\ 0 & \sin\psi & \cos\psi \end{bmatrix}$$

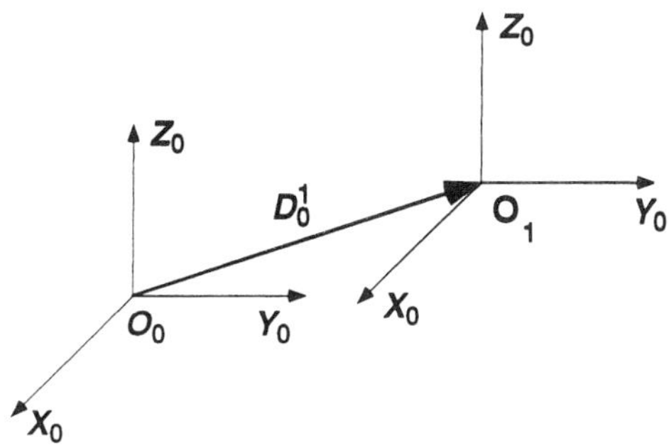

Figure 12.11. *Illustration of parallel translation.*

$$\mathbf{R}_0^1 = \begin{bmatrix} c_\phi c_\theta & -s_\phi c_\psi + c_\phi s_\theta s_\psi & s_\phi s_\psi + c_\phi s_\theta c_\psi \\ s_\phi c_\theta & c_\phi c_\psi + s_\phi s_\theta s_\psi & -c_\phi s_\psi + s_\phi s_\theta c_\psi \\ -s_\theta & c_\theta s_\psi & c_\theta c_\psi \end{bmatrix}. \tag{12.26}$$

Instead of yaw–pitch–roll relative to a fixed frame, the above transformation could also be interpreted as roll–pitch–yaw. Because of the fixed frame the result is the same.

12.3.7. Homogeneous transformations

Translations (linear motion) are as important as rotation in robotic systems. Consider a coordinate system $O_1x_1y_1z_1$ obtained from $Ox_0y_0z_0$ by a parallel translation of distance $|d|$, (i.e. $\boldsymbol{i}_0, \boldsymbol{j}_0, \boldsymbol{k}_0$ and $\boldsymbol{i}_1, \boldsymbol{j}_1, \boldsymbol{k}_1$ are parallel) (figure 12.11).

Any point p has a representation p_0 and p_1. Since the two coordinate systems are parallel (pure translation), any point p on one can be represented by a translation from one frame to another

$$\begin{aligned} &\boldsymbol{p}_0 = \boldsymbol{p}_1 + \boldsymbol{d}_0^1 \\ &p_{0x} = p_{1x} + d_{0x}^1 \qquad p_{0y} = p_{1y} + d_{0y}^1 \qquad p_{0z} = p_{1z} + d_{0z}^1. \end{aligned} \tag{12.27}$$

If $\mathbf{R}$ is orthogonal (i.e., a rotation matrix) than a relationship for rigid motion can be defined as follows

$$p_0 = \mathbf{R}p_1 + d. \tag{12.28}$$

For two rigid motions, the relationships are defined as follows

$$p_0 = \mathbf{R}_0^1 p_1 + d_0^1 \qquad p_1 = \mathbf{R}_1^2 p_2 + d_1^2. \tag{12.29}$$

The composition of these rigid motions defines a third rigid motion

$$p_0 = \mathbf{R}_0^1 \mathbf{R}_1^2 p_2 + \mathbf{R}_0^1 d_1^2 + d_0^1. \tag{12.30}$$

Motion between p_0, p_2 is also rigid thus

$$p_0 = \mathbf{R}_0^2 p_2 + d_0^2. \tag{12.31}$$

Comparing equation (12.30) to equation (12.31) yields the composite rotation matrix and translation vector

$$\mathbf{R}_0^2 = \mathbf{R}_0^1 \mathbf{R}_1^2 \qquad d_0^2 = d_0^1 + \mathbf{R}_0^1 d_1^2. \tag{12.32}$$

This shows that rotation matrices simply multiply, whereas translation is a combination of vector addition of the translations and the cross-product of translation and rotation. Translation and rotation can be combined to create a single matrix identity, referred to as the homogeneous transformation. The homogeneous transformation holds for multiple rigid motions. The homogeneous transformation is defined by the following matrix identity

$$\mathbf{H} = \begin{bmatrix} \mathbf{R} & d \\ \mathbf{0}^T & 1 \end{bmatrix}. \tag{12.33}$$

$\mathbf{R}$ is an orthogonal rotation matrix, thus the inverse of $\mathbf{H}$ can be found explicitly:

$$\mathbf{H}^{-1} = \begin{bmatrix} \mathbf{R}^T & -\mathbf{R}^T d \\ \mathbf{0}^T & 1 \end{bmatrix}. \tag{12.34}$$

The verification of the inverse will be left to the reader.

Multiple rigid motions are treated by multiplying the individual homogeneous transformations

$$\mathbf{H}_0^2 = \mathbf{H}_0^1 \mathbf{H}_1^2 = \begin{bmatrix} \mathbf{R}_0^1 & d_0^1 \\ \mathbf{0}^T & 1 \end{bmatrix} \begin{bmatrix} \mathbf{R}_1^2 & d_1^2 \\ \mathbf{0}^T & 1 \end{bmatrix} = \begin{bmatrix} \mathbf{R}_0^1 \mathbf{R}_1^2 & d_0^1 + \mathbf{R}_0^1 d_1^2 \\ \mathbf{0}^T & 1 \end{bmatrix}. \tag{12.35}$$

12.4. ROBOT MOTION

Robot motion must be controlled in order to be useful as a rehabilitation tool. The motion of a robot is complex and an understanding of the accelerations and velocities of various motions is important for safe and effective control. Rehabilitation robotics systems often work in close proximity to people and may make physical contact with the operator. This requires that only the necessary force can be applied to the person, and that velocities must be slow when the end-effector is around the person. Higher speeds can be used when the robot is not in close proximity to the person. Some tasks can also be performed at greater speed than others. For example, an empty glass can be moved faster than a full glass if spilling is not permitted.

12.4.1. Velocities and accelerations

Suppose that a rotation matrix $\mathbf{R}$ is time varying, so that $\mathbf{R} = \mathbf{R}(t)$ for all real t. Then the rate of change in the position matrix can be determined through matrix differentiation. Because of the properties of rotation matrices, their derivatives assume a special form

$$\dot{\mathbf{R}}(t) = \mathbf{S}(t)\mathbf{R}(t). \qquad (12.36)$$

$\mathbf{S}(t)$ is skew symmetric

$$\mathbf{S}(t)^T + \mathbf{S}(t) = 0.$$

$\mathbf{S}(t)$ is a function of the angular velocity of the rotating frame with respect to a fixed frame at each instant in time.

$$\mathbf{S}(t) \Leftrightarrow \mathbf{S}(\omega(t)). \qquad (12.37)$$

The matrix $\mathbf{S}(t)$ can be derived from the properties of $\mathbf{R}(t)$

$$\mathbf{R}(\psi(t))\mathbf{R}(\psi(t))^T = I. \qquad (12.38)$$

By differentiating both sides of equation (12.38) with respect to ψ

$$\frac{\mathrm{d}\mathbf{R}}{\mathrm{d}\psi}\mathbf{R}(\psi(t))^T + \mathbf{R}(\psi(t))\frac{\mathrm{d}\mathbf{R}^T}{\mathrm{d}\psi} = 0. \qquad (12.39)$$

Then $\mathbf{S}(t)$ can be defined as

$$\mathbf{S}(t) = \frac{\mathrm{d}\mathbf{R}}{\mathrm{d}\psi}\mathbf{R}(\psi(t))^T$$

which is skew symmetric. For example, examine the derivative of the basic rotation matrix about the x-axis

$$\mathbf{R}(t) = \mathbf{R}_{x,\psi(t)} \qquad \dot{\mathbf{R}}(t) = \frac{\mathrm{d}\mathbf{R}(t)}{\mathrm{d}t}. \qquad (12.40)$$

The chain rule can be used to determine the derivative of $\mathbf{R}(t)$:

$$\dot{\mathbf{R}}(t) = \frac{\mathrm{d}\mathbf{R}(t)}{\mathrm{d}\psi}\frac{\mathrm{d}\psi(t)}{\mathrm{d}t} = \dot{\psi}(t)\mathbf{S}(i)\mathbf{R}(t) = \mathbf{S}(\omega(t))\mathbf{R}(t). \qquad (12.41)$$

The angular velocity about the x-axis is equal to the derivative of the rotation angle

$$\omega(t) = \mathrm{i}\dot{\psi}(t). \tag{12.42}$$

Suppose p_1 represents a vector fixed in a coordinate frame $Ox_1y_1z_1$ and that this frame is rotating relative to $Ox_0y_0z_0$. Then p_1 in $Ox_0y_0z_0$ is given by

$$\begin{aligned} p_0 &= \mathbf{R}(t)p_1 \\ \dot{p}_0 &= \mathbf{S}(\omega(t))\mathbf{R}(t)p_1 = \mathbf{S}(\omega(t))p_0 = \omega(t) \times p_0. \end{aligned} \tag{12.43}$$

The explicit use of t can be dropped as it is implied. If p_1 is a constant vector representing rigid motion, then the velocity equation is as follows

$$\begin{aligned} p_0 &= \mathbf{R}p_1 + d \\ \dot{p}_0 &= \dot{\mathbf{R}}p_1 + \dot{d} = \mathbf{S}(\omega)\mathbf{R}p_1 + \dot{d} = \omega \times r + v. \end{aligned} \tag{12.44}$$

If both sides of the above equation are differentiated with respect to time once more, then the accelerations in space result

$$\begin{aligned} &\ddot{p}_0 - \ddot{d} = \ddot{\omega} \times \mathbf{R}p_1 + \omega \times \dot{\mathbf{R}}p_1 = \ddot{\omega} \times r + \omega \times (\omega \times r) \\ &\ddot{p}_0 = \ddot{\omega} \times r + \omega \times (\omega \times r) + a \\ &\ddot{p}_0 = \text{transverse acceleration} + \text{centripetal acceleration} + \text{linear acceleration}. \end{aligned} \tag{12.45}$$

The configuration of the robot, and the control scheme determine the relative balance between the components of the acceleration vector. Often simple movements (i.e., pure translation, pure rotation) are used to simplify the control problem.

12.4.2. Derivation of the Jacobian

The forward kinematics define a function between the space of Cartesian positions and orientations, and the space of joint positions. The velocity relationships are determined by a matrix valued function called the Jacobian. The Jacobian is an important relation in the control of robotics motion which arises in nearly every aspect of manipulation.

The transformation ($T_0^n(q)$) from the end-effector frame to the base frame with joint variables $q_1, \ldots, q_n$, will be a function of time. This transformation is necessary to relate linear and angular velocity of the end-effector to the vector of joint velocities.

$$\mathbf{T}_0^n(q) = \begin{bmatrix} \mathbf{R}_0^n(q) & d_0^n(q) \\ \mathbf{0} & 1 \end{bmatrix}. \tag{12.46}$$

The vector of joint variables is defined as $q = [q_1, \ldots, q_n]^T$. As the robot moves, the joint variables, the end-effector position and orientation will be functions of time. We want to relate the linear velocity and angular velocity of the end-effector to the vector of joint velocities ($\dot{q}$). The angular velocity (ω_0^n) of the end-effector is defined as follows:

$$\mathbf{S}(\omega_0^n) = \dot{\mathbf{R}}_0^n(\mathbf{R}_0^n)^T. \tag{12.47}$$

The linear velocity of the end-effector is defined by the derivative of the coordinate translation vector

$$v_0^n = \dot{d}_0^n. \tag{12.48}$$

To find the Jacobian, we need equations of the form:

$$v_0^n = \mathbf{J}_v\dot{q} \qquad \omega_0^n = \mathbf{J}_\omega\dot{q} \tag{12.49}$$

where each of the submatrices of the Jacobian are $3 \times n$ matrices. The Jacobian (J_0^n) is defined by combining the above two equations.

$$\begin{bmatrix} v_0^n \\ \omega_0^n \end{bmatrix} = \begin{bmatrix} \mathbf{J}_\omega \\ \mathbf{J}_v \end{bmatrix} \dot{q} = \mathbf{J}_0^n\dot{q}. \tag{12.50}$$

The Jacobian is a $6 \times n$ matrix, where n is the number of links.

12.4.3. Angular velocities

Vector addition can be used with angular velocities provided they are expressed relative to a common coordinate frame. Thus we can determine the angular velocity of the end-effector relative to the base by expressing the angular velocity of each link in the orientation of the base frame and then summing them. If the ith joint is revolute, then the ith joint variable q_i equals θ_i and the axis of rotation is the z_{i-1} axis. Thus the angular velocity of the link i expressed in the frame $i-1$ is given by,

$$\omega_{i-1}^i = \dot{q}_i k = \dot{\theta}_i k. \tag{12.51}$$

If the ith joint is prismatic, then the motion of frame i relative to frame $i-1$ is translation and

$$\omega_{i-1}^i = 0. \tag{12.52}$$

If joint i is prismatic, the angular velocity of the end-effector does not depend on q_i, which equals d_i. Therefore the overall angular velocity of the end-effector in the base frame is given by

$$\begin{aligned} \omega_0^n &= \omega_0^1 + \mathbf{R}_0^1\omega_1^2 + \mathbf{R}_0^2\omega_2^3 + \mathbf{R}_0^3\omega_3^4 + \ldots + \mathbf{R}_0^{n-1}\omega_{n-1}^n \\ \omega_0^n &= p_1\dot{q}_1\boldsymbol{k} + p_2\dot{q}_2\mathbf{R}_0^1\boldsymbol{k} + \ldots + p_n\dot{q}_n\mathbf{R}_0^{n-1}\boldsymbol{k} \\ \omega_0^n &= \sum_{i-1}^{n} p_i\dot{q}_i\mathbf{R}_0^{i-1}\boldsymbol{k}. \end{aligned} \tag{12.53}$$

When the joint is revolute $p_i = 1$, otherwise $p_i = 0$. Equation (12.53) can be simplified by defining a new variable z_i:

$$z_{i-1} = \mathbf{R}_0^{i-1}\hat{k} \quad \text{and} \quad z_0 = \hat{k} = (0, 0, 1)^T. \tag{12.54}$$

Using these relations, the angular velocity component of the Jacobian can be defined:

$$\mathbf{J}_\omega = \left[p_1 z_0 \ldots p_n z_{n-1}\right]. \tag{12.55}$$

12.4.4. Linear velocity

The linear velocity of the end-effector can be found by using the chain rule for differentiation:

$$\dot{d}_0^n = \sum_{i=1}^{n} \frac{\partial d_0^n}{\partial q_i}\dot{q}_i. \tag{12.56}$$

Thus the ith column of $\mathbf{J}_v$ is $\partial d_0^n/\partial q_i$. The ith column of the Jacobian is generated by holding all joints fixed except for the ith joint and then actuating the ith joint at unit velocity.

Two cases need to be considered:

CASE 1: If joint i is prismatic, then $\mathbf{R}_0^{i-1}$ is independent of $q_i = d_i$ for all i and

$$d_{i-1}^i = d_i\hat{k} + \mathbf{R}_{i-1}^i a_i\hat{i}. \tag{12.57}$$

If all joints are fixed, except for the ith, we have

$$\dot{d}_0^n = \mathbf{R}_0^{i-1}\dot{d}_{i-1}^i = \dot{d}_i\mathbf{R}_0^{i-1}\hat{k} = \dot{d}_i z_{i-1} \tag{12.58}$$

and

$$\frac{\partial d_0^n}{\partial q_i} = z_{i-1} \quad \text{when} \quad q_i = d_i.$$

CASE 2: If joint i is revolute, let O_k denote the vector d_0^k from the origin O_0 to the origin O_k for any k, and we write

$$d_0^n = d_0^{i-1} + \mathbf{R}_0^{i-1}d_{i-1}^n \tag{12.59}$$

or with the newly defined notation

$$O_n = O_{i-1} + \mathbf{R}_0^{i-1}d_{i-1}^n. \tag{12.60}$$

Both d_0^{i-1} and $\mathbf{R}_0^{i-1}$ are constant if only the ith joint is activated, therefore

$$\dot{d}_0^n = \mathbf{R}_0^{i-1}\dot{d}_{i-1}^n. \tag{12.61}$$

Since the motion of link i is a rotation θ_i about z_{i-1}, we have

$$\dot{d}_{i-1}^{n} = \dot{\theta}_i \hat{k} \times d_{i-1}^{n}. \tag{12.62}$$

Using this relationship, the linear velocity portion of the Jacobian can be calculated:

$$\begin{aligned} \dot{d}_{i-1}^{n} &= \mathbf{R}_0^{i-1}(\dot{q}_i \hat{k} \times d_{i-1}^{n}) = \dot{q}_i \mathbf{R}_0^{i-1} \hat{k} \times \mathbf{R}_0^{i-1} d_{i-1}^{n} \\ &= \dot{q}_i \dot{z}_{i-1} \times (O_n - O_{i-1}) \,. \end{aligned} \tag{12.63}$$

This leads to the following relationship, which defines the linear velocity Jacobian of the ith column:

$$J_{vi} = \frac{\partial d_0^n}{\partial q_i} = \dot{z}_{i-1} \times (O_n - O_{i-1}) \,. \tag{12.64}$$

The upper half of the Jacobian $\mathbf{J}_v$ is given by the following matrix:

$$\mathbf{J}_v = [J_{v1}, \ldots, J_{vn}] \,. \tag{12.65}$$

If joint i is prismatic, then the column vector is simpler.

$$\mathbf{J}_{vi} = z_{i-1} . \tag{12.66}$$

The total Jacobian is defined by the following matrix

$$\mathbf{J} = [J_1, J_2, \ldots, J_n] \tag{12.67}$$

with

$$\mathbf{J}_i = \begin{bmatrix} z_{i-1} \times (O_n - O_{i-1}) \\ z_{i-1} \end{bmatrix}$$

when the corresponding joint is revolute, and

$$\mathbf{J}_i = \begin{bmatrix} z_{i-1} \\ 0 \end{bmatrix} \tag{12.68}$$

when the corresponding joint is prismatic.

These points may be demonstrated with an example. Consider a two-link planar robotic arm (figure 12.12). The joint axes z_0, z_1, z_2 are normal to the surface of the page. The base frame is Ox_0y_0z. The origin is chosen as the point of intersection of the z_0 axis with the page. Once the base frame is

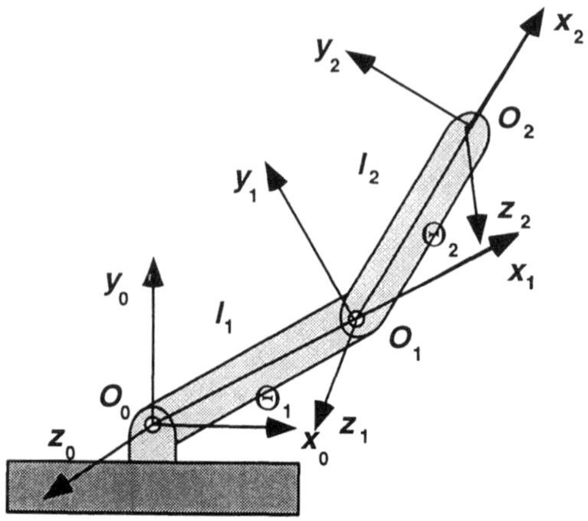

Figure 12.12. *Two-link revolute robotic arm.*

Table 12.1. *Link parameters for example robotic arm.*

Link	Label	ψ_l	θ_l	ϕ_l	d_i
1	l_1	0	0	ϕ_1	0
2	l_2	0	0	ϕ_2	0

established, the frames $Ox_1y_1z_1$ and $Ox_2y_2z_2$ are defined. All joints of this robotic arm are revolute.

A convenient way to analyze robot kinematics is to make a table defining the pertinent system variables, table 12.1. In this example, the rotations about the x, y, and z axes, the types of joints, and the number of links are most important.

The homogeneous transformation for position and orientation of the end-effector to the base can be derived by cascading the homogeneous transformation for each joint coordinate system.

$$\mathbf{H}_0^n = \mathbf{H}_0^1\mathbf{H}_1^2 \ldots \mathbf{H}_{n-1}^n. \tag{12.69}$$

In this example, there are two homogeneous transformations: one from the base frame to joint one, and another from joint one to joint two. Because of the structure of this robot, the transformations are similar in form, but this need not be the case.

$$\mathbf{H}_0^1 = \begin{bmatrix} c\phi_1 & -s\phi_1 & 0 & l_1 c\phi_1 \\ s\phi_1 & c\phi_1 & 0 & l_1 s\phi_1 \\ 0 & 0 & 1 & 0 \\ 0 & 0 & 0 & 1 \end{bmatrix}$$
$$\mathbf{H}_1^2 = \begin{bmatrix} c\phi_2 & -s\phi_2 & 0 & l_2 c\phi_2 \\ s\phi_2 & c\phi_2 & 0 & l_2 s\phi_2 \\ 0 & 0 & 1 & 0 \\ 0 & 0 & 0 & 1 \end{bmatrix}. \tag{12.70}$$

The homogeneous transformation form the end-effector to the base frame can be found by multiplying the joint homogeneous transformations.

$$\mathbf{H}_0^2 = \mathbf{H}_0^1 \mathbf{H}_1^2$$
$$\mathbf{H}_0^2 = \begin{bmatrix} c\phi_1 c\phi_2 - s\phi_1 s\phi_2 & -c\phi_1 s\phi_2 - s\phi_1 c\phi_2 & 0 & l_2 c\phi_1 c\phi_2 + l_2 s\phi_1 s\phi_2 + l_1 c\phi_1 \\ s\phi_1 c\phi_2 + c\phi_1 s\phi_2 & -s\phi_1 s\phi_2 + c\phi_1 c\phi_2 & 0 & l_2 s\phi_1 c\phi_2 + l_2 c\phi_1 s\phi_2 + l_1 s\phi_1 \\ 0 & 0 & 1 & 0 \\ 0 & 0 & 0 & 1 \end{bmatrix}. \tag{12.71}$$

Because of the simplicity of the geometry of this robotic arm, the homogeneous transformation from the end-effector to the base frame can be simplified by using relative angles.

$$\mathbf{H}_0^2 = \begin{bmatrix} c\phi_{12} & -s\phi_{12} & 0 & l_2 c\phi_{12} + l_1 c\phi_1 \\ s\phi_{12} & c\phi_{12} & 0 & l_2 s\phi_{12} + l_1 s\phi_1 \\ 0 & 0 & 1 & 0 \\ 0 & 0 & 0 & 1 \end{bmatrix}. \tag{12.72}$$

The coordinates for the O_2 frame base is given by the following equations.

$$O_2 = \begin{bmatrix} x \\ y \end{bmatrix} = \begin{bmatrix} l_1 c\phi_1 + l_2 c\phi_{12} \\ l_1 s\phi_1 + l_2 s\phi_{12} \end{bmatrix}. \tag{12.73}$$

Since both joints are revolute, the Jacobian matrix is revolute

$$\mathbf{J}(\phi) = \begin{bmatrix} z_0 \times (O_2 - O_0) & z_1 \times (O_2 - O_1) \\ z_0 & z_1 \end{bmatrix}. \tag{12.74}$$

From the coordinate frame transformation matrices, the joint coordinate frame locations with respect to the base frame can be determined.

$$O_0 = \begin{bmatrix} 0 \\ 0 \\ 0 \end{bmatrix} \quad O_1 = \begin{bmatrix} l_1 c\phi_1 \\ l_1 s\phi_1 \\ 0 \end{bmatrix} \quad O_2 = \begin{bmatrix} l_1 c\phi_1 + l_2 c\phi_{12} \\ l_1 s\phi_1 + l_2 s\phi_{12} \\ 0 \end{bmatrix}$$
$$z_0 = z_1 = \begin{bmatrix} 0 \\ 0 \\ 1 \end{bmatrix}. \tag{12.75}$$

The z vectors for both joints are the same because there is no rotation outside of the plane. The product of the appropriate z vector and joint coordinate location can be determined from the information provided above

$$\begin{aligned} z_0 \times (O_2 - O_0) &= \begin{bmatrix} l_1 s\phi_1 - l_2 s\phi_{12} \\ l_1 c\phi_1 + l_2 c\phi_{12} \\ 0 \end{bmatrix} \\ z_1 \times (O_2 - O_1) &= \begin{bmatrix} -l_1 s\phi_{12} \\ l_2 c\phi_{12} \\ 0 \end{bmatrix}. \end{aligned} \tag{12.76}$$

All of the necessary information has been calculated to construct the Jacobian for this two-link robotic manipulator example

$$\mathbf{J} = \begin{bmatrix} l_1 s\phi_1 - l_2 s\phi_{12} & -l_1 s\phi_{12} \\ l_1 c\phi_1 + l_2 c\phi_{12} & l_2 c\phi_{12} \\ 0 & 0 \\ 0 & 0 \\ 0 & 0 \\ 1 & 1 \end{bmatrix}. \tag{12.77}$$

The robot manipulator in this example has been characterized. The kinematics of the robot are important for its control. However, actuator and sensor characteristics must also be incorporated into the controller.

12.5. ROBOT CONTROL

Many robotic systems use permanent magnet d.c. servomotors as actuators. Although suitable for many activities and devices, they often require gearing to match the requirements for normal human functions. If the operating environment of the robot is to remain substantially optimized for human operation, it may be appropriate to develop actuators more similar in function to human muscle. Many activities performed under human control require mechanical compliance. Active compliance helps to prevent damage and to provide finer control. Robotic control systems can be designed to simulate human compliance. Human control systems are remarkably flexible and adaptive. For example, a handshake between two people may be viewed as an interactively determined trajectory of two actively compliant links that are coupled at the hand. Each party may have a particular handshake trajectory in mind, but the actual trajectory is a weighted version of each participant's desired trajectory.

Human arms routinely perform tasks using compliant control (i.e., they deviate from their path in response to contact force). Robots which use position control exhibit high mechanical stiffness to maintain accuracy. If the

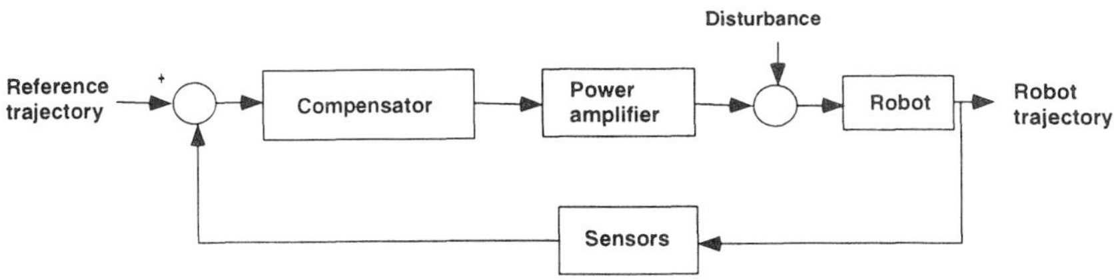

Figure 12.13. *Block diagram for a simple robot control system.*

control system simulates compliance then it may be able to perform personal care tasks safely and comfortably. Compliance within position control can be accomplished in a number of ways. Mechanical compliance, position/force control and software springs can be used to provide end-effector compliance.

The sophistication of the robot controller is an important factor in determining the system's immunity to failure. Redundant safety provisions, from mechanical design choices and encoder signal integrity to multiple level software for monitoring joint limits, are required for safe operation. A number of sophisticated control schemes have been developed and continue to be developed. This section will focus on basic, well developed techniques to introduce the reader to robotic system control.

12.5.1. Feedback control

Some robotic systems use an open loop (i.e., no direct feedback from the system state) controller. Stepper motors permit reasonably accurate position control without the use of a position feedback sensor. However, viable rehabilitation robotic systems incorporate feedback control to improve accuracy, reliability and precision. A simple robotic control system includes a compensator to attain desired system performance, a power servo amplifier to convert compensator signals to actuator power levels, a model of external disturbances to the systems (e.g., sensor error, unbalanced loads), a model of the robot and a model and input signals from various sensors (figure 12.13).

The compensator must be designed such that the robot trajectory (output) follows the desired trajectory and so the effects of the disturbances on the robot (plant) output are reduced. If the compensator is successful in suppressing the effect of the disturbances on the robot trajectory (output), it is said to reject disturbances. We have two primary objectives with robot control: disturbance rejection and tracking.

In order to achieve these objectives, models of the components of the system are required. These mathematical models are used by the control algorithm to predict and control the system's behavior in response to the operator's desires. Real systems are non-linear and of infinite order. However, many of the higher-order terms can be neglected for control purposes and linear models are quite accurate at predicting the gross behavior

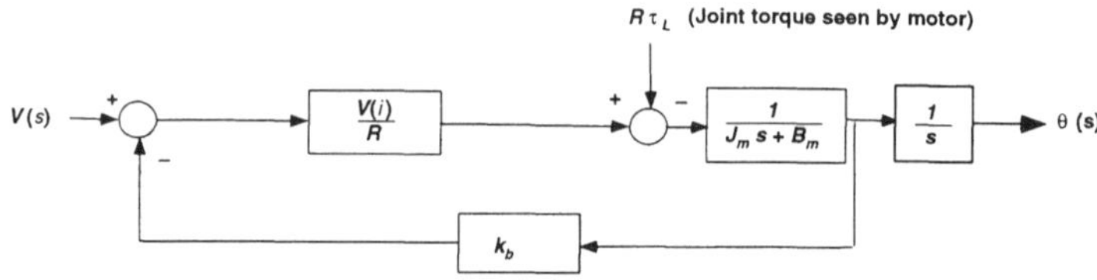

Figure 12.14. *Reduced-order model of motor and joint.*

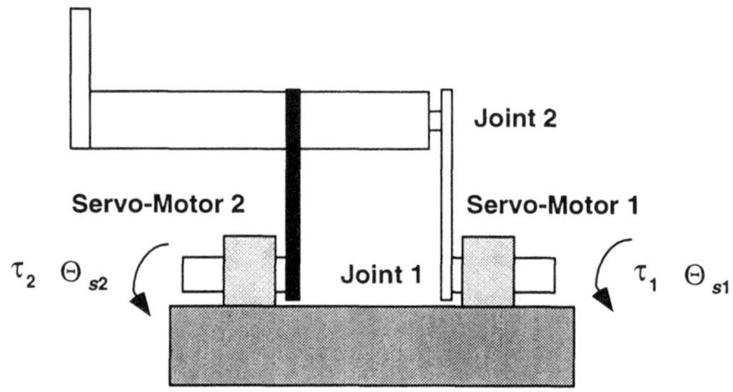

Figure 12.15. *Two-link remotely driven manipulator.*

of many system components. The appropriate model must be selected based upon the quality of the data available and the improvement in control gained with increasing model complexity. Often a reduced-order model of a motor and joint works well for many applications (figure 12.14).

For manipulators that use drive mechanisms such as belts, pulleys, chains, etc., then $r_k\theta_m$, the motor state variable, need not equal q_k, the joint variable. Therefore, in general, one must incorporate a transformation between joint variables and actuator variables into the manipulator dynamics. An example of a two-link remotely driven manipulator will illustrate this point (figure 12.15).

For this example, the joint variables and actuator variables are related as follows

$$q_1 = \theta_{s1} \qquad q_2 = \theta_{s1} + \theta_{s2}. \tag{12.78}$$

The joint torques and the actuator load torques are related by

$$\tau_{l1} = \tau_1 \qquad \tau_{l2} = \tau_1 + \tau_2. \tag{12.79}$$

The inverse transformations between joint position and servo motor position and joint torque and servo motor torque are found

$$\begin{aligned} \theta_{s1} &= q_1 & \theta_{s2} &= q_1 - q_2 \\ \tau_1 &= \tau_{l1} & \tau_2 &= \tau_{l1} - \tau_{l2}. \end{aligned} \tag{12.80}$$

12.5.2. Set-point tracking

Set-point tracking is used to define control when the objective is to have the system maintain or track some constant state (i.e., have the output follow some constant path). This type of control is adequate for applications where motions are slow. For many simple rehabilitation robotic applications set-point tracking provides adequate control of the system. More sophisticated methods can be found in the references for this chapter. Proportional, integral, derivative (PID) compensation is one of the most commonly implemented industrial and rehabilitation control application algorithms.

Set-point track requires a model of the actuator and manipulator kinematics. For simplicity the subscript s will be used to denote a servoactuator variable and the subscript m to denote a manipulator variable. The variable r_k represents the gear reduction for a particular joint-actuator pair k

$$q_k = \theta_{sk} = r_k \theta_{mk}. \tag{12.81}$$

The manipulator torque can be written in terms of the acceleration, angular velocity and orientation with respect to gravity of the joint and link

$$\tau_{lk} = \sum_{j=1}^{n} b_{jk}(\boldsymbol{q})\ddot{q}_{jk} + \sum_{i,j=1}^{n} c_{ijk}(\boldsymbol{q})\dot{q}_{ik}\dot{q}_{jk} + g_k(\boldsymbol{q}). \tag{12.82}$$

The equation of motion for each joint of the manipulator can be written in terms of the motor inertia (I_m), motor friction (B_m), the motor back electromagnetic force constant (k_b), motor torque constant (k_m), motor armature resistance (R), motor voltage (V) and the load torque (τ_k)

$$I_{mk}\ddot{\theta}_{mk} + \left(B_{mk} + \frac{\boldsymbol{k}_{bk}\boldsymbol{k}_{mk}}{R}\right)\dot{\theta}_{mk} = \frac{\boldsymbol{k}_{mk}}{RV_k} - r_k\tau_k \qquad k = 1, 2, \ldots, n. \tag{12.83}$$

The simplest approach is to treat τ_k as an external disturbance to the motor and design an independent controller for each joint using equation (12.83) for the actuator dynamics. The effect of the gear ratio is to reduce the coupling non-linearities. If an attempt is made to drive the robot fast or not use gear ratios, large tracking errors will result without sophisticated control.

The effective inertia and friction of the joint–actuator system can be used to simplify the equations describing the system dynamics

$$\begin{aligned} q_k &= r_k\theta_{mk} \\ I_{\text{eff}}\ddot{\theta}_{mk} &= \left(I_{mk} + r_k^2 b_{kk}(\boldsymbol{q})\right)\ddot{\theta}_{mk} \qquad k = 1, 2, \ldots, n \\ B_{\text{eff}} &= B_{mk} + \frac{\boldsymbol{k}_{bk}\boldsymbol{k}_{mk}}{R}. \end{aligned} \tag{12.84}$$

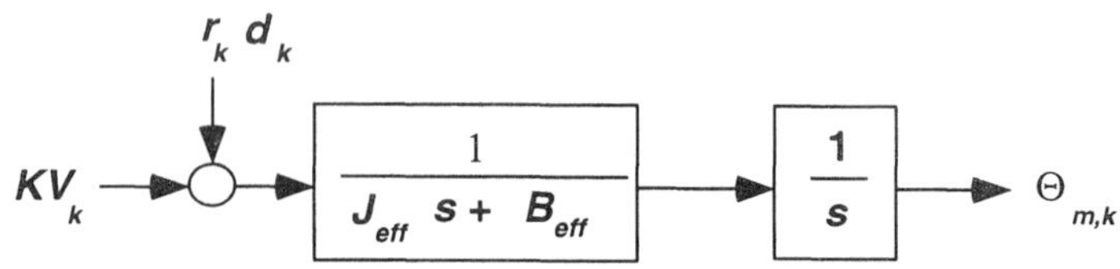

Figure 12.16. *Simplified open loop block diagram for a motor and joint.*

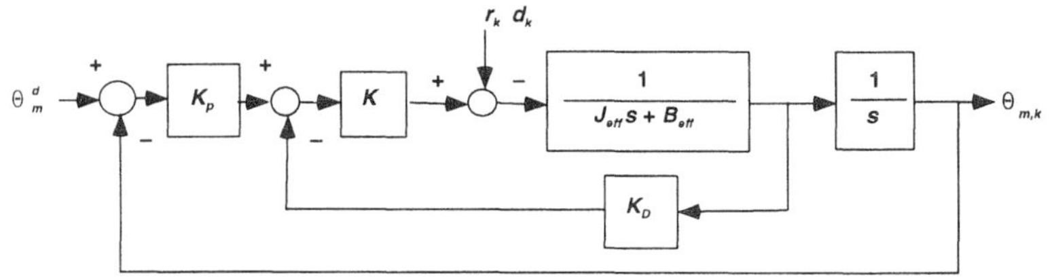

Figure 12.17. *Proportional and derivative control of a motor and single joint.*

The system dynamics can be described in terms of the effective inertia, and friction to simplify the analysis

$$I_{\text{eff}}\ddot{\theta}_{mk} + B_{\text{eff}}\dot{\theta}_{mk} = \left(\frac{k_{mk}}{R}\right) V_k - r_k \tau_k. \tag{12.85}$$

In this form, the problem can be posed to treat the load torque as a disturbance (i.e., $\tau_k = d_k$) to the actuator control system. This eliminates the need to directly account for the load torque. The set-point tracking problem with a constant step reference command will be used as an example of control of a single robot joint. The open loop model of the system is presented in figure 12.16.

A proportional and derivative compensator will be used to provide the desired response (figure 12.17). The angular position of the joint will be fed back and subtracted from the desired angle input. When the actual angle is equal to the desired angle, the error signal will be zero and the system will remain stationary. The appropriate proportional (K_p) and derivative (K_D) gains must be determined to achieve the desired response.

The input to the motor $V(s)$ is a function of the input ($\theta_m^d(s)$), output ($\theta_m(s)$) and PD compensator

$$V(s) = K_p\left(\theta_m^d(s) - \theta_m(s)\right) - K_D s\theta_m(s). \tag{12.86}$$

The dynamics of the system can, again, be determined using the effective inertia, effective resistance, gear ratio and joint angle

$$I_{\text{eff}}s^2\theta_m(s) + B_{\text{eff}}s\theta_m(s) = KV(s) - rD(s). \tag{12.87}$$

The function for $V(s)$ can be substituted into equation (12.87) to demonstrate the dependence of the system output upon the PD compensator.

$$I_{\text{eff}}s^2\theta_m(s) + B_{\text{eff}}s\theta_m(s) = K[K_p\left(\theta_m^d(s) - \theta_m(s)\right) - K_D s\theta_m(s)] - rD(s). \tag{12.88}$$

The poles, or roots, of the system characteristic equation determine its stability. In continuous time the real part of each pole must lie in the left hand of the complex plane. For discrete systems, the poles must lie within the unit circle in the complex plane. We can manipulate equation (12.88) to solve for the manipulator joint angle. First, the characteristic equation of the system must be found. The coefficients of the compensator must be chosen so that the system is bounded input, bounded output (BIBO) stable. The characteristic equation of this system is second order and the roots can be determined explicitly

$$\Omega(s) = I_{\text{eff}}s^2 + (B_{\text{eff}} + KK_D)\,s + KK_p. \tag{12.89}$$

Several methods exist to test the stability of higher order systems.

The closed loop system is stable for all positive values of K_p and K_D. The transfer function's relationship with the output, the input and the disturbance can be derived after some algebra

$$\theta_m(s) = \frac{KK_p}{\Omega(s)}\theta_m^d(s) - \frac{r}{\Omega(s)}D(s). \tag{12.90}$$

The tracking error is given by the difference between the actual angle and the desired angle

$$\begin{aligned} E(s) &= \theta_m^d(s) - \theta_m(s) \\ E(s) &= \frac{I_{\text{eff}}s^2 + (B_{\text{eff}} + KK_D)\,s}{\Omega(s)}\theta_m^d(s) + \frac{r}{\Omega(s)}D(s). \end{aligned} \tag{12.91}$$

The error will be bounded for a bounded input and disturbance. The proper selection of the suitable compensator coefficients can minimize the tracking error. The final value theorem can be used for this purpose. The steady-state error will be examined using an unit step function as the input and another as the disturbance

$$\theta_m^d(s) = \frac{\boldsymbol{k}_d}{s} \qquad D(s) = \frac{D}{s}. \tag{12.92}$$

This is appropriate for set-point tracking problems, more rich inputs and disturbances are required for contours tracking.

The steady-state error under these conditions can be estimated by using the Laplace transform final value theorem

$$e_{ss} = \lim_{s \Rightarrow 0} sE(s) = \frac{rD}{KK_p}. \tag{12.93}$$

The steady-state error is small for large gear reduction (i.e., small r) or high proportional gain (K_p). Other inputs can be evaluated as well. An integral component can also be added to the compensator to improve steady-state error.

12.5.3. Intelligent control

Many animals use genetic implant information that is passed on from generation to generation without communication. Robotic systems have great potential for genetic implantation. Information gained with any robot could be translated into the next version of software and be delivered intact to the next robot. All robots can be initiated with a past history.

There is considerable advantage in terms of system flexibility if the user has direct control over the robot's motion. The complexity and tedium of direct manipulator control must be reduced. Artificial intelligence or task-level control can allow the user to determine the tasks to be performed without being overly specific. The sophistication of the control software and the user control must be balanced to obtain acceptable performance. The use of more sophisticated control software often leads to the implementation of sensors to provide the robotics system with feedback from the environment.

Many industrial robots are programmed to accurately guide the end-effector through a desired sequence of configurations with a high degree of precision in a highly structured environment. The human controller is often ignored during this process and in many cases is excluded from operation of the robot. Human interaction may be limited to supervision of the robot's operation. This approach is not well suited for rehabilitation applications where the user is often in intimate contact with the robot.

One method for providing for functional interaction between the user and the robot is to adopt an interactive programming approach. The user is permitted to give simple commands to toggle the robot from automatic tasks to direct teleoperator-type control. This ability to switch modes of control is important for system flexibility and safety. The concept of task-level programming and controlling robots lends itself to the integration of visual and sensor information at a level which does not burden the user with technical details. Task-level commands can be used to read and interpret visual and sensor information. Task-level programming uses key words to perform high-level functions. Appropriately developed key task-level programming could allow people to interact with a robot using human-like language. Task-level control requires a high degree of artificial intelligence on the part of the robotic system, but does not relegate control by the person. For example, during meal preparation it would not be necessary for the robot to know the recipe or process for cooking the meal. The user would give high-level instructions to guide the robot through the tasks required to prepare the meal (table 12.2).

Table 12.2. *Example of task-level commands for meal preparation.*

Remove bowl from cupboard
Remove spoon from drawer
Remove milk from refrigerator
Remove cereal from cupboard
Pour cereal into bowl
Pour milk into bowl
Pick up spoon
Scoop cereal from bowl
Present spoon to me

The user must be able to string together groups of commands to further simplify common tasks. This allows the user to maintain his/her desired level of control over the robot system and to customize the system for his/her needs. Task-level programming allows the user to guide the robot through a large number of tasks which could not be envisaged by the system designer. Artificial intelligence is required to convert high-level language to low-level robot operations. This increases the need for computational power and sensory inputs. Task-level programming requires an interpreter which translates text from the application level into low-level robot and visual system commands. Models of the robot and environment must be maintained by the interpreter. System designers must have a clear understanding of the environment and desired tasks for the robotic system to be acceptable.

Task-level cannot be made practical without flexibility in low-level components to avoid obstacles and plan motion. Algorithms must be programmed into the robot to plan motions to avoid obstacles in the environment. This requires knowledge of fixed and movable objects in the environment. A functional approach is to have the robot maintain a 3D map (an obstacle-free space map (OFSM)) of locations and types of all objects in its environment. The interpreter uses the map to schedule joint-level motion sequences necessary to complete the requested task. The map is updated after each sequence. Information about the object is used to maintain the proper orientation of objects (e.g., a glass of liquid will need to remain upright) during path planning. Orientation maintenance reduces the number of degrees of freedom usable by the manipulator and complicates obstacle avoidance. A visual system can be used to provide feedback which can help avoid collisions. Interaction with the user may be required for obstacle avoidance within complex environments. Most obstacles can be avoided using a 2D map of the footprints of obstacles in the environment. A simple approach to obstacle avoidance is to determine the maximum clearance height of the object to be moved and compare that to the obstacles in the environment. Obstacle avoidance can then be treated as a 2D problem.

12.6. ROBOT SENSORS

Sensors may be used to make the robotic system responsive to the external world. The number and type of sensors depends upon the structure of the robot (e.g., mobile, stationary, wheelchair-mounted) and the tasks to be performed. Implementation of sensors improves control at the cost of greater system sophistication. Humans and animals use a wide range of senses to query the environment. A robotic system could eventually surpass humans and animals in their ability to sense the environment. This is because many sensing technologies exist which provide substantially better performance than human or animal sensing. In addition, robotic systems can be integrated with the best sensing capabilities for a variety of environmental factors. Visual, proximity, touch, and force sensing play a significant role in the performance of robotic systems. The function of robot sensors may be divided into two categories: internal state and external state.

Internal-state sensors provide information about joint and other variables related to the status of the robot internal hardware. External sensors are used to detect environmental variables such as range, proximity, object orientation and touch. External sensing is used for guidance, path planning, object identification, and object handling.

12.6.1. Robot vision

Vision provides a robot with a sophisticated sensing mechanism that permits the machine to respond to its environment in a flexible manner. Vision can increase the scope and effectiveness of robots. Hardware and software associated with vision systems are quite complex. Robotic vision systems offer tremendous potential improvement over human or animal vision. An eagle has about 100 000 pixels per square centimeter and can see up to 2000 feet. A person has about 30 000 pixels per square centimeter, but with the aid of a telescope can see much further than an eagle. A robotic system could incorporate lens systems to give it better vision than humans or animals. A cat uses a biological photomultiplier to improve night vision by amplifying the ambient light. Synthetic photomultipliers exist which are many times better than a cat's. Humans and animals use two eyes offset from one another to locate items visually in three-space. Multiple vision sensors have been implemented quite successfully. Snakes use infrared sensors on their heads to detect and locate prey in total darkness. Infrared sensors are readily available and have been successfully applied to provide vision to humans in darkness.

Robot vision can be divided into several processes. The process of attaining a visual image is referred to as sensing. Noise reduction and enhancement of details is achieved through preprocessing. Objects of interest are partitioned by a process called segmentation. Description encompasses the computation of features suitable for differentiating various objects. The process of identifying objects (e.g., cup, spoon, razor) is called recognition. Interpretation is the process of associating the various objects properly. Vision may be used to perform simple or complex tasks. Low-level vision may be used to create a reflex reaction to certain situations, whereas high-level vision is used more to interpret the environment and for motion planning. The degree of processing and the information content define the level of vision.

An image is acquired by sampling sensors spatially and quantizing the information. Television or video cameras are the most common sensors used to acquire a digital image. Charge-coupled devices (CCDs) form the basis for solid-state imaging technologies. CCD cameras can be divided into line scan sensors and area sensors. A line scan sensor uses a row of photosites (silicon imaging elements). Photons hitting the photosites pass through a transparent polycrystalline silicon gate structure and are absorbed into the silicon crystal creating electron–hole pairs. The charge collected by each photosite is proportional to the illumination intensity at that location. Each row of photosites is interfaced to two transfer gates used to clock the contents into transport registers. An output gate is used to transfer the contents of the transport gate to an output amplifier whose output voltage signal is proportional to the illumination intensity.

Charge-coupled arrays use photosites arranged in a matrix with a gate-transport register between columns of photosites. The contents of odd numbered photosites are sequentially gated into the vertical transport registers and then into the horizontal transport register. The content of the horizontal transport gate is fed into a charge amplifier to create a line of video. The procedure is repeated 1/60 of a second later on the even numbered photosites to yield a total scanning rate of 30 times per second. CCD devices commonly have resolutions ranging from 32×32 to 1024×1024 elements.

The image from a television or CCD camera can be represented by a two-dimensional array ($f(x, y)$ of N rows and M columns, where x and y denote spatial coordinates in the image plane. The value of the function $f(x, y)$ is proportional to the intensity of the image at that point. Each element in the array is called an image element, picture element, or pixel. An image capture board is often used to create a digital image. Digitization of spatial coordinates is referred to as image sampling, whereas amplitude digitization is called intensity or gray-level quantization

$$f(x, y) = \begin{bmatrix} f(0,0) & f(0,1) & \dots & f(0, M-1) \\ f(1,0) & f(1,1) & \dots & f(1, M-1) \\ \vdots & \vdots & \dots & \vdots \\ f(N-1,0) & f(N-1,1) & \dots & f(N-1, M-1) \end{bmatrix}. \tag{12.94}$$

The elements of $f(x, y)$ at each time instant represent the spatial location and intensity of each pixel. The element $f(0, 0)$ is located in the upper left corner of the image and represents the origin for the spatial coordinate system. Computation is simplified by having N, M and the range of discrete intensity levels to be powers of 2. The quality of the image is affected by each of these values.

Lighting is also critical to achieve a usable image. Lighting is used to reduce the complexity of the image by reducing extraneous details, reflections, and low-contrast. Diffuse lighting is used to determine surface characteristics. Back lighting is used when silhouettes of objects are required. Structured lighting can be used to minimize some external disturbances and to provide an indication of three-dimensional shape. Directional lighting can be used to detect the contour of surfaces.

The optimal position of the camera may be different for various tasks. This may require the orientation of the object in the camera reference frame to be transformed into another reference frame. This may be accomplished using the kinematic methods described earlier. In order for the camera to be used as a measuring tool it must be calibrated. Calibration requires knowledge of focal length, camera offsets, pan angle and tilt angle. The simplest way of measuring these parameters is to use the camera itself along with a set of image points with known world coordinates. In three-space, a minimum of six points is required for camera calibration.

Several basic concepts must be developed before explaining some simple methods for processing digital images. Each pixel p at coordinates (x, y) has four horizontal and vertical neighbors. The four neighbors are unit distance from (x, y). Some of the neighbors will be outside the digital image if the pixel is on the edge of the image. There are also four diagonal neighbors added to the vertical and horizontal neighbors, making up the eight neighbors of the pixel. Connectivity defines the intensity of the pixels determined to be connected. Typically, a vector (e.g., $V = (77, 78, 79)$) defines the intensity of the pixels which are defined to be connected.

A number of techniques are available for preprocessing general image data, but only a few techniques have the necessary speed and cost to be suitable for robotic applications. Images are processed using spatial-domain techniques and frequency-domain techniques based on Fourier analysis. Spatial domain methods operate directly on pixels. A spatial-domain preprocessor can be thought of as an operator on $f(x, y)$ to create a preprocessed image $g(x, y)$

$$g(x, y) = h\left[f(x, y)\right]. \tag{12.95}$$

One of the most frequently used spatial-domain techniques is based on convolution masks or templates. A template is a small two-dimensional array whose coefficients are chosen to detect a desired property in an image. The center of the template is moved around the image. Each pixel position within the template is multiplied by the corresponding template coefficient.

$$h = \begin{bmatrix} -1 & -1 & -1 \\ -1 & 8 & -1 \\ -1 & -1 & -1 \end{bmatrix}. \tag{12.96}$$

With the mask illustrated above, the center pixel is multiplied by 8, while its neighbors are multiplied by -1. If all pixels within the template have the same intensity, the sum will be zero. If the center of the template is located at one of the higher intensity points of interest, the sum will be different from zero. If one of the higher intensity points is located off-center, the sum will also be different from zero, but the magnitude of the result will be weaker. A threshold can be used to determine if a marker of interest is within the template and whether it is located at the center. A more general representation of the convolution mask is described by

$$\begin{aligned} h\left[f(x, y)\right] = & w_1 f(x-1, y-1) + w_2 f(x-1, y) + w_3 f(x-1, y+1) \\ & + w_4 f(x, y-1) + w_5 f(x, y) + w_6 f(x, y+1) \\ & + w_7 f(x+1, y-1) + w_8 f(x+1, y) + w_9 f(x+1, y+1). \end{aligned} \tag{12.97}$$

A template can include any number of terms depending upon the purpose of the image processing. Templates are simple, fast algorithms for processing well structured images. Frequency-domain methods based on Fourier transforms create an aggregate of complex pixels. Fourier transforms are composed of complex sinusoids which require extensive computational capability

$$\begin{aligned} F(u, v) &= \frac{1}{N} \sum_{x=0}^{N-1} \sum_{y=0}^{N-1} f(x, y) e^{-j2\pi(ux+vy)/N} \\ f(x, y) &= \frac{1}{N} \sum_{u=0}^{N-1} \sum_{v=0}^{N-1} F(u, v) e^{j2\pi(ux+vy)/N}. \end{aligned} \tag{12.98}$$

Fourier transform techniques are used for image enhancement and reconstruction.

The two-dimensional Fourier transform can be expressed as two one-dimensional summations similar in form to the one-dimensional Fourier transform. Fourier transforms can be used for edge detection and object motion. Optical means can be used to compute Fourier transforms in real time.

Noise and other spurious effects can be reduced by smoothing. Several fast techniques exist for smoothing image data for robotic systems. Neighborhood averaging, median filtering and image averaging can be used to reduce the effects of disturbances. Neighborhood averaging simply takes the average of the neighborhood intensity to represent the center pixel. Neighborhood averaging has the tendency to blur edges and sharp details. The median filter can reduce blurring by using the median of the neighborhood pixels in place of the mean. Image averaging is used to minimize the effects of additive noise by taking the average pixel intensity for a given position over several images. More sophisticated techniques are available and information about them can be found in the references.

12.6.2. Non-contact sensing

Non-contact sensors rely on the response of a detector to variations in acoustic or electromagnetic radiation. Non-contact sensors are available for measuring range, proximity and visual properties of objects. Many animals use sound to communicate and to locate prey. Owls and sharks use binaural hearing that they can aim to detect and locate prey. Multiple microphones have been used to isolate and locate objects in space quite effectively. Elephants have sensors in their cortexes to use infrasound to communicate over long distances without visual contact. Infrasound has long been used to communicate with submerged submarines. Infrasound has the ability to pass through openings and around barriers much better than ultrasound. Dolphins use ultrasound to communicate and to generate images with high resolution. Ultrasound has been used quite successfully to non-invasively generate images of the human heart and the wombs of pregnant women. Bats use radar to detect and isolate prey with amazing speed and accuracy. They use a frequency sweep from about 30 to 100 kHz with varying power. Bats also use varying sweep rates and in some cases use a whisper mode to avoid detection. Synthetic radar has been developed and used quite successfully. However, the algorithms used with robotic systems are often simplistic, using only a single frequency.

A range sensor measures distance from a reference point to objects in the field of sensor operation. Range sensors are used for robot navigation and obstacle avoidance. Techniques have also been developed to determine object shape using range sensors. Range is most simply measured using triangularization techniques. Triangularization uses a narrow beam focused onto the surface of an object. The reflection of the beam is measured by a detector. The distance between the detector and the emitter is fixed and only the angle of the detector varies to achieve highest intensity of the reflected signal. Using the known distance between the emitter and detector and the angle of the detector, simple geometry can be used to determine the distance to the object. A profile of the object can be estimated by moving the emitter–detector pair relative to the object.

Range can also be determined by using the time-of-flight. A common approach is to use a pulse laser and to measure the time the light takes to reach an object and be reflected coaxially from the object surface. The distance to the surface is given by the relationship $D = cT/2$, where D is the distance to the object, c is the speed of light and T is the pulse transit time. This method requires a highly accurate and stable clock. An alternative to the pulsed laser method is to use a continuous beam laser and measure the phase shift between the outgoing and returning laser beams. Since the wavelength of laser light is small (e.g., 632.8 nm for helium–neon) the amplitude is modulated by a waveform of much higher wavelength (i.e., recall $c = f\lambda$, therefore for $f = 10$ MHz, $\lambda = 30$ m). The modulated laser light of effective wavelength λ is split into two beams. The reference beam travels a distance L from the laser through a beam splitter to a phase measurement device, the measurement beam travels a distance $D' = 2D+L$, to the object and back through the beam splitter to the phase measurement device. If the distance to the object (D) is zero, then $D' = L$ and both beams will arrive simultaneously at the phase detection device without phase shift. As D increases, the phase shift between the two beams increases:

$$D = L + \frac{\theta}{360}\lambda = \text{distance to phase device} + \frac{\text{phase shift}}{360}\text{ wavelength.} \tag{12.99}$$

Continuous systems can provide intensity information as well as range information. However, they require considerably more power than pulsed systems. Both pulsed and continuous systems are susceptible to noise, therefore several measurements are averaged to reduce uncertainty.

Proximity sensors use a binary output to indicate the presence of an object within a specified distance. Robots use proximity sensors for near-object sensing and obstacle avoidance. Proximity sensors can be based on change in inductance due to the presence of a metallic object. A coil and magnet are integrated into a single housing. The magnet produces constant magnetic flux lines which pass through the coil. If a metallic object is in proximity to the sensor, the flux lines change and a current pulse is induced in the coil proportional to the change in the flux. The amplitude of the current pulse is related to the speed with which the object enters or leaves the magnetic field. The polarity of the current pulse refers to whether the object is entering or leaving the field. The sensitivity of inductive sensors shows a log–linear relationship with distance to the object.

A Hall-effect sensor relates the voltage between two points in a conducting or semiconducting material to a magnetic field across the material. Hall-effect sensors are used in conjunction with permanent magnets so that together they form a sensor capable of detecting the presence of ferromagnetic materials. When a ferromagnetic object comes close to a Hall-effect sensor the magnetic field through the Hall effect element weakens as the field lines are pulled through the metal.

Capacitive sensors are capable of sensing the presence of nearly any solid or liquid material. A capacitive transducer consists of a sensing electrode and a reference electrode. Dry air is used to provide isolation. When an object comes close to the sensor, the capacitance between the sensing and reference electrode changes. The change in capacitance is affected by distance as well as the type of material placed in proximity to the sensor. A simple method for detecting proximity, uses the capacitive sensor as part of an oscillator circuit designed so that oscillations only occur when the capacitance of the sensor exceeds a predefined threshold.

Ultrasonic sensors show a greatly reduced dependence upon the material being detected. The basic element of an ultrasonic sensor is a piezoelectric electroacoustic transducer. A layer of resin is often used to isolate the transducer from humidity and dust and to act as an acoustical impedance matcher. The transducer is decoupled from its housing and placed in an acoustical damping medium to provide fast damping of the energy. This permits the same transducer to be used as transmitter and receiver. Several methods have been developed to determine the proximity of objects with ultrasonic transducers. Most methods operate on the timing of transmitted and received pulses within a known medium.

Optical sensors can also be used to detect proximity. A simple approach is to use a light emitting diode (LED) and a photodiode. The LED transmits infrared light to the photodiode receiver. Interruption of the volume of infrared light between the LED and the photodiode will produce a change in output of the photodiode. This can be used to detect the presence of objects.

12.6.3. Tactile sensation

Tactile sensation is important to many animals and is equally so for robotic systems. The whiskers of some animals (e.g., cats, seals) can be used to navigate and sense prey. A number of technologies have been developed to simulate touch sensing. Touch information is used for object location, recognition, and to control force exerted by a manipulator on an object. Touch sensors are either binary, which respond to the presence or absence of an object, or analog which provide a signal proportional to the local force.

Binary sensors are contact devices. Microswitches are, perhaps, the simplest and most widely used binary touch sensor. Microswitches may be placed on the end-effector to determine when contact has been made with an object and on the work surface to indicate which objects are in their home positions. Multiple binary sensors can be used to provide tactile information on the location of objects within the end-effector. They may also be used to indicate whether an object is slipping within the end-effector's grasp.

Some degree of compliance is required to develop an analog tactile sensor whose output is proportional to local force. Simple mechanisms based upon spring-loaded plungers equipped with rotary or linear position sensors (e.g., potentiometers, encoders, LVDTs) can be used as tactile sensors. Force is proportional to displacement, thus both are given when the spring constant is known. This method is rather simplistic.

Arrays of tactile sensors have been developed to provide more complete information regarding grasping. A single array of electrodes in electrical contact with a compliant conductive material (e.g., conductive polymer, graphite-based substances) whose resistance varies with compression can be used to provide a more effective and versatile grip. An object pressing against the array causes local pressure variations which are measured and transmitted to the control unit. Tactile arrays can be made by electrically isolating passive elements and reading each sensor simultaneously. Active arrays can be used where sensors are excited sequentially and a sweep is used to sample each sensor in the array. Another method is to use anisotropic materials which are conductive in only one direction.

Force and torque sensing are very important for rehabilitation applications. Because of the close proximity often required of rehabilitation robotic systems, force and torque must be limited to prevent harm to the user. Moreover, force and torque must be sufficient to produce the desired effect (i.e., clean the teeth, wash the face, lift the plate). Force and torque sensors can be designed to measure the reaction force with the end-effector. Wrist sensors are mounted between the tip of the robot arm and the end-effector. Strain gages which measure the mechanical deflection of beams are often used as force transducers.

Wrist sensors must be sensitive, lightweight and compact with a dynamic range of up to 100 pounds. Hysteresis and accuracy can be improved by making the sensor out of a single piece of metal. To measure six degrees of freedom (e.g., three forces and three moments), a minimum of six properly oriented sensors is required. Strain gages are used to measure the small deflections of the sensors. If the strain gages are oriented in full-bridge configurations then the sensitivity is increased and temperature compensation is improved. Wrist sensors must have high stiffness, compact design, good linearity and low hysteresis. Natural frequency of a mechanical sensor is related to its stiffness. High stiffness ensures disturbing forces will damp-out quickly, helping to ensure an accurate and rapid response. A compact wrist sensor helps to ensure that the dextrous and reachable space of the manipulator will not be limited. It is important to place the sensor as close as possible to the applied load to minimize the moment arm of the load through the end-effector. Good linearity permits resolving the elements of the applied forces and torques via matrix operations. Hysteresis is a function of sensor internal friction, which reduces the device sensitivity.

12.6.4. Robot navigation

A variety of transducers can be used to navigate mobile rehabilitation robots through their operating environment. Robot navigation consists of three basic platforms: obstacle avoidance, path planning and operator tracking. Path-planning often uses a software map of the environment to plan an optimal path between any two points (i.e., the current location and the desired location). Path-planning algorithms must have the ability to account for fixed, semi-fixed and variable obstacles. If the path is interrupted, then the robot must be able to determine a new path to reach its target without input from the user. Ultrasonic sensors can be used to locate a local coordinate reference frame. This helps the robot to update its absolute position, rather than to rely purely on relative position feedback mechanisms (i.e., optical encoders on the wheels or tracks). Ultrasonic sensors can also be attached to the robot to detect when obstacles appear or move in the environment. This information is used to continuously update the map. Birds have the ability to sense fluctuations in magnetic fields which they use for navigation. The compass provides the same information to robotic systems. Flies use a second set of undeveloped wings as gyroscopes for inertial navigation. Inertial guidance and control is well developed and has been applied to a number of vehicles. Many animals mark their territory to define their feeding and breeding grounds. Any number of transponders or other sensors can be used to define a robot's operating region. This may help to prevent robots from operating outside of a desired area or within a restricted area. Many retail stores use passive transponders to prevent unauthorized goods from being removed from the store.

Companion tracking systems consist of a non-physical link between the robot and operator. The link must be able to keep the robot in contact with the operator despite background clutter and other people being in the area (including other robot operators). Companion tracking may operate in lead or track mode depending upon the environment and the needs of the user. An infrared beacon combined with a global positioning system can be used to maintain contact. Other sensors which convey proximity and direction information can be used to provide companion tracking as well.

Obstacle avoidance often involves heavy computational cost and slow time response. Ultrasonic sensors are used in many obstacle avoidance applications because of their low price and ease of operation. However, most ultrasonic sensors exhibit poor directionality, high sensitivity to ultrasonic noise and cross-talk and some surfaces project ultrasonic sound randomly. The use of multiple sensors with overlapping fields can improve directional sensitivity. Selecting proper sampling rate and sequencing with multiple sensors can reduce noise and cross-talk problems. Histogram signal conditioning methods with compensation for field overlap can be applied to develop a computer map of a dynamic environment. The probability distribution property of the histogram helps to reduce errors due to individual sensors. In some instances, the robot may be required to ask the operator for assistance with obstacle avoidance.

12.7. HUMAN INTERFACES TO ROBOTIC SYSTEMS

Mechanical and mechatronic systems have been more successful than general robotic systems in assisting people with disabilities. Mouthsticks, page turners, and powered prostheses provide the user with several

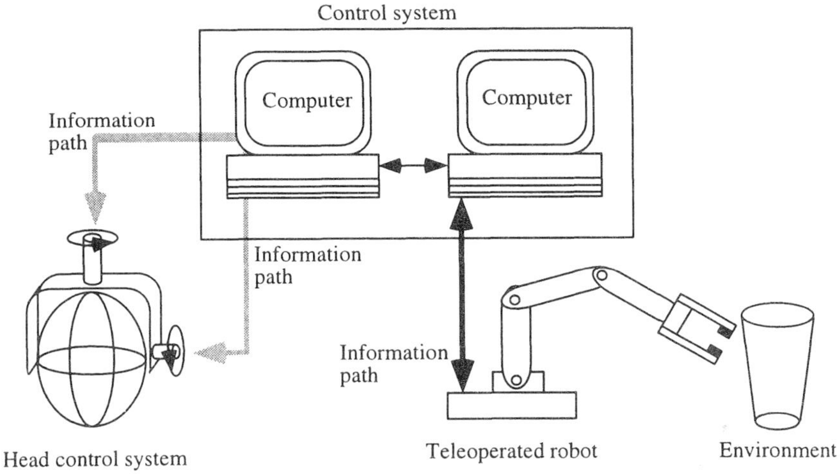

Figure 12.18. *Head-mounted control system for teleoperation of a robotic system. Computers are used to obtain command signals, to control the robot and to provide feedback. Such systems could permit people with severe disabilities to effectively manipulate their environment with greater autonomy.*

feedback paths containing information about position, force, speed and other important variables. Robotic systems have primarily relied on joysticks, voice recognition and simple switches. These devices, by themselves, may not provide adequate feedback for successful robotic control in an assistive device. Robotic systems have also been used to perform multiple complex tasks at the initiation of a few simple commands by the user. However, during our daily lives manipulation of the environment is more interactive. Therefore, teleoperators which provide multiple forms of feedback to the user may find greater acceptance. Telemanipulation has the ability to perform as an extension of the user by simulating proprioception and force feedback.

Robotic systems are intended to provide manipulation of the environment for people with very severe physical impairments. It must be assumed that individuals who will use robotic systems will have very limited voluntary movement. Therefore sites with voluntary control must be used to direct the robotic system. Head motion shows promise for providing a functional input for robotic systems. Head-operated systems have been successfully used for power wheelchair and specialized environmental control units. A method of providing head control with sufficient feedback is to use a head-mounted interface (e.g., halo) which measures head position (figure 12.18).

Small motors or brakes and position sensors can be integrated into the head input device. The position sensors would direct the robot's position. Tactile feedback from the robot would provide information about force applied to an object or make the robot aware of an obstacle in its environment. The robot tactile feedback could be translated to an increased resistance to

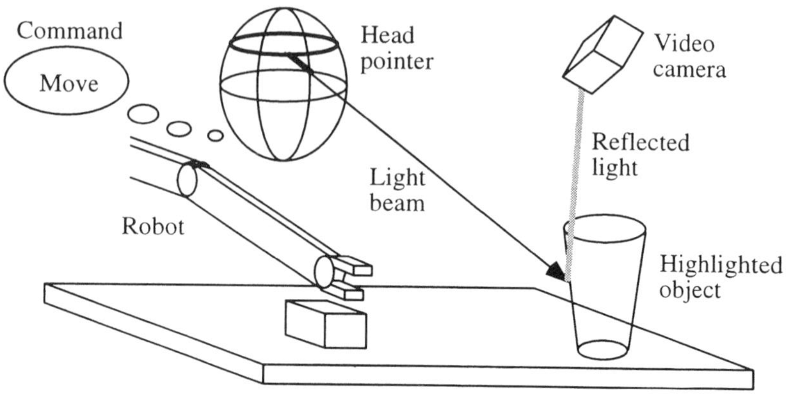

Figure 12.19. *Schematic diagram of a combined pointing and voice recognition system for assisting in the identification and manipulation of objects.*

motion at the head interface. The user could alter the force application based upon his/her intention and knowledge of the environment. Both force and position feedback provide the user greater control. More effective control can be used with multi-modal interfacing. Multi-modal interfacing combines a number of interfacing technologies into a single complex interface. For example, voice control can be used to select a mode among the robot's preprogrammed routines, pointing can be used to select an object to be manipulated and a head control interface can direct the manipulation of the object. For some individuals, shoulder motion can be used as an interface device rather than head control. The interface would be identical in function, although dissimilar in appearance.

Pointing systems may be useful to direct robots to objects to be manipulated. A head or shoulder interface can use relative motion much like the arms and hands. However, to move or manipulate an object, people use a combination of their visual and motor processing systems. A pointing system provides a 'visually based' input system for robotic manipulation (figure 12.19). The pointing system identifies the initial and final position of the object to be manipulated. Pointing systems will probably work best in structured lighting. A method that has been recently developed, which may work well, is to place highly reflective paint or tape on objects to be manipulated. The head pointer could emit a pulse of infrared light at the user's command. This would illuminate the area marked on the object. One or more video cameras would see the reflection and through triangulation or precalibration would know the location of the object. If different distinct markers were placed on the items then they could be identified through this process as well. A voice or switch command can be used to signal to the robot which is the initial and which is the final position. Once the object is in the robot's grasp, a head or shoulder control system can guide the path and or orientation of the object to its final position. The use of pointing, voice and a body interface should improve the effectiveness and acceptance of robotic systems.

FURTHER READING AND REFERENCES

Adjouadi M 1992 A man-machine vision interface for sensing the environment *Int. J. Rehab. Dev.* **29** pp 57–76

Andersson M and Fahlen L E 1993 A virtual environment user interface for a robotic assistive device *Proc. 2nd Eur. Conf. Advancement Rehabilitation Technol. (Stockholm, 1993)* p 29.3

Asfahl C R 1985 *Robots and Manufacturing Automation* (New York: Wiley)

Asimov I 1950 *I, Robot* (New York: Ballantine)

Birch G E, Fengler M and Gosine R G 1993 Regenesis robotic appliance evaluation progress report *Proc. RESNA 16th Ann. Conf. (Las Vegas, NV, 1993)* pp 501–3

Bolt R A 1980 'Put-that-there': voice and gesture at the graphics interface *Comput. Graph.* **14** 262–70

Borenstein J and Koren Y 1989 Real time obstacle avoidance for fast mobile robots *IEEE Trans. Systems Man. Cybern.* **19** 1179–87

Brooks R A 1986 A robust layered control system for a mobile robot *IEEE Trans. Robot. Automat.* **RA-2** 14–23

Cammoun R, Detriche J M, Lauture F and Lesigne B 1993 Improvements of the master man-machine interface *Proc. 2nd Eur. Conf. Advancement of Rehabilitation Technol. (Stockholm, 1993)* pp 24.2

Chiachhio P, Pierrot F, Sciavicco L and Sicialiano B 1993 Robust design of independent joint controllers with experimentation on a high-speed paralleled robot *IEEE Trans. Ind. Electron.* **40** 393–403

Cooper R A 1993 An introductory laboratory exercise in rehabilitation robotics *Proc. RESNA 16th Ann. Conf. (Las Vegas, NV*, 1993) pp 492–4

Crochetiere W J and Johnson T L 1983 Coordinated compliant motion *Proc. IEEE Conf. Decision Control* **3** pp 1464–5

Crochetiere W J and Mahoney R M 1990 Task-oriented control of a robot manipulator part i: the concept *Proc. RESNA 13th Ann. Conf. (Washington, DC, 1990)* pp 299–300

Dallaway J L and Jackson R D 1993 The raid workstation for office environments *Proc. RESNA 16th Ann. Conf. (Las Vegas, NV, 1993)* pp 504–6

Dallaway J L, Mahoney R M and Jackson R D 1993 CURL—A robot control environment for Microsoft windows *Proc. RESNA 16th Ann. Conf. (Las Vegas, NV, 19)* pp 510–1

Dally J W, Riley W F and McConnel K G 1993 Instrumentation for Engineering Measurements (New York: Wiley)

Detriche B and Lesigne B 1991 Man machine interface and programming aspects of the robotisized system 'Master' for disabled people *Med. Biol. Eng. Comput.* **29**

Drotning W, Christensen and Thunborg S 1992 Graphical model based control of intelligent robot systems *IEEE Control Systems Mag.* **12** 13–8

Eftring H and Bolmsjoe G 1993 RAID—A robotic workstation for the disabled *Proc. 2nd Eur. Conf. Advancement Rehabilitation Technol. (Stockholm, 1993)* p 24.3

Erlandson R F and Phelps J A 1993 Mechatronic systems as vocational enablers for persons with severe multiple handicaps *Proc. RESNA 16th Ann. Conf. (Las Vegas, NV, 1993)* pp 495–7

Fengler M and Cameron W 1988 Clinical testing of a low cost robotic arm for the severely disabled *Proc. First Int. Workshop on Robotic Applications in Medicine and Health Care (Ottawa, 1988)*

Fijany A and Bejczy A K 1989 Efficient Jacobian inversion for the control of simple robot manipulators *NASA Tech. Brief* **13** Item #125

Foulds R (ed) 1986 *Interactive Robotic Aids—One Option for Independent Living: An Int. Perspective (Monograph 37)* (New York: World Rehabilitation Fund)

Fu K S, Gonzalez R C and Lee C S G 1987 *Robotics Control, Sensing, Vision, and Intelligence* (New York: McGraw Hill)

Gilbert M and Trefsger J (ed) 1990 *Proc. Int. Conf. Rehabilitation Robotics, Applied Science and Engineering Laboratories, University of Delaware/Alfred I. duPont Institute (Wilmington, Delaware, 1990)*

Gordon G B, Roark J C and Schleifer A 1993 ORCA optimized robot for chemical analysis *Hewlett Packard J.* June 6–19

Gosine R G, Harwin W S, Furby L J and Jackson R D 1988 An intelligent end-effector for a rehabilitation robot *J. Med. Eng. Technol.* **13** 37–43

Griffin S 1985 *PRO-ARM RS2200 User's Manual* Version 1 (Kennewick, WA: Marcraft Int. Corporation)

Guittet J, Kwee H and Quetin N 1979 The SPARTACUS telethesis: manipulator control studies *Bull. Prosthet. Res.* **10** 69–105

Hammel J, Hall K, Lees L, Leifer L, Van der Loos M, Perkash I and Crigler R 1989 Clinical evaluation of a desktop robotic assistant *Int. J. Rehab. Dev.* **26** 1–16

Hammel J M and Van der Loos H F M 1991 Factors in the prescription and cost-effectiveness of robot systems for high-level quadriplegics *Proc. RESNA '91 (Kansas City, MO, 1991)* pp 16–8

Hammel J M, Van der Loos H F M and Perkash I 1992 Evaluation of a vocational robot with a quadriplegic employee *Arch. Phys. Med. Rehab.* **73** 683–93

Hillman M R, Gammie A R and Pullin G M 1990 The development of a robot workstation for the disabled *Proc. RESNA 13th Ann. Conf. (Washington, DC, 1990)* pp 287–288

Hogan N 1984 Impedance control of industrial robots *Robot. Comput.-Integrated Manuf.* **1** 97–113

Holmberg L 1993 The installation of a robotized workstation at Samhall-Hadar *Proc. 2nd Eur. Conf. Advancement Rehabilitation Technol. (Stockholm, 1993)* p 29.2

Jaros L A, Raschke U, Levine S P and Borenstein J 1991 Mobile robotic platform capable of following a human companion *Proc. RESNA 14th Ann. Conf. (Kansas City, MO, 1991)* pp 11–2

Kazerooni 1988 Direct drive active compliant end-effector (Active RCC) *IEEE Trans. Robot. Automat.* **4** 324–33

Kazerooni H and Mahoney S L 1991 Dynamics and control of robotic systems worn by humans *J. Dyn. Syst. Meas. Control* **113** 379–87

Komeda T, Uchida T, Matsouka H, Koyama H, Miyagi M, Beomonte-Zobel P and Funakubo H 1993 Mobile robot system to aid the daily life of bedridden persons in the private house *Proc. 2nd Eur. Conf. Advancement Rehabilitation Technol. (Stockholm, 1993)* p 24.4

Kuc R and Barshan 1992 Bat-like sonar for guiding mobile robots *IEEE Control Syst. Mag.* **12** 4–12

Kwee H, Tramblay R, Barbier R, Dupeyroux M, Vincenneux M F, Semoulin P and Pannier S 1983 First experimentation of the SPARTACUS telethesis in a clinical environment *Paraplegia* **21** 275–83

Lee H S and Leifer L J 1987 Automatic robotic grasping by use of optical proximity sensors *Proc. 10th Ann. Conf. Rehabilitation Technol. (San Jose, 1987)* pp 790–2

Lee H and Wade G 1985 *Imaging Technology* (New York: IEEE)

Leifer L J 1981 Rehabilitative robots *Robot. Age* 4–15

Leifer L 1992 RUI: factoring the robot user interface *Proc. RESNA Int. '92 (Toronto, Ont., 1992)* pp 581–3

Lieberman L I and Wesley M A 1977 AUTOPASS: an automatic programming system for computer-controlled mechanical assembly *IBM J. Res. Dev.* **21** 321–33

Lin S H, Tosunoglu S and Tesar D 1991 Control of a six degree of freedom flexible industrial manipulator *IEEE Controls Syst.* **11** 24–30

Lozano-Perez T 1983 Robot programming *Proc. IEEE* **71** 821–41

Lozano-Perez T 1987 A simple motion-planning algorithm for general robot manipulators *IEEE Trans. Robot. Automat.* **RA-3** 224–38

Madarasz RL, Lorren C, Cromp RF and Mazur NM 1986 The design of an autonomous vehicle for the disabled *IEEE Trans. Robot. Automat.* **RA-2** 117–16

Mason M T 1981 Compliance and force control for computer-controlled manipulators *IEEE Trans. Syst. Man Cybern.* **11** 418–32

Mataric M J 1992 Integration of representation into goal-driven behavior-based robots *IEEE Trans. Robot. Automat.* **RA-8** 304–12

McGorry R 1989 Active dynamometry in quantitative evaluation and rehabilitation of musculoskeletal dysfunction *Assistive Technol.* **1** 91–9

Mills J K and Goldenburg A A 1989 Force and position control of manipulators during constrained motion tasks *IEEE Trans. Robot. Automat.* **5** 30–46

Minsky M 1985 *Robotics* (New York: Omni)

Napper ST and Seaman RL 1989 Applications of robots in rehabilitation *Robot. Auton. Syst.* 227–39

Neveryd H and Bolmsjoe G 1993 Mobile robot systems for the disabled *Proc. 2nd Eur. Conf. Advancement Rehabilitation Technol. (Stockholm, 1993)* p 24.1

Oderud T, Bastiansen J E and Tyvand S 1993 Experiences with the MANUS wheelchair-mounted manipulator *Proc. 2nd Eur. Conf. Advancement of Rehabilitation Technol. (Stockholm, 1993)* p 29.1

Ouerfelli M, Harwin W and Kumar V 1992 A pneumatic actuation system for a wheelchair-mounted robot arm *Prod. RESNA Int. '92 (Toronto, Ont., 1992)* pp 331–3

Peremans H, Audenaert K and Van Campenhout J M 1993 A high-resolution sensor based on tri-aural perception *IEEE Trans. Robot. Automat.* **RA-9** 36–48

Prior S 1990 A survey of potential users of an electric wheelchair-mounted robotic arm *Proc. RESNA 13th Ann. Conf. (Washington, DC, 1990)* pp 297–8

Prior S D and Warner P 1991 Low cost pneumatic actuators for applications in rehabilitation robotics *Proc. RESNA 14th Ann. Conf. (Kansas City, MO, 1991)* pp 19–20

Pruski A 1992 Multivalue coding: application to autonomous robots *Robotica* **10** 125–33

Rahman T and Harwin W S 1992 Bilateral control in teleoperation of a rehabilitation robot *Prod. RESNA Int. '92 (Toronto, Ont., 1992)* pp 325–7

Raibert M H and Craig J J 1981 Hybrid position/force control of manipulators *ASME Trans. Dyn. System Meas. Control* **102** 126–33

Regalbuto M A, Cheatham J B and Krouskop T A 1990 A framework for a practical mobile robotic aid for the severely physically disabled *Proc. RESNA 13th Ann. Conf. (Washington, DC, 1990)* pp 127–8

Regalbuto M A, Krouskop T A and Cheatham J B 1992 Toward a practical mobile robotic aid system for people with severe physical disabilities *Int. J. Rehab. Dev.* **29** 19–26

Regalbuto M A, Krouskop T and Cheatham J 1993 Applying robotic technology to aid people with severe disabilities *J. Assistive Technol.* **4** 87–94

Sandorf M, Sample W, Edwards R and Engelhardt K J 1990 Development of a robotic vocational workstation *Proc. RESNA 13th Ann. Conf. (Washington, DC, 1990)* pp 123–4

Seamone W and Schmeisser G 1985 Early clinical evaluation of a robot arm/worktable system for spinal-cord-injured persons *Int. J. Rehab. Dev.* **22** 38–57

Spong M W and Vidyasagar M 1989 *Robot Dynamics and Control* (New York: Wiley)

Stripling T, Fonseca J E, Tsou V and Copperthite A 1983 A demographic study of spinal cord injured veterans *J. Am. Paraplegia Soc.* **6** 62–5

Van der Loos H F M, Hammel J, Lees D, Chang D, Perkash I and Leifer L 1990 A voice-control robot system as a quadriplegic programmer's assistant *Proc. RESNA 13th Ann. Conf. (Washington, DC, 1990)* pp 129–30

van Vliet P and Wing A 1992 A new challenge—robotics in the rehabilitation of the neurologically motor impaired *Phys. Ther.* **71** 39–47

Whitney D E 1987 Historical perspective and state of the art in robot force control *Int. J. Robot. Res.* **6** 3–14

Wolovich W A 1987 *Robotics Basic Analysis and Design* (New York: Holt, Rinehart and Winston)

INDEX